AF775141

BIOINSPIRATION AND BIOMIMICRY IN CHEMISTRY

BIOINSPIRATION AND BIOMIMICRY IN CHEMISTRY

REVERSE-ENGINEERING NATURE

Edited by

Gerhard F. Swiegers

WILEY

A JOHN WILEY & SONS, INC., PUBLICATION

Copyright © 2012 by John Wiley & Sons, Inc. All rights reserved

Published by John Wiley & Sons, Inc., Hoboken, New Jersey
Published simultaneously in Canada

No part of this publication may be reproduced, stored in a retrieval system, or transmitted in any form or by any means, electronic, mechanical, photocopying, recording, scanning, or otherwise, except as permitted under Section 107 or 108 of the 1976 United States Copyright Act, without either the prior written permission of the Publisher, or authorization through payment of the appropriate per-copy fee to the Copyright Clearance Center, Inc., 222 Rosewood Drive, Danvers, MA 01923, (978) 750-8400, fax (978) 750-4470, or on the web at www.copyright.com. Requests to the Publisher for permission should be addressed to the Permissions Department, John Wiley & Sons, Inc., 111 River Street, Hoboken, NJ 07030, (201) 748-6011, fax (201) 748-6008, or online at http://www.wiley.com/go/permission.

Limit of Liability/Disclaimer of Warranty: While the publisher and author have used their best efforts in preparing this book, they make no representations or warranties with respect to the accuracy or completeness of the contents of this book and specifically disclaim any implied warranties of merchantability or fitness for a particular purpose. No warranty may be created or extended by sales representatives or written sales materials. The advice and strategies contained herein may not be suitable for your situation. You should consult with a professional where appropriate. Neither the publisher nor author shall be liable for any loss of profit or any other commercial damages, including but not limited to special, incidental, consequential, or other damages.

For general information on our other products and services or for technical support, please contact our Customer Care Department within the United States at (800) 762-2974, outside the United States at (317) 572-3993 or fax (317) 572-4002.

Wiley also publishes its books in a variety of electronic formats. Some content that appears in print may not be available in electronic formats. For more information about Wiley products, visit our web site at www.wiley.com.

Library of Congress Cataloging-in-Publication Data:

Bioinspiration and biomimicry in chemistry : reverse-engineering nature /
edited by Gerhard F. Swiegers.
p. cm.
Includes bibliographical references and index.
ISBN 978-0-470-56667-1 (cloth)
1. Biomimicry. 2. Biomimetics. 3. Biomedical engineering. 4. Biomedical materials. I. Swiegers, Gerhard F.
QP517.B56B478 2012
610.28–dc23

2011049801

10 9 8 7 6 5 4 3 2 1

Dedicated to Crawford Long, William Thomas Green Morton, and Wilhelm Röntgen

CONTENTS

FOREWORD

The highest level of complexity of matter is that expressed in living matter, the substances and processes supporting life. In the course of evolution from nonliving to living matter, more and more complex forms of matter have been generated. Life has funneled molecular systems into specific types and improved their functions toward efficiency and selectivity as high as required for the operation of the full living organism.

Describing these highly efficient and selective systems and understanding their functioning is a challenge for chemistry. It involves designing mimics that help to unravel how these natural systems work. But, as important and in fact of wider significance is to go beyond models and implement on the wider scene the knowledge gained through mimicry to explore on one hand how similar functional features may be borne by different structures and, on the other, to show that novel functions of similar or even higher efficiencies and selectivities may be evolved in synthetic, nonnatural systems. Thus, mimicry of biological processes is crucial in first progressing toward understanding them and then going beyond.

Chemistry and in particular supramolecular chemistry entertain a double relationship with biology. Numerous studies are concerned with substances and processes of a biological or *biomimetic* nature. The scrutinization of biological processes by chemists has led to the development of models for understanding them on a molecular basis and of suitably designed effectors for acting on them.

On the other hand, the challenge for chemistry lies in the development of *abiotic*, nonnatural systems, figments of the imagination of the chemist, displaying desired structural features and carrying out functions other than those present in biology with comparable efficiency and selectivity. Not limited by the constraints of living organisms, abiotic chemistry is free to invent new substances and processes. The field of chemistry is indeed broader than that of the systems actually realized in Nature.

Supramolecular chemistry has been following both paths. Molecular recognition, catalysis, and transport processes are the basic functions investigated on both the biomimetic and abiotic fronts over the years. As recognition implies information, supramolecular chemistry has brought forward the concept that chemistry is also an information science, information being stored at the molecular level and processed at the supramolecular level. On this basis, supramolecular chemistry is actively exploring systems undergoing *self-organization*, that is, systems capable of generating, spontaneously but in an information-controlled manner, well-defined functional architectures by self-assembly from their components, thus behaving as *programmed chemical systems*.

The realization that supramolecular chemistry is intrinsically a *dynamic chemistry* in view of the lability of the interactions connecting the molecular components of a supramolecular entity led to the emergence of the concept of *constitutional dynamic chemistry* (CDC) that extended these dynamic features also to the molecular level. Dynamic entities are thus able to exchange their components by reversible formation or breaking of noncovalent interactions or of reversible covalent bonds, therefore allowing a continuous change in constitution by reorganization and exchange of building blocks.

CDC introduces a paradigm shift with respect to constitutionally static chemistry and takes advantage of dynamic diversity to allow variation and selection. The implementation of selection in chemistry introduces a fundamental change in outlook. Whereas *self-organization by design* strives to achieve full control over the output molecular or supramolecular entity by explicit programming, *self-organization with selection* operates on dynamic constitutional diversity in response to either internal or external factors to achieve *adaptation* in a Darwinian way. Synthetic systems are thus moving toward an *adaptive* and *evolutive chemistry*.

Along the way, the chemist finds illustration, inspiration, and stimulation in biological processes, as well as confidence and reassurance since they are proof that such fantastic complexity of structure and function can be achieved on the basis of molecular components. The mere fact that biological systems exist demonstrates that such a complexity can indeed exist in the world of molecules, despite our present inability to understand how it operates and how it has come about. Indeed, the molecular world of biology is only one of all the possible worlds of the universe of chemistry, that await to be created at the hands of the chemist!

It has been my privilege and pleasure to have participated in the development of bioinspiration and biomimicry in chemistry, and in the steps beyond, over the last 40 years. This field has made striking progress, but it still has much to teach us. I recommend it to you, the reader, for the promise and stimulation it holds. I wish to warmly congratulate the authors of this volume for their efforts in presenting the realizations and the perspectives of this most inspiring frontier of science.

JEAN-MARIE LEHN

FOREWORD

In the years since *Biomimicry: Innovation Inspired by Nature* chronicled the rise of a new design discipline,[1] the number of bioinspired patents, products, and practitioners has steadily risen. Each year, new biomimetic research centers open, more students take biomimetics courses, and more Fortune 500 companies invite biomimics to their design tables. In a study of U.S. patents between 1985 and 2005, Richard Bonser of the University of Bath found that patents with "biomimetic" or "bioinspired" in the title increased by a factor of 93, against a 2.7 times rise in other patents.[2] Why this surge of interest in Nature's designs?

I believe our species has begun to sense and respond to the same set of selection pressures that other organisms have faced for 3.8 billion years. As energy prices climb, chemists are asked to dial back temperatures and pressures while minimizing processing steps. Peaking supplies of nonrenewable feedstocks prompt calls for higher selectivity and atom economy, while focus shifts to renewable and waste-derived feedstocks. Meanwhile, regulatory laws oblige companies to minimize hazardous emissions and, in some countries, to take responsibility for long-term toxicological effects. In this perfect storm for change, conscientious consumers, governments, and corporations are demanding safer and more sustainable chemistry.

Life on earth has operated under these strict guidelines for billions of years. Organisms don't have the luxury of buying their chemicals from a manufacturing facility; they *are* the facility. Chemistry is performed in or near an organism's living tissues, and the by-products are released not just to any environment, but to the very habitat that must nurture the organism's offspring.

Life has had to perform this in situ chemistry without high temperatures, organic solvents, hazardous reagents, or extremes of pH. The feedstocks of choice are renewable or waste derived, procured locally and used judiciously. Compared to industry's use of the entire periodic table (even the toxic elements), the rest of life uses only a small subset of elements as grist for an astounding variety of functional molecules, structures, and materials. The feedstocks are few, the reactions are aqueous and elegant, and recyclability is built in though a process of anabolism and catabolism. Life's processes are proof that chemistry can occur under mild, life-friendly conditions, with an impressive degree of efficiency, selectivity, chemical yield, purity, and end of life reuse.

This realization dawns at an important moment in the history of sustainable chemistry. The first decades of safer chemistry featured lists of substances to avoid and challenged chemists to find alternatives to individual compounds. With more than 100,000 synthetic chemicals on the market, this compound-by-compound substitution has not kept pace. To overcome this limitation, bioinspired chemists should spend the coming decades moving upstream in the design process, finding alternatives for whole *families* of chemical reactions, not just compounds.[3] Rather than designing for acceptable risk, or writing containment protocols for questionable substances, young chemists should look forward to a career-long challenge of replacing industry's recipe book with Nature's own.

Pledging to work as Nature does—within planetary boundaries—is in no way a limit on creativity. In fact, the relatively unexplored space of biological chemistry—the process strategies of 30 million species—is broad and inspiring. A design brief that specifies no "heat, beat, and treat," no waste, and no rare or toxic materials serves as a creative frame, allowing us to achieve what we might not have imagined.

One example is a kiln-free route to high-tech ceramics. During the oil shocks of the 1970s, Jeffrey Brinker of Sandia National Labs was asked by his supervisor if he could make ceramics without fossil fuels. Brinker's research led him to mimic nacre, the iridescent lining of the abalone's shell. This layered nanocomposite is twice as tough as our jet engine ceramics thanks to the inclusion of polymer interlayers between the calcium carbonate layers. After nucleating crystal formation, the polymer allows the nacre to slide like a metal under compression, and under tension, the polymer stretches and self-heals. Our conventional kiln-based processes would have burned off this essential organic component, and with it, step changes in performance and functionality. In the same way, Nature's habit of "building from the bottom up" confers a strategic advantage. Templated self-assembly gives rise to long-range, hierarchical order, with surprising ancillary effects such as functional gradients and built-in redundancy from molecule to biosystem. Building to shape rather than subtractive cutting and grinding is inherently waste-free, a welcome change in an economy where most manufactured products yield 93% waste and only 7% product.[4]

Biomimetic companies are beginning to reverse this equation in several breakthrough products. Novomer has designed a photosynthesis-inspired catalyst that combines CO_2 and limonene to create biodegradable polycarbonates in a low-temperature process.[5] Calera has borrowed the recipe from corals to turn flue-gas CO_2 and seawater into a cement alternative that sequesters a half ton of CO_2 for every ton of cement.[6] Biomatrica has mimicked the anhydrobiosis chemistry of tardigrades to create a new way of storing biologicals without refrigeration, significantly reducing energy use in research labs, hospitals, and vaccine cold chains.[7] AQUAporin is making desalination membranes studded with life's water-escorting aquaporin molecules to increase rates of permeability by 100 times.[8] Donlar Corporation's TPA product reduces mineral scaling in pipes by borrowing the principles of mollusk stop proteins which limit seashell size.[9] Mussel glue has also been mimicked, allowing Columbia Forest Products to market a plywood resin that replaces more than 47 million pounds of formaldehyde-based adhesive annually.[10] Biosignal researchers found a resistance-free way to prevent biofilms by mimicking furonones—compounds that red algae use to interrupt bacterial signaling.[11] Several companies are working to replicate the self-cleaning properties of lotus leaves, and Big Sky Technologies has learned to make a lotus fabric coating with a minimum of fluorinated compounds.[12]

The products in the research pipeline are just as impressive. Labs around the world are studying photosynthesis to create an artificial leaf that turns photons into fuel, mimicking the water splitting and CO_2 reducing parts of photosynthesis.[13] Others are mimicking the active site at the heart of the hydrogenase protein to create an inexpensive substitute for platinum in the anodes of fuel cells. Materials researchers are studying biosilification to one day create computer chips and other silica compounds in water, at room temperature, using the process chemistry learned from diatoms and sponges.[14] One of the holy grails for spider researchers is to recreate the processing conditions of the spider's abdomen and spinnerets to impart superlative fiber properties to conventional silkworm silk.[15]

Behind all these brilliant ideas, there is a larger, more ubiquitous pattern that will hopefully guide biomimetic chemistry in the 21st century. For organisms of all species, the measure of success is simple and consistent—it's the continuation of an individual's genetic material thousands and thousands of generations from now. The only way to take care of an offspring that far into the future is to take care of the place that will take care of your offspring. Well-adapted organisms have therefore evolved to meet their needs in ways that also build soil, clean air, filter water, support biodiversity, and so on. On a planetary level, life creates conditions conducive to life.

Luckily, in this time of unprecedented need, the researchers in this volume have realized that we are surrounded by a world that works. They are in the vanguard of a growing movement to learn not just how to do smarter chemistry, but how to create conditions conducive to life. There is no more exciting or important work.

JANINE BENYUS

REFERENCES

1. Benyus, J. *Biomimicry: Innovation Inspired by Nature*, William Morrow & Company Inc., New York, 1997.
2. Bonser, R. H. C. "Patented biologically-inspired technological innovations: A twenty year view," *Journal of Bionic Eng*. **2006**, *39,* 39–41.
3. Geiser, K. *Making Safer Chemicals*, 2004, pp. 1–15.
4. Crystal Faraday Partnership; `http://www.crystalfaraday.org/`.
5. `http://www.novomer.com`.
6. `http://www.calera.com`.
7. `http://www.biomatrica.com`.
8. `http://www.aquaporin.com`.
9. `http://www.donlar.com`.
10. `http://www.columbiaforestproducts.com`.

11. http://www.biosignal.com.
12. http://www.bigskytechnology.com.
13. Schwartz, S.; Masciangiol, T.; Boonyaratanakornkit, B. *Bioinspired Chemistry for Energy: A Workshop Summary to the Chemical Sciences Roundtable*, National Academies Press, National Academy of Science USA, Washington DC, 2008.
14. Foo, C. W. P.; Huang, J.; Kaplan, D. L. "Lessons from seashells: Silica mineralization via protein templating," *Trends Biotechnol*. **2004**, *22,* 577.
15. Vollrath, F.; Madsen, B.; Shao, Z. "The effect of spinning conditions on the mechanics of a spider's dragline silk," *Proc. R. Soc. London Ser. B: Biol. Sci*. **2001**, *268* (1483), 2339.

CONTRIBUTORS

Pilar Aranda, Materials Science Institute of Madrid, ICMM-CSIC, c/Sor Juana Inés de la Cruz 3, 28049 Madrid, Spain

Katsuhiko Ariga, World Premier International (WPI) Research Center for Materials Nanoarchitectonics (MANA), National Institute for Materials Science (NIMS), 1-1 Namiki, Tsukuba 305-0044, Japan

Christopher R. Benson, Department of Chemistry, Indiana University, 800 East Kirkwood Avenue, Bloomington, Indiana 47405, USA

Wolfgang H. Binder, Lehrstuhl Makromolekulare Chemie, Fakultät f. Naturwissenschaften II, Institut f. Chemie, Martin-Luther University Halle-Wittenberg, Von Danckelmannplatz 4, D-06120 Halle, Germany

Clyde W. Cady, Department of Chemistry and Chemical Biology, Rutgers The State University of New Jersey, 610 Taylor Road, Piscataway, New Jersey 08854, USA

Jun Chen, Intelligent Polymer Research Institute and ARC Centre of Excellence for Electromaterials Science, University of Wollongong, Wollongong, NSW 2522, Australia

Jack K. Clegg, School of Chemistry and Molecular Biosciences, The University of Queensland, Brisbane St Lucia, QLD 4072, Australia

Liming Dai, Department of Macromolecular Science and Engineering, Case School of Engineering, Case Western Reserve University, 10900 Euclid Avenue, Cleveland, Ohio 44106, USA

Andrea M. Della Pelle, Department of Chemistry, University of Massachusetts–Amherst, 710 N. Pleasant Street, Amherst, Massachusetts 01003, USA

Gianfranco Ercolani, Dipartimento di Scienze e Tecnologie Chimiche, Università di Roma Tor Vergata, Via della Ricerca Scientifica, 00133 Roma, Italy

Francisco M. Fernandes, Materials Science Institute of Madrid, ICMM-CSIC, c/Sor Juana Inés de la Cruz 3, 28049 Madrid, Spain

Amar H. Flood, Department of Chemistry, Indiana University, 800 East Kirkwood Avenue, Bloomington, Indiana 47405, USA

Zhong-Ze Gu, State Key Laboratory of Bioelectronics, Southeast University, Nanjing, Peoples Republic of China 210096

Timothy W. Hanks, Department of Chemistry, Furman University, 3300 Poinsett Highway, Greenville, South Caralina 29613, USA

Florian Herbst, Lehrstuhl Makromolekulare Chemie, Fakultät f. Naturwissenschaften II, Institut f. Chemie, Martin-Luther University Halle-Wittenberg, Von Danckelmannplatz 4, D-06120 Halle, Germany

Jonathan P. Hill, World Premier International (WPI) Research Center for Materials Nanoarchitectonics (MANA), National Institute for Materials Science (NIMS), 1-1 Namiki, Tsukuba 305-0044, Japan

Sabine Himmelein, Organic Chemistry Institute and Graduate School of Chemistry, Westfälische Wilhelms-Universität Münster, Corrensstrasse 40, 48149 Münster, Germany

Reinhard W. Hoffmann, Fachbereich Chemie der Philipps Universität, Hans Meerwein Strasse, D-35032 Marburg, Germany

Ivan Jabin, Laboratoire de Chimie Organique, Université Libre de Bruxelles (U.L.B.), Av. F. D. Roosevelt 50, CP160/06, B-1050 Brussels, Belgium

Stéphane Le Gac, UMR CNRS 6226-Institut des Sciences Chimiques de Rennes, 263 Avenue du Général Leclerc-CS 74205, Université de Rennes 1, 35042 Rennes Cedex France

Yan Li Center of Advanced Science and Engineering for Carbon (Case4Carbon), School of Chemistry, Beijing Institute of Technology, Beijing 100081, Peoples Republic of China

Leonard F. Lindoy, School of Chemistry, The University of Sydney, Sydney, NSW 2006, Australia

Fabio Nudelman, Laboratory of Materials and Interface Chemistry and Soft Matter CryoTEM Unit, Eindhoven University of Technology, P.O. Box 513, 5600 MB, Eindhoven, The Netherlands

Bhanuprathap Pulamagatta, Lehrstuhl Makromolekulare Chemie, Fakultät f. Naturwissenschaften II, Institut f. Chemie, Martin-Luther University Halle-Wittenberg, Von Danckelmannplatz 4, D-06120 Halle, Germany

Liangti Qu, Center of Advanced Science and Engineering for Carbon (Case4Carbon), School of Chemistry, Beijing Institute of Technology, Beijing 100081, Peoples Republic of China

Bart Jan Ravoo, Organic Chemistry Institute and Graduate School of Chemistry, Westfälische Wilhelms-Universität Münster, Corrensstrasse 40, 48149 Münster, Germany

Olivia Reinaud, Laboratoire de Chimie et de Biochimie Pharmacologiques et Toxicologiques, CNRS UMR 8601, PRES Sorbonne Paris Cité, Université Paris Descartes, 45 rue des Saints Péres, 75006 Paris, France

Christopher Richardson, School of Chemistry, University of Wollongong, Wollongong, NSW 2522, Australia

David M. Robinson, Department of Chemistry and Chemical Biology, Rutgers The State University of New Jersey, 610 Taylor Road, Piscataway, New Jersey 08854, USA

Eduardo Ruiz-Hitzky, Materials Science Institute of Madrid, ICMM-CSIC, c/Sor Juana Inés de la Cruz 3, 28049 Madrid, Spain

Luca Schiaffino, Dipartimento di Scienze e Tecnologie Chimiche, Università di Roma Tor Vergata, Via della Ricerca Scientifica, 00133 Roma, Italy

Marlen Schunack, Lehrstuhl Makromolekulare Chemie, Fakultät f. Naturwissenschaften II, Institut f. Chemie, Martin-Luther University Halle-Wittenberg, Von Danckelmannplatz 4, D-06120 Halle, Germany

Andrew I. Share, Department of Chemistry, Indiana University, 800 East Kirkwood Avenue, Bloomington, Indiana 47405, USA

Paul F. Smith, Department of Chemistry and Chemical Biology, Rutgers The State University of New Jersey, 610 Taylor Road, Piscataway, New Jersey 08854, USA

Nico A. J. M. Sommerdijk, Laboratory of Materials and Interface Chemistry and Soft Matter CryoTEM Unit, Eindhoven University of Technology, P.O. Box 513, 5600 MB, Eindhoven, The Netherlands

Gerhard F. Swiegers, Intelligent Polymer Research Institute and ARC Centre of Excellence for Electromaterials Science, University of Wollongong, Wollongong, NSW 2522, Australia

Sankaran Thayumanavan., Department of Chemistry, University of Massachusetts–Amherst, 710 N. Pleasant Street, Amherst, Massachusetts 01003, USA

Pawel Wagner, Intelligent Polymer Research Institute and ARC Centre of Excellence for Electromaterials Science, University of Wollongong, Wollongong, NSW 2522, Australia

Bernd Wicklein, Materials Science Institute of Madrid, ICMM-CSIC, c/Sor Juana Inés de la Cruz 3, 28049 Madrid, Spain

Cun Zhu, State Key Laboratory of Bioelectronics, Southeast University, Nanjing, Peoples Republic of China 210096

CHAPTER 1

Introduction: The Concept of Biomimicry and Bioinspiration in Chemistry

TIMOTHY W. HANKS

Department of Chemistry, Furman University, Greenville, South Carolina, USA

GERHARD F. SWIEGERS

Intelligent Polymer Research Institute and ARC Center of Excellence for Electromaterials Science, University of Wollongong, Wollongong, NSW 2522, Australia

1.1 WHAT IS BIOMIMICRY AND BIOINSPIRATION?

The idea of looking to Nature to solve problems is undoubtedly as old as humanity itself. Observations of Nature, particularly of its biological face, have impacted the development of every facet of human society, from basic survival tactics to art, and from fashion to philosophy. Indeed, as a part of the biosphere ourselves, we cannot help but frame our conceptual understanding of ourselves and our environment in terms of biology. *Bioinspiration* and *biomimicry*, then, are ancient processes that take advantage of millions of years of evolutionary experimentation to help us address the many challenges that affect human well-being.

The term *biomimetics* was suggested by Schmitt in the early 1960s and was listed in Webster's dictionary as early as 1974. Webster's dictionary defined the concept as "The study of the formation, structure, or function of biologically produced substances and materials (as enzymes or silk) and biological mechanisms and processes (as protein synthesis or photosynthesis) especially for the purpose of synthesizing similar products by artificial mechanisms that mimic natural ones."[1]

While there are many historical examples that fit this definition, the formalization of the concept occurred only in the late 20th century. This formalization was

Bioinspiration and Biomimicry in Chemistry: Reverse-Engineering Nature, First Edition. Edited by Gerhard F. Swiegers.
© 2012 John Wiley & Sons, Inc. Published 2012 by John Wiley & Sons, Inc.

significant in that it arguably represented a key paradigm shift in which the chemistry community changed its focus from molecular composition to the morphology and function of molecular and supramolecular structures.

While *biomimicry* formally involves a direct replication of processes or techniques that are employed by Nature, *bioinspiration* involves a more indirect "drawing of ideas" from Nature. Here Nature serves as a rich and readily accessible source of new concepts and approaches. Of particular interest are approaches that have the potential to help solve intractable and challenging problems. *Bioinspiration* is mostly concerned with understanding the principles that underlie natural processes and then applying these principles in nonbiological settings. Benson, Share, and Flood describe the principle as follows in Chapter 4, "Bioinspired Molecular Machines":

> Bioinspiration is described as understanding the fundamental aspects of some biological activity and then recasting it in another form. Consider the Wright brothers' research program, where lift, control, and propulsion were all accepted elements of bird flight. The first two elements were recast in similar forms as wing shape and wing warp, whereas the latter was completely replaced with an engine-driven propeller. It is illustrative that propulsion was generated using very different means.

The distinction between biomimicry and bioinspiration is, however, not clear-cut. There are many shades of overlap between these two concepts. For example, a deliberate and systematic mimicry of techniques employed by Nature within systems that are far removed from Nature could be considered to be either biomimicry or bioinspiration. A good illustration of this is given by Hoffmann in his masterly exposition in Chapter 14, "Biomimicry in Organic Synthesis." He says:

> When the targets of natural product synthesis become even more complex in the 21st century, it is evident that the strategies and methods used in the last century reach their limits. Hence, organic chemistry is faced in the 21st century with the necessity to substantially increase the efficiency of syntheses by turning to new strategies. Combined with better synthesis methods, this should reduce the number of steps necessary to reach complex target structures. . . . Natural products are synthesized by Nature in the living cells from simple starting materials. . . . When new strategies for synthesis of such compounds are needed, it is obvious and advantageous to ask how Nature synthesizes such molecules in the process of biosynthesis. This raises the hope that Nature has found, through the process of evolution, an efficient route for the synthesis of a particular natural product, a route that could serve as a model for in vitro synthesis. Thus, knowledge of a biosynthetic pathway for a natural product of interest could serve as a guideline to develop a "biomimetic" synthesis. This line of thought could be expected to open reasonable approaches to the synthesis of a natural product, or at least provide a much better synthetic route than used before.

The formal distinctions between biomimicry and bioinspiration can therefore blur and become difficult to separate. For this reason, this book assigns the same weight and importance to both topics. It is left up to the reader to decide whether a particular experiment is best considered as biomimicry or bioinspiration.

1.2 WHY SEEK INSPIRATION FROM, OR REPLICATE BIOLOGY?

1.2.1 Biomimicry and Bioinspiration as a Means of Learning from Nature and Reverse-Engineering from Nature

Perhaps the key reason for studying biomimicry and bioinspiration is to learn from Nature. Biological entities and processes have evolved over billions of years to achieve forms and functions that are often remarkable, both for their efficacy and their efficiency. Humanity has a lot to learn from Nature.

Zhu and Gu in Chapter 10, "Bioinspired Surfaces II: Bioinspired Photonic Materials," put it very succinctly:

> Nature provides inexhaustible wealth to humankind [and this is the reason to learn from it].

In Chapter 6, "Bioinspired Materials Chemistry II: Biomineralization as Inspiration for Materials Chemistry," Nudelman and Sommerdijk state it thus:

> Living organisms are well known to exploit the material properties of amorphous and crystalline minerals in building a wide range of organic–inorganic hybrid materials for a variety of purposes, such as navigation, mechanical support, protection of the soft parts of the body, and optical photonic effects. The high level of control over the composition, structure, size, and morphology of biominerals results in materials of amazing complexity and fascinating properties that strongly contrast with those of geological minerals and often surpass those of synthetic analogs. It is no surprise, then, that biominerals have intrigued scientists for many decades and served as a source of inspiration in the development of materials with highly controllable and specialized properties. Indeed, by looking at examples from the biological world, one can see how organisms are capable of manipulating mineral formation so as to produce materials that are tailor-made for their needs.

Finally, Benson and colleagues make the amusing note that we do not need an alien civilization to land on Earth in order to undertake technological development by reverse-engineering. We can reverse-engineer from Nature. That is, indeed, the very basis of biomimicry and bioinspiration. They state in Chapter 4, "Bioinspired Molecular Machines":

> A variation on this last notion of bioinspiration has a healthy life in our fertile cultural imagination—revisited in fiction and urban legend alike. The proposition has been made that the explosion in technological development over the past century or so came about when humanity reverse-engineered technology that was originally fabricated by advanced alien species. While absurd as an account of modern civilization, this sequence of events is somewhat analogous to chemistry's use of bioinspiration, which takes cues from Nature's mature "technology."

1.2.2 Biomimicry and Bioinspiration as a Test of Our Understanding of Nature

It has often been said that one only truly understands a principle or a system if one is able to apply it in a functionally operational way, in a setting of one's own making. Much of the work described in this book is dedicated to this concept. It asks: Do we properly understand Nature's principles? If we do, then we should be able replicate, in at least some small measure, the feats of biology. If we cannot, then our understanding is necessarily and unambiguously incomplete. The experiment leaves little leeway for self-delusion. As noted by Benson, Share, and Flood in Chapter 4:

> Here, the direct question to be answered once the machine has been made is: "Does it move?" Or, in the parlance of the Wright brothers, "Does it fly?"

Seen in this light, bioinspiration and biomimicry can also be considered to be a test of our understanding of Nature. Indeed, every experiment is, effectively, a measure of our understanding. Swiegers, Chen, and Wagner have stated it thus in Chapter 7, "Bioinspired Catalysis":

> Every winged aircraft and putative aircraft ever built comprises nothing less than a test of the builder's understanding of the underlying principle by which birds fly, namely, the law of the aerofoil.

1.2.3 Going Beyond Biomimicry and Bioinspiration

A question that arises is: what, in the fullness of time, is the ultimate purpose of biomimicry and bioinspiration? According to several commentators, this "ultimate purpose" is not merely to emulate Nature or achieve capacities similar to those enjoyed by Nature, but rather to go beyond Nature into a man-made realm that surpasses Nature. Nobel Laureate Jean-Marie Lehn is perhaps the foremost proponent of this approach. He describes it thus in his Foreword to this book:

> Chemistry and in particular supramolecular chemistry entertain a double relationship with biology. Numerous studies are concerned with substances and processes of a biological or *biomimetic* nature. The scrutinization of biological processes by chemists has led to the development of models for understanding them on a molecular basis and of suitably designed effectors for acting on them. On the other hand, the challenge for chemistry lies in the development of *abiotic*, nonnatural systems, figments of the imagination of the chemist, displaying desired structural features and carrying out functions other than those present in biology with comparable efficiency and selectivity. Not limited by the constraints of living organisms, abiotic chemistry is free to invent new substances and processes. The field of chemistry is indeed broader than that of the systems actually realized in Nature.

1.3 OTHER MONIKERS: BIOUTILIZATION, BIOEXTRACTION, BIODERIVATION, AND BIONICS

Bioinspiration and biomimicry however, are arguably not the only descriptors of our interaction with Nature. There are several distinct approaches for making use of facts learned by observing the biosphere. The most obvious is to use natural materials directly; what we might call *bioutilization*. When the natural component of interest is too dilute for our purposes as harvested, such as natural products to be used in pharmaceuticals, they must be *bioextracted*. This technique has long been a major approach to exploiting the bounty of the biosphere and will continue to play a major role in society.

It is, moreover, often the case that a product of Nature does not meet our needs in the initially extracted form or that the extraction process may not be economically feasible. *Bioderived* materials are the result of modifying Nature's offerings to provide enhanced performance. The optimization and production of bioderived products has arguably been the key tool for the transformation of human society for centuries. For example, the development of organic chemistry from its origins in dye chemistry to its current key role in the pharmaceutical, plastics, and many other industries is largely a result of the modification of products found in Nature.

In addition to extracting and modifying natural materials for our own purposes, we have long strived to reproduce biological form and function. There are many examples of such efforts, including attempts by the Chinese to make artificial silk more than 3000 years ago, the invention of Velcro based on the hooked seeds of the burdock plant, and dry adhesive tape based on the surface morphology of gecko feet.[2] The term *bionics* was introduced by Steele, in late 1958, to promote the study of biological systems for solving physical problems. Bionics was originally defined as "the science of systems which have some function copied from Nature," but perhaps as a result of the TV series *The Six Million Dollar Man*, and recent interest in the brain/machine interface, the term has largely come to mean "biological electronics." While specific interfaces between living systems and electronics may indeed have some of the features of the original definition, we will largely avoid the use of the term here to avoid confusion.

1.4 BIOMIMICRY AND SUSTAINABILITY

In order to rationally exploit the products and processes of Nature for our own purposes, it is necessary to deconstruct very complex systems in order to decipher the underlying physical, chemical, and biological processes that result in the natural phenomena we wish to emulate. This process of deducing and exploiting the fundamental laws that govern the universe has proved to be a powerful strategy for technological development. Indeed, while modern science and technology has its origins in Nature, many of the products we surround ourselves with show little, or only superficial, resemblance to naturally occurring materials. The sheer number of humans on the planet and our ability to manipulate energy on a scale unlike anything found in the biosphere means that we have created (and continue to create) environments that are radically different from those produced by Nature. All biological systems impact their surroundings, but the unprecedented scale and rate of our activities has outstripped the capacity of the biosphere to adapt using its evolutionary approach. Our efforts to provide ourselves with comfort, security, and even amusement are often highly detrimental to the rest of the biosphere and ultimately to ourselves. Plastics are generally not degraded by the usual biological processes and their mass is not readily recycled. Sediment disruption from mining and concentration of particular elements in fabrication processes can lead to areas that are highly toxic to life forms, including our own. Pesticides, industrial waste, and pharmaceutical products can make their way into the environment, causing mutations or cellular disruptions in plants and animals. It has been clear now for some decades that the industrialization of society with scant regard for the larger biosphere has serious consequences.

The term biomimicry has been used since at least 1976 as a synonym for biomimetic,[3] but it has more recently been linked to environmentalism with the publication of *Biomimicry: Innovation Inspired by Nature*[4] by Janine Benyus and through the popularization of the idea through the work of the Biomimicry Institute.[5] Benyus's book focuses on nine core concepts derived from the study of the natural world:

Nature runs on sunlight.

Nature uses only the energy it needs.

Nature fits form to function.

Nature recycles everything.

Nature rewards cooperation.

Nature banks on diversity.

Nature demands local expertise.

Nature curbs excesses from within.

Nature taps the power of limits.

From this perspective, biomimicry becomes a strategy for not only taking advantage of Nature to produce novel structures and processes, but also as a way to combat the negative environmental impacts of current practices. New developments toward sustainable agriculture practices parallel these ideas, but there is movement within the science and engineering communities that embraces these ideas as well. A recent review[6] highlights some of the activities in the chemical engineering research and education establishments to develop programs that not only take advantage of the technological insights afforded by Nature, but also strategies for integrating industrial processes with those of the biosphere. Likewise, recent texts have explored the role that biomimicry might play in architecture[7, 8] and urban planning.[9] As human population continues to increase and resources become scarce, a biomimetic approach to organizing our cities offers a strategy for long-term survival.

In the interests of providing a balanced view, we should note that the "green" biomimetic approach described above is not without critics. Kaplinsky argues that humans too are part of Nature and that our technical achievements and physical constructs are not only on par with those of evolution, but are "natural" in the same way that the building of shelters by other animals are natural.[10] The interdependence of Nature is such that the activities of one species necessarily impact the environment of others, and while the activities of humans are dramatically larger than those of any other species, the basic principle is the same. Kaplinsky agrees that there is much to be learned from Nature, but he points out that biological designs are by no means completely optimized, even for the unique microenvironment of a given species. Evolution has produced amazing structures and strategies over the eons, but the process is exceedingly slow. Conversely, humans are able to learn, adapt, and innovate on a time scale that is very brief compared to evolutionary processes.

Kaplinsky takes issue with other ideas of the green biomimicry viewpoint. In effect, he proposes that it is possible to get carried away with the wonders of Nature, while ignoring the less palpable aspects. For example, at the risk of being overly cynical, he notes that "the fossil fuels that supply our energy are, after all, nothing but waste products of Nature that escaped its supposedly miraculous recycling process." Moreover, while Nature may "reward cooperation," it also rewards competition, parasitism, violence, and some of the most underhanded, nefarious behaviors imaginable. Indeed, the entire biosphere is a battle zone of species engaged in all-out physical, chemical, and biological warfare in a relentless struggle for resources. This battle is carried out over multiple size and temporal regimes where the primary difference between winners and losers is reproduction and whether the "recycling" commences soon or somewhat later.

Clearly, Nature is not inherently benign—a fact not lost on the defense establishment, which is concerned not only with the implications of bioweapons, but also about the ways in which biomimetics will impact areas of the warfare system from fuels to robotics.[11] Biomimicry offers tremendously powerful strategies, but also demands responsible development in order to provide benefits while mitigating potential damage. The biomimetic approach does, however, inherently encourage an examination of how a particular structure or process fits into its surroundings and may thereby assist in the development of sustainable approaches to technological and industrial development.

1.5 BIOMIMICRY AND NANOSTRUCTURE

The concept of biomimicry has been explored in a wide range of fields and attempts have been made to apply the "lessons of Nature" in a number of ways, some of them in unexpected fields. For example, Thompson uses biomimicry to propose approaches to personnel management[12] and a recent report describes a bioinspired approach to credit risk analysis.[13] While computational models have been applied extensively to biological systems, biomimetic principles have also been successfully directed toward problems in computer science, such as systems management,[14] control systems and robotics,[15] and distributed computing algorithms.[16] However, by far the most active fields making use of bioinspiration and biomimicry are those of chemistry and materials science.[17] This comes as no surprise, since there has always been a close relationship between biology and chemistry. What has changed in recent years, and is reflected in the content of this book, is the level of complexity that is involved in the biomimicry. This complexity shows itself in many ways, but particularly in material morphology across multiple size regimes—structural hierarchy-and in the new field of nanotechnology.

In 1994, the U.S. National Research Council issued a report outlining the potential offered by biological hierarchical structures to materials scientists.[18] They noted that while Nature has a relatively limited range of materials to work with, composites with astoundingly diverse properties result through structural control over multiple length scales.

Hierarchical materials systems in biology are characterized by:

- Recurrent use of molecular constituents (e.g. collagen), such that widely variable properties are attained from apparently similar elementary units
- Controlled orientation of structural elements
- Durable interfaces between hard and soft materials
- Sensitivity to—and critical dependence on—the presence of water
- Properties that vary in response to performance requirements
- Fatigue resistance and resiliency
- Controlled and often complex shapes
- Capacity for self-repair

The report goes on to describe specific examples of natural materials with unique properties and technological challenges that could potentially be met by mimicking key features. Yet the actual realization of the examples offered is difficult, as it requires not only understanding the material's composition and properties at the different length scales, but also the ways in which they work together to provide the properties of interest.

In 2010, the U.S. National Nanotechnology Initiative reached its 10th anniversary, with more than $14 billion directed toward the development of new technologies.[19] Worldwide, more than $50 billion (U.S.) has been spent by the public and private sectors, with many nations instituting formal nanotechnology programs. The global focus on nanotechnology has accelerated the ongoing development of imaging and analytical tools that bridge the gap between the traditional chemistry size regime and that of biology. From the "top–down" perspective, these tools permit ever-higher resolution for probing of material structure. From the "bottom–up" perspective, they give insight into the organization of molecules into increasingly larger and more elaborate assemblies.

Optical and electron microscopes provide striking and appealing images of natural structures that can take us from very large to very small (nanometer) length scales. At the small end though, the scanning probe microscope (SPM) family of instruments are key tools that help nanoscience and biology combine to provide a unique biomimetic perspective.[20]

Beginning with the scanning tunneling microscope and later the more biologically relevant atomic force microscope (AFM), SPMs involve the rastering of a very sharp tip (on the order of 10 nm in radius of curvature) across a surface. The tip is affixed to a cantilever, which undergoes deflection in response to surface topography (in the case of simple AFM) or other forces. A recent review on the use of AFM in the study of amyloids illustrates the power of scanning probe technologies to provide a variety of detailed information.[21]

AFM and other SPM technologies are tremendously powerful tools for examining the surfaces and interfaces found in both synthetic and biological materials. It is the surface of a material, or a component within a composite, that determines whether another environmental actor will adhere or simply slip away. Surfaces are responsible for the ways in which light is absorbed and reflected, giving an object its color. Surfaces are where an object is first subject to wear and corrosion. In atomically homogenous nanoparticles, the surface atoms experience forces different from those in the bulk and may have distinctly different chemical behavior.

In Chapters 9 and 10, inspiration is taken from different types of biological surfaces. In a sweeping and detailed exposition, Qu, Li, and Dai examine, in Chapter 9, the issue of dry adhesion using the gecko foot as inspiration. They discuss recent progress and the potential of synthetic mimics of this incredible structural design. In Chapter 10, Zhu and Gu consider the phenomenon of structural color, which involves the use of nanopatterned surfaces to generate bright and vividly colored surfaces. Their inspiration is the wings of the Morpho butterfly and related structures, which achieve vibrant color by means of interference effects due to their surface and near-surface structures.

1.6 BIOINSPIRATION AND STRUCTURAL HIERARCHIES

Throughout Chapters 9 and 10, the importance of structural hierarchy on surface properties is demonstrated. The gecko's toes, for example, are covered arrays of hair-like structures called *setae*, which are in turn split into even finer structures. This concept of increasing effective surface area is not restricted to increasing adhesive forces. In Chapter 13, Della Pelle and Thayumanavan present examples where functional arrays can be used for light-harvesting and drug delivery. Some arrays may be thought of as large two-dimensional surfaces that are roughened into the third by the attachment of ever smaller structures. Dendridic structures, also discussed in Chapter 13, are better conceptualized as polymers that grow from simple molecules into increasingly bifurcated three-dimensional arrays through the coupling of monomers with connectivity greater than two.

In Chapter 8 Himmelein and Ravoo look at amphiphilic bilayer "surfaces" that have effectively been bent until they form hollow vesicles. At their most basic, these vesicles are composed of a homogenous collection of amphiphiles—molecules containing a hydrophilic head group and a lipophilic tail. At their most complex level, they are the elaborate architectures that define the cell walls in living organisms. The phospholipid-based cell wall is a highly sophisticated, dynamic structure complete with functional components that enable the cell not only to retain its contents but also to transport nutrients and waste, to respond to chemical and physical stimuli, and to perform other functions. Synthetic vesicles used in commercial applications are far less ambitious in their function, mainly serving to encapsulate drugs or other species. However, through biomimicry, more complex structures are being developed by adding molecular recognition elements to the surface, introducing subcompartments, and introducing "smart" stimulus–response capability. The relative ease with which different regions of the vesicle may be modified makes these structures interesting platforms for the development of nanoscale devices.

Nature produces much more than interesting surfaces and pseudosurfaces. There is a tremendous interest in bioinspired composite materials in which the synergism between materials with different physical properties and different size scales leads to useful macroscopic physical properties, as well as to important biological and chemical features.[22] For example, both the aging of the world's population and ongoing violent conflicts are driving the search for synthetic materials that can be used to replace human tissue. The challenges of tissue engineering and regenerative medicine are as great as the need for high volume abiological replacements.[23] Some applications in this field require materials with good mechanical strength, while others demand constructs that are soft and extensively vascularized. The majority of materials must be biocompatible, meaning not only nontoxic and acceptable to the immune system, but also with the proper mechanical properties to interface with natural tissue. Sometimes the requirements for a particular application seem almost absurd in light of previous generations of synthetic materials, yet Nature shows they are possible. For example, an implanted neural electrode should be very soft and highly hydrated, yet capable of conducting electricity. Ideally, it would act as a cellular scaffold that minimizes the inflammatory response generated by the insertion of the electrode and would encourage the directional growth of neurons through the controlled release of chemical, electrical, and perhaps viscoelastic cues. Biocompatible hydrogels are under development that may be able to fulfill all of these functions.[24]

Chapters 5 and 6 review biomimetic materials in which the inorganic aspects of biology are exploited. In Chapter 5, Aranda, Fernandes, Wicklein, Ruiz-Hitzky, Hill, and Ariga discuss the formation, properties, and applications of organic–inorganic hybrid materials, which can provide strength *and* fracture resistance due to clever structural hierarchy and control of component interfaces. In Chapter 6, Nudelman and Sommerdijk present a class of synthetic materials inspired by biomineralization. There are countless examples in Nature where organisms extract inorganic ions from their environment to create relatively hard structures with both striking macroscopic shapes and microscopic structures that provide properties critical to the organism. Sommerdijk illustrates how lessons learned from these structures can be applied to the construction of new ceramics and semiconductors. Throughout this chapter, an emphasis is placed on the importance of considering not only the structures of biological models, but also the processes that lead to their formation.

1.7 BIOINSPIRATION AND SELF-ASSEMBLY

Biological processes generally take place under mild conditions and in aqueous solution. Not only are these conditions quite different from those of traditional materials synthesis, the dynamical behavior of the resulting products is also quite different. Synthetic structures are generally conceived as being in their final, complete form at the end of the fabrication process, while supramolecular biological structures derive much of their functionality from their spatial organization. They are also dynamic, responding to environmental cues to change both shape and activity. To achieve this, biological systems rely on a combination of relatively strong covalent bonds for their primary structure and both directional and nondirectional weak interactions for higher level structure and assemblies.[25] The primary mechanism for the construction (and deconstruction) of biological entities is one of self-assembly, where the basic building blocks of a superstructure are guided into place by strategic positioning of the functional groups that give rise to the weak interactions. The ability to build structures with atomic precision is also a goal of nanoscience and considerable effort is being applied toward designing the self-assembling building blocks that lead to useful superstructures.

Self-assembly inevitably generates defects in a structure. While "defects" are the origin of a property of interest in some materials, even in those cases it is necessary to be able to control the number and locations of defects. Fully reversible systems operate under thermodynamic control, allowing defects to be repaired, but this is a slow process and only provides access to structures at the global thermodynamic minimum. The first limitation can be problematic in biology, but is even more so in the industrial world, where high throughput is not only desirable, but may determine the ultimate feasibility of a given process. The second limitation is also important, because many interesting structures lie at local thermodynamic minima. Biology shows that such structures may be accessed by "assisted self-assembly," where reaction conditions or biocatalysts provide viable pathways to kinetic structures.[26]

An alternative to thermodynamic self-assembly is "kinetic" or "nonequilibrium" self-assembly. Here the system cascades through a series of steps to end up at the kinetically favored product, which would typically not lie at the global thermodynamic minimum. In such processes, each step sets up the next, leaving the system with little option but to traverse a pathway that seems almost predetermined, much like the pathway that is followed when a line of dominos is toppled. Nonequilibrium processes of this type are believed to be common in biology. For example, the remarkable rapidity with which proteins fold is consistent with this process being largely a nonequilibrium one.

Self-assembly is a general theme that necessarily runs throughout this text, but the topic is addressed in detail in Chapters 2 and 3. In Chapter 2, Lindoy, Richardson, and Clegg provide an overview of self-assembly in polymeric, metal-organic, and other nonbiological systems that generate structures that have "biological" features and functions. The authors provide specific examples of self-assembled structural elements that may lead to novel applications. In Chapter 3, Ercolani and Schiaffino discuss the role of cooperativity in biological and abiological systems.

Cooperativity is an important feature in molecules that display allosteric responses and can provide selectivity in binding events for sensors and stimuli-responsive constructs. Cooperativity is particularly important in nonequilibrium self-assembly processes, which are path and time dependent.

In Chapter 11, Binder, Schunack, Herbst, and Pulamagatta expand on the self-assembly theme in a discussion of the dynamic behavior of biomimetic polymers. As with biological polymers such as proteins, these synthetic polymers display dynamic changes in their higher order structure, including folding and coiling into predefined shapes. Elements of self-assembly and molecular recognition are also found in Chapter 4, by Benson, Share, and Flood, with an examination of bioinspired molecular machines. Here, self-assembly is required for both the initial formation of the machines and to drive the switching between individual states. The harnessing of biology to design nanoelectromechanical systems has the potential to lead to systems with not only hierarchical structure but also hierarchical mechanical motion.[27]

1.8 BIOINSPIRATION AND FUNCTION

In biology, structure is intimately coupled to function. Natural structures are exquisitely engineered to operate within the chemical and energetic constraints of the biological environment and therefore often incorporate highly efficient or even unexpected (from the synthetic viewpoint) functions. For example, polymer science has provided us with products that have excellent mechanical properties, but when polymeric products fail they are usually nonrecyclable (at least by biological mechanisms) and their primary purpose is irreversibly compromised. Conversely, biological composites feature mechanisms for restoring functionality after sustaining damage. Recent efforts to develop self-healing polymers and polymer composites are taking the initial steps toward low-maintenance, high-durability products and devices.[28] The challenge in this area is to refine the biological inspiration so that it will work with synthetic processes. Biological repair requires a dynamic and relatively elaborate support infrastructure; it is often slow (days to months) compared to the needs of synthetic materials (minutes to perhaps hours).[29]

Bioinspired functionality is a theme woven throughout this text, but possibly the field where Nature's functional molecules inspire the greatest respect from those developing their synthetic counterparts is that of catalysis. Enzymes are highly efficient and can display extraordinary selectivity by orienting substrates, stabilizing intermediates, and other processes that are not yet fully understood.[30] In Chapter 7, Swiegers, Chen, and Wagner explain how the conformational dynamics of enzymes is an integral part of their catalytic function and how biomimetic catalysts can make use of conformational flexing to replicate natural efficiencies.

Natural compounds have long been a source of inspiration for the pharmaceutical and related health care industries. Tremendous effort has gone into the total synthesis of natural products as well as into preparing derivatives that might show superior performance.[31] In Chapter 14, Hoffmann looks at functionality from the perspective of process rather than from a largely structure/function viewpoint. Biomimetic and cascade reactions in synthetic organic chemistry, for instance, are able to produce target molecules with high step, atom, and redox efficiencies. This approach to the production of pharmaceutically and industrially important compounds ties into the green promise of biomimicry. Comparisons of our current abilities with those of biological systems gives us a benchmark of our progress and ideas for further refinement.

Biological organisms live in complex environments and survive by collecting quantitative and qualitative information about the changes around (and within) them. Like biological structures, the sensing function is hierarchical and takes place on the subcellular level on up to macroscopic sensors with sensing processes triggering responses across different size and temporal regimes.[32]

In Chapter 12, Le Gac, Jabin, and Reinaud use the example of synthetic receptors based on calix[6]arene-based receptors as biomimics of molecular and ion-pair recognition elements. The low toxicity and versatility of this platform places it alongside crown ethers and cyclodextrins as some of the most important classes of macrocycle with a myriad of potential uses.[33]

1.9 FUTURE PERSPECTIVES: DRAWING INSPIRATION FROM THE COMPLEX SYSTEM THAT IS NATURE

In the final chapter of this work, Cady, Robinson, Smith, and Swiegers briefly explore some future perspectives in the field of bioinspiration and biomimicry in chemistry. These include an examination of the big picture of life itself, its origin and its character. They show that life in Nature comprises an extraordinarily complicated web of interconnected interactions that displays properties which are characteristic of so-called complex systems, including emergence, evolution, autonomy, and others.

The field of complex systems science studies the way in which multiplicities of independent elements interact with each other to create chains of action and reaction that lead to amplified and/or unique outcomes. Examples include family trees (chains of procreation), weather systems (chains of interacting weather events), traffic patterns on intersecting highways (chains of automobile movements), and economic behavior (in, for example, the chains of mutually beneficial transactions on stock exchanges). The most important complex system, at least to us, involves the way that biochemical entities interact with each other to create life itself (chains of biochemical events). A future perspective that is just beginning to emerge in bioinspiration and biomimicry is to understand and replicate the processes at play. In biology, this field is called *systems biology*. The corresponding new and emerging field of *systems chemistry* aims to study and apply the same concepts to chemistry.

The significance of these studies is that they go beyond mere chemistry and have implications in a host of other fields, including some of those mentioned previously, like information technology (self-improving computer programs), social interactions and human behavior (e.g., criminology, sociology, ethics), and economics (the phenomenon of "economic growth"). As such, they offer the prospect of unifying science and improving the human experience.

This book provides a perspective on how the study of Nature has had a profound impact on the disciplines of chemistry and materials science. It is a story that is thousands of years old, yet we are still in the introductory chapter. The inspiration that will be gleaned from the earthly biosphere over the coming years is vast and we may never discover all of its secrets, much less elucidate the web of synergistic interactions that makes it all work. It is breathtaking to realize that our world is but one among a vast number of likely worlds, many of which will surely have evolved their own biospheres with their own unique materials and interconnected processes. In the fullness of time bioinspiration and biomimicry may ultimately grow to encompass an interplanetary aspect. Perhaps this will one day turn out to be the best justification for humankind to reach for the stars.

REFERENCES

1. Harkness, X. X. *IEEE Eng. Med. Biol.* **2004**, *23,* 20.
2. Vincent, J. F. V.; Bogatyreva, O. A.; Bowyer, A.; Paul, A.-P. *J. R. Soc. Interface* **2006**, *3,* 471.
3. Busch, D. H. *Abs. Papers Am. Chem. Soc.* **1976**, *Suppl. 1*, 1.
4. Benyus, J. M. *Biomimicry: Innovation Inspired by Nature*, William Morrow, New York, 1997.
5. http://www.biomimicryinstitute.org/.
6. Garcia-Serna, J.; Perez-Barringon, L.; Cocero, M. J. *Chem. Eng. J.* **2007**, *133,* 7.
7. Ginatta, C. *ARCHITECTURE Without Architecture: Biomimicry Design*, VDM Verlag Dr. Müller GmbH & Co. KG, Saarbrucken, Germany, 2010.
8. Gruber, P. *Biomimetics in Architecture: Architecture of Life and Buildings*, Springer, New York, 2011.
9. Spiegelhalter, T.; Arch, R. A. *The Sustainable City VI: Urban Regeneration and Sustainability* (Eds. Brebbia, C. A.; Hernandez, S.; Tiezzi, E.), Wit Press, Southampton, UK, 2010.
10. Kaplinsky, J. *Architectural Design* **2006**, *76,* 66.
11. *Bio-inspired Innovation and National Security* (Eds. Armstrong, R. E.; Drapeau, M. D.; Loeb, C. A.; Valdes, J. J.), National Defense University Press, Washington DC, 2010.
12. Thompson, K.; Bonk, C. J.; Cross, J. *Bioteams: High Performance Teams Based on Nature's Most Successful Designs*, Meghan Kiffer Press, Tampa, FL, 2008.
13. Yu, L.; Wang, S.; Lai, K. K.; Zhou, L. *Bio-Inspired Credit Risk Analysis: Computational Intelligence with Support Vector Machines*, Springer, New York, 2008.
14. Nakrani, S.; Tovey, C. *Bioinspir. Biomim.* **2007**, *2,* S182.
15. Passino, K. M. *Biomimicry for Optimization, Control and Automation*, Springer-Verlag, London, 2005.
16. Afek, Y.; Alon, N.; Barad, O.; Barkai, N.; Bar-Joseph, Z. *Science* **2011**, *331,* 183.

17. Sanchez, C.; Arribart, H.; Guille, M. M. G. *Nature Materials* **2005**, *4,* 277.

18. Tirrell, D. A. (coord.) *Hierarchical Structures in Biology as a Guide for New Materials Technology*, National Material Advisory Board, The National Academic Press, Washington DC, 1994.

19. Sargent, J. F. Jr. *The National Nanotechnology Initiative: Overview, Reauthorization, and Appropriations Issues*, Congressional Research Service, Washington DC, 2011.

20. Casuso, I.; Rico, F.; Scheuring, S. *J. Mol. Recog*. **2011**, *24,* 406.

21. Gosal, W. S.; Myers, S. L.; Radford, S. E.; Thomson, N. H. *Prot. Pept. Lett*. **2006**, *13,* 261.

22. *Biomimetics, Learning from Nature* (Ed. Mukherjee, A.), InTech, Vukovar, Croatia, 2010.

23. Schenke-Layland, K. *Adv. Drug Deliv. Rev*. **2011**, *63,* 193.

24. Guiseppi-Elie, A. *Biomaterials* **2010**, *31,* 2701.

25. Mohammed, J. S.; Murphy, W. L. *Adv. Mater*. **2009**, *21,* 2361.

26. Hirst, A. R.; Roy, S.; Arora, M.; Das, A. K.; Hodson, N.; Murray, P.; Marshall, S.; Javid, N.; Sefcik, J.; Boekhoven, J.; van Esch, J. H.; Santabarbara, S.; Hunt, N. T.; Ulijn, R. V. *Nature Chemistry* **2010**, *2,* 1089.

27. Huang, T. J.; Flood, A. H.; Brough, B.; Liu, Y.; Bonvallet, P. A.; Kang, S.; Chu, C.-W.; Guo, T.-F.; Lu, W.; Yang, Y.; Stoddart, J. F.; Ho, C.-M. *IEEE Trans. Autom. Sci. Eng*. **2006**, *3,* 254.

28. Blaiszik, B. J.; Kramer, S. L. B.; Olugebefola, S. C.; Moore, J. S.; Sottos, N. R.; White, S. R. *Annu. Rev. Mater. Res*. **2010**. *40,* 179.

29. Vincent, J. F. V. *Proc. Inst. Mech. Eng. H: J. Eng. Med*. **2009**, *223,* 919.

30. Deuss, P. J.; den Heeten, R.; Laan, W.; Kamer, P. C. J. *Chem. Eur. J*. **2011**, *17,* 4680.

31. Beghyn, T.; Deprez-Poulain, R.; Willand, N.; Folleas, B.; Deprez, B. *Chem. Biol. Drug Des*. **2008**, *72,* 3.

32. Johnson, E. A. C.; Bonser, R. H. C.; Jeronimidis, G. *Philos. Trans. R. Soc. A* **2009**, *367,* 1559.

33. Sansone, F.; Baldini, L.; Casnati, A.; Ungaro, R. *New J. Chem*. **2010**, *34,* 2715.

CHAPTER 2

Bioinspired Self-Assembly I: Self-Assembled Structures

LEONARD F. LINDOY

School of Chemistry, The University of Sydney, Sydney, NSW 2006, Australia

CHRISTOPHER RICHARDSON

School of Chemistry, University of Wollongong, Wollongong, NSW 2522, Australia

JACK K. CLEGG

School of Chemistry and Molecular Biosciences, The University of Queensland, Brisbane St Lucia, QLD 4072, Australia

2.1 INTRODUCTION

Self-assembly processes are ubiquitous in Nature where, typically, multifunctional building blocks are assembled into larger molecular entities showing considerable sophistication in both their function and form. These natural self-assembly processes are often of exquisite subtlety and include, among many others, protein folding, the assembly of DNA, and the formation of bilayers, micelles, and vesicles. In this chapter we discuss selected self-assembled synthetic structures, in general prepared by the "bottom–up" approach. These structures mimic or were inspired to a greater or lesser degree by aspects of natural systems. This discussion is intended, in part, to introduce the reader to certain of the structural types that will be discussed in greater detail in later chapters.

Since the rise of supramolecular chemistry as a defined subdiscipline of chemistry over the past four decades or so—recognized by the award of the Nobel Prize to D. J. Cram, J.-M. Lehn, and C. J. Pedersen in 1987—there has been steady progress in understanding the rules involved in self-assembly processes and applying them to the construction of synthetic assemblies. This has led to generally increased appreciation of the latent steric and electronic information inherent in a wide range of molecular or ionic building blocks that are either already available

Bioinspiration and Biomimicry in Chemistry: Reverse-Engineering Nature, First Edition.
Edited by Gerhard F. Swiegers.
© 2012 John Wiley & Sons, Inc. Published 2012 by John Wiley & Sons, Inc.

or are able to be synthesized. A large number of structures incorporating metal ions as structural elements have now been reported. This is perhaps not surprising since each metal ion type may be considered to be a package of structural and electronic information whose unique identity can potentially be harnessed to achieve a particular supramolecular outcome. As for natural systems, the bottom–up approach has been successful in constructing a range of new nanoscale materials displaying a wide range of properties, often related to those of Nature's materials.

As in Nature, the "glue" between the ionic and/or molecular building blocks in these synthetic assemblies consists of weaker noncovalent interactions. These include hydrogen bonding, dispersion forces, π-π stacking, π-cation and π-anion interactions, as well as the full complement of electrostatic (ion–ion, ion–dipole, dipole–dipole) interactions. Metal-donor atom coordination bonds of generally moderate strength can also be employed (giving rise to metallosupramolecular systems).[1, 2] Overall, hydrogen bonding is a key interaction across many self-assembled systems, reflecting the directionality and specificity of this interaction type. This is not surprising in the case of Nature's assemblies since classical natural building blocks such as the carbohydrates, amino acids, and nucleic acids all incorporate numerous hydrogen bond donors and acceptors.

While the level of complexity generally achieved so far for synthetic systems falls considerably short of that routinely exhibited by natural systems,[3] steady progress toward more complex systems is being made. Since the same rules apply to Nature's assemblies as for synthetic systems, there is every reason to anticipate continuing progress in this respect in the future. Nevertheless, there can be no doubt that the design and preparation of more complex, functional synthetic systems remains a considerable challenge at both the intellectual and the practical level—requiring not only the appropriate choice of building blocks but also keen insight into the skilled application of the weak noncovalent interactions mentioned above. A good measure of creativity on the part of the practitioners is also clearly an important ingredient!

Supramolecular assemblies, whether natural or synthetic, show a common feature in being capable of self-repair, implying inherent reversibility with respect to their formation processes. This is a direct consequence of the use of weak (or moderate, in the case of metal-donor) interactions in their construction. Apart from providing a mechanism for error correction during their formation (see below), such reversibility is also sometimes crucial to the function of the resulting assembly. For example, in molecular and ionic transport processes the host (receptor) entity must first take up the substrate, then release it under the influence of an external stimulus (such as a pH change or the presence of a concentration differential). Hence, strong thermodynamic binding of a suitable guest molecule or ion in such cases is not necessarily advantageous.

Electronic and steric complementarity is another feature of importance in the design and assembly of supramolecular systems. Such complementarity may be considered to act at two levels. First, it may refer to the complementarity between adjacent building blocks in constructing a given molecular assembly. Second, on a larger scale, a degree of host–guest complementarity between a cleft- or

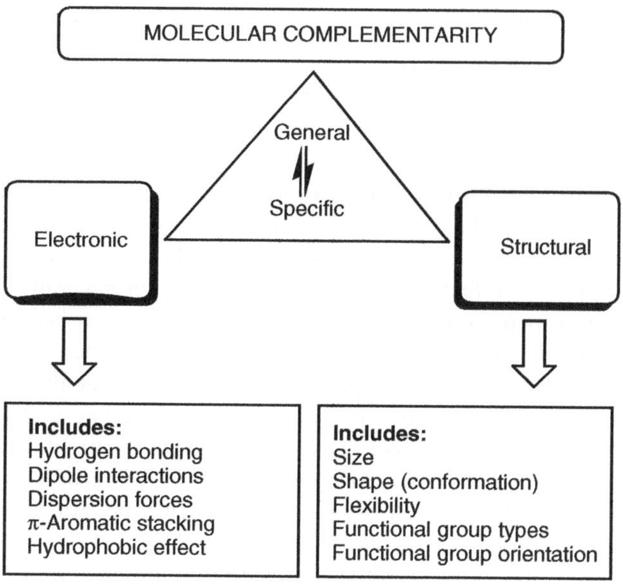

Figure 2.1 Parameters that may influence molecular and ionic complementarity.

cavity-containing assembly and its potential guest is necessary for guest encapsulation to occur. A range of parameters, including appropriate use of selected noncovalent weak intermolecular interactions of the types mentioned above, may typically contribute to achieving such complementarity (see Figure 2.1).

The preparation of bioinspired systems has, so far, proceeded along two general lines. First, fully synthetic components have been constructed and assembled into systems that mimic, at least in part, particular functions of the natural systems. A large number of published structures fall into this category. Second, there have been a smaller number of studies that employ naturally occurring molecules, with their built-in propensities for self-assembly (and in some cases functionality), to produce new synthetic molecular assemblies that "borrow" some of their characteristics from the natural systems. Particularly notable in the latter case is the use of DNA-containing synthetic assemblies, often displaying aesthetic nuances, as exemplified by the research of the Seeman group.[4] In addition to being very challenging science, the results of such a hybrid approach have so far been quite impressive.

2.2 MOLECULAR CLEFTS, CAPSULES, AND CAGES

Nature makes good use of molecular (and ionic) encapsulation in a range of important biological systems that include enzymes, proteasomes, and viral capsids. Natural cage and cleft systems are typically characterized by the presence of deep cavity-like structures whose biological function is associated with the encapsulation of guest species. Such an inclusion process may be associated with some combination of the following:

- Selective guest uptake (sometimes associated with chiral recognition)
- Induced high catalytic activity
- Concentration and guest storage of particular guests (including toxic species)
- Molecular/ionic guest transport processes

The guest can range from a single molecule (or ion) to considerably larger entities. In many cases, where guest uptake occurs, this involves transfer from the bulk surroundings into the microenvironment in the host's cavity. Such enclosure typically has a marked effect on the guest, with inclusion generally resulting in a range of altered properties. In a good many instances this has been demonstrated to include a marked alteration of guest reactivity (even when the cavities lack specifically tailored functional groups of the type present at the active sites of enzymes).[6, 8, 13, 16–20] To give just one example, a carboxylate group is unable to hydrolyze a polysaccharide in water but can do so when enclosed at the active site of *lysozyme* since in the latter case the "protecting" solvating water shell has been effectively stripped in the hydrophobic pocket of the enzyme, effectively enhancing the nucleophilicity of the carboxylate group.

The spontaneous and often selective inclusion of numerous neutral and charged (both cationic and anionic) guests in the cavities of a wide range of cage-like synthetic hosts (often referred to as "container" molecules or assemblies)[9, 11–14] has now, of course, been the subject of a vast number of studies, with many systems of this type showing parallels in their behavior and properties to those observed in Nature's encapsulating systems.[15] For example, like particular natural cage and cage-like systems, the reaction of molecules held in the restricted cage environments of particular synthetic hosts has been documented to differ significantly from their reaction in bulk solution.[6, 8, 13, 16–20] Encapsulations giving rise to novel stereo- and regioselective reactions are well documented; for example, Diels–Alder reactions displaying unusual regioselectivities have been demonstrated under such conditions.[16] Once again, factors such as the degree of complementarity present between host and guest together with solvation/desolvation differences, the restriction of possible guest conformations in the cavity, and the forced presence of reagent proximity may all contribute to such a reactivity difference.

Over the past two decades there has been increased effort to devise syntheses for larger cage-like molecules and assemblies, a number of which exhibit polyhedral shapes corresponding to those of Platonic or Archimedean solids.[21] While the majority of these larger structures fall within the realm of metallo-supramolecular chemistry,[22–27] a range of such larger metal-free structures has now also been reported.[8, 14, 28–32] A number of the larger polyhedral structures[28, 33] mirror the shapes of particular naturally occurring systems such as the virus capsids and the protein enzyme *lumazine synthase*. The capsids represent a well-defined group of biostructures[34] that serve as the protein shells of viruses and consist of oligomeric structural subunits that enclose the genetic material of the virus. Collectively, they are assembled from common protein building blocks whose shapes and connection points give rise to their shell-like structures. The faces of these unusual cages are composed of one or more proteins and are held together by weak noncovalent interactions. Overall, capsids prototypically display icosahedral (12 vertices and 20 equilateral triangular faces) structures—although a number of other closed polyhedral shapes, including dodecahedra, truncated-icosahedra, and pentakis-dodecahedra, also occur.[35] Similarly, *lumazine synthase*, an enzyme that catalyzes the penultimate step in the conversion of a precursor to riboflavin,[36] has a shape composed of protein subunits in the form of a large icosahedral assembly (60 subunits ≡ 12 pentamer). Examples of larger synthetic cages whose shapes resemble these unusual natural systems are presented later in this chapter.

2.2.1 Organic Cage Systems

As intimated earlier, the range of purely organic cage and cage-like structures now synthesized is quite large. Individual systems of great structural diversity have been investigated across both the fully covalent and self-assembled categories. For example, cage-like systems that range from simple self-assembling micelles that can act as primitive hosts, to large container systems capable of including a nanoscale guest or multiple smaller guests have now been reported.[37] Indeed, numerous organic cages and clefts of varying size have been synthesized over the past 40 years or so. These include the well-known cryptand and fullerene categories of closed cages as well as a range of fully organic "closed" and "open" cages based on preformed macrocyclic scaffolds such as the natural cyclodextrins or synthetic cyclic derivatives based on calixarene, resorcinarene, and curcurbit[*n*]uril rings.[29, 38, 39] Cram's spherands, cavitands, and related derivatives also fall into this category.[40, 41]

Fully covalent larger organic cages and cage-like systems not derived from the preformed ring systems just mentioned have also been synthesized via conventional direct organic synthesis starting from a range of precursors.[42–44] However, in part as a consequence of the nonreversibility of the majority of most classical organic reaction types, such cavity-containing products have tended to be obtained in quite low overall yield.[42] For example, the interesting cage trinacrene **1** (Scheme 2.1), which was proposed as a suitable host for a range of organic, organometallic, and

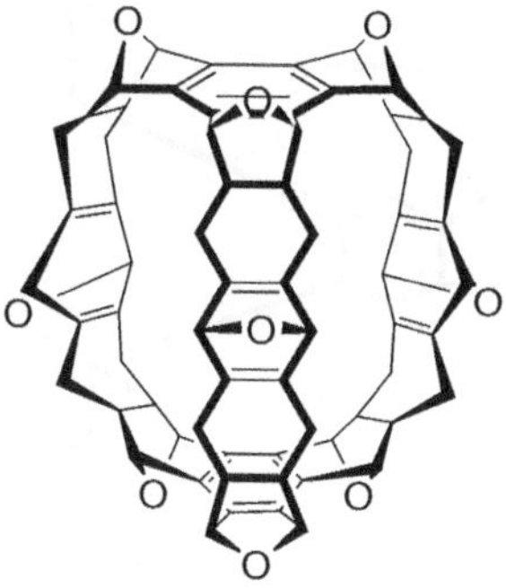

1

Scheme 2.1

inorganic guests,[45] was prepared in a conventional four-step synthesis starting from furan and hexabromobenzene in an overall yield of less than 0.01%![42, 46]

In view of the above propensity for conventional synthetic procedures to result in low overall yields, the use of a template strategy provides a potential means for assisting the formation of a particular cavity-containing product in higher yield. With regard to this, it is noted that templation is a common process employed by Nature for directing the formation of covalent bonds—as exemplified by the building of protein chains on messenger RNA via catalysis involving the ribosome. Indeed, the use of templation to direct the formation of self-assembled cages and capsules has been a moderately common procedure over many years.[47] For example, in a recent study, it was demonstrated that the water soluble cavitand **2** (Scheme 2.2) spontaneously forms a self-assembled dimeric capsule in the presence of different hydrophobic templating guests that include highly complementary rigid steroids as well as, perhaps surprisingly, flexible straight-chain hydrocarbons.[48] The templating action in the latter case is presumably driven by weak C-H-π interactions of the hydrocarbon guests with **2** as well as through the operation of the hydrophobic effect.

In contrast to the above self-assembled (supramolecular) systems, the use of templates has been less frequently exploited for the preparation of fully covalent cage and capsule products,[41] especially for synthesizing larger (nanoscale) cages.

Although in principle more restricted in scope, the rise of dynamic covalent chemistry[49, 50] over recent years has provided a successful approach for the synthesis of a number of larger all-covalent organic cages. The use of dynamic covalent chemistry mimics the reversible behavior associated with many biosynthetic

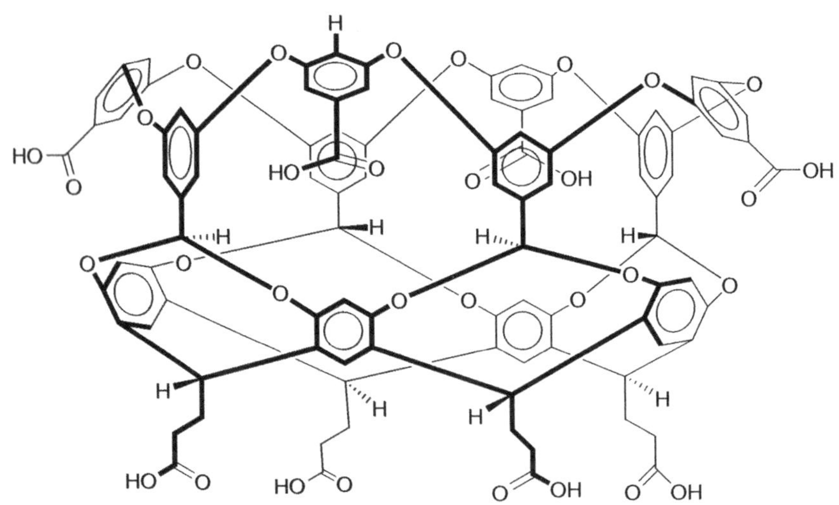

2

Scheme 2.2

processes in that it relies on reversible covalent bond formation and hence promotes the formation of the target cage in higher yield through the provision of "error correction" (provided the cage corresponds to the most stable thermodynamic product under the conditions employed).[50, 51]

For the majority of studies of the above type, the synthetic procedure has been centered on the use of reversible imine (Schiff base) bond formation.[42, 52] Imine linkages are one of the few organic groups that can exist in solution in equilibrium with their precursors—namely, the corresponding carbonyl (aldehyde or ketone) and amine derivatives—where the equilibrium corresponds to the reversible hydrolysis or solvolysis of the imine linkage. In an increasing number of instances such an approach has allowed the generation of cages in fewer steps and in higher yield than occurs on employing a more traditional synthetic approach such as that exemplified by the synthesis of **1** (Scheme 2.1).

The power of dynamic covalent synthesis is convincingly demonstrated by its use in a one-pot synthesis to yield imine-linked, fully covalent nanocubes of type **3** (Scheme 2.3). Both cubes were obtained in approximately 90% yield.

In these products the tritopic C_3-triformylcyclobenzylene unit **4** (Scheme 2.3) was employed to form each corner of the cube, with the corners being linked via linear diamines to yield the sides. Overall, this involved the formation of 24 imine bonds. The tri-aldehyde **4** was chosen as a suitable precursor for the corners since the linked aryl units holding the aldehydes are essentially orientated mutually orthogonally. The mixing of **4** with the required diamine (1,4-phenylenediamine or benzidine), in an 8:12 ratio in chloroform containing trifluoroacetic acid as catalyst, led in each case to the formation of the corresponding cube of type **3**.

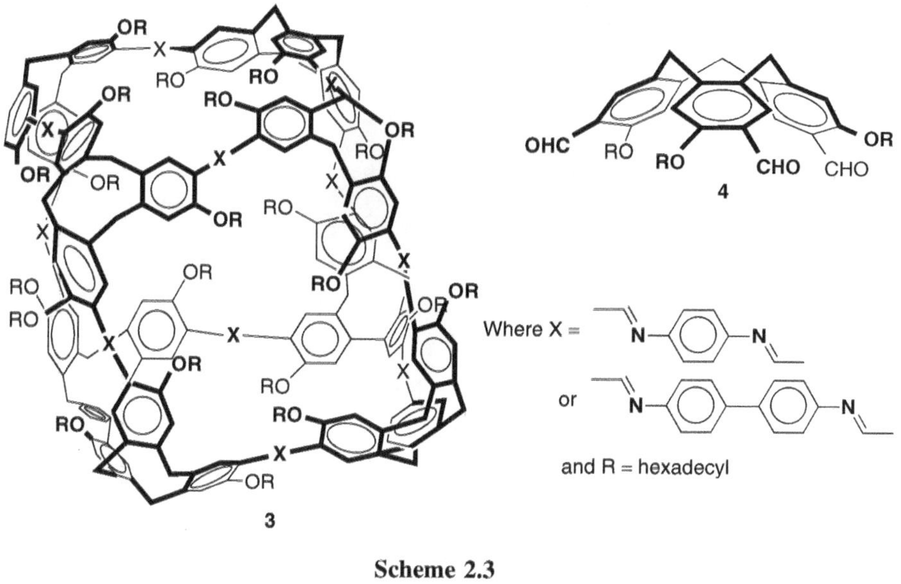

Scheme 2.3

2.2.2 Metallosupramolecular Cage Systems

There are now a very considerable number of metallosupramolecular cage and capsule systems formed by self-assembly processes in which metal ions are incorporated in the framework of the cage structures and which show inclusion behavior that resembles that exhibited by biological systems. As some of these systems have recently been reviewed,[6, 16, 22, 26, 27, 53] only a selected few examples will be mentioned here. In the first of these, Fe(II) was demonstrated to interact in acetonitrile with the quaterpyridine-derived ligand **5** (Scheme 2.4), leading to the assembly of an 8+ charged tetrahedral shaped cation of type $[Fe_4L_6]^{8+}$ which was found to spontaneously encapsulate the polyatomic anions BF_4^-, PF_6^-, and $[FeCl_4]^-$ from solution.[25, 54]

There is evidence that the smaller BF_4^- anion undergoes fast exchange in and out of the cavity while the larger PF_6^- anion is slow to enter the cavity and, once inside, showed no tendency for exchange under ambient conditions. The structure of the $[FeCl_4]^-$ inclusion complex is depicted in Figure 2.2.[24] In this case the cage

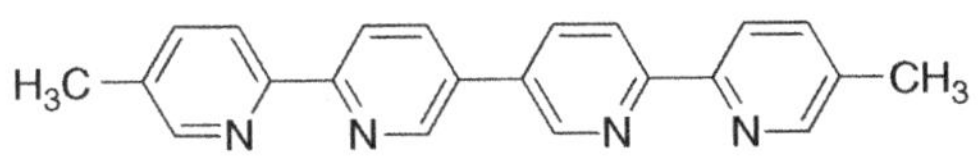

5

Scheme 2.4

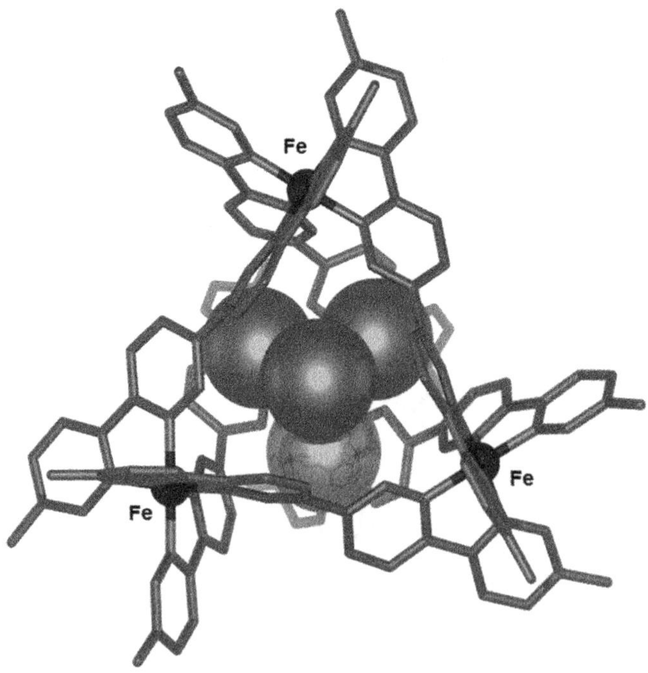

Figure 2.2 X-ray structure of the cationic metallosupramolecular cage $[Fe_4L_6(FeCl_4)]^{5+}$ in which an $[FeCl_4]^-$ anion occupies the center of the tetrahedron.

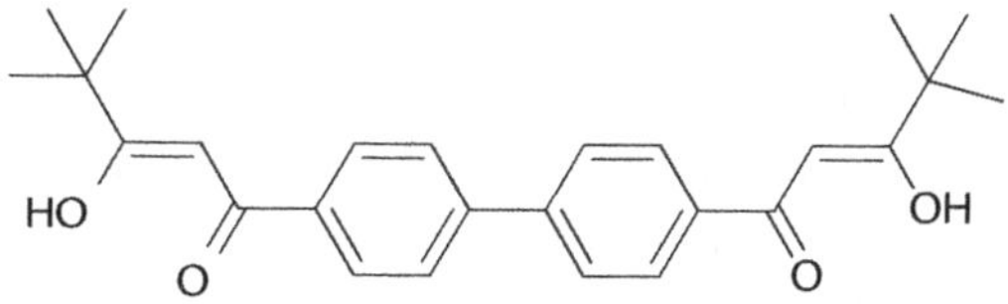

6

Scheme 2.5

was shown to be selective for the singularly charged Fe(III) species, $[FeCl_4]^-$, over its doubly charged Fe(II) analog, $[FeCl_4]^{2-}$.

In contrast to the above cationic cage, the extended neutral tetrahedral cage assembled from the doubly deprotonated form of **6** (Scheme 2.5) (incorporating 4,4′-biphenylene rather than 1,4-phenylene spacers) and Fe(III) in tetrahydrofuran (THF) was shown by X-ray diffraction to have the structure illustrated by Figure 2.3.

This metallocage has an enlarged internal volume of 844 Å^3 and was found to encapsulate four tetrahydrofuran solvent molecules in the solid state,[55] contrasting

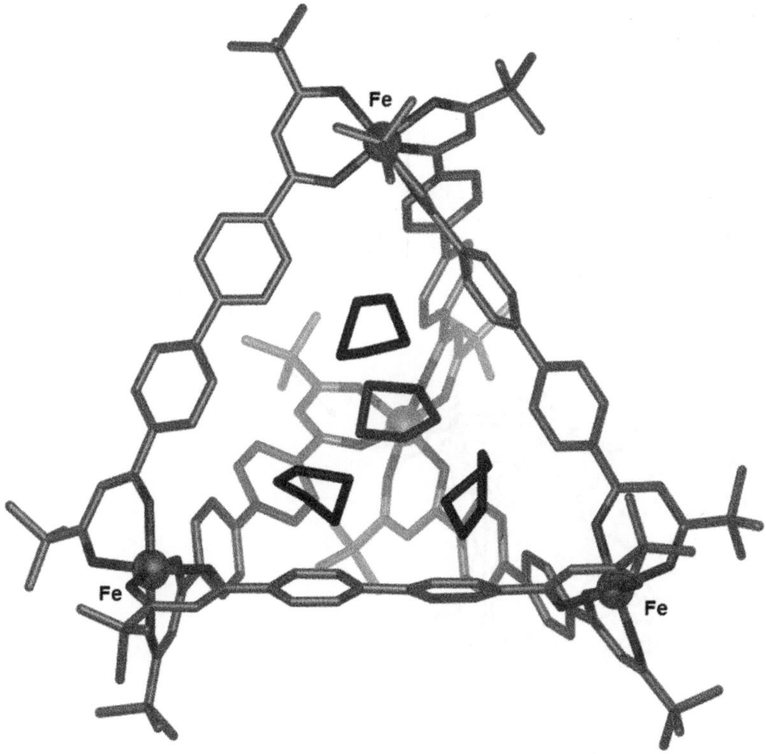

Figure 2.3 X-ray structure of the neutral metallosupramolecular cage $[Fe_4L_6(THF)_4]^{5+}$ in which four tetrahydrofuran molecules occupy the cavity of the tetrahedron.

with the 174 Å^3 volume observed for the smaller phenylene-spaced analog which included a single tetrahydrofuran molecule.

As mentioned already, many examples of larger self-assembled metallocages displaying other polyhedral shapes have also been reported by various groups. The design and successful synthesis of new "box-like" (cubic) cages based on the subcomponent Schiff base assembly reaction shown in Figure 2.4 has been reported.[56] In this, the reaction of the Ni(II) or Zn(II) tetrakis(4-aminophenyl)porphyrin complexes of type **7** (or the corresponding diprotonated free ligand–not shown in Figure 2.4) with 2-formylpyridine and Fe(II) trifluoromethanesulfonate (triflate, OTf^-) in dimethylformamide (DMF) yielded the corresponding tetra-Schiff base derivative intermediates of type **8** (ML, where M = Ni or Zn). Reaction of these with Fe(II) resulted in metal-directed assembly to form the corresponding cubic assembly.

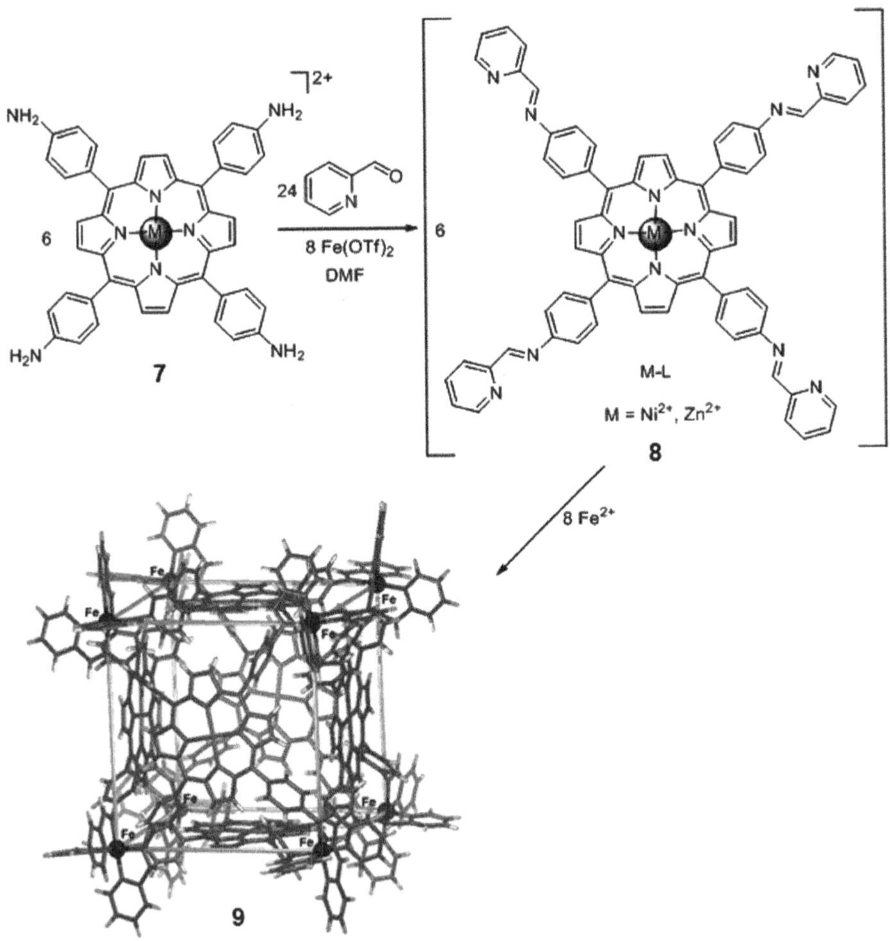

Figure 2.4 Synthesis and structure of metallo-cubes of type **9**.

The X-ray structure of the guest-free $[Fe_8(M\text{-}L)_6]^{16+}$ (M = Ni) cage incorporating Ni(II) in the respective porphyrin rings of each L is given as **9** in Figure 2.4. Each face of the cube is formed by a porphyrin moiety with each corner occupied by a low-spin Fe(II) metal center that has an octahedral N_6-donor coordination shell derived from three pyridylimine fragments from three different M-L ligands. Single cubes display homochirality with respect to their Fe(II) centers, while their single crystals are racemic with equal numbers of each chiral form present. The internal volume of the cage is 1340 $Å^3$ in this case. An integral part of the design of this cage type was the incorporation of large areas of π-electron density in the walls of the cavity since it was a goal of this investigation to produce a receptor cage that would include large unsaturated organic species such as (flat) coronene ($C_{18}H_{12}$) as well as individual spherical fullerene molecules.

Accordingly, the $[Fe_8(M\text{-}L)_6]^{16+}$ (M = Ni) cage was found to uptake three coronene molecules while both C_{60} and C_{70} were demonstrated to form 1:1 inclusion complexes, with the latter fullerene showing higher binding affinity, perhaps reflecting the less symmetrical nature of this guest, allowing greater $\pi-\pi$ interaction with the cavity walls. NMR evidence suggested that the three coronene molecules adopt a triple-sandwich π-stacked arrangement. Because the pore sizes in the walls of the fully formed cage type are quite small (with a spherical guest requiring a radius of no greater than ~1 Å to enter the cage without distorting the pore), it was postulated that guest exchange requires decomplexation of at least one of the pyridylmethyl imine arms from a Fe(II) center. Thus, a "gate mechanism," which in principle can be controlled chemically, for example, by pH variation, appears to be a feature of (large) guest binding by this cage type.

Over recent years a series of $M_{12}L_{24}$ spherical assemblies have been generated using self-assembly procedures, with such species typically having diameters of several nanometers. An early example of such a self-assembled metallocage was reported in 2001.[57] The self-assembly of isophthalic acid (1,3-benzenedicarboxylic acid) with Cu^{2+} yielded such a product composed of 12 dinuclear copper paddle-wheel cluster units and 24 bridging dicarboxylate ligands in the shape of a discrete polyhedron with an internal spherical cavity estimated to be close to 1600 $Å^3$. At about the same time a similar structure was found to self-assemble from the interaction of 5-hydroxy-isophthalic acid with Cu^{2+}.[58] The kinetics of formation of this functionalized derivative has since been reported[59] and its incorporation into polymers investigated.[60] The cage displays cuboctahedral symmetry and is remarkably stable.[61] The large internal cavity of this polyhedron is maintained by the rigid metal–organic scaffold.

A further example of such a highly symmetric assembly from the Fujita group is shown in Figure 2.5.[62] This structure also self-assembles from a total of 36 components (12 Pd(II) and 24 L) and once again displays cuboctahedral symmetry (diameter of 2.6 nm). Unlike the formation of infinite metal–organic framework structures by linear rod-like ligand systems, the slight bend in the present ligand (see Figure 2.5) results in a constant radius of curvature as the structure assembles, ultimately leading to the observed finite spherical network whose impressive capsid-like structure was confirmed by an X-ray diffraction study.

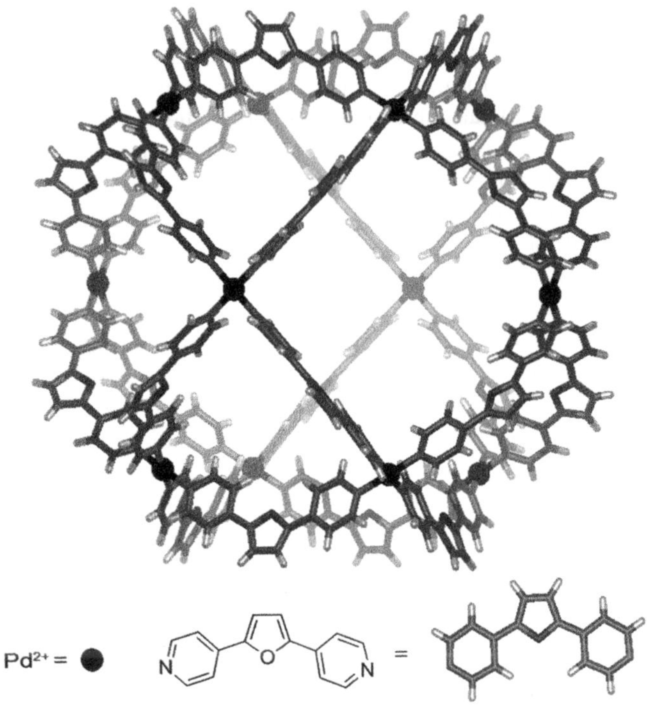

Figure 2.5 The X-ray crystal structure of the nanocage of composition $[Pd_{12}L_{24}]^{24+}$; counterions and solvent molecules are not shown.

Most recently, these structures have been extended to $M_{24}L_{48}$ spherical assemblies with a mass of >20,000 Da. Not surprisingly, such species have a substantially larger diameter, approaching 4.0 nm.[23]

2.3 ENZYME MIMICS AND MODELS: THE EXAMPLE OF CARBONIC ANHYDRASE

As already mentioned, the ubiquitous enzymes are, of course, especially characterized by the selective uptake of a guest substrate into an encapsulating cavity or pocket, coupled with high induced catalytic reactivity. As for "simple" cage and cage-like systems of the type just discussed, the characteristic substrate selectivity shown by the enzymes is clearly again strongly associated with the nature of the cavity; including its dimensions, lipophilicity, and the presence or absence of complementary functional groups. The presence of a portal and/or sufficient flexibility of the host cage to permit ingress and egress of a substrate is also an important feature.[63]

In general, two goals have predominated in the investigation of synthetic enzyme mimics: first, as an aid for defining the active site of a natural system as well as for probing its potential mode of action; and second, to probe the possibility of producing a functioning catalyst system for application in the "real world." However, in general, the catalytic activity of most synthetic systems tends to be low to moderate in comparison to that of the native enzymes—in part, likely due to the common absence of some form of polyfunctional activation in the former, relative to the situation known to occur in most natural enzymes. The following discussion focuses on the well-studied zinc enzyme *carbonic anhydrase*.

Carbonic anhydrase II is an enzyme that reversibly converts carbon dioxide into the bicarbonate ion. The active site incorporates a zinc ion bound to three histidines and a water molecule, such that a distorted tetrahedral coordination geometry is present. A feature of the system is that the zinc-bound water molecule has a pK_a of near 7 and hence is readily deprotonated under physiological conditions (see Figure 2.6). Deprotonation is promoted by the Lewis acidity of the zinc center but also appears to be assisted by the bound water being part of a hydrogen bonded network within the cavity (not shown).

There have been many studies employed to model the active site of *carbonic anhydrase*, with most of these involving the assembly of simple zinc complexes using a selection of ligand types that were perceived to be suitable for modeling the nature of the histidines in the natural system. Most of the studies have focused on the role of the Zn(II)–OH^- group for the hydration of CO_2 (as well as on the reverse process, the dehydration of HCO_3^-).[64] For example, in a series of early studies, tripodal (substituted) tris(pyrazolyl)borate ligand derivatives were employed to form complexes of type $[ZnL(OH)]^+$ as mimics of the active site of carbonic anhydrase. The resulting complex assemblies showed similarities to the natural system.[65] First, the coordinated zinc is tetrahedrally coordinated. Second, the low pK_a of the zinc-bound water was demonstrated (~6.5 in one instance) and, third, it was also shown that the bound hydroxide ion reacts rapidly and reversibly with CO_2.

In other, related biomimetic studies, Kimura and co-workers have investigated the zinc complex, $[ZnL(H_2O]^{2+}$, where L is the triaza-macrocycle,1,5,9-triazacyclododecane **10** (Scheme 2.6). The pK_a of the zinc-bound water molecule measured by potentiometric means was 7.3—again close to that of the natural system.[64, 66] This zinc complex system was demonstrated to induce catalytic

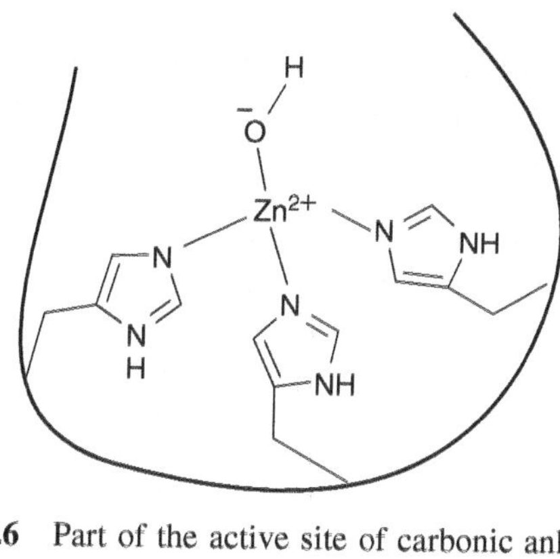

Figure 2.6 Part of the active site of carbonic anhydrase II.

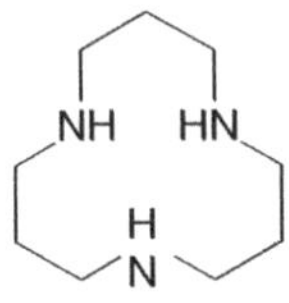

10

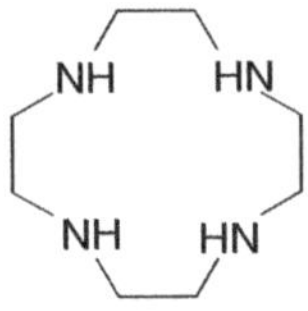

11

Scheme 2.6

enhancement of both the hydration of CO_2 and the dehydration of HCO_3^-. The reaction profile was in keeping with the coordinated OH^-/H_2O being the active species in these reactions.[67] Thus, the pH dependence of the reactions clearly showed that it is only the OH^- form of the complex that catalyzes the hydration reaction (via CO_2 uptake by coordinated hydroxide) and it is only the aqua complex that catalyzes the dehydration of HCO_3^- (via a ligand substitution process).

The rate constant for the hydration of CO_2 is approximately four orders of magnitude less than for the native enzyme; the observed lower rate constant was suggested to reflect the absence of the hydrophobic cavity that occurs in the natural system (the cavity may enhance catalysis by preassociating CO_2 or by aiding proton transfer).[64] As is the case for carbonic anhydrase, this system was found to be effective in catalyzing the hydration of acetylaldehyde as well as the hydrolysis of methyl acetate and *p*-nitrophenyl acetate. The pH profiles of the rates of the latter reactions were also in keeping with a nucleophilic attack mechanism involving the zinc-bound OH^- group acting as the nucleophile.

On changing the triaza macrocycle for its tetraaza analog (**11**, cyclen, Scheme 2.6) it was found that the binding of the additional nitrogen resulted in the pK_a of the coordinated water being raised to 8.0, suggesting that the "low" pK_a of the Zn(II)–OH_2 group in the native enzyme reflects, at least in part, the presence of a four-coordinate zinc center.[66]

2.4 SELF-ASSEMBLED LIPOSOME-LIKE SYSTEMS

Bilayer structures are mimics of natural membranes and as such duly receive a great deal of attention, with self-assembly processes playing a critical role in the formation of both natural and synthetic bilayers. The ability to control bilayer composition and assembly will be crucial to advances in, for example, the delivery of both therapeutic (drugs, genes) and diagnostic agents. The self-assembly of liposomes and caposomes using polymers is currently at the cutting edge in bionanoscience[68] and the impressive results in this field have recently been reviewed.[69] In this section we highlight recent advances in the formation of self-assembled bilayer systems as mimics of liposomes that are constructed from small components.

A powerful demonstration of forming bilayer structures through the use of small amphiphilic components was reported in 2010.[70] In this study small dendrimeric amphiphiles comprising a core, branching hydrophilic arms, and branching

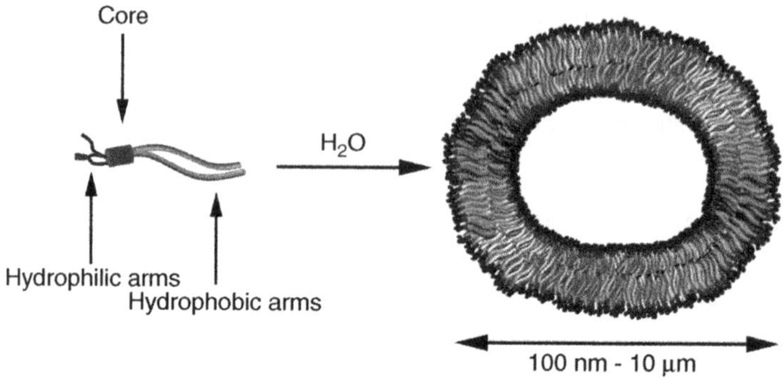

Figure 2.7 Schematic representation of an amphiphilic dendrimer with a cross-sectional view of a spherical dendrimersome.

hydrophobic arms were synthesized in a modular fashion. In water the amphiphiles were observed to form discrete hollow bilayer structures, which the researchers named dendrimersomes (Figure 2.7). Different amphiphiles give differently sized dendrimersomes, with the largest of these being an impressive 10 μm in diameter. Indeed, cryo-TEM measurements indicated that the dendrimersomes formed a variety of shapes and sizes depending on the amphiphile employed, indicating that the amphiphilic component plays a significant role in determining the structure of the resulting dendrimersome.

Advantages of dendrimersomes include significant mechanical strength relative to other micellar systems and narrow size distributions. Both of these features bestow benefits for use in delivery applications. Significantly, these systems have the added advantages of flexibility and tailorability, which can be manipulated through synthetic chemistry. The ability to prepare highly specific groups for attachment to the amphiphile core (employing a modular strategy), thereby yielding the desired components, makes this approach a powerful enabling strategy.

Other advantages that stem from this kind of methodology include deliberate selection of components that vary the thickness of the membrane. This allows the use of these synthetic nanoparticles in conjunction with molecules and systems that occur in natural membranes, or are membrane-bound (e.g., proteins, pumping systems, cholesterol), for generating future technology in which synthetic liposome-like systems work with Nature to breach new frontiers in in vivo nanomedicine. Toward this goal, new drug delivery nanosystems combining liposomal and dendrimeric technology (liposomal locked-in dendrimers) for cancer therapy are currently being pursued.[71]

One of the problems associated with self-assembling systems is their propensity to disassemble when exposed to new environmental conditions. This is of course highly undesirable in many instances. To improve the stability of liposomic systems (such as dendrimersomes), the next step is to prepare systems that can react on their outer surface with a polymerization reagent to form a layer around the liposome, thereby protecting it. It is noteworthy, in this regard, that natural systems can withstand considerable internal and external pressures before cellular

rupture. Forming such a protective layer may provide a membrane boundary with increased toughness and hence enhance the stability of the system, allowing its use under more challenging conditions. Potentially, the surface layer could then undergo appropriate chemical modification for use in specific applications.

2.5 ION CHANNEL MIMICS

Ion channels continue to be desirable targets for biomimicry. The passage of ions across cell membranes is crucial to cellular vitality. Mastery of this flow could provide medical advances toward a number of channelopathies. In a sustained research effort, Gokel and co-workers[72] have explored self-assembled pyrogalloarene cages and nanotubes as ion channel mimics. Pyrogalloles are conveniently formed in one-pot reactions from alkyl aldehydes and pyrogallol in modest yields. They self-assemble into a variety of structures, primarily dependent on the crystallization conditions.[73] Some of the most interesting are the metallocages formed on coordination of pyrogallole with metal ions.[26, 74] With sufficiently long alkyl chains extending from the metallocage core, these assemblies display an excellent ability to insert into phospholipid bilayers and act as protein channel mimics. Typically, the bilayer can be investigated by conductance measurements in planar bilayers through the voltage-clamp method. For a metallocage with pendant dodecane alkyl groups (**12** in Figure 2.8),[75] the measurements showed ion channel behavior toward Na^+, K^+, and Cs^+ ions with the greatest selectivity toward Na^+. This was consistent with the expected size of the hydrophobic channel presented by the membrane-bound pyrogallole cage.

The formation of other large cage polyhedra has been carried out by employing several different 5-substituted isophthalic acids to give complexes with the general formula of $M_{12}L_{24}$.[76] A notable example is a structure with dodecyloxy chains extending from the 5-position of the coordinated isophthalate units (**13** in Figure 2.8).[76, 77] With this extension the size of the resulting nanoparticle is approximately 5 nm. The structure of **13** has pore windows with triangular and square shapes with "diameters" of 3.8 and 6.6 Å,[77, 78] respectively. These appear large enough for small molecules to exchange in and out of the cavity. The potential of this ideally sized nanoparticle to span a phospholipid membrane was investigated (Figure 2.8).[78] Conductance and fluorimetry measurements were used and showed that this system can act as a transmembrane ion gate with preference for K^+ over Na^+. The authors proposed this to be a selective system.

An appealing prospect of porous systems is the potential to temporarily house molecules in the pores, especially with regard to applications like drug delivery. Zhou and co-workers[79] seized this opportunity and first investigated functionalizing a $Cu_{12}L_{24}$ polyhedron with pendant alkyne groups after its formation by attaching PEG chains via Hüigsen azide-alkyne cycloaddition chemistry to produce a material with improved water solubility. A choice was made in selecting a small drug molecule, 5-fluorouracil, as potential guest which could potentially fit through the pore windows of the structure and reside in the cavity during delivery, then exit when the cage was membrane bound. It was shown that 4.38 weight percent of the drug was carried by the PEGylated polyhedron, indicating multiple drug

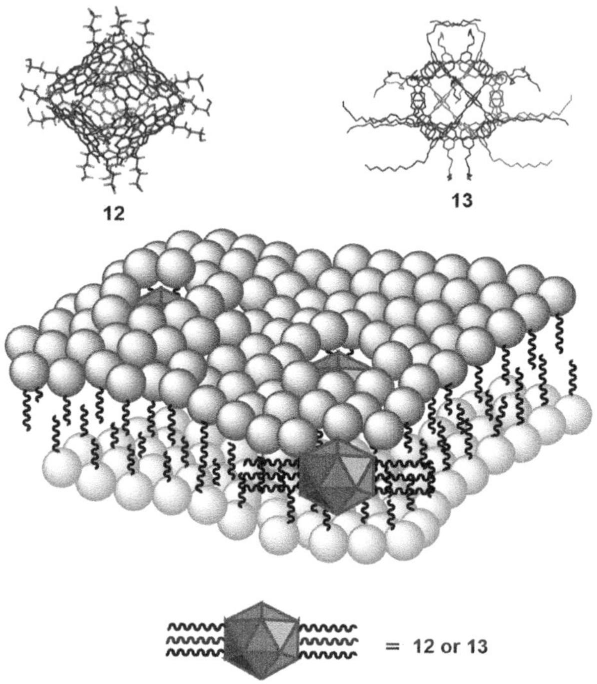

Figure 2.8 A representation of metallocage compounds 12 and 13 embedded in a bilayer, where each can individually act as ion transporters; 12 is the pyrogallole metallocage and 13 is the functionalized isophthalate metallocage.

molecules are associated with this cage. Although the association of the drug to the cage was not definitively established, it seems possible that a combination of hydrophobic effects in the interior cavity and the formation of coordinate bonds from 5-fluorouracil with the copper atoms of the polyhedron are involved in the uptake of this drug.

Other examples of ion channel mimics based on the use of amphiphilic palladium complexes with long alkyl groups on the ligands have been investigated (Figure 2.9).[80] Octacationic molecular squares form when the amphiphilic palladium complex is combined with the linear 4,4′-bipyridine ligand. This system mimics the shape of naturally occurring G-quadruplexes and its behavior was also investigated in bilayers. The approach here again has drawn upon metallosupramolecular chemistry with the use of directional coordinate bonding to assemble defined metallosupramolecular shapes.[81]

Recent work from the Fujita group has demonstrated the assembly of perfectly monodisperse nanosized metallosupramolecular assemblies that are adorned with groups that will encourage interactions with proteins[82] and DNA.[83] The studies reported so far point to a rich future for ion-channel mimics that combine natural membrane-bound or membrane-like organic components, which provide the mechanism for membrane association, with the shapes and sizes of core components derived from metallosupramolecular chemistry.

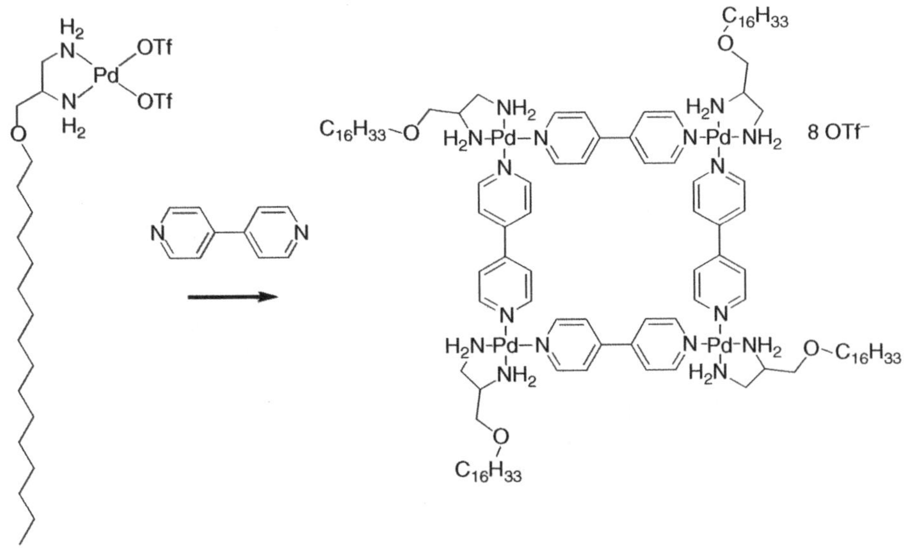

Figure 2.9 The formation of a cationic molecular square from an amphiphilic palladium complex and 4,4′-bipyridine.

2.6 BASE-PAIRING STRUCTURES

The structure of duplex DNA is one familiar to biologists and chemists alike and is of undoubted beauty. Many efforts continue to be made to utilize this simple-yet-complex motif. The process of transcription shows that DNA can be thought of as an information dense material. To artificially encode readable information within DNA strands represents a significant goal. Watson–Crick base pairing is central to the structure and function of DNA and because of the reliable nature of the base-pairing interactions and the strength of the double helical structure over nanometer dimensions, DNA and other forms of nucleic acids are now being used by nanobiotechnologists to self-assemble dedicated nanostructures. It is here where mimicry of base pairing is a target for preparing materials with designed function for applications in nanomedicine[84] and information storage,[85] as examples.

A popular target in DNA structures is to mimic base pairs through metal–mediated interactions.[86] This is where metal-ligand bonding, rather than hydrogen bonding, directs the formation of metal-coordinated base pairs and hence the structure of the resultant DNA. Metal–mediated base pairs have long been targeted but the results so far have often tended to be inconclusive. Crucial to the success of this approach is to match the coordination geometry of the metal ion to the desired structure. A successful example of employing NMR spectroscopy to obtain structural information was reported recently.[87] A palindromic sequence of 17 nucleotides with three artificial imidazole nucleosides in the central positions was used for improved metal–ligand bonding and positioned so as to result in consecutive metal-mediated base pairs. The use of Ag^+ with its known tendency

for linear or near-linear two-coordinate ligation was an apt choice for the metal to mediate the base pairing through imidazole–Ag–imidazole bonds; the structure was unambiguously determined through $^{107/109}$Ag–^{15}N heteronuclear correlation NMR experiments. Significantly, the metallated sequence forms a double helix (duplex) (Figure 2.10), while in the unmetallated case a hairpin structure is formed, showing that the mediation templates the formation of a double helix. The duplex and hairpin structures were deduced through a combination of NMR spectroscopy and structural computation.

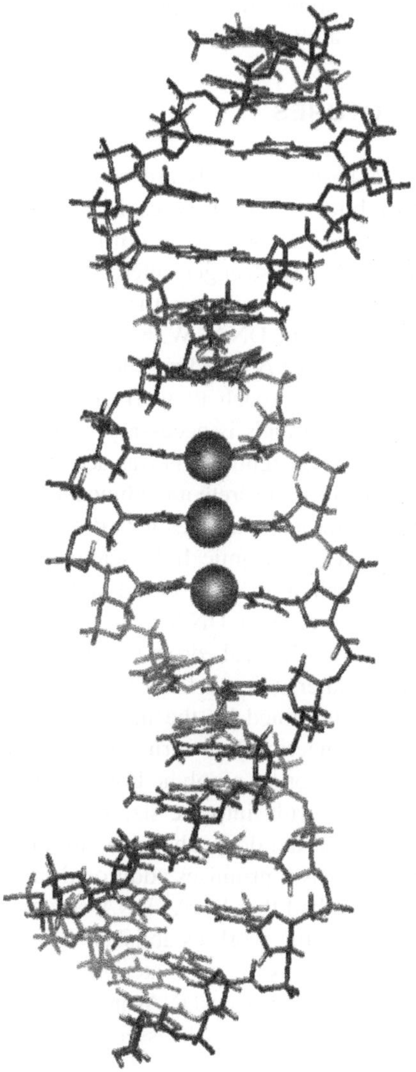

Figure 2.10 The duplex structure of a synthetic nucleotide showing the coordination of Ag^+ by the central imidazole nucleosides.

Built upon the discoveries of the 20th century involving DNA materials, this example shows superb mimicry in delivering geometrically defined sequence-specific nanomaterials. The inclusion of metal ions in this type of assembly promises DNA materials exhibiting new properties; studies into how such artificial duplex structures may be used in nanomedicine and information storage are already underway. The next generation of this material type will surely see metal-mediated base pairs at different positions along a DNA backbone giving rise to start and stop positions between which information is encoded. Of course, congruent with this aim must be the development of molecular technology capable of reading (by way of transcription or some similar process) or relaying the information. However, this avenue of research is far less advanced than this at present.

2.7 DNA–RNA STRUCTURES

As indicated earlier, DNA-based materials are being thoroughly investigated for constructing discrete assemblies of various types. However, intrinsically more interesting may be the use of RNA since it occurs in different forms that show different functions.[88] In general, RNA has stronger interactions between base pairs than DNA and can give more thermally robust structures. RNA is also slightly more stable to acidic environments than DNA. With RNA possessing functional roles, the combination of DNA and RNA building blocks in functional assemblies is extremely appealing. For this approach to work, matching of helical lengths of the individual DNA and RNA strands is necessary in order to favor the formation of DNA–RNA heteroduplexes. Recently, this was accomplished, with DNA–RNA heteroduplexes being used to form discrete nanodimensional dodecahedra.[89] Highly symmetric hollow dodecahedral nanoparticles with dimensions of $\sim$21 nm (as measured through a combination of dynamic light scattering and cryo-EM methods) form in a one-pot synthesis when a tRNA strand and a cyclic trimer of DNA are combined with a third short strand of DNA in water. A bent triangular unit that becomes a vertex point of the dodecahedron is formed with each of its "sticky ends" able to engage three other units to form the polyhedron (Figure 2.11). The size of the polyhedron is determined by the number of turns the linking strands take along the edges. By engineering four turns, the bent units face the same way, thereby encouraging the formation of a polyhedron rather than a planar array. This discrete cage assembly is now truly into the size regime of virus-sized nanoparticles. Although the formation of the dodecahedra is not quantitative ($\sim$40% in this case), the union of DNA and RNA promises additional functionality to this kind of assembly. The authors established that DNA–RNA co-assembly is versatile, being able to be applied to much of the work established for DNA alone. This significant discovery now provides access to nanomaterials for RNA-derived applications using already generated knowledge from DNA-only assemblies.

In an exciting recent outcome, the self-assembly of RNA was demonstrated to yield large and uniquely shaped square antiprisms.[90] The prisms were formed from the assembly of eight, three-armed tRNA monomers. Each monomer unit features fine control over the angles between the arms, with the distinct lengths

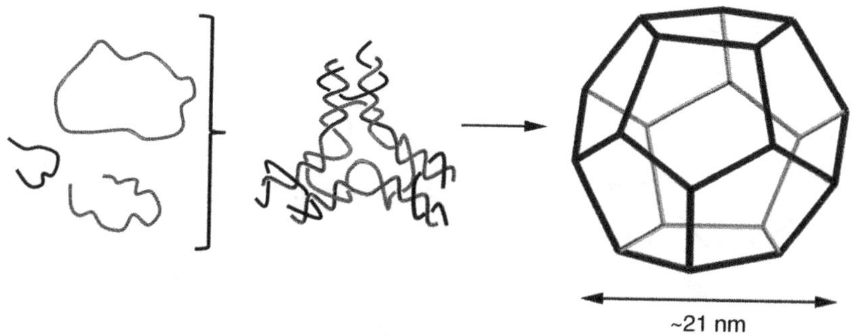

Figure 2.11 Schematic representation of tRNA and DNA heteroduplex formation to give a triangular unit that assembles into a large discrete dodecahedron.

of the base-pair arms determining the prism dimensions (as large as 8 × 14 nm); defined roles exist for the arms in the assembly process. All of these result in "programmed" assembly of the prisms. By combining known protein binding ligands with the tRNA monomers, antiprisms were formed that were shown to host the small globular protein streptavidin (~5 nm). Thus, the significance of these kinds of assemblies is that they can bind and position molecules of considerable size and act as scaffolds for the protection or delivery of biomolecules.

Exciting breakthroughs often occur at the union of traditionally separated research areas. An example of self-assembly through a union of metallo-supramolecular chemistry and DNA nanotechnology is the production of a metal–DNA cage with the impressive internal dimensions, estimated through modeling, to be 25–30 nm^3.[91] The inspiration to create such structures clearly comes from the structure and function of metalloenzymes, where, as discussed earlier, the metal is held in a defined position within the active site.

The incorporation of metal ions in synthetic DNA assemblies brings with it the possibility of additional functionality for the resulting assembly (including catalysis, unusual magnetism, or luminescence) that can, in principle, be controlled by the particular metal ion chosen—while at the same time maintaining the inherent biocompatibility due to the presence of DNA. Thus, this hybrid approach shows much promise and clearly will lead to new challenging, worthwhile science, requiring understanding from multiple viewpoints, in order to appropriately control the assembly parameters required for generating a given target structure. The strategy employed for obtaining the above metal–DNA cage hinged on the creation of two chelating 1,10-phenanthroline metal binding sites within a single oligonucleotide strand. Through base-pairing interactions, three strands self-assemble in a triangular shape, with the metal binding phenanthroline domains from different strands held in close proximity such that metal ion binding is preorganized. Two triangular units are then brought together through three linking strands, which are then structurally reinforced by formation of a duplex with a second oligonucleotide sequence to give a trigonal prismatic shape to the final polygon (Figure 2.12).

The above approach is flexible as the triangular units could be either left unmetallated or reacted with Cu^+ or Ag^+ before prism formation. The formation of the

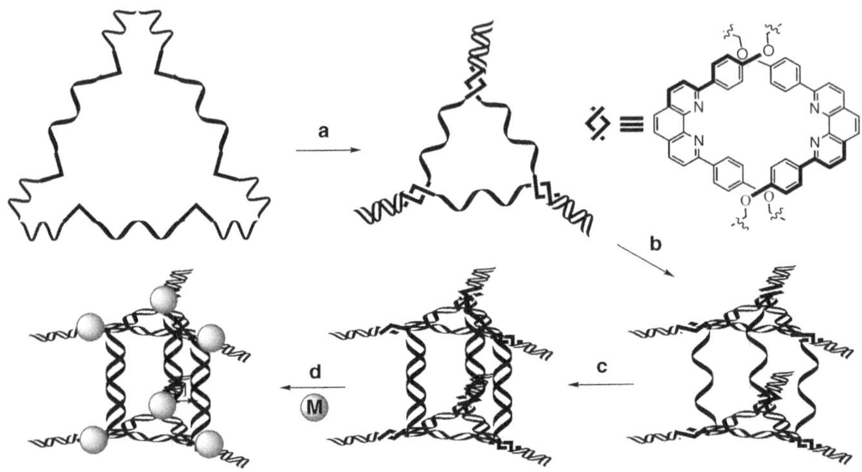

Figure 2.12 (a) Formation of a triangle through base-pairing interactions; (b) linking strands bringing two units together to form the prism; (c) reinforcement of the linking strands; (d) metallation of the prism.

prism was quantitative and the stability of the system at each step of the assembly was probed through exposure to denaturing enzymes (to which the prism showed good resistance). The significant feature of this assembly is the size of the cavity (25–30 nm^3). Such large cavities may well be able to host large guests such as proteins and could possibly utilize the spatially separated metal ions held in the rigid structure for another purpose such as for catalysis and/or generating luminescence. Applications in nanomedicine and sensing are obvious reasons for further research in this general area. In addition, this kind of assembly may prove to be of importance to synthetic biology; biomolecules may be hosted in predetermined orientations within such a cage, thus allowing site-specific reactions to take place. This impressive study demonstrates that the union of DNA technology with metallosupramolecular chemistry can be a powerful strategy for forming useful nanostructured vessels. There is no doubt that continuing research in this area will provide exciting discoveries in the near future.

2.8 BIOINSPIRED FRAMEWORKS

The translation from discrete porous structures in solution into the solid state becomes increasingly important when considering how technologies based on porous solid materials might be commercialized. In some cases immobilization on a solid support can be employed. More commonly, the solid materials can be prepared directly via precipitation or crystallization from solution. Metal–organic frameworks, or porous coordination polymers as they are alternatively known, are currently an intensively investigated class of such porous solids of the latter type, with applications that include molecular storage, sequestration,[92] and catalysis.[93]

More recently, applications in medicine have become apparent.[94, 95] Such metal–organic frameworks are composed of organic ligands that link metal centers to produce porous infinite arrays. Network structures of this type have the potential to include therapeutic agents in their porous structures or act as diagnostics or sensors for use in medicine. An important consideration for any compound used medicinally is its inherent toxicity as well as that of its decomposition products. For metal–organic frameworks this means that both the metal and the bridging ligand need to be considered.[95] The use of bridging ligands endogenous to the body provides a bioinspired means of removing toxicological concerns about this component. In this regard, frameworks constructed using adenine bridging ligands have recently been reported.[96]

Other bioinspired examples of this type include incorporating amino acids as ligands. However, these are flexible and typically they have been combined with rigid bridging ligands, such as 4,4′-bipyridine, which help "stiffen" the porous frameworks.[97]

Relatively recently, there have been reports of metal–organic framework materials spawned from conformationally flexible peptide bridging ligands. The inspiration to utilize such ligands was influenced by their inherent chirality, biocompatibility, and availability, with a view to imparting desirable features to the framework solid. However, perhaps the reason that peptidyl bridging ligands have not found widespread use to date is the anticipated difficulty of producing phase-pure porous frameworks. Certainly, the combination of multiple competing binding sites for the metal centers and the conformational flexibility of the bridging ligands undoubtedly conspire against assembling predictable porous structures with this ligand type.

One of the first examples of the assembly of a metal–peptide framework was reported in 2008.[98] The Cbz protected peptidyl ligand Z-L-Val-L-Val-L-Glu(OH)OH was reacted with Ca^{2+} and with Cu^{2+} in the presence of aqueous ammonia and fibrous solids precipitated in each case. Although analysis by single crystal studies could not be performed, the structure of the copper-containing material was solved and refined from powder diffraction studies. The structure shows that the peptide ligand chelates to copper through its glutamate carboxylates, with square planar coordination of the copper being completed by two ammonia ligands. A prominent feature in the extended structure is the alignment into β-sheet-like layers through familiar $NH\cdots O{=}C$ hydrogen bonds, indicating significant direction of the lattice structure by the coordinated peptidyl ligands. Individual metal coordination units are involved in hydrogen bonded interactions with water molecules in the lattice, as shown in Figure 2.13. The magnetic properties of this material were investigated and were in accord with the expected result for one-dimensional $S = \frac{1}{2}$ chains of Cu^{2+} ions. Thus, this example represents a nice illustration of a bio-motif displaying further structural and functional features gained from metal ion complexation.

The complexity of dealing with peptidyl ligands is exemplified by the structure of $[Zn(Gly\text{-}Ala)_2]$.[99] In this case the Zn^{2+} ions are tetrahedrally coordinated by four Gly-Ala ligands with two coordinating by oxygen atoms from the C-terminal

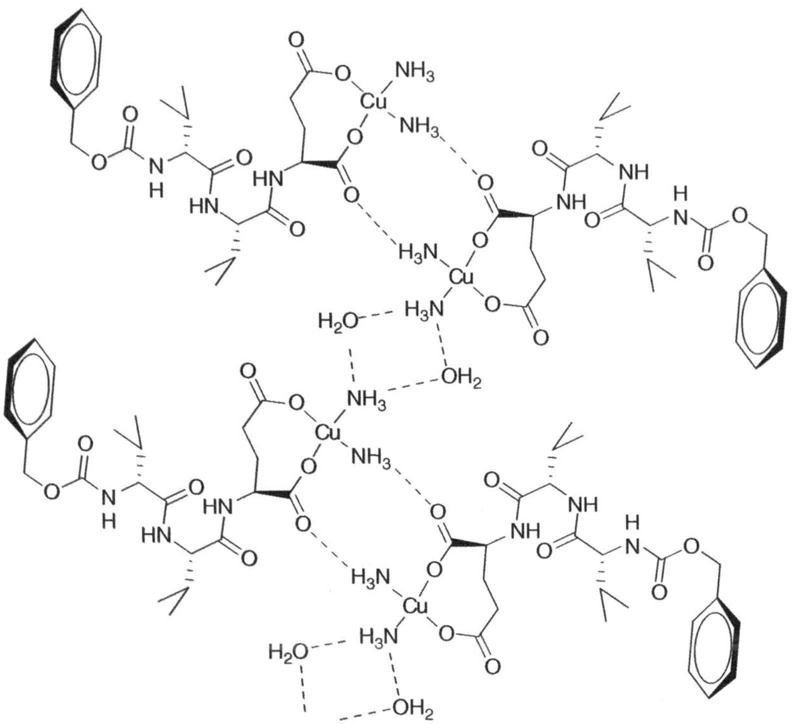

Figure 2.13 The hydrogen bonding motif between lattice water molecules and the metal coordination unit.

Ala carboxylate termini and two by nitrogen atoms from the N-terminal Gly amine groups. A sheet structure reminiscent of a β-sheet is formed through which small square-shaped pores exist by alignment of the sheets (Figure 2.14). Even with this small and simple dipeptide (consisting of one glycine and one alanine residue), the structural intricacies were complex, with a combination of structural techniques coupled with molecular dynamics calculations required to assign plausible mechanisms for pore opening and closing by this material.

When guests are removed from this network, the alanine methyl groups, which project into the pores, rotate to close the space and the material becomes impermeable to gas uptake with respect to H_2 and N_2. When exposed to polar adsorbates such as CO_2, MeOH, and water, the material takes up these molecules under the influence of a temperature–pressure gate. The material behaves cooperatively and is able to change the orientation of the pore-blocking methyl groups in each layer during readsorption. This kind of response is uniquely accessed as a result of the low-energy conformational changes possible with the Gly-Ala ligand. This work certainly points to the potential for designing more complex ligands for incorporation into metal–organic frameworks. Such frameworks may then become capable of addressing more efficiently such challenging and complex tasks as molecular sorting.

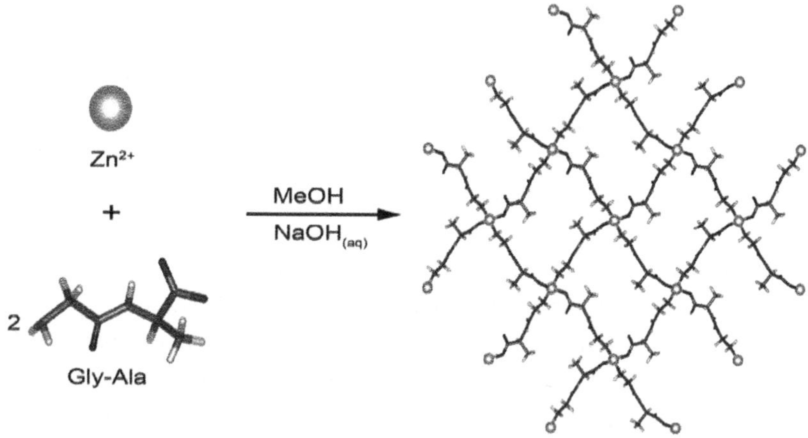

Figure 2.14 A view of the sheet structure of $[(Gly\text{-}Ala)_2Zn]$ with the alanine methyl groups projecting into the square-shaped pores.

2.9 CONCLUSION

Nature provides numerous examples of the efficacy of the bottom–up approach for assembling new biomaterials that often display great sophistication in both their form and function. Clearly, it is a strategy that offers the opportunity for achieving control over both the structure and properties of a material, and can also be effective in adding designed functionality to the assembled structure. In this chapter, we have introduced a cross section of synthetic supramolecular structures that are either directly bioinspired or that mimic in some way selected biostructures. Clearly, much scope remains for the design and construction of further synthetic assemblies that display both greater complexity and functionality at levels that approach those routinely found in Nature. This remains the challenge for the future!

REFERENCES

1. Lehn, J.-M. *Science* **2002**, *295,* 2400.
2. Lindoy, L. F.; Atkinson, I. M. *Self-Assembly in Supramolecular Chemistry*, Royal Society for Chemistry, Cambridge, UK, 2000 (a monograph of 224 pages).
3. Williams, R. J.; Smith, A. M.; Collins, R.; Hodson, N.; Das, A. K.; Ulijn, R. V. *Nature Nanotechnology* **2009**, *4,* 19.
4. Seeman, N. C. *Annual Review of Biochemistry*, 2010, 79, 65.
5. Ajami, D.; Tolstoy, P. M.; Dube, H.; Odermatt, S.; Koeppe, B.; Guo, J.; Limbach, H. H.; Rebek, J. *Angew. Chem. Int. Ed*. **2011**, *50,* 528.
6. Pluth, M. D.; Bergman, R. G.; Raymond, K. N. *Acc. Chem. Res*. **2009**, *42,* 1650.
7. (a) Horiuchi, S.; Nishioka, Y.; Murase, T.; Fujita, M. *Chem. Commun*. **2010**, *46,* 3460; (b) Yoshizawa, M.; Fujita, M. *Bull. Chem. Soc. Jpn*. **2010**, *83,* 609.

8. Rebek, J. *Acc. Chem. Res*. **2009**, *42,* 1660.
9. Breiner, B.; Clegg, J. K.; Nitschke, J. R. *Chem. Sci*. **2011**, *2,* 51.
10. Yoshizawa, M.; Tamura, M.; Fujita, M. *Angew. Chem. Int. Ed*. **2007**, *46,* 3874.
11. Cram, D. J. *Nature* **1992**, *356,* 29.
12. Cram, D. J.; Cram, J. M. *J. Am. Chem. Soc*. **1995**, *117,* 2680.
13. Fiedler, D.; Leung, D. H.; Bergman, R. G.; Raymond, K. N. *Acc. Chem. Res*. **2005**, *38,* 349.
14. Laughreya, Z.; Gibb, B. C. *Chem. Soc. Rev*. **2011**, *40,* 363.
15. Schneider, H.-J.; Yatsimirsky, A. K. *Chem. Soc. Rev*. **2008**, *37,* 263.
16. Yoshizawa, M.; Klosterman, J. K.; Fujita, M. *Angew. Chem. Int. Ed*. **2009**, *48,* 3418.
17. Mal, P.; Breiner, B.; Rissanen K.; Nitschke, J. R. *Science* **2009**, *342,* 1697.
18. Leung, D. H.; Bergman, R. G.; Raymond, K. N. *J. Am. Chem. Soc*. **2008**, *130,* 2798.
19. Koblenz, T. S.; Wassenaar, J.; Reek, J. N. H. *Chem. Soc. Rev*. **2008**, *37,* 247.
20. Yoshizawa, M.; Sato, N.; Fujita, M. *Chem. Lett*. **2005**, *34,* 1392.
21. (a) MacGillivray, L. R.; Atwood, J. L. *Angew. Chem. Int. Ed*. **1999**, *38,* 1018; (b) Leininger, S.; Olenyuk, B.; Stang, P. J. *Chem. Rev*. **2000**, *100,* 853; (c) Caulder, D. L.; Raymond, K. N. *Acc. Chem. Res*. **1999**, *32,* 975; (d) Cram, D. J.; Cram, J. M.; *Container Molecules and Their Guests*, Royal Society of Chemistry, Cambridge, UK, 1994.
22. Ward, M. D. *Chem. Commun*. **2009**, 4487.
23. Sun, Q.-F.; Iwasa, J.; Ogawa, D.; Ishido, Y.; Sato, S.; Ozeki, T.; Sei, Y.; Yamaguchi, K.; Fujita, M. *Science* **2010**, *328,* 1144.
24. (a) Mirtschin, S.; Slabon-Turski, A.; Scopelliti, R.; Velders, A. H.; Severin, K. *J. Am. Chem. Soc*. **2010**, *132,* 14004; (b) Seidel, S. R.; Stang, P. J. *Acc. Chem. Res*. **2002**, *35,* 972; (c) Albrecht, M.; Froehlich, R. *Bull. Chem. Soc. Jpn*. **2007**, *80,* 797; (d) Liu, X. J.; Liu, Y.; Li, G.; Warmuth, R. *Angew. Chem. Int. Ed*. **2006**, *45,* 901; (e) Hristova, Y. R.; Smulders, M. M. J.; Clegg, J. K.; Breiner, B.; Nitschke, J. R. *Chem. Sci*. **2011**, *2,* 638; (f) Riddell, I. A; Smulders, M. M. J.; Clegg, J. K.; Nitschke, J. R. *Chem. Commun*. **2011**, *47,* 457; (g) Oppel (nee) Müller, I. M.; Focker, K. *Angew. Chem. Int. Ed*. **2008**, *47,* 402; (h) Glasson, C. R. K; Meehan, G. V.; Motti, C. A.; Clegg, J. K.; Turner, P.; Jensen, P.; Lindoy, L. F. *Dalton Trans*. **2011**, *40*, 10481.
25. Glasson, C. R. K.; Clegg, J. K.; McMurtrie, J. C.; Meehan, G. V.; Lindoy, L. F.; Motti, C. A.; Moubaraki, B.; Murray, K. S.; Cashion, J. D. *Chem. Sci*. **2011**, *2,* 540.
26. Dalgarno, S. J.; Power, N. P.; Atwood, J. L. *Coord. Chem. Rev*. **2008**, *252,* 825.
27. Saalfrank, R. W.; Maid, H.; Scheurer, A. *Angew. Chem. Int. Ed*. **2008**, *47,* 8794.
28. Liu, Y.; Hu, C.; Comotti, A.; Ward, M. D. *Science* **2011**, *333,* 436.
29. Avram, L.; Cohen, Y.; Rebek, J. *Chem. Commun*. **2011**, *47,* 5368.
30. Rebek, J. *Angew. Chem. Int. Ed*. **2005**, *44,* 2068.
31. Heaven, M. W.; Cave, G. W. V.; McKinlay, R. M.; Antesberger, J.; Dalgarno, S. J.; Thallapally, P. K.; Atwood, J. L. *Angew. Chem. Int. Ed*. **2006**, *45,* 6221.
32. Purse, B. W.; Butterfield, S. M.; Ballester, P.; Shivanyuk, A.; Rebek, J. *J. Org. Chem*. **2008**, *73,* 6480.

33. (a) Stephenson, A.; Argent, S. P.; Riis-Johannessen, T.; Tidmarsh, I. S.; Ward, M. D. *J. Am. Chem. Soc.* **2011**, *133,* 858; (b) Li, D.; Zhou, W.; Landskron, K.; Sato, S.; Kiely, C. J.; Fujita, M.; Liu, T. B. *Angew. Chem. Int. Ed.* **2011**, *50,* 5182 and references therein.

34. (a) Katen, S.; Zlotnick, A. *Methods Enzymol*. **2009**, *455,* 395; (b) Wikoff, W.R.; Liljas, L.; Duda, R. L.; Tsuruta, H.; Hendrix, R.W.; Johnson, J. E. *Science* **2000**, *289,* 2129.

35. Vernizzi, G.; Olvera, de la Cruz, M. *Proc. Natl. Acad. Sci. USA* **2007**, *104,* 18382.

36. Fornasari, M. S.; Laplagne, D. A.; Frankel, N.; Cauerhff, A. A.; Goldbaum, F. A.; Echave, J. *Mol. Biol. Evol*. **2004**, *21,* 97.

37. Olenyuk, B.; Whiteford, J. A.; Fechtenkötter, A.; Stang, P. J. *Nature* **1999**, *398,* 796.

38. (a) Ajami, D.; Rebek, J. *J. Org. Chem*. **2009**, *74,* 6584; (b) Lledó, A.; Hooley, R. J.; Rebek, J. *Org. Lett*. **2008**, *10,* 3669.

39. Kim, K.; Selvapalam, N.; Ko, Y. H.; Park, K. M.; Kim, D.; Kim, J. *Chem. Soc. Rev*. **2007**, *36,* 267.

40. von dem Bussche-Hunnefeld, C.; Bühring, D.; Knobler, C. B.; Cram, D. J. *J. Chem. Soc. Chem. Commun*. **1995**, 1085.

41. Helgeson, R. C.; Paek, K.; Knobler, C. B.; Maverick, E. F.; Cram, D. J. *J. Am. Chem. Soc*. **1996**, *118,* 5590.

42. Mastalerz, M. *Angew. Chem. Int. Ed*. **2010**, 49, 5042.

43. Liu, S.; Gan, H.; Hermann, A. T.; Rick, S. W.; Gibb, B. C. *Nature Chemistry* **2010**, *2,* 847.

44. Young, P. G.; Clegg, J. K.; Bhadbhade, M.; Jolliffe, K. A. *Chem. Commun*. **2011**, *47,* 463.

45. Ashton, P. R.; Isaacs, N. S.; Kohnke, F. H.; d'Alcontes, G. S.; Stoddart, J. F. *Angew. Chem. Int. Ed. Engl*. **1989**, *28,* 1261.

46. Ashton, P. R.; Girreser, U.; Giuffrida, D.; Kohnke, F. H.; Mathias, J. P.; Raymo, F. M.; Slawin, A. M. Z.; Stoddart, J. F.; Williams, D. J. *J. Am. Chem. Soc*. **1993**, *115,* 5422.

47. (a) Mullen, K. M.; Beer, P. D. *Chem. Soc. Rev*. **2009**, *38,* 170; (b) Linton, B.; Hamilton, A. D. *Chem. Rev*. **1997**, *97,* 1669; (c) Tokunaga, Y.; Rudkevich, D. M.; Santamaria, J.; Hilmersson, G.; Rebek, J. *Chem. Eur. J*. **1998**, *4,* 1449.

48. Gibb, C. L. D.; Gibb, B. C. *Chem. Commun*. **2007**, 1635; (b) Gibb, C. L. D.; Gibb, B. C. *J. Am. Chem. Soc*. **2006**, *128,* 16498 ; (c) Gibb, C. L. D.; Gibb, B. C. *J. Am. Chem. Soc*. **2004**, *126,* 11408.

49. (a) Sadownik, J. W.; Ulijn, R. V. *Curr. Opin. Biotechnol*. **2010**, *21,* 401; (b) Rowan, S. J.; Cantrill, S. J.; Cousins, G. R. L.; Sanders, J. K. M.; Stoddart, J. F. *Angew. Chem. Int. Ed*. **2002**, *41,* 898.

50. Lehn, J.-M. *Chem. Eur. J*. **1999**, 2455.

51. Corbett, P. T.; Leclaire, J.; Vial, L.; West, K. R.; Wietor, J.-L.; Sanders, J. K. M.; Otto, S. *Chem. Rev*. **2006**, *106,* 3652.

52. (a) Xu, X. N.; Wang, L.; Wang, G. T.; Lin, J. B.; Li, G. Y.; Jiang, X. K.; Li, Z. T. *Chem. Eur. J*. **2009**, *15,* 5763; (b) Tartaggia, S.; Scarso, A.; Padovan, P.; De Lucchi, O.; Fabris, F. *Org. Lett*. **2009**, *11,* 3926; (c) Liu, X.; Sun, J.; Warmuth, R.

Tetrahedron **2009**, *65,* 7303; (d) Steinmetz, V.; Couty, F.; David, O. R. P. *Chem. Commun*. **2009**, 343; (e) Mastalerz, M. *Chem. Commun*. **2008**, 4756; (f) Kuhnert, N.; Le-Gresley, A. *Tetrahedron Lett*. **2008**, *49,* 1274; (g) Xu, D.; Warmuth, R. *J. Am. Chem. Soc*. **2008**, *130,* 7520; (h) Liu, Y.; Liu, X.; Warmuth, R. *Chem. Eur. J*. **2007**, *13,* 8959; (i) Meyer, C. D.; Joiner, C. S.; Stoddart, J. F. *Chem. Soc. Rev*. **2007**, *36,* 1705; (j) Ro, S.; Rowan, S. J.; Pease, A. R.; Cram, D. J.; Stoddart, J. F. *Org. Lett*. **2000**, *2,* 2411.

53. Glasson, C. R. K.; Lindoy, L. F.; Meehan, G. V. *Coord. Chem. Rev*. **2008**, *252,* 940.
54. Glasson C. R. K.; Meehan, G. V.; Clegg, J. K.; Lindoy, L. F; Turner, P.; Duriska, M. B.; Willis, R. *Chem. Commun*. **2008**, 1190.
55. Clegg, J. K.; Li, F.; Jolliffe, K. A.; Meehan, G. V.; Lindoy, L. F. *Chem. Commun*. **2011**, *47,* 6042.
56. Meng, W.; Breiner, B.; Rissanen, K.; Thoburn, J. D.; Clegg, J. K.; Nitschke, J. R. *Angew. Chem. Int. Ed*. **2011**, *50,* 3479.
57. Eddaoudi, M.; Kim, J.; Wachter, J. B.; Chae, H. K.; O'Keeffe, M.; Yaghi, O. M. *J. Am. Chem. Soc*. **2001**, *123,* 4368.
58. Abourahma, H.; Coleman, A. W.; Moulton, B.; Rather, B.; Shahgaldian, P.; Zaworotko, M. J. *Chem. Commun*. **2001**, 2380.
59. Larsen, R. W. *J. Am. Chem. Soc*. **2008**, *130,* 11246.
60. Tonigold, M.; Hitzbleck, J.; Bahnmüller, S.; Langstein, G.; Volkmer, D. *Dalton Trans*. **2009**, 1363.
61. Atwood, J. L.; Barbour, L. J.; Dalgarno, S. J.; Hardie, M. J.; Raston, C. L.; Webb, H. R. *J. Am. Chem. Soc*. **2004**, *126,* 13170.
62. (a) Masatoshi, I.; Takashi, K.; Sachiko M.; Kiotaka, S.; Sota, S.; Fujita, M. *Chem. Sci*. **2010**, *1,* 68; (b) Tominaga, M.; Suzuki, K.; Kawano, M.; Kusukawa, T.; Ozeki, T.; Shakamoto, S.; Yamaguchi, K.; Fujita, M. *Angew. Chem. Int. Ed*. **2004**, *43,* 5621.
63. (a) Pluth, M. D.; Raymond, K. N. *Chem. Soc. Rev*. **2007**, *36,* 161; (b) Ringe, D.; Petsko, G. A. *Science* **2008**, *320,* 1428.
64. Krishnamurthy, V. M.; Kaufman, G. K.; Urbach, A. R.; Gitlin, I.; Gudiksen, K. L.; Weibel, D. B.; Whitesides, G. M. *Chem. Rev*. **2008**, *108,* 946.
65. (a) Looney, A.; Han, R.; McNeill, K.; Parkin, G. *J. Am. Chem. Soc*. **1993**, *115,* 4690; (b) Rombach, M.; Maurer, C.; Weis, K.; Keller, E.; Vahrenkamp, H. *Chem. Eur. J*. **1999**, *5,* 1013 and references therein.
66. Kimura, E.; Shiota, T.; Koike, T.; Shiro, M.; Kodama, M. *J. Am. Chem. Soc*. **1990**, *112,* 5805.
67. Zhang, X.; van Eldik, R.; Koike, T.; Kimura, E. *Inorg. Chem*. **1993**, *32,* 5749.
68. (a) Chandrawati, R.; Hosta-Rigau, L.; Vanderstraaten, D.; Lokuliyana, S. A.; Stadler, B.; Albericio, F.; Caruso, F. *ACS Nano* **2010**, *4,* 1351; (b) Hosta-Rigau, L.; Staedler, B.; Yan, Y.; Nice, E. C.; Heath, J. K.; Albericio, F.; Caruso, F. *Adv. Funct. Mater*. **2010**, *20,* 59.
69. (a) Brizard, A. M.; van Esch, J. H. *Soft Matter* **2009**, *5,* 1320; (b) Monnard, P.-A.; DeClue, M. S.; Ziock, H.-J. *Curr. Nanosci*. **2008**, *4,* 71; (c) Chang, T. M. S. *Nature Rev. Drug Discov*. **2005**, *4,* 221.

70. Percec, V.; Wilson, D. A.; Leowanawat, P.; Wilson, C. J.; Hughes, A. D.; Kaucher, M. S.; Hammer, D. A.; Levine, D. H.; Kim, A. J.; Bates, F. S.; Davis, K. P.; Lodge, T. P.; Klein, M. L.; DeVane, R. H.; Aqad, E.; Rosen, B. M.; Argintaru, A. O.; Sienkowska, M. J.; Rissanen, K.; Nummelin, S; Ropponen, J. *Science* **2010**, *328,* 1009.

71. Gardikis, K.; Hatziantoniou, S.; Bucos, M.; Fessas, D; Signorelli, M.; Felekis, T.; Zervou, M.; Screttas, C. G.; Steele, B. R.; Ionov, M.; Micha-Screttas, M.; Klajnert, B.; Bryszewska, M.; Demetzos, C. *J. Pharm. Sci*. **2010**, *99,* 3561.

72. (a) Negin, S.; Daschbach, M. M.; Kulikov, O. V.; Rath, N.; Gokel, G. W. *J. Am. Chem. Soc*. **2011**, *133,* 3234; (b) Li, R.; Kulikov, O. V.; Gokel, G. W. *Chem. Commun*. **2009**, 6092; (c) Wang, W.; Li, R.; Gokel, G. W. *Chem. Commun*. **2009**, 911; (d) Gokel, G. W.; Carasel, I. A. *Chem. Soc. Rev*. **2007**, *36,* 378.

73. Jin, P.; Dalgarno, S. J.; Warren, J. E.; Teat, S. J.; Atwood, J. L. *Chem. Commun*. **2009**, 3348.

74. Jin, P.; Dalgarno, S. J.; Atwood, J. L. *Coord.Chem. Rev*. **2010**, *254,* 1760.

75. Kulikov, O. V.; Li, R.; Gokel, G. W. *Angew. Chem. Int. Ed*. **2009**, *48,* 375.

76. Li, J.-R.; Zhou, H.-C. *Nature Chemistry* **2010**, *2,* 893.

77. Furukawa, H.; Kim, J.; Plass, K. E.; Yaghi, O. M. *J. Am. Chem. Soc*. **2006**, *128,* 8398.

78. Jung, M.; Kim, H.; Baek, K.; Kim, K. *Angew. Chem. Int. Ed*. **2008**, *47,* 5755.

79. Zhou, D.; Tan, S.; Yuan, D.; Lu, W.; Rezenom, Y. H.; Jiang, H.; Wang, L.-Q.; Zhou, H.-C. *Adv. Mater*. **2011**, *23,* 90.

80. Fyles, T. M.; Tong, C. *New J. Chem*. **2007**, *31,* 655.

81. Fujita, M.; Tominaga, M.; Hori, A.; Therrien, B. *Acc. Chem. Res*. **2005**, *38,* 369.

82. Kamiya, N.; Tominaga, M.; Sato, S.; Fujita, M. *J. Am. Chem. Soc*. **2007**, *129,* 3816.

83. Kikuchi, T.; Sato, S.; Fujita, M. *J. Am. Chem. Soc*. **2010**, *132,* 15930.

84. (a) Campolongo, M. J.; Tan, S. J.; Xu, J.; Luo, D. *Adv. Drug Deliv. Rev*. **2010**, *62,* 606; (b) Luo, D. *Adv. Drug Deliv. Rev*. **2010**, *62,* 591; (c) Yang, D.; Campolongo, M. J.; Tran, T. N. N.; Ruiz, R. C. H.; Kahn, J. S.; Luo, D. Wiley Interdisciplinary Reviews: *Nanomedicine and Nanobiotechnology* **2010**, *2,* 648.

85. (a) Mansuripur, M.; Khulbe, P. K.; Kuebler, S. M.; Perry, J. W.; Giridhar, M. S.; Erwin, J. K.; Seong, K.; Marder, S. R.; Peyghambarian, N. *Proc. SPIE–The Int. Soc. Opt. Eng*. **2003**, *231,* 5069; (b) Duran, C.; Appleby, N.; Edwards, D.; Batley, J. *Curr. Bioinformatics* **2009**, *4,* 16.

86. (a) Bandy, T. J.; Brewer, A.; Burns, J. R.; Marth, G.; Nguyen, T. N.; Stulz, E. *Chem. Soc. Rev*. **2011**, *40,* 138; (b) Tanaka, K.; Shionoya, M. *Coord. Chem. Rev*. **2007**, *251,* 2732; (c) Mueller, J. *Eur. J. Inorg. Chem*. **2008**, *24,* 3749; (d) Yang, H.; Rys, A. Z.; McLaughlin, C. K.; Sleiman, H. F. *Angew. Chem. Int. Ed*. **2009**, *48,* 9919.

87. Johannsen, S.; Megger, N.; Böhme, D.; Sigel, R. O. K.; Müller, J. *Nature Chemistry* **2010**, *2,* 229.

88. Gou, P. *Nature Nanotechnology* **2010**, *5,* 833.

89. Ko, S. H.; Su, M.; Zhang, C.; Ribbe, A. E.; Jiang, W.; Mao, C. *Nature Chemistry* **2010**, *2,* 1050.

90. Severca, I.; Geary, C.; Chworos, A.; Voss, N.; Jacovetty, E.; Jaeger, L. *Nature Chemistry* **2010**, *2,* 772.

91. Yang, H.; McLaughlin, C. K.; Aldaye, F. A.; Hamblin, G. D.; Rys, A. Z.; Rouiller, I.; Sleiman, H. F. *Nature Chemistry* **2009**, *1,* 390.

92. (a) Czaja, A. U.; Trukhan, N.; Müller, U. *Chem. Soc. Rev*. **2009**, *38,* 1284; (b) Murray, L. J.; Dinca, M.; Long, J. R. *Chem. Soc. Rev*. **2009**, *38,* 1294; (c) Clegg, J. K.; Iremonger, S. S.; Hayter, M. J.; Southon, P. D.; MacQuart, R. B.; Duriska, M. B.; Jensen, P.; Turner, P.; Jolliffe, K. A.; Kepert, C. J.; Meehan, G. V.; Lindoy, L. F. *Angew. Chem. Int. Ed*. **2010**, *49,* 1075.

93. (a) Lee, Y.-Y.; Omar K.; Farha, O. K.; Roberts, J.; Scheidt, K. A.; Nguyen, S. T.; Hupp, J. T. *Chem. Soc. Rev*. **2009**, *38,* 1450; (b) Corma, A.; Garcia, H.; Llabres i Xamena, F. X. *Chem. Rev*. **2010**, *110,* 4606.

94. (a) Jung, S.; Kim, S. J.; Kwon, T. H. Huh, S.; Park, S. *Chem. Commun*. **2011**, 2904; (b) Barron, P. M.; Wray, C. A.; Hu, C.; Guo, Z.; Choe, W. *Inorg. Chem*. **2010**, *49,* 10217; (c) Horcajada, P.; Serre, C.; Grosso, D.; Boissière, C.; Perruchas, S.; Sanchez, C.; Férey, G. *Adv. Mater*. **2009**, *21,* 1931.

95. McKinlay, A. C.; Morris, R. E.; Horcajada, P.; Férey, G.; Gref, R.; Couvreur, P.; Serre, C. *Angew. Chem. Int. Ed*. **2010**, *49,* 6260.

96. (a) An, J.; Shade, C. M.; Chengelis-Czegan, D. A.; Petoud, S.; Rosi, N. L. *J. Am. Chem. Soc*. **2011**, *133,* 1220; (b) An, J.; Geib, S. J.; Rosi, N. L. *J. Am. Chem. Soc*. **2010**, *132,* 38; (c) An, J.; Geib, S. J.; Rosi, N. L. *J. Am. Chem. Soc*. **2009**, *131,* 8376.

97. (a) Vaidhyanathan, R.; Bradshaw, D.; Rebilly, J.-N.; Barrio, J. P.; Gould, J. A.; Berry, N. G.; Rosseinsky, M. J. *Angew. Chem. Int. Ed*. **2006**, *45,* 6495; (b) Vaidhyanathan, R; Bridges, C. A; Bradshaw, D; Rosseinsky, M. J. *Crystal Growth and Design* **2010**, *10,* 4348.

98. Mantion, A.; Massuger, L.; Rabu, P.; Palivan, C.; McCusker, L. B.; Taubert, A. *J. Am. Chem. Soc*. **2008**, *130,* 2517.

99. Rabone, J.; Yue, Y.-F.; Chong, S. Y.; Stylianou, K. C.; Bacsa, J.; Bradshaw, D.; Darling, G. R.; Berry, N. G.; Khimyak, Y. Z.; Ganin, A. Y.; Wiper, P.; Claridge, J. B; Rosseinsky, M. J. *Science* **2010**, *329,* 1053.

CHAPTER 3

Bioinspired Self-Assembly II: Principles of Cooperativity in Bioinspired Self-Assembling Systems

GIANFRANCO ERCOLANI and LUCA SCHIAFFINO
Dipartimento di Scienze e Tecnologie Chimiche, Università di Roma Tor Vergata, 00133 Roma, Italy

3.1 INTRODUCTION

Cooperativity plays a crucial role in biology as a regulation mechanism for a large variety of processes, such as multiple molecular recognition, enzyme catalysis, membrane transport, and self-assembly.[1] Not surprisingly, one of the main goals of supramolecular chemistry is to implement cooperativity in artificial systems with the aim of better understanding the mechanisms involved in the natural processes, as well as of preparing functional materials and devices that benefit from such an efficient regulation.[2] In spite of the importance of the concept, there is widespread confusion about the definition and quantification of cooperativity, especially as far as self-assembly is concerned.[3–6] Over time, it has become clear that different types of cooperativity mechanisms must be considered to truly understand the behavior of biological, as well as bioinspired, self-assembling systems.[7] The aim of this chapter is to illustrate these mechanisms and, at the same time, to provide the physicochemical background for their comprehension.

Multiple binding events may be cooperative, rather than independent, meaning that the binding at one site may influence the binding at another, causing an increase (positive cooperativity) or a decrease (negative cooperativity) of the binding strength. Cooperativity is thus the deviation of the behavior of a real system of multiple binding events from the hypothetical system in which binding events occur independently of each other. The assessment of cooperativity requires, therefore,

Bioinspiration and Biomimicry in Chemistry: Reverse-Engineering Nature, First Edition.
Edited by Gerhard F. Swiegers.
© 2012 John Wiley & Sons, Inc. Published 2012 by John Wiley & Sons, Inc.

the following three-step procedure: (1) evaluation of the isolated binding events;*[8] (2) development of a model (the noncooperative model) where each binding event of the system behaves as if it were isolated from the others; and (3) testing of the behavior of the real system against that predicted by the noncooperative model: any deviation, positive or negative, of the real system is by definition the unambiguous mark of cooperativity. This procedure will be followed to analyze the three basic types of cooperativity mechanisms: (1) allosteric cooperativity, arising from the interplay of intermolecular binding events; (2) chelate cooperativity, due to the mere presence of one or more intramolecular binding interactions; and (3) interannular cooperativity, occurring when intramolecular binding events are not independent of other.[7]

The structure of this chapter is as follows. In Section 3.2, a detailed derivation and explanation of statistical factors is presented, since they are essential components in describing the equilibria of multivalent interactions. In Section 3.3, allosteric cooperativity, which is the type of cooperativity exhibited by hemoglobin, is briefly discussed, mainly to illustrate the above three-step procedure. In Section 3.4, the concept of effective molarity (*EM*) is introduced, as this physicochemical parameter is of fundamental importance for the understanding and discussion of any intramolecular interaction. In Sections 3.5 and 3.6, the two mechanisms of chelate and interannular cooperativity, respectively, are presented. Finally, in Section 3.7, the stability of a self-assembling system is discussed, taking into account inter- and intramolecular interactions and their possible interplay.

3.2 STATISTICAL FACTORS IN SELF-ASSEMBLY

An accurate and consistent evaluation of statistical factors in self-assembly processes is crucial to predict the expected stability constant in the absence of cooperative effects and, therefore, to spotlight the emergence of cooperativity. The evaluation of statistical factors has recently been critically reexamined.[9] Two methods are most useful: the symmetry number method and the direct count method. The two methods if properly applied give the same results; however, the symmetry number method is generally of easier and faster application, and therefore we recommend its use. Only in case of doubt, we recommend the use of the direct count method as an independent check. For brevity reasons, we only present here an outline of the symmetry number method.

The observed equilibrium constant, K_{obs}, of a generic equilibrium (Eq. 3.1) can be regarded as given by the product of a microscopic or "chemical" constant K and a statistical factor, K_σ.

$$a\mathrm{A} + b\mathrm{B} \xrightleftharpoons{K_{\text{obs}}=K_\sigma K} c\mathrm{C} + d\mathrm{D} \tag{3.1}$$

*In the case of multiple binding of the same type of ligand (homotropic cooperativity), only a single reference interaction is needed, whereas in the case of binding of different types of ligands (heterotropic cooperativity), more than a single reference interaction must be defined. The binding of oxygen to hemoglobin is the classical example of homotropic cooperativity. For a recent example of heterotropic cooperativity.

According to the symmetry number method, pioneered by Benson,[10] K_σ is given by the ratio of symmetry numbers for reactant and product species in equilibrium (Eq. 3.2).

$$K_\sigma = \frac{\sigma_{\text{reactants}}}{\sigma_{\text{products}}} = \frac{\sigma_A^a \sigma_B^b}{\sigma_C^c \sigma_D^d} \tag{3.2}$$

The factor σ is the product of the external, σ_{ext}, and internal, σ_{int}, symmetry numbers. The external symmetry number is defined as the number of different but indistinguishable atomic arrangements that can be obtained by rotating a given molecule as a whole. It is found, in practice, by multiplying the order of the independent simple rotational axes of the point group to which the molecule belongs (axes of infinite order are not considered because they do not generate different atomic arrangements). External symmetry numbers for the several point groups are shown in Table 3.1.

The internal symmetry number is defined as the number of different but indistinguishable atomic arrangements that can be obtained by internal rotations around single bonds, or, in the case of fluxional molecules, by inversion, pseudorotation, or other intramolecular processes. It is implied that the processes giving rise to the internal symmetry number are fast with respect to the time scale in which the equilibrium in Eq. 3.1 is attained and measured. For example, ethane has $\sigma = 18$ that is the product of $\sigma_{\text{ext}} = 6$ because of D_{3d} symmetry (in this point group there are a threefold axis and an independent twofold axis) and $\sigma_{\text{int}} = 3$, due to the internal rotation of one methyl group with respect to the other; ammonia has $\sigma = 6$ that is the product of $\sigma_{\text{ext}} = 3$ because of C_{3v} symmetry (only a threefold rotational axis) and $\sigma_{\text{int}} = 2$, due to the process of pyramidal inversion. Equation 3.2 has its physical basis on the fact that the symmetry number of a molecule affects its rotational entropy by a factor $-R \ln \sigma$. A different type of correction accompanies a chiral molecule present at equilibrium as a racemic mixture; its symmetry number must be divided by 2 to account for the entropy of mixing of the two enantiomers.[9, 10]

A complication arises from flexible molecules in which internal rotations produce distinct conformations of different energy; in these cases the only viable and consistent approximation is one that considers all the internal rotations as

TABLE 3.1 External Symmetry Numbers for Various Point Groups

Point Group	σ_{ext}
$C_1, C_i, C_s, C_{\infty v}, R_3$	1
$D_{\infty h}$	2
C_n, C_{nv}, C_{nh}	n
D_n, D_{nd}, D_{nh}	$2n$
S_n (n even)	$n/2$
T_d	12
O_h	24
I_h	60

free, implying that all torsion angles have equal probability of occurrence. Under this approximation, it can be demonstrated that the external symmetry number of a molecule is equal to that of the most symmetrical of its conformations.[9] For example, in the case of a normal alkane, the most symmetrical conformation is the extended one (all dihedrals in anti) whose point group is $C_{2h}(\sigma_{\text{ext}} = 2)$. Since $\sigma_{\text{int}} = 3^2$, because of the internal rotations of the two terminal methyl groups, the overall symmetry number of a normal alkane is $\sigma = 18$. In the case of a cycloalkane with n carbons, the most symmetrical conformation is the planar one (D_{nh} symmetry) for which $\sigma = \sigma_{\text{ext}} = 2n$, even though this conformation does not even correspond to a minimum. It goes without saying that the symmetry number obtained under this approximation has no bearing on the rotational entropy of the molecule.

3.3 ALLOSTERIC COOPERATIVITY

Allosteric cooperativity arises from the interplay of intermolecular binding interactions. It can be homotropic or heterotropic, depending on whether the binding to a multivalent receptor involves the same or different types of ligands. Homotropic cooperativity, exemplified by the binding of oxygen to hemoglobin,[1] is the most interesting and the most difficult to realize in artificial receptors; thus, we will limit the discussion to this type of cooperativity.

Consider, for example, the stepwise binding of a monovalent ligand to a divalent receptor as depicted in Figure 3.1. The observed stepwise constants are given by the two microscopic association constants K_1 and K_2 multiplied by the statistical factors 2 and $\frac{1}{2}$, respectively. The statistical factors are easily understood considering that the unbound receptor has $\sigma = 2$, the half-bound receptor has $\sigma = 1$, and the fully bound receptor has $\sigma = 2$. In general, for the interaction of an n-valent receptor with a monovalent ligand, the statistical factor for the ith stepwise equilibrium is given by $(n - i + 1)/i$. The reference constant K can be evaluated by studying the binding of the monovalent ligand B to a monovalent model, A, of the receptor, or alternatively, directly taking the value of the constant K_1 as the reference constant. Thus, in the absence of cooperativity, all the microscopic stepwise equilibrium constants must be equal to the reference value K (for the case in Figure 3.1, $K_2 = K_1 = K$). A way to quantify allosteric cooperativity is to evaluate the cooperativity factor α given by the ratio of the overall experimental constant to the hypothetical overall noncooperative constant. For the general case of the

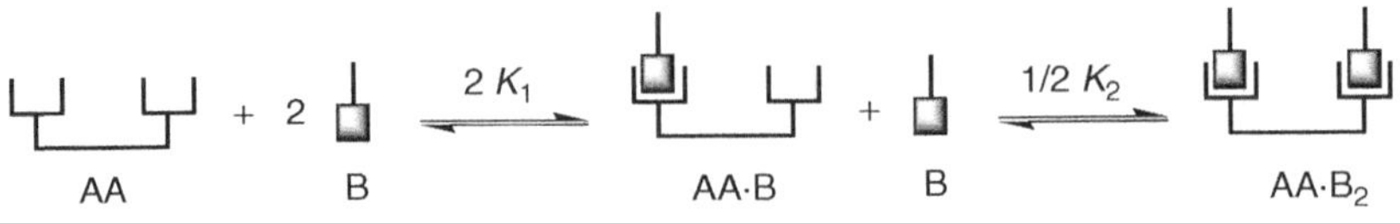

Figure 3.1 Stepwise binding of a monovalent ligand B to a divalent receptor AA.

interaction of a monovalent ligand with an n-valent receptor, α is given by Eq. 3.3:

$$\alpha = \frac{\prod_i K_i}{K^n} \tag{3.3}$$

The factor α is a dimensionless constant larger than 1 in the case of positive cooperativity, equal to 1 in the case of noncooperativity, and smaller than 1 in the case of negative cooperativity. It can be viewed as the equilibrium constant for the conversion of the hypothetical noncooperative complex (independent binding sites) into the cooperative complex (interacting binding sites) (Figure 3.2).

There are other equivalent tests to assess allosteric cooperativity, mainly graphical ones, based on the calculation of the occupancy, r, that is to say the average number of occupied sites of the n-valent receptor.[2a, 11] It is easy to show that the occupancy is given by Eq. 3.4:

$$r = \frac{\sum_{j=1}^{n} j\beta_j[\mathrm{B}]^j}{1 + \sum_{j=1}^{n} \beta_j[\mathrm{B}]^j} \tag{3.4}$$

where β_j are the cumulative binding constants given by Eq. 3.5 and [B] is the free ligand concentration.

$$\beta_j = \prod_{i=1}^{j} \frac{(n-i+1)}{i} K_i \tag{3.5}$$

If the binding is noncooperative ($K_i = K$ for all i values), it can be demonstrated that Eq. 3.4 reduces to the binding isotherm shown in Eq. 3.6, where $y(= r/n)$ is the degree of saturation of the receptor.[2a, 11]

$$y = \frac{K[\mathrm{B}]}{1 + K[\mathrm{B}]} \tag{3.6}$$

Plots that deviate from the above equation are diagnostic for cooperativity. In particular, a sigmoid plot is clear-cut evidence of marked positive cooperativity. However, such deviations are not always easily recognizable. A much better diagnostic can be obtained by putting Eq. 3.6 into linear form because deviations from a

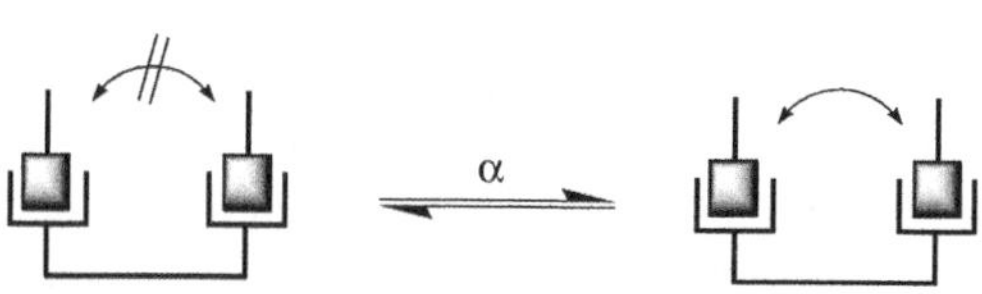

Figure 3.2 The cooperative factor α is the equilibrium constant for the conversion of the hypothetical noncooperative complex into the cooperative complex. The figure depicts the case of a saturated divalent receptor.

straight line are more easily detectable. Two popular linear forms are the Scatchard equation (Eq. 3.7) and the Hill equation (Eq. 3.8):

$$\frac{r}{[\mathrm{B}]} = -Kr + nK \tag{3.7}$$

$$\log \frac{y}{1-y} = \log[\mathrm{B}] + \log K \tag{3.8}$$

A Scatchard plot is a plot of $r/[\mathrm{B}]$ as a function of r; it is linear in the case of noncooperativity, whereas it presents a concave downward curve in the case of positive cooperativity, or a concave upward curve in the case of negative cooperativity. A Hill plot is a plot of $\log y/(1-y)$ versus log [B]; apart from the case of noncooperativity that is evidenced by a straight line of unit slope, cooperativity manifests itself as two lines of unit slope connected by an S-shaped curve. The value of the slope in the central region of the curve is called the Hill coefficient (n_H). It can vary between 0 and n; values larger than 1 are diagnostic for positive cooperativity whereas values lower than 1 are diagnostic for negative cooperativity.[2a, 11]

It is useful to remark that all of these methods to assess cooperativity provide meaningful results only for a collection of intermolecular binding events.[12]

3.4 EFFECTIVE MOLARITY

Intramolecular reactions are often more favored than analogous intermolecular reactions. This advantage, known as the proximity effect or the chelate effect, is measured by the effective molarity (*EM*), a physicochemical parameter that has units, as the name implies, of mol L^{-1}.[13, 14] To illustrate the concept of *EM*, let us envisage a solution of a chain molecule with two end groups –A and –B, as schematically illustrated in Figure 3.3. The functional group –A can react reversibly with the functional group –B to form a new bond AB; if –A and –B belong to the same chain the reaction is intramolecular and leads to the formation of a ring, whereas if the two groups belong to different chains the reaction is intermolecular. The equilibrium constants for the two processes are, respectively, K_intra that is dimensionless, and K that has units of mol^{-1} L. There are a number of equivalent definitions of *EM*; the first in order of time, and somewhat outdated, considers the *EM* as the molar concentration of one chain end experienced by the other end of the same chain. Although formally correct, this definition is not physically reliable for short chains because the *EM* can assume unreasonably high values such as 10^8 mol L^{-1}. However, this definition is a useful starting point to obtain other more physically accurate definitions. On the basis of Figure 3.3, it appears that if the concentration of the bifunctional chain A–B is equal to the *EM*, the intermolecular process proceeds with the same extent of reaction of the intramolecular process. This observation is very important because the *EM* can now be viewed as the limit concentration of the chain A–B below which intramolecular processes are more favored than intermolecular processes. Under the condition that $[\mathrm{A{-}B}] = EM$,

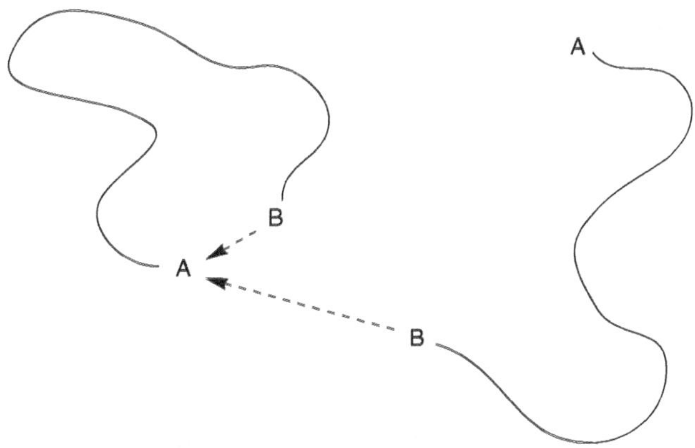

Figure 3.3 Schematic representation of the competition between an intramolecular process and an intermolecular process aimed at illustrating the concept of *effective molarity*.

the intramolecular constant K_{intra} is equal to the apparent intermolecular constant $K[A\text{–}B] = K \cdot EM$, from which an operative definition of *EM* is immediately obtained (Eq. 3.9):

$$EM = K_{intra}/K \tag{3.9}$$

The constant K_{intra} can thus be viewed as the product of *K* representing the inherent chemical reactivity of end groups, and *EM* representing a connection factor that accounts for the fact that the two reactive groups are connected to each other. The *EM* has both an enthalpic and an entropic component ($EM = EM_H\, EM_S$). In most of the cases the enthalpic component only depends on the strain energy of the ring, so EM_H is lower than 1 unless a strainless ring is formed, in which case, it is equal to 1. For a strainless ring the *EM* is solely dependent on entropy. The value of EM_S decreases on increasing the number of rotatable bonds of the chain connecting the end groups because internal rotations become more restricted upon ring closure. Thus, the maximum *EM* value can be reached when the end groups are connected by a rigid structure, preorganized to form a strainless ring. The maximum *EM* value has been estimated by Page and Jencks.[15] According to their classical analysis, bringing two molecules together is accompanied by a negative change in entropy because of the reduced volume of space available to the reactants. Mechanically, a free molecule has three degrees of translational freedom and three degrees of overall rotational freedom. When two molecules condense to form one, three degrees of each are converted into six degrees of vibrational freedom that have a lower entropy content. The more severely the geometrical relationship between the reactant molecules is defined the greater is the loss of entropy. For bond-forming reactions commonly encountered in organic chemistry, this entropy change is about -150 J K^{-1} mol^{-1} at a standard state of 1 mol L^{-1}; it makes intermolecular reaction equilibria unfavorable by a factor of about 10^8 mol L^{-1}. The formation of weaker bonds such as hydrogen bonds and metal–ligand bonds, often encountered in self-assembling systems, involves a smaller entropy change (about -40 J K^{-1} mol^{-1} at a standard state of 1 mol L^{-1}) that makes intermolecular reaction equilibria unfavorable by a factor of about 10^2 mol L^{-1}. These changes do not occur in intramolecular reactions and thus 10^8 mol L^{-1} and 10^2 mol L^{-1} are the maximum *EM* values that can be found for intramolecular processes involving the formation of tight and loose bonds, respectively. Introduction of rotors in the chain connecting the end groups reduces the value of the *EM* from these maxima. In Figure 3.4 are reported the *EM* values for covalent[13b–d] and noncovalent[16] cyclization processes versus the number of rotors in the linking chain (r). The limits to which the experimental results tend in the absence of rotors are in substantial agreement with the analysis of Page and Jencks.[15]

The theory for predicting the *EM* is well developed for long flexible chains, say more than 25–30 skeletal bonds, yielding strainless rings; according to the theory, the *EM* of an r-meric chain is proportional to $r^{-3/2}$.[17] This factor is related to the probability that a Gaussian chain of r repeating units has its ends coincident. Thus, for long flexible chains, the original definition of the *EM* as the molar concentration

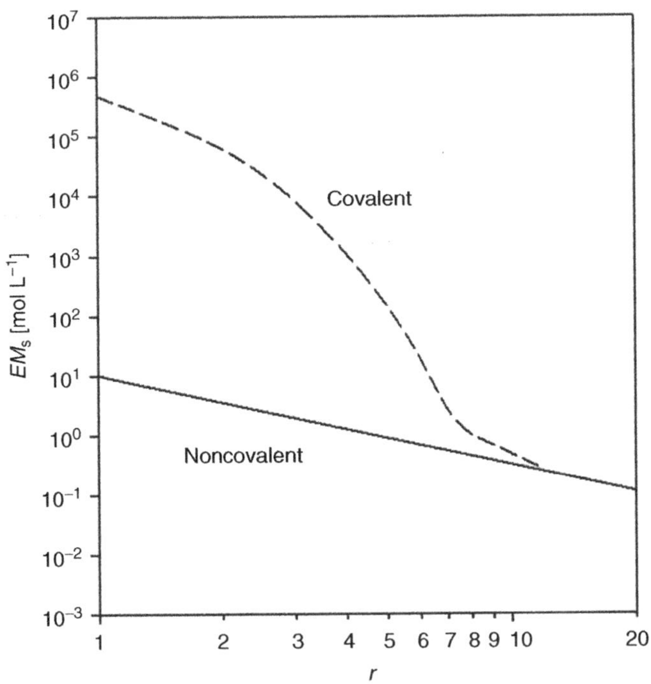

Figure 3.4 Log–log plot of the entropic component of the effective molarity (EM_S) for covalent (dashed line) and noncovalent (solid line) cyclization processes and the number of rotors in the linking chain (r). The covalent EM_S values are from Refs. 13b–d; the noncovalent EM_S values, according to Ref. 16, satisfy the linear relationship log $EM_S \approx 10 - \frac{3}{2} \log r$.

of one chain end experienced by the other end of the same chain makes physical sense. Hunter and co-workers have recently shown that the same relationship also holds for cyclizations of shorter chains provided that the newly formed intramolecular bond is noncovalent.[16] This finding is evidenced in Figure 3.4 by the straight line of slope $-\frac{3}{2}$ observed for noncovalent cyclizations. Rather interestingly, the covalent curve approaches the noncovalent one when the orientational correlations between the end groups are less severe because of the increased chain length.[17c]

From the analysis of Page and Jencks,[15] as well as from the experimental data of Hunter and co-workers,[16] it appears that noncovalent *EM*s never reach the very high values observed for covalent processes, which places limitations on the magnitudes of the effects that one is likely to achieve through the use of chelate interactions in supramolecular assembly. On the other hand, the decrease in *EM* due to the introduction of conformational flexibility is less dramatic than one might expect based on the behavior of covalent systems, which limits the losses in binding affinity caused by poor preorganization of the interaction sites.

3.5 CHELATE COOPERATIVITY

Chelate cooperativity arises from the presence of one or more intramolecular binding interactions, that is, as a consequence of the chelate effect, also called multivalency. While allosteric cooperativity is well recognized, the assessment of chelate cooperativity is still the object of debate.[3–7] Let us consider the archetypal case that involves the binding of a divalent ligand BB to a divalent receptor AA (Figure 3.5). The ligand is present in a large excess relative to the receptor so that complexes involving more than one receptor molecule can be neglected, and $\alpha = 1$ to exclude allosteric cooperativity. Under the given conditions there are only four states for the receptor: free AA, the partially bound 1:1 open complex *o*-AA·BB, the fully bound 1:1 cyclic complex *c*-AA·BB, and the 1:2 complex $AA{\cdot}(BB)_2$. The intramolecular binding constant is expressed as the product $\frac{1}{2}\,K\,EM$, where $\frac{1}{2}$ is the statistical factor for the cyclization process, K is the microscopic intermolecular constant expressing the strength of the binding interaction, and *EM* is the microscopic effective molarity.*

An important characteristic of chelate cooperativity is that, in contrast with allosteric cooperativity, it depends on ligand concentration. To illustrate this behavior, in Figure 3.6 are reported the speciation profiles for the equilibria shown in Figure 3.5 assuming $K\cdot EM = 50$. Positive cooperativity is characterized by a low concentration of partially bound species. In the most extreme cases only the unbound and fully bound receptor are significantly populated and the system exhibits an "all-or-none" behavior. From Figure 3.6 it appears that the presence of the chelate interaction leads to a low concentration of the partially bound intermediate complex *o*-AA·BB favoring the fully bound cyclic complex *c*-AA·BB.

*Sometimes the statistical factor for the cyclization process is incorporated into the value of the *EM*. We use the term microscopic effective molarity to indicate that the symbol *EM* does not account for the statistical factor.

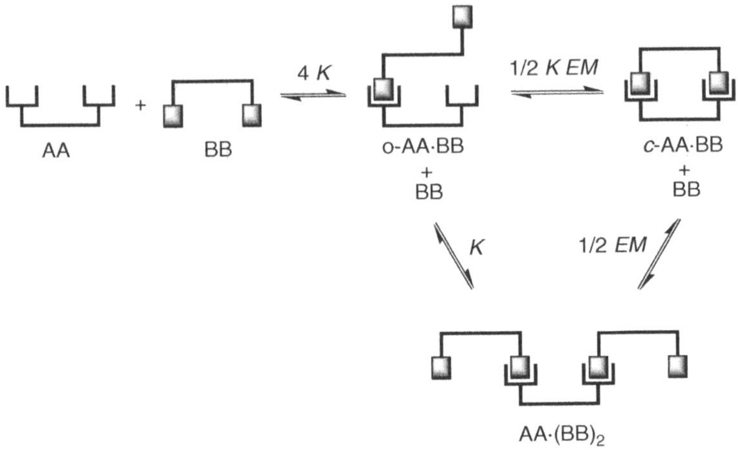

Figure 3.5 Binding of a divalent ligand BB to a divalent receptor AA, assuming $[BB]_0 \gg [AA]_0$ and $\alpha = 1$.

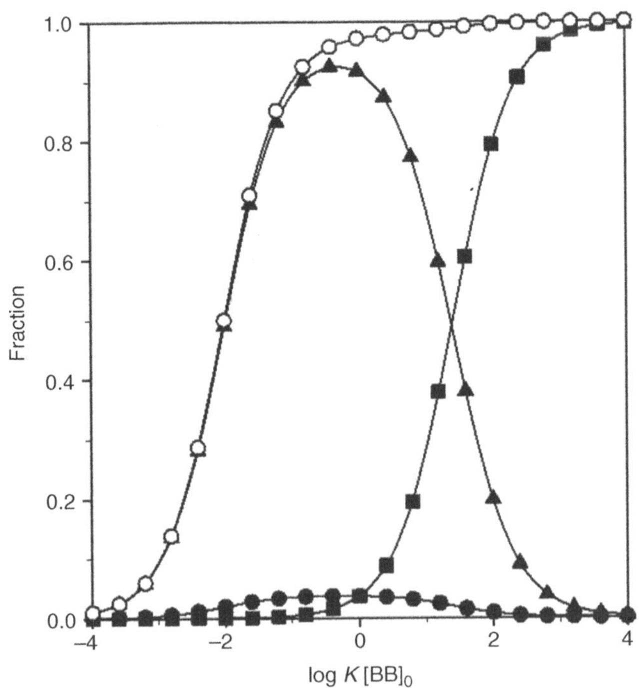

Figure 3.6 Speciation profiles for the equilibria shown in Figure 3.5 in the case where $K \cdot EM = 50$. The concentration scale on the abscissa is normalized by multiplying by K. Symbol legend: *o*-AA·BB (filled circles); $AA{\cdot}(BB)_2$ (squares); *c*-AA·BB (triangles); degree of saturation of the receptor AA (empty circles).

However, although the overall degree of saturation of the receptor tends to an "all-or-none" process, the speciation profile of the chelate complex c-AA·BB is bell-shaped, suggesting that the intramolecular process can be more properly regarded as "none-all-none." This is due to the fact that at low concentration the cyclic complex is disfavored with respect to the unbound receptor, whereas at high concentration it suffers from the competition with the fully bound 1:2 open complex. The ligand concentration at which the switch between the cyclic complex and the fully bound open complex occurs is easily obtained by considering the equilibrium between these two species illustrated in Figure 3.5 and the corresponding equilibrium constant defined by Eq. 3.10:

$$\frac{EM}{2} = \frac{[c - \mathrm{AA} \cdot \mathrm{BB}][\mathrm{BB}]}{[\mathrm{AA} \cdot (\mathrm{BB})_2]} \tag{3.10}$$

When [BB] $= EM/2$ the concentrations of the two species are identical. Thus, it appears that *EM* is the threshold concentration of ligand binding groups above which the intramolecular process loses the competition with the intermolecular one. The conclusion is that the advantage provided by the chelate interaction is dissipated at high ligand concentrations. The dependency of chelate cooperativity on ligand concentration has been often overlooked, leading to inconsistencies. For example, it has been advocated that chelate cooperativity manifests itself when the intramolecular constant ($\frac{1}{2}K \cdot EM$ in Figure 3.5) is larger than the intermolecular constant ($4K$ in Figure 3.5)*.[5] This comparison does not make sense because the intramolecular and the intermolecular constants have different units, specifically, the former is unitless whereas the latter has units of $\mathrm{mol^{-1}\ L}$. Another inconsistency has been pointed out by Jencks[18] with reference to the binding of a divalent asymmetric ligand AB to a complementary receptor as illustrated in Figure 3.7: a common pitfall is to consider that in the absence of chelate cooperativity the observed free energies of binding for the individual parts A and B are additive in the molecule AB, so that $\Delta G_{\mathrm{AB}}{}^{\circ} = \Delta G_{\mathrm{A}}{}^{\circ} + \Delta G_{\mathrm{B}}{}^{\circ}$. Jencks pointed out that there is no basis for this assumption: the addition of Gibbs energies is equivalent to the multiplication of binding constants and, if K_{A}, K_{B}, and K_{AB} are measured in $\mathrm{mol^{-1}\ L}$, the equation $K_{\mathrm{AB}}\ (\mathrm{mol^{-1}\ L}) = K_{\mathrm{A}}K_{\mathrm{B}}\ (\mathrm{mol^{-1}\ L})^2$ is meaningless. The correct way to address the problem of additivity of binding energies is to add

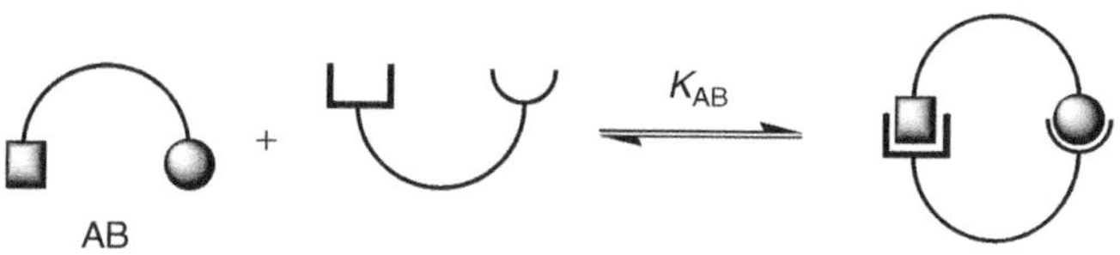

Figure 3.7 Binding of a divalent asymmetric ligand AB to a complementary receptor.

*The concept was expressed in terms of dissociation constants, that is, $K_{\mathrm{d2}} < K_{\mathrm{d1}}$, where $K_{\mathrm{d2}} = 2(K \cdot EM)^{-1}$ and $K_{\mathrm{d1}} = (4K)^{-1}$. See the caption of Figure 1(b) of Ref. 5.

a connection Gibbs energy that represents the change in the probability of binding that results from the connection of A and B in AB. This connection Gibbs energy is in fact the free energy associated to the *EM*, $\Delta G_{EM}{}^{\circ}$. Accordingly, $\Delta G_{\mathrm{AB}}{}^{\circ} = \Delta G_{\mathrm{A}}{}^{\circ} + \Delta G_{\mathrm{B}}{}^{\circ} + \Delta G_{EM}{}^{\circ}$, that when translated into binding constants gives a dimensionally correct equation $K_{\mathrm{AB}}(\mathrm{mol}^{-1}\,\mathrm{L}) = K_{\mathrm{A}}K_{\mathrm{B}}\cdot EM\ (\mathrm{mol}^{-1}\ \mathrm{L})$. In analogy with the definition of the allosteric cooperativity factor α, it is tempting to assume the *EM* as the measure of chelate cooperativity; however, this assumption is inconsistent because, in contrast with the factor α which is dimensionless, the *EM* has units of concentration, and its numerical value depends on the choice of the standard state. Moreover, as pointed out above, the ligand concentration must also be taken into account for the correct assessment of chelate cooperativity.

Recently, it has been suggested the product $K \cdot EM$ is a measure for chelate cooperativity.[6] Again, this factor does not account for the dependency of chelate cooperativity on ligand concentration. Moreover, the product $K{\cdot}EM$ tends to zero in the absence of cooperativity, whereas a consistent cooperativity factor must tend to 1. One of us has previously shown that the product $K{\cdot}EM$ is related to the maximum amount of a cyclic or multicyclic supramolecular assembly in solution.[14, 19] However, although $K{\cdot}EM$ is the driving force for self-assembly, it has no bearing on chelate cooperativity.

To assess this type of cooperativity, we follow the three-step procedure detailed in the introduction. In the absence of chelate cooperativity, the overall constant for the fully bound receptor, as given by Eq. 3.11, is used as the reference intermolecular binding:

$$4K^2 = \frac{[\mathrm{AA}\cdot(\mathrm{BB})_2]}{[\mathrm{AA}][\mathrm{BB}]^2} \tag{3.11}$$

In the presence of chelate cooperativity, the overall apparent constant for the fully saturated receptor is given by Eq. 3.12:

$$\frac{([\mathrm{AA}\cdot(\mathrm{BB})_2] + [c-\mathrm{AA}\cdot\mathrm{BB}])}{[\mathrm{AA}][\mathrm{BB}]^2} = 4K^2 + 2K^2\frac{EM}{[\mathrm{BB}]} \tag{3.12}$$

The two terms on the right-hand side of Eq. 3.12 represent the contributions of intermolecular and intramolecular binding, respectively. The chelate cooperativity factor, β, is the ratio of the intramolecular binding to the reference intermolecular binding as given by Eq. 3.13:

$$\beta = \frac{EM}{2[\mathrm{BB}]} \tag{3.13}$$

It represents the apparent equilibrium constant for the conversion of the complex $\mathrm{AA{\cdot}(BB)_2}$ into the cyclic complex *c*-AA·BB as shown in Figure 3.8. The value of β depends on ligand concentration. When $\beta = 1$, $[\mathrm{BB}] = EM/2$; at this ligand concentration, the fully bound open complex and the cyclic complex are equally populated. At lower ligand concentrations, the cyclic complex is dominant ($\beta > 1$,

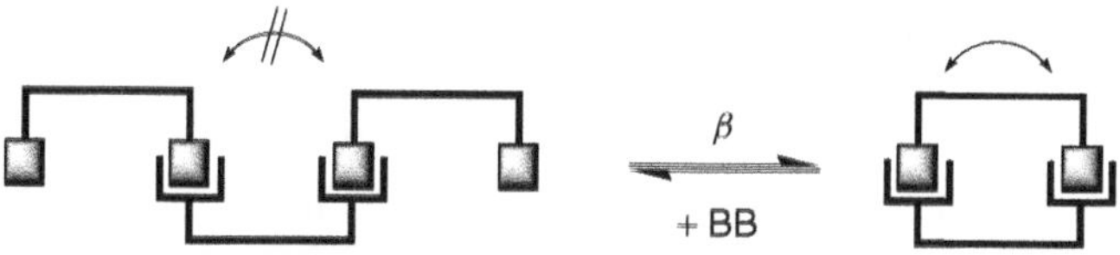

Figure 3.8 The cooperative factor β is the apparent equilibrium constant for the conversion of the noncooperative open complex into the chelate cooperative complex. The factor β depends on ligand concentration.

positive chelate cooperativity), whereas the opposite occurs at higher ligand concentrations. Indeed, as shown in Figure 3.6, the process of disassembly of the cyclic complex is due to a decrease in chelate cooperativity, which changes from positive to negative on increasing ligand concentration.

These results can be generalized to the case of the binding of an n-valent ligand ^{n}B to an n-valent receptor ^{n}A to form a 1:1 multicyclic complex c-^{n}A·^{n}B, the constituent rings of which have identical structure so that their *EM* values are also identical. It is assumed that the ligand is present in large excess relative to the receptor and that $\alpha = 1$. In Figure 3.9 is shown the equilibrium between the family of fully saturated open complexes and the multicyclic complex c-^{n}A·^{n}B, in the case $n = 3$. The apparent constant for such equilibrium is the chelate cooperativity factor β. Equation 3.13 can easily be generalized to the multivalent case giving Eq. 3.14, in which it appears that β depends on the ratio of the *EM* to the ligand concentration, raised to the degree of cyclicity of the assembly.[7]

$$\beta = \frac{2}{n^n}\left(\frac{EM}{[{}^{n}\mathrm{B}]}\right)^{n-1} \tag{3.14}$$

As Eq. 3.14 suggests, chelate cooperativity becomes very sensitive to ligand concentration when $\beta \approx 1$; that is, when $[{}^{n}\mathrm{B}] \approx (2/n^n)^{1/(n-1)}EM$. This is nicely illustrated by the fraction of the chelate complex c-^{n}A·^{n}B calculated in the case

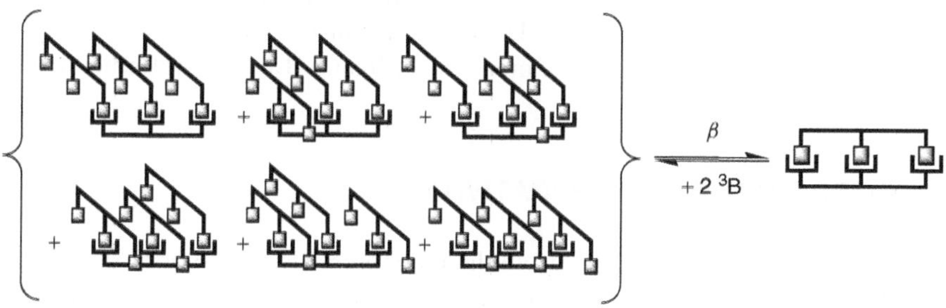

Figure 3.9 In the multivalent case, the cooperative factor β is the apparent equilibrium constant for the conversion of the family of noncooperative open complex into the multicyclic chelate cooperative complex. In the figure is illustrated the case of a trivalent ligand interacting with a trivalent receptor.

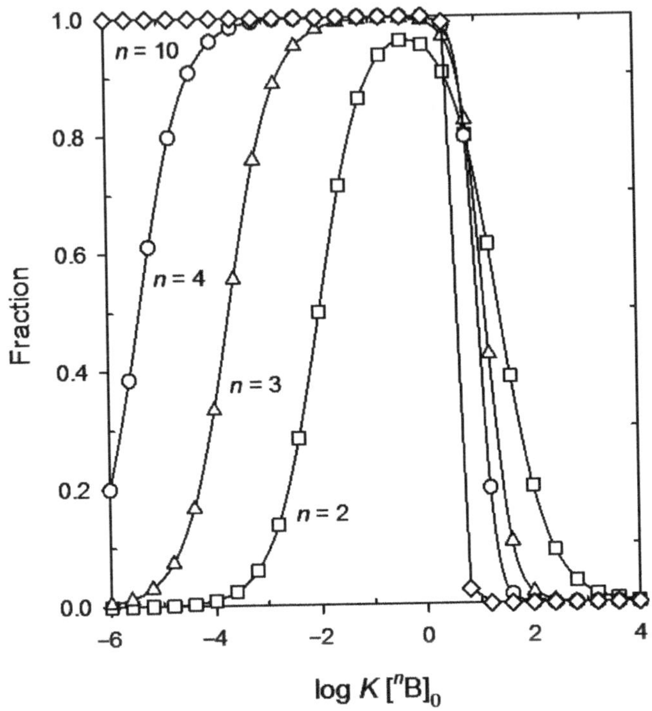

Figure 3.10 Fraction of the chelate complex $c\text{-}^{n}A\cdot{}^{n}B$ for the case $K \cdot EM = 50$ and $n = 2$ (squares), 3 (triangles), 4 (circles), and 10 (diamonds). The concentration scale on the abscissa is normalized by multiplying by K.

$K \cdot EM = 50$ and $n = 2$, 3, 4, and 10 (Figure 3.10), showing that disassembly (or denaturation) becomes sharper and sharper as n increases, because of an increasingly sharp drop of chelate cooperativity. This drop is accompanied by a complementary sharp increase of the fractional amount of the 1:n fully bound open complexes. The apparent cooperativity of the process of denaturation is in fact due to dissipation of chelate cooperativity caused by the increasing ligand concentration.

Of course, the physical basis of denaturation remains the same if it is carried out, as more usual, by addition of a monovalent ligand B to the preformed chelate complex.[7]

3.6 INTERANNULAR COOPERATIVITY

While chelate cooperativity arises from the mere presence of one or more independent intramolecular interactions, interannular cooperativity arises from the interplay of two or more such interactions. Thus, interannular cooperativity necessarily implies the presence of chelate cooperativity but the opposite is not true. A simple system to illustrate interannular cooperativity is that constituted by a tetravalent receptor ^{4}A, in which one pair of binding sites can freely rotate with respect to

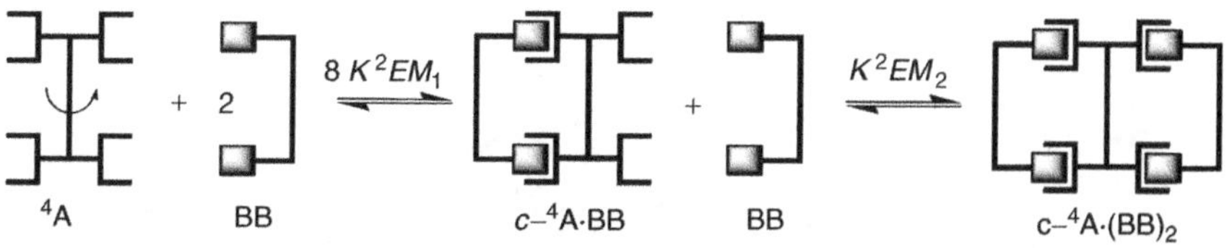

Figure 3.11 Binding of a divalent ligand BB to a tetravalent receptor ^{4}A with a free internal rotation, assuming $[BB]_0 \gg [^4A]_0$ and $\alpha = 1$.

the other pair, interacting with a divalent ligand BB (Figure 3.11). The ligand is present in a large excess relative to the receptor so that complexes involving more than one receptor molecule can be neglected, but not in such large excess so that disassembly of the bicyclic complex c-^{4}A·(BB)$_2$ occurs; $\alpha = 1$ to exclude allosteric cooperativity. The initially free internal rotation of the receptor is restricted by the binding of the first ligand molecule. Thus, part of the free energy of binding is spent to compensate for the corresponding entropy loss. Binding of the second ligand molecule is stronger because the entropic cost for the freezing of internal rotation has already been paid by the first ligand molecule. This type of cooperativity is due not to an increase of the affinity of the binding sites of the receptor but to an increase of the *EM* of the second ring with respect to that of the first ring. The equilibrium constants in Figure 3.11 are expressed in terms of a statistical factor, a microscopic intermolecular constant K, and the microscopic effective molarities EM_1 and EM_2. The statistical factors are easily calculated considering that the unbound tetravalent receptor has $\sigma = 8$ ($\sigma_{ext} = 4$ and $\sigma_{int} = 2$), the ligand and the half-bound receptor have $\sigma = 2$, and the fully bound receptor has $\sigma = 4$. The reference constant K can be evaluated by studying the binding of a monovalent model of the ligand, B, to a monovalent model of the receptor, A, while the reference *EM* can be evaluated by studying the binding of the ligand BB to a divalent receptor model AA, as illustrated in Figure 3.12. In the absence of interannular cooperativity, $EM_2 = EM_1 = EM$.

To quantify interannular cooperativity, we define the cooperativity factor γ that, with reference to Figure 3.11, is given by the ratio of the overall experimental constant to the hypothetical overall noncooperative constant; that is, $\gamma = EM_1\,EM_2/EM^2$. It is worth noting that interannular cooperativity, in contrast with chelate cooperativity, does not depend on ligand concentration.

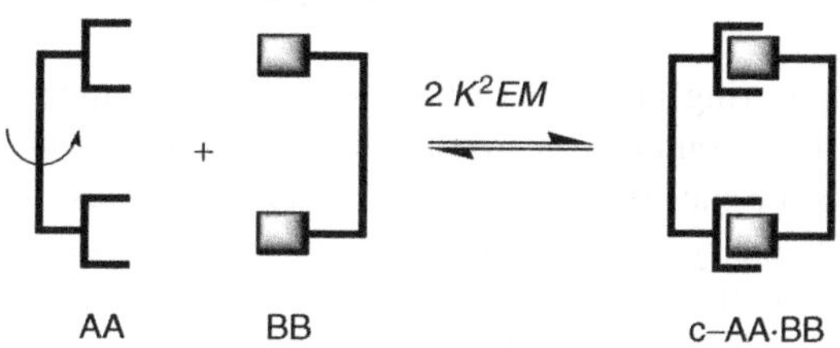

Figure 3.12 Binding of a divalent ligand BB to a divalent receptor model AA to obtain the reference *EM* value.

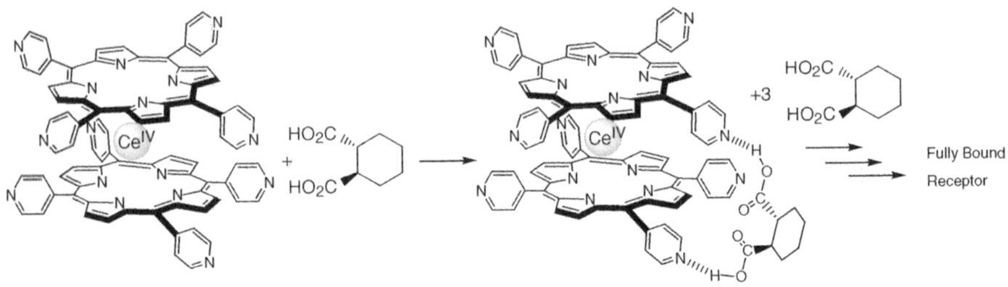

Figure 3.13 Hydrogen bonding by the first ligand molecule suppresses internal rotation of the double-wheel receptor, making the binding of successive ligand molecules easier.

An example of interannular cooperativity is provided by the double-wheel receptor of Shinkai and co-workers shown in Figure 3.13.[2b, 20] The receptor consists of two porphyrin "wheels," each of which bears four pyridinyl binding sites; the two wheels are connected by a cerium "axle," so that they can rotate relative to each other. Simultaneous binding by hydrogen bonding of a first ditopic ligand, such as $(1R, 2R)$-cyclohexane-1,2-dicarboxylic acid, to both the wheels suppresses their internal rotational freedom so that the successive ligands are bound more efficiently.

Of course, freezing of torsional motion is just one of the possible mechanisms of interannular cooperativity; other mechanisms can involve either attractive or repulsive interligand interactions.

3.7 STABILITY OF AN ASSEMBLY

Self-assembly consists of the spontaneous generation of ordered supramolecular architectures from a given set of components under thermodynamic equilibration. The overall equilibrium constant for the formation of an assembly, K_{sa}, depends, in general, on a plurality of intermolecular and intramolecular interactions; thus, it is very useful to establish theoretical models that allow the prediction of K_{sa} on the basis of the knowledge of the single elementary interactions. Taking into account the various types of cooperativity discussed so far, a master equation for the stability of an assembly can be formulated as Eq. 3.15.[7]

$$K_{sa} = \alpha\gamma K_\sigma K^b EM^c \tag{3.15}$$

The parameters appearing in Eq. 3.15 have the following meanings: α and γ are cooperativity factors accounting for overall allosteric and interannular cooperativity, respectively; K_σ is the statistical factor of the assembly process, easily evaluated on the basis of the symmetry numbers of the assembly and of its constituent building blocks; K and EM are the key reference parameters, the first measuring the strength of the single binding interaction and the second, in the case of a cyclic or multicyclic assembly, measuring the ease of formation of the reference cyclic structure; b is the number of binding interactions joining the building blocks together; and c is the degree of cyclicity of the assembly given by $b - i + 1$, where i is the number of building blocks. Depending on the values of the parameters α, γ, and c, several models can result.

1. *The Noncooperative Model.* ($\alpha = \gamma = 1, c = 0$). This model applies to assemblies involving only intermolecular interactions without any allosteric effect. The occupation of the various binding sites of the receptor is only dictated by statistics. It provides the reference model to spot the presence of allosteric effects in real systems. It also applies to the formation of a given oligomer in isodesmic polymerizations. This process is exemplified by a monomer A–B, which undergoes a reversible polymerization in which all of the stepwise association constants are identical and equal to K. The formation constant of each oligomer $(\text{A–B})_i$ is given by Eq. 3.15 in which $\alpha = \gamma = 1$, $c = 0$, $K_\sigma = 1$, and $b = i - 1$.[14, 21]

2. *The Allosteric Cooperative Model.* ($\alpha \neq 1, \gamma = 1, c = 0$). This model is typical of cooperative systems involving only intermolecular interactions such as hemoglobin. It also applies to the reversible formation of linear oligomers under the condition of different stepwise association constants, as in the case of nucleation-growth polymerizations.[21]

3. *The Chelate Cooperative Model.* ($\alpha = \gamma = 1, c > 0$). This model applies to cyclic and multicyclic assemblies in which the constituent cyclic units are identical. Every cyclic or multicyclic assembly benefits from chelate cooperativity, measuring the stability of the assembly with respect to the corresponding fully saturated open receptor. The chelate cooperative model depends on two reference parameters, K and EM, that allow the calculation of the hypothetical self-assembly constant by using Eq. 3.15 and assuming $\alpha = \gamma = 1$. The experimental stability constants of a number of assemblies of different topology, that is, helicates,[22] ladders,[23] D_{3h},[24] and D_{4h}[25] symmetrical cages, are consistent with those calculated by the chelate cooperative model, evidencing the absence of other cooperative effects.[9, 12, 14] This result makes the chelate cooperative model a powerful instrument for predicting that which can be considered the most usual self-assembly behavior.

4. *The Allosteric–Chelate Cooperative Model.* ($\alpha \neq 1, \gamma = 1, c > 0$). In this model both allosteric and chelate cooperativity are taken into account. The factor α can be evaluated by studying the interaction of the receptor with a monovalent ligand. An example of an assembly conforming to the allosteric–chelate cooperative model is offered by the supramolecular complex **4** (Figure 3.15), whose formation is discussed below.

5. *The Interannular–Chelate Cooperative Model.* ($\alpha = 1, \gamma \neq 1, c > 0$). This model takes into account both interannular and chelate cooperativity. Provided that allosteric cooperativity can be excluded a priori or by the study of the interaction of the receptor with a monovalent ligand, the factor γ can be obtained as the ratio of the experimental self-assembly constant to that of the chelate cooperative model. A typical assembly to which this model can be applied is that shown in Figure 3.13 reported by Shinkai and co-workers.[2b, 20] Another clear-cut example has been reported by Wilson and Anderson.[26]

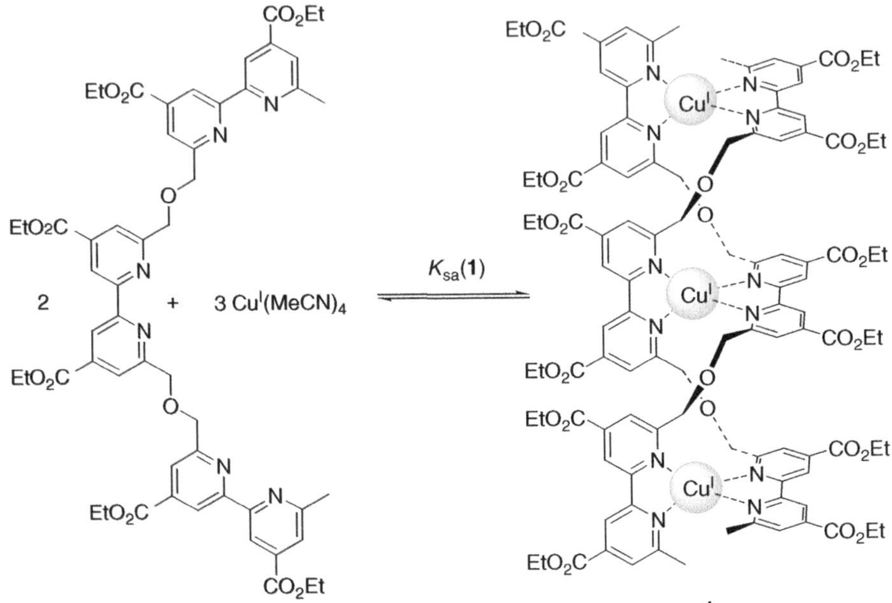

Figure 3.14 Self-assembly equilibrium of the double stranded copper(I) trihelicate **1**.

6. *The Allosteric–Interannular–Chelate Cooperative Model.* ($\alpha \neq 1, \gamma \neq 1, c > 0$). This is the most general model, in which all the possible types of cooperativity are taken into account. To separate the product $\alpha\gamma$ into its components, the factor α must be evaluated separately by studying the interaction of the receptor with a monovalent ligand. At present there are no clear-cut examples in which all three types of cooperativity have been evidenced and quantified, although there are hints suggesting that the double stranded copper(I) trihelicate **1**, due to the ingenuity of Lehn and co-workers (Figure 3.14), could be one of such cases. In this respect, it is instructive to illustrate the application of Eq. 3.15 to the helicate **1** that can be considered the archetypal bioinspired self-assembling system.[22]

The self-assembly equilibrium constant for the helicate **1**, log $K_{sa}(\mathbf{1}) = 18.6 \pm 0.1$, has been measured at 25 °C in $MeCN/CH_2Cl_2$ 50 : 50 (v/v).[22, 27a] To dissect the stability of the assembly **1** into the various contributions, we consider the model systems shown in Figure 3.15. The following stability constants at 25 °C in $MeCN/CH_2Cl_2$ 50:50 (v/v) have been reported (unfortunately without standard deviations): log $K_{sa}(\mathbf{2}) = 4.5$, log $K_{sa}(\mathbf{3}) = 8.6$, and log $K_{sa}(\mathbf{4}) = 14$.[27b] Considering the binding of a bipyridine ligand to a Cu^+ ion as the reference intermolecular process, we can extract the intermolecular constant, log $K = 3.4$, from $K_{sa}(\mathbf{2})$, since according to Eq. 3.15, $K_{sa}(\mathbf{2}) = K_\sigma(\mathbf{2})\ K$. The statistical factor for the formation of **2**, $K_\sigma(\mathbf{2}) = 12$, is easily calculated by Eq. 3.2 considering that the bipyridine ligand and the complex **2** have C_{2v} symmetry ($\sigma = 2$), and that Cu^+, which forms

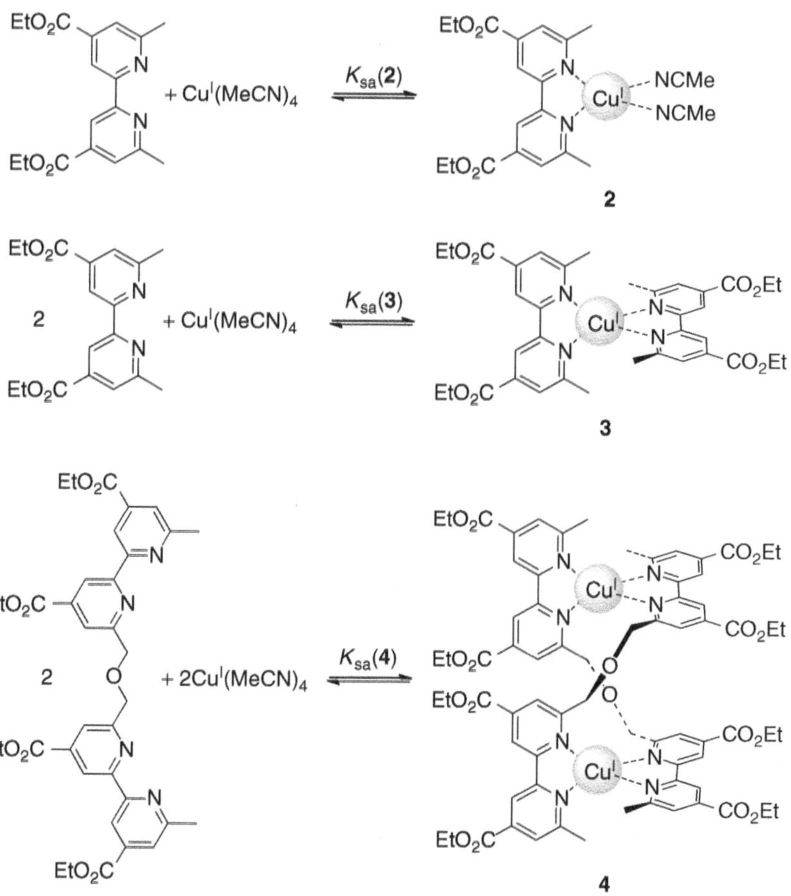

Figure 3.15 Self-assembly equilibria used as model systems to dissect the stability of the double-stranded copper(I) trihelicate **1** into various contributions.

in solution a tetrahedral complex with four MeCN molecules,[28] belongs to the T_d group ($\sigma = 12$).[29]

Knowing log K, we are now able to evaluate the allosteric cooperative factor $\alpha(\mathbf{3})$ for the successive binding of two bipyridine ligand molecules to Cu^+, yielding complex **3**. Indeed, according to Eq. 3.15, the constant $K_{sa}(\mathbf{3})$ can be factored out as $K_{sa}(\mathbf{3}) = \alpha(\mathbf{3})K_\sigma(\mathbf{3})K^2$. The statistical factor for the formation of **3**, $K_\sigma(\mathbf{3})$, is also equal to 12, since the bipyridine ligand has $\sigma = 2$, $[Cu(MeCN)_4]^+$ has $\sigma = 12$, and the complex **3** (point group D_{2d}) has $\sigma = 4$. The calculated value of the allosteric cooperative factor, $\alpha(\mathbf{3}) = 4.8$, indicates that formation of complex **3** is cooperative.

Now let us consider the formation of the assembly **4**. Since the assembly is cyclic, we can anticipate the presence of chelate cooperativity that depends on the corresponding *EM* value and on ligand concentration. The components of the assembly **4** are held together by four bipyridine–Cu(I) interactions that are two-by-two cooperative; thus, the allosteric cooperative factor $\alpha(\mathbf{4})$ is given by the square of $\alpha(\mathbf{3})$, that is, $\alpha(\mathbf{4}) = 23$. The self-assembly constant $K_{sa}(\mathbf{4})$ can be factored out as $K_{sa}(\mathbf{4}) = \alpha(\mathbf{4})\, K_{\sigma}(\mathbf{4})\, K^4\; EM$. The statistical factor for the formation of **4**, $K_{\sigma}(\mathbf{4}) = 288$, is calculated considering that the bis-bipyridine ligand has $\sigma = 2$ (point group C_{2v}), $[Cu(MeCN)_4]^+$ has $\sigma = 12$, and the assembly **4** has $\sigma = 4$ (point group D_2). Since **4** is chiral and present at equilibrium as a racemic mixture, its symmetry number must be divided by 2. The calculated value of EM is 3.2×10^{-4} mol L^{-1}. The significance of this datum can be appreciated by considering that there are 8 rotatable bonds in the structure of **4**, that are constituted by the four σ bonds per strand connecting the bipyridine units. An interpolation from the noncovalent plot in Figure 3.4 gives a predicted strainless EM value of 0.44 mol L^{-1}. The lower experimental EM value is evidence of a ring strain energy of 4.3 kcal mol^{-1} at 25 °C. No doubt a significant part of this strain is due to the electrostatic repulsion between the two positively charged copper ions. Note, therefore, that the EM already accounts for the electrostatic repulsion between nearby charged ions.

Now we are ready to analyze the stability of the double-stranded copper(I) trihelicate **1**. It is a bicyclic assembly formed by two identical rings with the same EM. Its components are held together by six bipyridine–Cu(I) interactions that are two-by-two cooperative; thus, according to Eq. 3.15, the self-assembly constant $K_{sa}(\mathbf{1})$ can be factored out as $K_{sa}(\mathbf{1}) = \alpha(\mathbf{1})\gamma(\mathbf{1})K_{\sigma}(\mathbf{1})K^6\; EM^2$, where $\alpha(\mathbf{1}) = \alpha(\mathbf{3})^3 = 110.6$. The statistical factor for the formation of **1**, $K_{\sigma}(\mathbf{1}) = 3456$, is calculated considering that the tris-bipyridine ligand has $\sigma = 2$ (point group C_{2v}), $[Cu(MeCN)_4]^+$ has $\sigma = 12$, and the helicate **1** has $\sigma = 4$ (point group D_2) and is chiral. The calculated value for the interannular cooperativity factor, $\gamma(\mathbf{1}) = 0.3$, suggests some anticooperativity. Although a small degree of anticooperativity could be justified by the electrostatic repulsion between the first and the third copper ion along the helicate that is not accounted for by the EM, we are inclined to think that the value of $\gamma(\mathbf{1})$ must be considered equal to 1 within the experimental errors. Indeed, even taking into account an optimistic error of ± 0.1 for the calculated values of $\log \alpha(\mathbf{1})$, $\log K$, and $\log EM$, one obtains, by considering error propagation, that $\log \gamma(\mathbf{1}) = -0.5 \pm 1$. In conclusion, the formation of Lehn's helicate displays allosteric cooperativity, chelate cooperativity, and a small degree, if any, of interannular anticooperativity. The original claim of cooperativity made by Pfeil and Lehn[22] on the basis of the curvatures of the Scatchard plot and the Hill plot is clearly not significant, since these tests are valid indicators of cooperative behavior only in the cases of pure allosteric cooperativity.[12]

A somewhat different factorization scheme for the formation of self-assembling polynuclear complexes such as metallohelicates has been proposed by Piguet and co-workers (Eq. 3.16).[2e,g]

$$K_{sa} = \left[\prod_k \exp(-\Delta E_k^{M,M}/RT) \prod_l \exp(-\Delta E_l^{L,L}/RT)\right] K_{\sigma} K^b EM_S^c \qquad (3.16)$$

In their model, dubbed the "extended site binding model," K_σ and K have the same meaning of the parameters in Eq. 3.15, whereas only the entropic component of the effective molarity, EM_S, is considered; $\exp(-\Delta E_k^{M,M}/RT)$ is a Boltzmann factor accounting for intermetallic interactions arising from Coulombic effects between metal ions in solution, and $\exp(-\Delta E_l{}^{L,L}/RT)$ is a Boltzmann factor accounting for the fact that ligands have different binding affinities for successive attachments to the same metal ion. In their model, the presence of ring strain due to the classical strain sources (angle strain, torsional strain, and transannular strain) is not considered. Since the various factors, apart from $K_{\sigma,}$ are generally obtained by a best-fit procedure, the lack of consideration of classical ring strain might lead to estimated factors whose magnitude has an uncertain physical meaning. Another limitation of the extended site binding model is that it cannot account for other cooperativity effects beyond those explicitly considered by the two Boltzmann factors. Despite these limitations, the extended site binding model has been used extensively to rationalize the stability of polynuclear assemblies with particular regard to metallohelicates.[2e,g, 30]

3.8 CONCLUSION

Summing up, we have listed three types of cooperativity that should be considered: allosteric cooperativity (α), chelate cooperativity (β), and interannular cooperativity (γ). While the presence of chelate cooperativity is immediately evident on the basis of the cyclic or multicyclic nature of the assembly, allosteric and interannular cooperativity need reference models or traditional plots to be spotted (binding isotherm, Scatchard plot, and Hill plot). In any case, Eq. 3.15 provides the conceptual framework to quantitatively assess cooperativity in self-assembly. The concepts illustrated here should pave the way for further advancements related to the theoretical prediction of the key cooperative factors α, β, and γ.

REFERENCES

1. (a) Perutz, M. F. Q. *Rev. Biophys*. **1989**, *22,* 139; (b) Fersht, A. *Structure and Mechanism in Protein Science*, Freeman, New York, 1999; (c) Ben-Naim, A. *Cooperativity and Regulation in Biochemical Processes*, Kluwer, New York, 2001; (d) Focus Issue on Cooperativity, *Nat. Chem. Biol*. **2008**, *4,* 433–507.
2. (a) Schneider, H.-J.; Yatsimirsky, A. K. *Principles and Methods in Supramolecular Chemistry*, Wiley, Chichester, UK, 2000, Sections A9 and D1.3.2; (b) Shinkai, S.; Sugasaki, A.; Ikeda, M.; Takeuchi, M. *Acc. Chem. Res*. **2001**, *34,* 494; (c) Mulder, A.; Huskens, J.; Reinhoudt, D. N. *Org. Biomol. Chem*. **2004**, *2,* 3409; (d) Badjic, J. D.; Nelson, A.; Cantrill, S. J.; Turnbull, W. B.; Stoddart, J. F. *Acc. Chem. Res*. **2005**, *38,* 723–732; (e) Hamacek, J.; Borkovec, M.; Piguet, C. *Dalton Trans*. **2006**, 1473–1490; Steed, J. W.; Atwood, J. L. *Supramolecular Chemistry*, 2nd ed., Wiley, Chichester, 2009; (g) Piguet, C. *Chem. Commun*. **2010**, *46,* 6209.
3. Mammen, M.; Choi, S.-K.; Whitesides, G. M. *Angew. Chem. Int. Ed. Engl*. **1998**, *37,* 2754.

4. Calderone, C. T.; Williams, D. H. *J. Am. Chem. Soc*. **2001**, *123,* 6262.
5. Whitty, A. *Nat. Chem. Biol*. **2008**, *4,* 435.
6. Hunter, C. A.; Anderson, H. L. *Angew. Chem. Int. Ed*. **2009**, *48,* 7488.
7. Ercolani, G.; Schiaffino, L. *Angew. Chem. Int. Ed*. **2011**, *50,* 2.
8. For a recent example of heterotropic cooperativity, see: Deutman, A. B. C.; Monnereau, C.; Moalin, M.; Coumans, R. G. E.; Veling, N.; Coenen, M.; Smits, J. M. M.; de Gelder, R.; Elemans, J. A. A. W.; Ercolani, G.; Nolte, R. J. M.; Rowan, A. E. *Proc. Natl. Acad. Sci. USA*. **2009**, *106,* 10471.
9. Ercolani, G.; Piguet, C.; Borkovec, M. Hamacek, J. *J. Phys. Chem. B* **2007**, *111,* 12195.
10. (a) Benson, S. W. *J. Am. Chem. Soc*. **1958**, *80,* 5151; (b) Benson, S. W. *Thermochemical Kinetics*, 2nd ed., Wiley-Interscience, Hoboken, NJ, 1976, pp. 37–39; (c) Bailey, W. F.; Monahan, A. S. *J. Chem. Educ*. **1978**, *55,* 489.
11. (a) Connors, K. A. *Binding Constants: The Measurement of Molecular Complex Stability*, Wiley, Hoboken, NJ, 1987, Chapter 2; (b) Perlmutter-Hayman, B. *Acc. Chem. Res*. **1986**, *19,* 90.
12. Ercolani, G. *J. Am. Chem. Soc*. **2003**, *125,* 16097.
13. (a) Kirby, A. J. *Adv. Phys. Org. Chem*. **1980**, *17,* 183. (b) Mandolini, L. *Adv. Phys. Org. Chem*. **1986**, *22,* 1; (c) Galli, C., Mandolini, L. *Eur. J. Org. Chem*. **2000**, 3117. (d) Cacciapaglia, R.; Di Stefano, S.; Mandolini, L. *Acc. Chem. Res*. **2004**, *37,* 113.
14. Ercolani, G. *Struct. Bonding (Berlin)* **2006**, *121,* 167.
15. (a) Page, M. I.; Jencks, W. P. *Proc. Natl. Acad. Sci. U.S.A*. **1971** *68,* 1678; (b) Page, M. I. *Chem. Soc. Rev*. **1973**, *2,* 295; (c) Jencks, W. P. *Adv. Enzym*. **1975**, *43,* 219.
16. Misuraca, M. C.; Grecu, T.; Freixa, Z.; Garavini, V.; Hunter, C. A.; van Leeuwen, P. W. N. M.; Segarra-Maset, M. D.; Turega, S. M. *J. Org. Chem*. **2011**, *76,* 2723.
17. (a) Kuhn, W. *Kolloid. Z*. **1934**, *68,* 2; (b) Jacobson, H.; Sockmayer, W. H. *J. Chem. Phys*. **1950**, *18,* 1600; (c) Flory, P. J.; Suter, U. V.; Mutter, M. *J. Am. Chem. Soc*. **1976**, *98,* 5733.
18. Jencks, W. P. *Proc. Natl. Acad. Sci. U.S.A*. **1981**, *78,* 4046.
19. (a) Ercolani, G. *J. Phys. Chem. B* **1998**, *102,* 5699; (b) Ercolani, G. *J. Phys. Chem. B* **2003**, *107,* 5052.
20. (a) Takeuchi, M.; Sugasaki, A.; Ikeda, M.; Shinkai, S. *Acc. Chem. Res*. **2001**, *34,* 865; (b) Ercolani, G. *Org. Lett*. **2005**, *7,* 803.
21. (a) De Greef, T. F. A.; Smulders, M. M. J.; Wolffs, M.; Schenning, A. P. H. J.; Sijbesma, R. P.; Meijer, E. W. *Chem. Rev*. **2009**, *109,* 5687; (b) de Greef, T. F. A.; Ercolani, G.; Ligthart, G. B. W. L.; Meijer, E. W.; Sijbesma, R. P. *J. Am. Chem. Soc*. **2008**, *130,* 13755.
22. Pfeil, A.; Lehn, J.-M. *J. Chem. Soc. Chem. Commun*. **1992**, 838.
23. Taylor, P. N.; Anderson, H. L. *J. Am. Chem. Soc*. **1999**, *121,* 11538.
24. Ballester, P.; Oliva, A. I.; Costa, A.; Deyà, P. M.; Frontera, A.; Gomila, R. M.; Hunter, C. A. *J. Am. Chem. Soc*. **2006**, *128,* 5560.
25. Baldini, L.; Ballester, P.; Casnati, A.; Gomila, R. M.; Hunter, C. A.; Sansone, F.; Ungaro, R. *J. Am. Chem. Soc*. **2003**, *125,* 14181.
26. Wilson, G. S.; Anderson, H. L. *Chem. Commun*. **1999**, 1539.

27. (a) Fatin-Rouge, N.; Blanc, S.; Pfeil, A.; Rigault, A.; Albrecht-Gary, A. -M.; Lehn, J.-M. *Helv. Chim. Acta* **2001**; *84,* 1694; (b) Pfeil, A.; Lehn, J.-M., unpublished results cited in Ref. 27a.

28. Cotton, F. A.; Wilkinson, G. *Advanced Inorganic Chemistry*, 5th ed., Wiley, Hoboken, NJ, 1988, p. 758.

29. The internal symmetry number of $Cu(MeCN)_4$ due to the internal rotations of the methyl groups is not considered because such rotations in part are maintained in the complex **2** and in part are converted into external rotations of the freed MeCN molecules.

30. (a) Hamacek, J.; Borkovec, M.; Piguet, C. *Chem. Eur. J.* **2005**, 11, 5217; (b) Hamacek, J.; Borkovec, M.; Piguet, C. *Chem. Eur. J.* **2005**, 11, 5227; (c) Canard, G.; Piguet, C. *Inorg. Chem.* **2007**, *46,* 3511; (d) Dalla-Favera, N.; Hamacek, J.; Borkovec, M.; Jeannerat, D.; Ercolani, G.; Piguet, C. *Inorg. Chem.* **2007**, *46,* 9312; (e) Dalla-Favera, N.; Hamacek, J.; Borkovec, M.; Jeannerat, D.; Gumy, F.; Bünzli, J. -C. G.; Ercolani, G.; Piguet, C. *Chem. Eur. J.* **2008**, 14, 2994; (f) Dalla Favera, N.; Guénée, L.; Bernardinelli, G.; Piguet, C. *Dalton Trans.* **2009**, 7625; (g) Riis-Johannessen, T.; Dalla Favera, N.; Todorova, T.; Huber, S. M.; Gagliardi, L.; Piguet, C. *Chem. Eur. J.* **2009**, *15,* 12702; (h) Dalla Favera, N.; Kiehne, U.; Bunzen, J.; Hytteballe, S.; Lützen, A.; Piguet, C. *Angew. Chem. Int. Ed.* **2010**, *49,* 125; (i) Lemonnier, J.-F.; Guénée, L.; Bernardinelli, G.; Vigier, J.-F.; Bocquet, B.; Piguet, C. *Inorg. Chem.* **2010**, *49,* 1252.

CHAPTER 4

Bioinspired Molecular Machines

CHRISTOPHER R. BENSON, ANDREW I. SHARE, and AMAR H. FLOOD

Department of Chemistry, Indiana University, Bloomington, Indiana, USA

4.1 INTRODUCTION

At their most basic level, mechanical machines generate motion, defined as the coherent change in location of one body with respect to another. Humanity has long been aware of the usefulness of movement and has harnessed animals and made wheels to facilitate its production and use. Engineered motors have since replaced horsepower and automated movements now fill the modern industrialized world. When these machines are at their best, they are labor-saving devices that perform work of various kinds. Frequently, this work is demanding, repetitive, or in environments where humans cannot act, such as at the nanoscale. At this small scale, synthetic molecular machines[1] are envisioned[2] to be the functional subunits for action and the inspiration for their design frequently reflects biological precedents.

Bioinspiration is described as understanding the fundamental aspects of some biological activity and then recasting it in another form. Consider the Wright brothers' research program, where lift, control, and propulsion were all accepted elements of bird flight (Figure 4.1). The first two elements were recast in similar forms as wing shape and wing warp, whereas the latter was completely replaced with an engine-driven propeller. It is illustrative that propulsion was generated using very different means. So it is with molecular machinery; where biology shows us that rotary motors[3–5] and walking machines[6, 7] are possible, access to the same functionality may be realized in different forms and with different building blocks.[1] It is also important to note that inspiration also stems from human-made machines and from the chemist's synthetic tools, which are the ones employed during the process of recasting biology's principles in new molecular formats. As Richard Feynman observed,[8] "there's plenty of room at the bottom," and given the burgeoning exploration of this idea in recent years, opportunities for the discovery and

Bioinspiration and Biomimicry in Chemistry: Reverse-Engineering Nature, First Edition.
Edited by Gerhard F. Swiegers.
© 2012 John Wiley & Sons, Inc. Published 2012 by John Wiley & Sons, Inc.

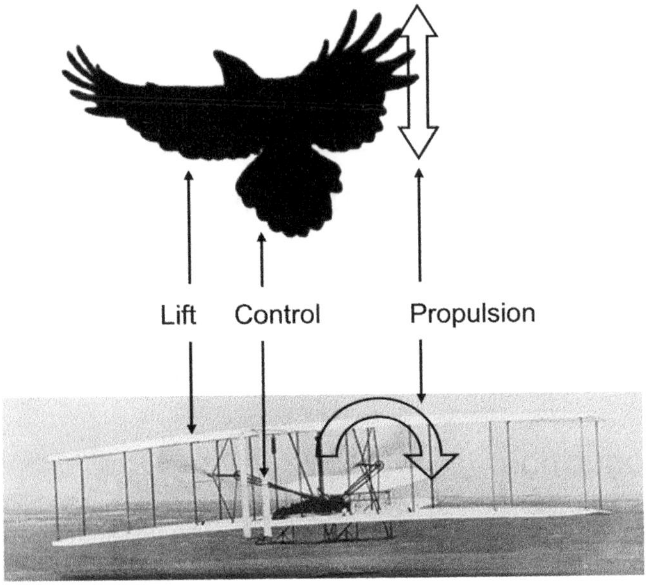

Figure 4.1 Archetypal bioinspiration showing the fundamental principles that allow bird flight and that were integrated together to achieve the first flight by a heavier-than-air craft on December 17, 1903. While lift and control are clear emulations, airplanes do not propel through the skies like birds.

creation of controllable molecular motion for the continued mechanization of our world are very rich.

4.1.1 Inspirational Antecedents: Biology, Engineering, and Chemistry

Knowledge of controllable movements at molecular length scales began with observations of biology. Microscopy allowed the very small micrometer-scaled world to be seen and experimentation confirmed the principles of motion. It is perhaps the first drawing by Leeuwenhoek (1670s) of his "little animals" moving from point C to point D (Figure 4.2a)[9] and then the chemotaxis of cells[10] that clearly show active and directed motion. More recently, the subcellular activities of motor proteins have been convincingly observed, such as muscle myosin,[7] the rotary movement of ATPase,[3] and the long walk of the kinesin.[6] These exemplars, represented also in Figures 4.2a,[9] 4.2b, and 4.2c, provide proof of motion from molecular entities and serve as direct inspiration for controllable movements at the nanoscale and for motility (Figures 4.2d[11] and 4.2e[12]), sliding (Figures 4.2f[13] and 4.2g[14]), and walking (Figures 4.2h[15] and 4.2i[16]) machines, respectively.

While biology is a primary source of inspiration for the creation of soft machines in molecular[1, 2] and other[17] formats, the hard machines in the world around us also provide some lessons. Planes, trains, and automobiles quickly capture the imagination and motivate future applications of machines for chemists as well as

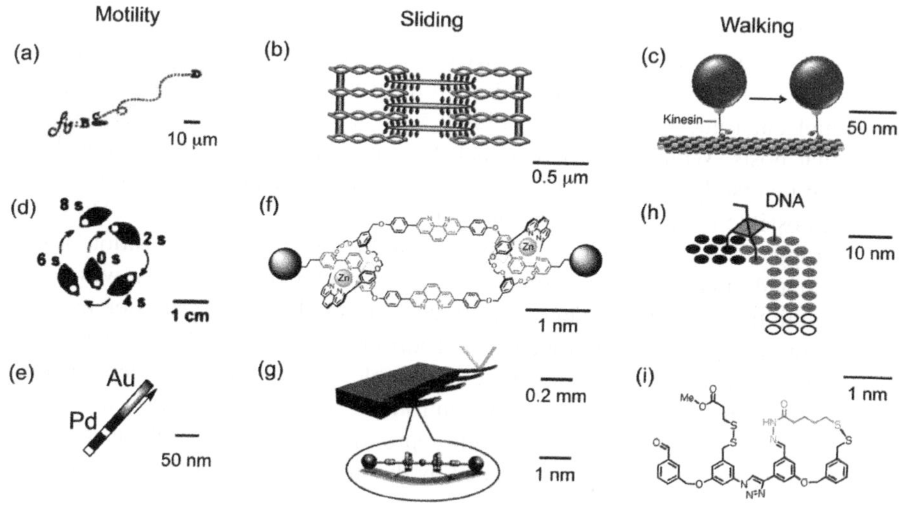

Figure 4.2 Examples of different types of motion in increasingly smaller biological systems that serve as sources of inspiration. (a) Leeuwenhoek's "little animals,"[9] (b) skeletal muscle's sarcomere, and (c) kinesin. They have inspired (d) Whitesides's autonomous swimmers,[11] (e) Mallouk's remote-controlled catalytic nanorods,[12] (f) Sauvage's molecular muscles,[13] (g) Stoddart's nanoelectrochemical system (NEMS),[14] (h) Stojanovic's DNA robot,[15] and (i) Leigh's synthetic walker,[16] respectively. (Images reproduced with permission. Copyright National Academy of Science, U.S.A.: Ref. 11. Copyright the American Chemical Society: Ref. 14.)

nonchemists. These devices from our everyday world also provide descriptions that allow a rapid delivery of ideas to audiences outside chemistry and biochemistry. However, the chemist has a responsibility to point out that machines at the meter scale follow Newtonian mechanics whereas those made from molecules follow the rules of chemistry.[18] Considering applications, engineering provides a basis for answering when and how molecules of all types have been successfully integrated into modern technologies, such as liquid crystalline displays (LCDs) and organic light-emitting diodes (OLEDs). The challenges faced are broadly called the interface problem[2]—how to connect a molecular world to the macroscopic one where the human operators, humming power supplies, and electrical control systems exist on a larger scale. Nature may have solved this problem, for instance, with the action and organization of its many myosins in skeletal muscle movements.[7] Not all applications require a bridge from the nano-to-macro, however, in much the same way that walking kinesin[4] operates at the single molecule level.

The expansive view of the molecular machine offered above raises important definitional questions. For instance, do the inter- and intramolecular rearrangements that occur during traditional chemical reactions also constitute movements? The short answer is yes, with the caveat that up until recently, the focus has instead been on the production of new value-added chemicals (e.g., pharmaceuticals, dyes and paints, polymers, and OLEDs), where movement was not a sought-after outcome.

With the birth of supramolecular chemistry[19, 20] in the 1960s and 1970s,[21–24] the focus shifted from strong and highly energetic covalent bonds to weak noncovalent bonds. Initial characteristics of movement were intermolecular in character when molecular guests entered into (Figure 4.3a) and moved out of complexes with molecular hosts. With the discovery that these complexes could be disrupted and reformed reversibly (Figure 4.3b) upon the addition and removal of stimulants,[25] it became possible to move molecules around from one position to another. However, the isotropy of simple diffusion quickly leads to the loss of relative position of host and guest and impairs concerted directional movement. Consequently, it was necessary to limit the translational degrees of freedom through spatial restriction by interlocking the hosts and guests together as rotaxanes[26] and catenanes,[27] facilitating predictable movements. Thus, dynamic, reversible, and stimuli-dependent changes (Figure 4.3c) lay the foundations for controllable movements within interlocked molecules, concepts that have now been appropriated in covalent chemistry.[28, 29] Thus, shape changes, which are mediated by the movement of one component with

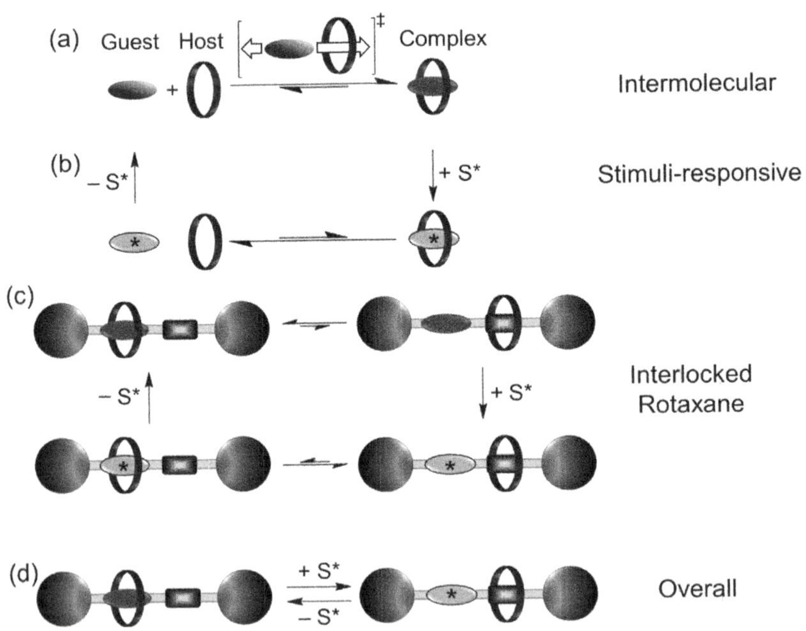

Figure 4.3 From supramolecular to molecular machines. (a) A macrocyclic guest binds to a host molecule to create a host–guest complex. (b) When a stimuli (S*) is applied to the encapsulated guest molecule, the host–guest interaction weakens, and the complex dissociates to its constituent parts. Removing the stimuli allows the host–guest complex to reform. (c) A stepwise square scheme representing the movements of a bistable mechanically interlocked rotaxane where adding or removing a stimulus will cause the macrocycle to move away (+S∗) or toward (−S∗) the oval station on the dumbbell. (d) The same switching diagram as in (c) with the intermediates removed for clarity.

respect to another, are veritable topics of study and deliberate designs have led to the elaboration, amplification, and harnessing of "moving molecules" that constitute all sorts of weird and wonderful varieties akin to the early exploration of flight.[30]

Synthesis, the classic domain of chemistry, permits practitioners to recast the principles of biological machines into different forms. With this tool, the pragmatic issue of what can and cannot be made is met. In the discipline of natural product chemistry,[31] where a natural organic molecule is being reproduced in the laboratory, students are trained in the art of *synthesis*, where the placement of every atom in stereochemical space matters. The creation of molecular machinery,[1, 2] however, is a training ground for the *design and synthesis* of functional molecules. Here, the direct question to be answered once the machine has been made is: "Does it move?" Or, in the parlance of the Wright brothers, "Does it fly?"

Chemical synthesis of molecular machines, therefore, is driven by the expediency of testing performance. As the syntheses and retrosyntheses are laid, another layer of inspiration emerges that is better described as *variation by realization*. New routes toward new targets as well as new chemical reactions in the canon of covalent and noncovalent synthesis provide opportunities to design machines that had not heretofore been considered or viable. As much as "chemistry creates its own object,"[32] this truism of synthesis leads to novel structures that emerge as a direct result of what could be created by the tools at hand.

All the examples considered in this chapter give a nod to all three sources of inspiration: Nature shows us what can be made, engineered machines verify what can be achieved by humanity and the increasing complexity of synthetic molecules (and supermolecules) continually reveal the edge of chemical possibility. Therefore, the creation of synthetic machines matches well with their sources of inspiration, motivation, and realization. Some synthetic systems are designed to test one or more of the concepts that are thought to operate in purely biological machines on the basis that if you can make it, then you understand it. Some examples of these will be presented here.[15, 16] If it can be made, therefore, it offers the chance for application. Thus, some molecular machines have been created for a proof-of-principle demonstration of a functional outcome—usually mechanical in character.[14] Here, one is motivated by the promise of nanoscaled technologies that are reliant upon moving parts coordinated by designed molecular behaviors. Then, there are the machines that are a direct result of synthetic capabilities, whether for the formation of covalent bonds or noncovalent interactions. By creating its own object, this approach often opens up new areas for design and also offers faster and more effective ways to make molecular machines.

4.1.2 Chemical Integration

Finally, all machines are integrated systems,[33] where multiple components come together to generate a whole that is greater than the sum of the parts.[20] Our definition of a mechanical movement, where parts are moving relative to each other, necessitates the involvement of more than one component (Figure 4.4a). Add controllability of the movements and the degree of integration grows (Figure 4.4b).

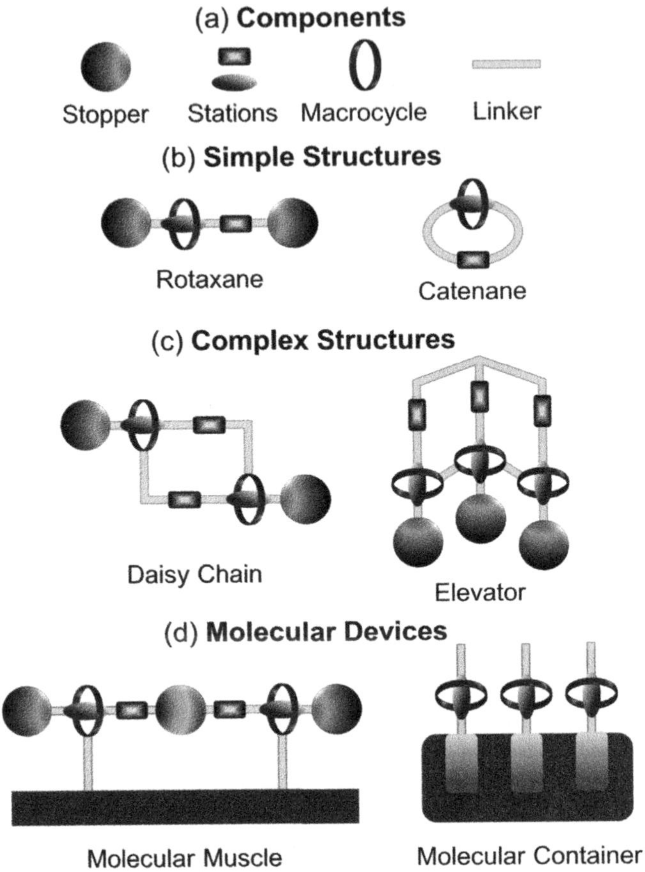

Figure 4.4 Chemically integrated systems. (a) Elementary components of many molecular machines can be synthetically combined into (b) simple molecular machines like rotaxanes or catenanes as well as (c) more complex structures like daisy chains and molecular elevators. (d) These machines can be integrated with support components to create molecular devices.

While chemically integrated systems are present in other areas (e.g., medicinal chemistry), within molecular machines there are often many components by dint of having reversible and repetitive moving parts. The tools available to handle these multicomponent systems are synthesis and measurement and, correspondingly, they predicate different working conditions by comparison to natural and engineered machines. Unlike biology, there is no natural selection to help optimize synthesis, and unlike engineered machines, it is not possible to hold objects in one's hand to aid in the integration of many components. Theory also has a role to play, particularly as the capabilities for modeling larger and multicomponent systems continues to grow.

Achieving chemically integrated molecular machines necessitates developmental phases of synthesis and characterization that have had consequences on the way in which this field has grown. Consider the fact that progress is made by the incremental accumulation of knowledge at the end of each scientific study, for which the conclusions are hard won. This reality has tended to focus research programs on the design and construction of similar molecules in order to mitigate the risk of the machine's performance (or failure!) by the predictive power of prior successes. Counterbalancing this approach is the fact that the practice of exploiting integration may depend on the use of previously untested components, particularly if they are the best ones to use. Therefore, it is becoming important to expand the scope of research efforts on molecular machines to address the scientific questions related to "integration." For instance, what components were integrated together? Why were they chosen? How were they integrated together? Furthermore, integration seeks to impart greater value, which is often cited as allostery, cooperativity, and synergism.[20] Therefore, to what extent has this added value been realized using the integration scheme? Take, for example, the cooperativity displayed in machines with multiple moving parts (Figure 4.4c) and in molecular devices (Figure 4.4d). Do they actually perform acts greater than the sum of their parts? Where appropriate, this theme will be explored in the content to follow.

4.1.3 Chapter Overview

This chapter is organized to provide an overview of mechanical effects in bioinspired molecular machines. We have omitted catalysis while recognizing that its emulation served as early[24] and continuing[34, 35] sources of inspiration for machine-like behaviors. We also leave the exciting field of mechanochemistry for others to review.[36] We begin in Section 4.2 with a brief description of biological machines (myosin, kinesin, ATPase), how they operate, their commonalities, and their places in various biologically integrated systems. In Section 4.3, we lay out theoretical considerations that address why and how it is possible to make a molecular motor. We then consider simple synthetic actuators. The first set of linear sliding machines in Section 4.4, typified by rotaxanes, provide a basis to explore ways of achieving controllable movements. These rotaxanes also serve as the building blocks for bioinspired machines that can do work, of which molecular muscles (Section 4.4.4) are an attractive, but not the only, target. The next series of elemental movements come from rotary motors (Section 4.5), where the holy grail of unidirectional motion first gets mastered in synthetic systems. Those synthetic systems that are capable of working together, like myosins, to move larger objects are given review in Section 4.6 followed by single-molecule kinesin-like walkers in Section 4.7. In Section 4.8, we present ingenious machines that are inspired by macroscopic ones and seek motility at the nanoscale and movements at solid surfaces. Unique among these are potential drug delivery devices (Section 4.9). We close with some perspectives and a conclusion in Sections 4.10 and 4.11 to provide some context for this field in the years ahead.

4.2 MECHANICAL EFFECTS IN BIOLOGICAL MACHINES

4.2.1 Skeletal Muscle's Structure and Function

Perhaps the most obvious mechanical movements serving as bioinspiration are the articulated folding and unfolding of our skeleton and the contraction and relaxation of our muscles that help make it happen. While hidden from view, there is also the nervous system that starts and then regulates this motion and the circulatory system that helps fuel the activity. Locomotion is achieved, therefore, in a highly integrated system that employs many materials and diverse functioning elements that are organized spatiotemporally across multiple scales. Such a lofty bioinspired production may be beyond our current capabilities, yet, likenesses find common currency in robotic analogs.[37] More prevalent in synthetic analogs perhaps is the inspiration provided by the individual components and the elementary levels of integration displayed in the biological systems.

The simplest contraction–extension unit of skeletal muscle is the sarcomere (Figure 4.5). Sarcomeres are composed of interdigitated actin and myosin filaments that interact with each other according to a cross-bridge model[7] to facilitate sliding motions; their lengths contract and extend during muscle action. Each myosin

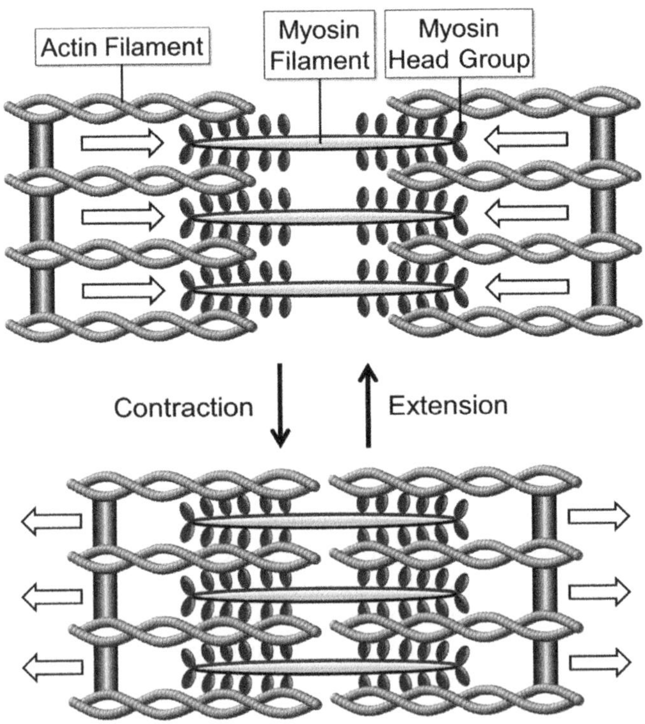

Figure 4.5 The contraction and extension of a sarcomere is induced by the myosin head groups acting on actin filaments.

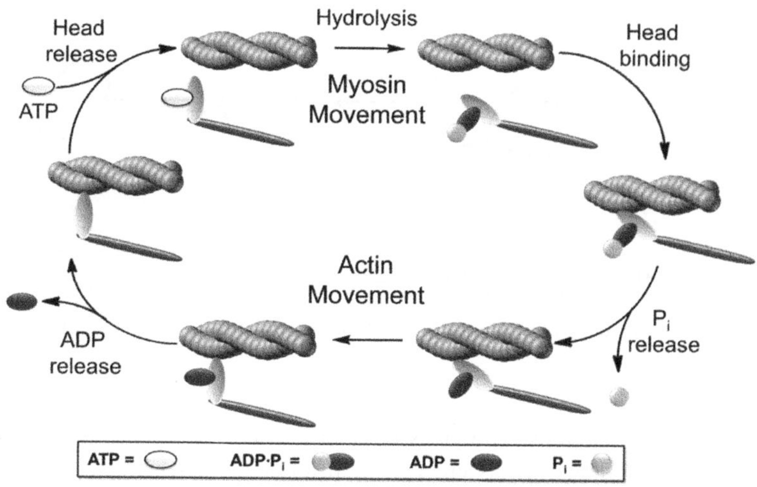

Figure 4.6 The individual states of the myosin unit as it moves an actin filament along.

filament is composed of protein bundles from which emerge head groups at regular intervals. These head groups are responsible for binding to actin filaments along their lengths and pulling the interdigitated components past each other. Thus, the myosin–actin complexes serve as the motor that drives muscle contraction.

The chemically energized contractile cycle begins with the binding of ATP to trigger the release of the myosin head from the actin filament (upper left arrow in Figure 4.6). Hydrolysis to ADP + P_i results in a conformational change that "cocks" the myosin head in preparation for binding to actin. Once an actin binding site is available, myosin forms a weak association with actin (relative to myosin with no bound nucleotide), releases P_i, and subsequently another conformational change takes place in the neck domain of nearly 90° that is transferred into a displacement of the actin filament. Release of the ADP is followed by the binding of a new ATP molecule to reset the cycle. At any stage, it is thought[7] that about one in twelve head groups are engaged with the actin. Thus, many myosins are used in muscle contraction.

These biomotor constructs and the contraction–extension cycle of the sacromeres are being emulated to different degrees by synthetic molecular machines (Section 4.4.4). Even though the specific hierarchical organization of the sarcomere has not yet been emulated, the long-range ordering of many machines may be the common factor represented in future systems of molecular muscles. In this chapter, we contend with small molecules of increasing complexity that have been deliberately designed to display muscle-like movement and highlight the few instances where many of them are employed to act together.

4.2.2 Kinesin

Kinesin,[4] by contrast, operates by itself. It uses the energy from ATP hydrolysis to walk in a hand-over-hand fashion along microtubules, while hauling vesicle cargo

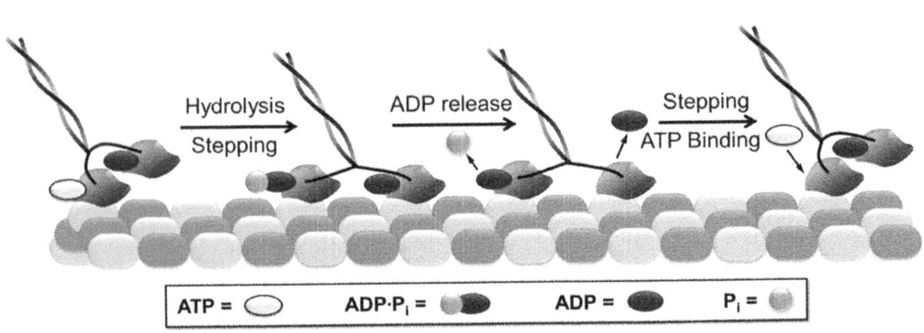

Figure 4.7 Kinesin walks hand-over-hand along a microtubule utilizing ATP hydrolysis to power its movement.

essential for life's functions. The kinesin dimer consists of two heads that can bind to microtubules. The two heads emerge from a coiled dimer stalk, which in turn can bind to a vesicular load (Figure 4.2c). Each cycle of ATP hydrolysis causes kinesin to take a single 8.3 nm step along the microtubule. There is still discussion as to the exact mechanism of kinesin movement, but based on the available data, Carter and Cross[38] have proposed the mechanism depicted in Figure 4.7. ATP/ADP binding and then hydrolysis in the head that is bound to the microtubule allow the unbound head bearing ADP to take a step forward. The step involves a diffusional search by the unbound head for the adjacent tubulin unit to settle on to. The front head then releases its ADP, causing it to bind more strongly to the microtubule. The back head subsequently releases P_i, allowing it to lift off from the microtubule in a sequence that may be gated by the preceding event. ATP can then bind to the head attached to the microtubule, bringing the motor back to its initial state. Applying a load to kinesin causes the dimer to walk at a slower rate. When kinesin is performing a diffusional search, a load increases the likelihood of the unbound head to take a backward step, rather than a forward step. Even so, unidirectionality is the general rule in kinesin's function, and a theme in biological rotary motors as well.

4.2.3 F_1-ATP Synthase

In contrast to myosin or kinesin, the mechanical movements of F_1-ATP synthase[3] (henceforth ATPase) are not coupled to any large-scale work-generating process, nor is its biological operation principally driven by ATP hydrolysis. Instead, its processive unidirectional rotation (Figure 4.8a) is powered by harnessing a chemical potential created by a pH gradient and using it to induce rotational motion accompanied by the synthesis of ATP. Although this latter feature is what gives it import in the study of biology, we are most interested in it as an example of the smallest known unidirectional rotary motor featured in biological systems.

A functional ATPase is composed of a 1 nm central rotor (γ subunit) embedded within a proton-conducting cylinder connected to a larger barrel-like structure (trimer of α/β subunits) approximately 5 nm in diameter. Upon binding free ADP

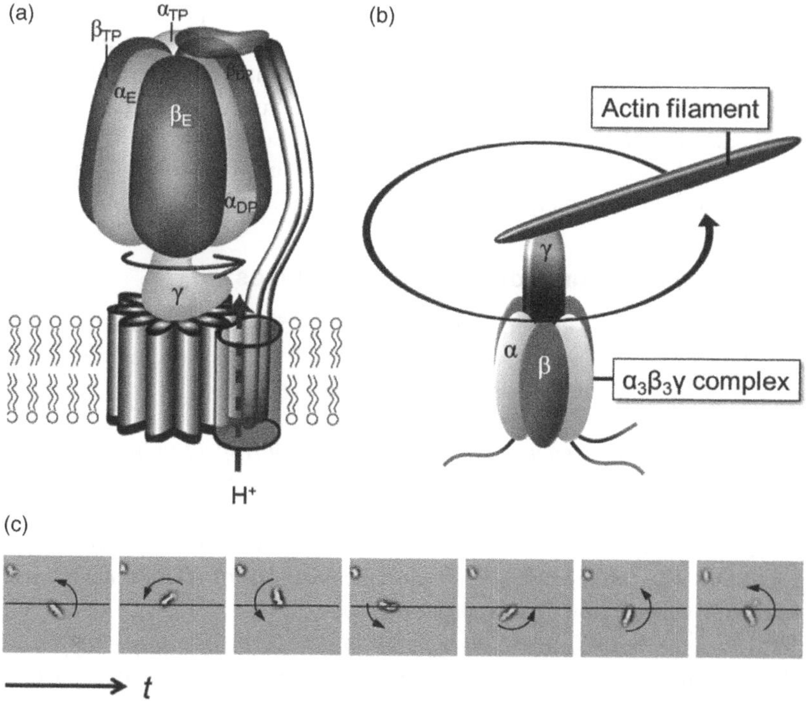

Figure 4.8 (a) Free ADP binds in empty α/β subunits (α/β_E), forming a complex with them (α/β_{DP}). Rotation of the ATPase γ subunit is driven by H^+ transport across a membrane, permitting the release of synthesized ATP from the α/β_{TP} subunit. (Reproduced with permission. Copyright Elsevier: Ref. 39.) (b) Surface-mounted ATPase rotates an actin filament as captured by (c) fluorescence images. (Adapted with permission. Copyright Nature Publishing Group: Ref. 40.)

and P_i, the α/β subunit catalyzes the formation of ATP, but still requires an energy input to liberate the bound nucleotide. This comes in the form of the γ subunit, which rotates within the α/β-subunit barrel, inducing conformational changes that permit the release of ATP. It is generally thought[39] that the exchange of three or four protons is the chemical toll required to produce a single ATP molecule.

One of the most fascinating explorations of the mechanical properties of ATPase was featured in a 1997 paper by Noji et al.[40] Here, researchers tagged ATPase machines with fluorescently labeled actin filaments and affixed the resulting ATPase–actin composites onto a glass substrate (Figure 4.8b). Upon the addition of ATP, unidirectional rotation of the actin filament was triggered and captured in a series of photographs (Figure 4.8c). It is noteworthy that this rotation would have proceeded in reverse relative to the protein's behavior in vivo. Additionally, the authors also determined that the force generated by this rotary motion was in excess of 45 pN, representing surprisingly powerful mechanical properties when compared with myosin II and kinesin, both of which operate with approximately 5 pN.

Taken together, these archetypical molecular machines provide a clear source of inspiration in addition to serving as a chemical yardstick for measuring the functionality of synthetic machines. A more subtle point, however, is their capacity to demonstrate the feasibility of creating molecular machines that are capable of various modes of activity across many length scales. Looking to the Wright brothers as an analogy, biological machines not only make us wish to "fly" but can also remind us that there are many ways to get "airborne."

4.2.4 Common Features of Biological Machines

With myosin–actin, kinesin, and ATPase (Figures 4.5–4.8), there are features common to all of them: movements, solid supports, ATP hydrolyses, binding/debinding events, use of thermal energy, and chemical cycles.

1. *There is clear movement of one component with respect to another.* Some of these movements are inferred from images obtained using electron microscopy and structures solved by X-ray diffraction. The head group of muscle myosin II has been imaged in different conformational states,[7] the walking myosin V has been imaged to look like a telegraph skier,[41] and the rotations of ATPase have been seen to rotate with various time-resolved photographs.[40]

2. *The motors and machines are usually interfaced with a large support material.* Depending on one's perspective, myosin head groups are attached to fibers and they act on fibrous actin to amplify motion along a single coordinate.[7] Kinesins follow microtubule pathways[4] that direct them from one place to another. ATPase is embedded in a membrane[3] to harness chemical potential differences on either side of the membrane. Supporting materials have also enabled the use of these biomotors in artificial systems. For instance, when ATPase is attached to glass supports it is still capable of rotations,[40] and inverse kinesin assays[42] are common ways to move microtubule "trains" in bioengineered constructs.

3. *Fuel in the form of ATP hydrolysis powers the movements.* Biochemical experiments show that ATP hydrolysis is coupled to the movement. It is also typical that the rates of the steps in the cycle of movement, including ATP binding and hydrolysis followed by ADP and P_i release, are all tightly coupled to the movements of the biomotor, whether a conformational change or a diffusive step. In some cases,[43] the rates of the reactions are dependent on the biomotor's mechanical state.[18]

4. *Sequential binding and debinding events are also involved.* For the myosin and kinesin families, protein–protein recognition events are inherent to their activity, while for ATPase it involves the binding, transport, and release of protons. For instance, the myosin–actin binding events are of different strengths and rates depending on the state of the system. In addition, diffusive searching of components (e.g., kinesin head groups) through space takes place to find one binding site after leaving another.

5. *The machines use ambient thermal energy and Brownian motion to move.* The beginning and end points of motion are preset and the machines utilize random

thermal energy to get from one location to the other. Therefore, the trajectory of motion is not preset and could actually have many false starts before success is met. Use of Brownian motion is essential for overcoming barriers to movement.

6. *The biomotor performs a cycle of motion that is intrinsically governed by chemistry.* All movements, ATP hydrolyses, and the binding events are individual chemical reaction steps. As a whole, biomotors are catalysts[18] that allow the machine to perform a complete chemical cycle during which chemical energy is converted into motion. It is obvious that other enzymes also have changes in structure that can be viewed as movement and they have led to bioinspiration as well.[35] This situation is best represented by ATPase in the cell,[3] where the rotations provide the basis for ATP synthesis rather than movement. Finally, the orders of events in the cycles of motion are complex and still ongoing topics of investigation.

4.2.5 Variation in Biomotors

Diversity is a feature of biology. For biomotors, this also includes variation within the same class; for example, myosin II spends more time off its actin track in our muscles to allow fast responses, while this behavior would spell failure for myosin V, which has to walk progressively down an actin filament. In an interesting article,[44] it has been proposed that the rates of motor-track binding/debinding have evolved to allow these two different functions. Such variation is alive and well in the different types of synthetic molecular machines described here.

4.2.6 Descriptions and Analogies of Molecular Machines

The fact that molecular machines invariably follow the rules of chemistry is at odds with descriptions of them based on macroscopic machines. While large machines capture the image and relate well to familiar human experience, as analogies they can lead one away from the chemical reality. For instance, consider that much attention has been focused on how biomachines resist the randomizing character of Brownian motion, which can be thought of as solvent buffeting the machines as they try to complete their tasks.[18] The fact that motion occurs successfully at the nanoscale in this seemingly chaotic environment is actually a result of thermal energy allowing energy barriers to be overcome. If analogies to macroscale machines were valid, where one expects large machines to be knocked off course by similar types of collisions, then molecular machines would never stand a chance!

4.3 THEORETICAL CONSIDERATIONS: FLASHING RATCHETS

Biomotors display properties that sometimes appear to disagree with the accepted laws of chemistry, such as directional movements. That is, the pathway forward is different from the pathway backward. This behavior *appears* to violate the principle of microscopic reversibility, which states that "in a reversible reaction,

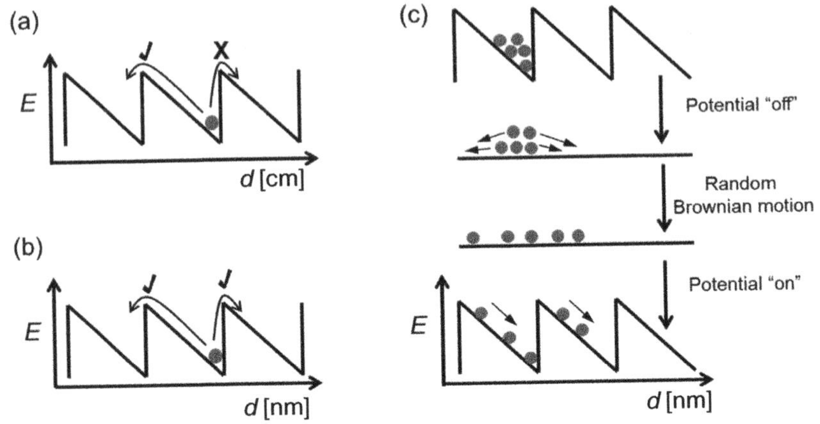

Figure 4.9 (a) The energy profile of movement in a centimeter-scale macroscopic ratchet that achieves directional movement. (b) The energy profile of a nanoscale ratchet that does not. (c) A flashing ratchet along with Brownian motion circumvents the principle of microscopic reversibility to drive movement to the right.

the pathway forward is the same as the pathway backward." This principle is gracefully sidestepped in biomotors because the state of the system is different in the forward and backward steps. This change can be induced in many ways: binding, ATP hydrolysis, redox changes, light absorption, or protonations. Thus, if motion goes in one direction along one pathway, the system switches states to ensure that the returning pathway is blocked while another is opened (i.e., gating).

The feasibility of unidirectional motion led to Feynman's thought experiment of a molecular scale ratchet.[45] It was based on a macroscopic ratchet with a nonsymmetric energy–distance potential surface (Figure 4.9a). At the molecular length scale (Figure 4.9b), the Curtin–Hammett principle dictates that a molecule is equally as likely to jump over the steep slope to the right as the shallow slope to the left because the energy barriers are equal, negating the chance for unidirectional motion. Feynman's molecular ratchet could, however, be achieved with an input of energy.

One approach to this ratchet suggests that, under Brownian conditions, unidirectional motion can be generated using a flashing ratchet design.[18] While this may be simpler than how biomotors operate, it shows mechanistically that unidirectional motion is feasible. Here, energy is added when an external stimulus "flashes" the system between sawtooth and flat potential energy surfaces (Figure 4.9c). When a particle is on the sawtooth energy surface, it will go to the lowest available well. When the potential energy surface is switched to a simple flat line, the particle can diffuse randomly along the one-dimensional coordinate. When the sawtooth potential is reapplied, the particle will be captured again and will always go to the bottom of the energy well. Given that diffusion is a probabilistic phenomenon, particles moving randomly in the thermal chaos on the flat surface will most likely

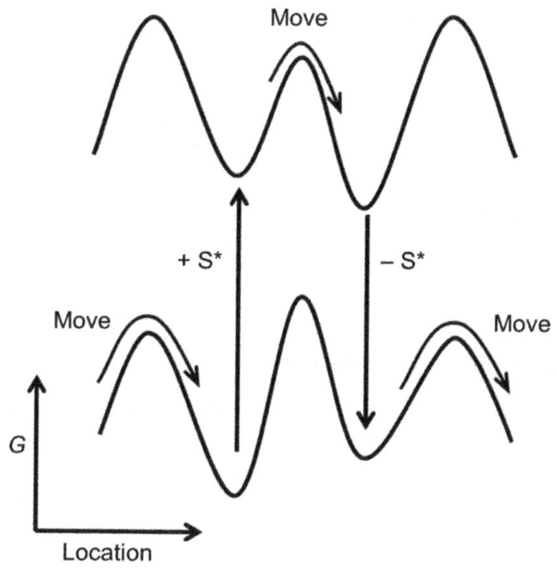

Figure 4.10 A hypothetical flashing ratchet on a curved potential energy surface causes directional motion of particles to the right.

translate evenly to the left and right. To the left, they get recaptured in the original well when the sawtooth is reinstated. But to the right, they could progress to the adjacent location once the sawtooth is flashed on again. Thus, "flashing" the potential energy surface "on" and "off" on appropriate time scales permits limited periods of diffusion, where probability dictates the particle will most likely maintain its position or move unidirectionally to the right.

At the molecular level, flat potential surfaces are extremely rare, and so, unidirectional motion can be achieved by switching between two different double-well potentials.[46] In Figure 4.10, there are two double-well potentials. The stimulus acts on the thermodynamics of the energy wells by turning states "off" and "on," a property known as bistability. At the same time, different pathways are switched "off" and "on" by making the kinetic barriers larger and smaller, a property defined as bilability.[47] Thus, unidirectional motion can be attained following (1) stimulation, (2) move right, (3) remove stimulation, (4) move right, and continue cycling periodically.

It is likely that biomotors operate under similar conditions with the critical difference being that the complexity is greater. There are a large number of different states the system can be "flashed" between that are accessed depending on whether ATP, ADP, and P_i are bound or unbound, and when different protein–protein interactions are present/absent. It is reasonable to believe[18] that the rates of different processes can change depending on which of these states the motor is in. For bioinspired machines, it is prudent to start emulating motion at simpler levels.

4.4 SLIDING MACHINES

Having now explored the world of the biological machine, we turn our attention to synthetic ones by first addressing linear machines capable of sliding motions, typified by the behavior of rotaxanes (Figure 4.3) and reminiscent of the activity of the sarcomere (Figure 4.5). This section seeks, more generally, to explore the idea of reversible length changes, and as such, it shall also include machines that demonstrate greater faithfulness to the activity of biological muscle than to its structure.

4.4.1 Linear Machines: Rotaxanes

Bistable rotaxanes[26] are elementary machines that display linear motion. They are composed (Figure 4.3) of a ring interlocked around a dumbbell. The dumbbell incorporates two stations defining the two locations between which the ring can move. The ring's affinity for one station is designed to respond to stimulation by photons, electrons, protons, or other chemistries, in order to move the ring controllably from one station to the other.[1, 2] When the stimulation is removed, the ring moves back again. Essentially, ATP has been replaced with a different fuel source. Degenerate rotaxanes,[26] where the stations are equivalent, can display shuttling movements back and forth. The repetitive motion is reminiscent of how biological machines reset after each cycle. Additionally, the molecular movements are governed by the same diffusional searching found in biological machines. To the best of our knowledge, there are no biological examples of this exact motif; however, the goal of linear motion at the molecular scale is common to myosin and kinesin and of the translational motion of ribosomes. Simple rotaxanes, however, often lack the intrinsic directional motion that biological machines possess. Also, the first examples of rotaxanes were not integrated into any larger devices; therefore, the work they performed was lost as thermal energy. These limitations have changed in more recent studies (vide infra). Nevertheless, rotaxanes serve as the formative and ongoing proof-of-principle demonstrations of controllable molecular motions at the nanoscale.

Stoddart and co-workers have pioneered a class of bistable rotaxane[33] (Figure 4.11) that is highly modular. They have been employed as the elemental switching gear in numerous molecular machines and larger constructs.[33] In this particular case, the cyclobis(paraquat-*p*-phenylene) ($CBPQT^{4+}$) macrocycle is initially centered on the tetrathiafulvalene (TTF) unit. This coconformation is stabilized by charge transfer interactions between the tetracationic ring (electron acceptor) and the TTF unit (electron donor) as well as by hydrogen bonds from the glycolic linkers. Oxidation of the TTF unit,[48] minimally to the TTF^{+} monocation as well as the TTF^{2+} dication, repels the positively charged macrocycle, driving it thermodynamically to move away to either the left or right. This moving away is almost certainly an activated process, and just as with most chemical events it has some finite time to occur as characterized by an activation barrier. It is then able to translate along the length of the dumbbell by random thermal motion,

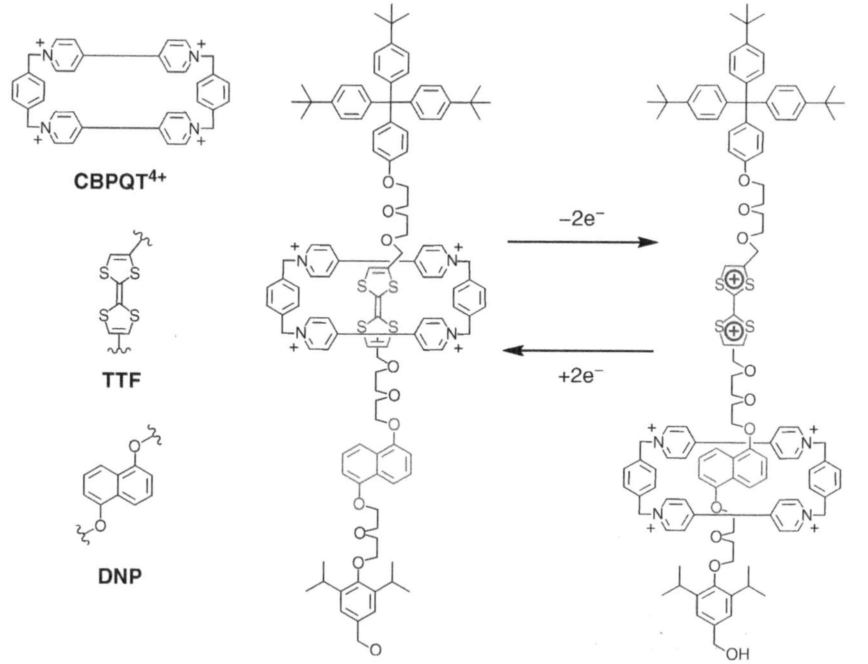

Figure 4.11 The movement in this TTF-based rotaxane is driven by the oxidation and reduction of the TTF unit.

where a right-hand escape successfully allows it to settle on the weaker electron donating dioxynaphthalene (DNP) unit in the new lowest energy coconformation. Reduction of the $TTF^{2+/+}$ cation causes the macrocycle to reequilibrate back to the thermodynamically most stable position around the neutralized TTF unit. Thus, a simple redox cycle can drive the machine's motion forward and backward using either chemical or electrochemical means. As an integrated system, this rotaxane displays controllable one-dimensional motion along its central rod in a manner that could not be replicated by the sum of the components. Moreover, by operating the switch as a molecular entity, it can be diluted into solutions or deployed on surfaces without its components flying apart from each other, which is a capability that then allows it to be incorporated into integrated molecular devices (vide infra).

Sauvage and co-workers developed another elementary bistable switching motif[49] that employs the different stereoelectronic preferences of copper in its two oxidation states: Cu(I) prefers four coordinate environments while Cu(II) prefers five. Thus, bistable copper-based rotaxanes (see Figure 4.12 for a recent example[50]) employ an interlocked macrocyclic ring that delivers two of the nitrogen donor atoms. To provide the other two and three donors, the dumbbell incorporates bidentate and terdentate stations often in the form of phenanthroline

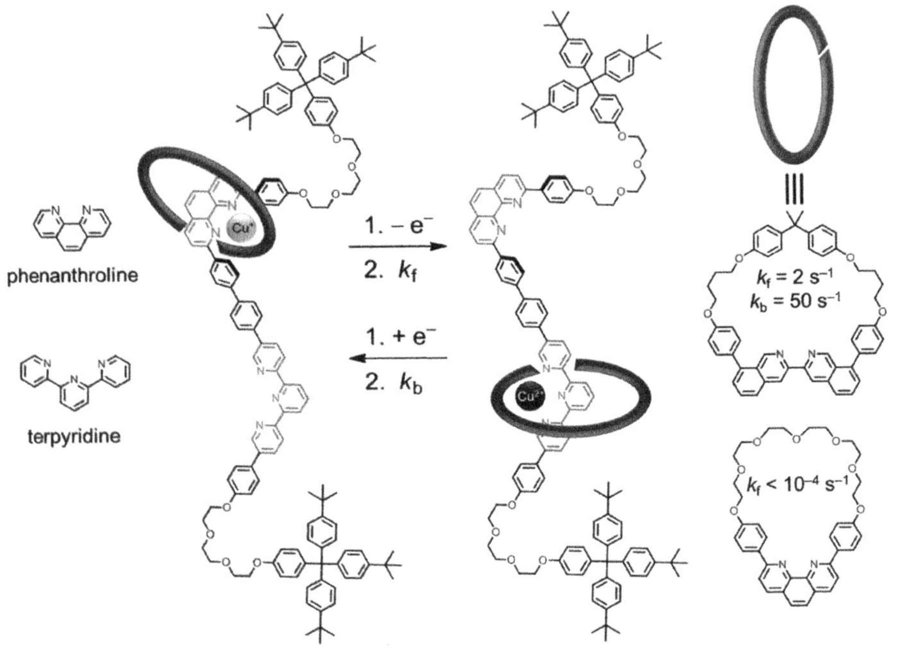

Figure 4.12 Oxidation and reduction of copper causes the movement of the macrocycle, which is based on the stereochemical preferences of Cu(I) and Cu(II). The larger macrocycle translates more rapidly across the dumbbell than the smaller one.

(or bipyridine) and terpyridine, respectively. Thus, the Cu(I)-macrocycle resides at the phenanthroline station until oxidation ($-e^-$) drives the Cu(II)-macrocycle moiety to the terpyridine station. Again, according to the rules of activated motion, reduction ($+e^-$) drives it back again. The example shown uses a biisoquinoline-based macrocycle to hasten the linear movements forward (k_f) and backward (k_b) compared to the original phenanthroline macrocycle.

The speed of each movement[50] refers to overcoming the barrier to motion. While the translational motion has an analogy to an abacus-like movement from the world around us, there is always a rate-limiting step operating at the molecular level. There are two movements investigated in these structures—forward and backward—and their origins are only now being unraveled.[51] Once the natures of these barriers are understood, they offer the chance for modulation either up or down. In the present case, the rigid spacer between the phenanthroline and terpyridine stations along the dumbbell only allows for a dissociative or solvent-assisted pathway to facilitate the movement of the copper-macrocycle. In these studies, the oxidation of Cu(I) to Cu(II) did not cause[50] the rotaxane with the smaller macrocycle to move from the phenanthroline to the terpyridine station. In the rotaxane with the larger macrocycle, however, the redox changes at the copper ion caused rapid switching of the macrocycle between the two stations. This structure–property study confirms the importance of sterics for the mechanism of translational movement.

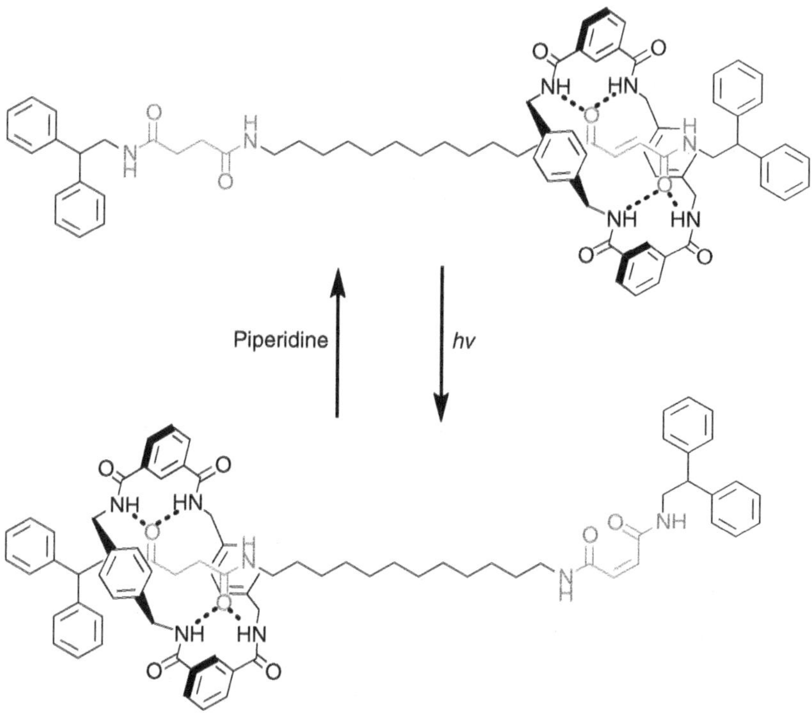

Figure 4.13 Isomerization of the double bond weakens the hydrogen bonds between the macrocycle and the station on the right. The macrocycle translates across the dumbbell to the other station with a better hydrogen bonding environment. Addition of piperidine reisomerizes the double bond, which drives the macrocycle back to the starting state.

Leigh and co-workers designed a class of linear molecular machines that are based on changing the strength of hydrogen bonds that occur between a tetraamide macrocycle and two different stations on a dumbbell (Figure 4.13).[52] The four amide NH groups on the macrocycle engage in hydrogen bonds with the two carbonyl oxygens of either station. In the example featured here, the lowest energy coconformation is where the macrocycle is situated around the station with the two carbonyls that are held in place by the *trans* alkene double bond. Photoisomerization of the double bond from *trans* to *cis* destabilizes this station because the macrocycle can no longer form hydrogen bonds to both carbonyls. The macrocycle then equilibrates to the other binding site. Reisomerizing the double bond with piperidine base restores the original station as the lowest coconformational energy state and the macrocycle moves back across the dumbbell.

4.4.2 Mechanistic Insights: Ex Situ and In Situ (Maxwell's Demon)

An important aspect of harnessing gliding motions and incorporating them into larger systems is coordinating each step in the mechanism, for example, delivering fuel when it is needed at the completion of a molecular movement and not before nor too long afterwards. This coordination was noted as a common aspect occurring in the cycles of motion in biomachines (vide supra, Section 4.2.4). While Nature achieves these aspects by evolution,[44] studies of the mechanism of movements are critical for determining the sequence of events in synthetic machines. Fortunately, once understood, they offer the power of modular synthetic control. A variety of studies on donor–acceptor machines have investigated the enthalpy and entropy of the kinetics of motion for the reverse[53] or the forward[54] movements. Copper-based machines have had some of their rate constants of movement measured for the forward and backward movements.[49] Efforts to understand the origins of the barriers to motion have come through structure–property studies.[50, 51] As integrated and modular systems, it has been relatively straightforward to swap one component for another to facilitate these studies. However, many movements are believed to involve a random Brownian walk once they have escaped the embrace of their parent station. Unlike macroscopic machines, therefore, they are not "set in motion" by getting a "push" along their rod following stimulation. This reality begs the question: How does the ring move from one side of a rotaxane to the other? While the answer offered is usually by a "random walk," such a process has only recently been quantified with one of Leigh's hydrogen-bonded rotaxanes.[55]

Leigh was able to measure the two types of movements (activated and Brownian) in a series of rotaxanes (Figure 4.14), where the length of the alkyl spacer, $-(CH_2)_n-$, between stations was varied, $n = 5$, 9, 12, and 16. In this rotaxane, the secondary naphthalimide binding site becomes the primary one when it gets reduced in a process driven by photoinduced electron transfer.[56] It was observed that the rates of movement from the original bisamide station to the reduced naphthalimide decreased as the length of the chain was extended. The model that accurately fit the data involved an initial escape over an activation barrier ($\Delta G^{\ddagger}$), followed by a biased diffusion along the alkyl spacer. This random walk was found to have a smaller probability for the macrocycle to take a step forward ($p = 0.45$) versus a step backward ($p = 0.55$) along the chain section. The longer spacer slows the rate of forward motion because the probability for the ring to take forward steps becomes smaller when the number of steps gets larger. This study is critical for opening up the design space to include kinetic effects on the biased diffusion along the rod. There is a similarity to biomachines. For example, in the walking myosin V,[41] the rate of this diffusive searching for the next binding site along the microtubule track is believed to have evolved[44] to be faster than the debinding of the back foot from the track; without this rate difference, the walking machine would fall off its track!

An important feature of biological molecular machines is the ability to drive movements in directions that are uphill in energy; that is, we can walk even in the face of gravity. In the previous synthetic examples, the input of energy drove the molecular machines to a thermodynamic minimum. Leigh was the first to design a dynamically controlled rotaxane that could increase the thermodynamic energy of the rotaxane. Consider the bistable rotaxane in Figure 4.15,[55] where the ring is more stable at the left-hand station. By using an information ratchet strategy,[1]

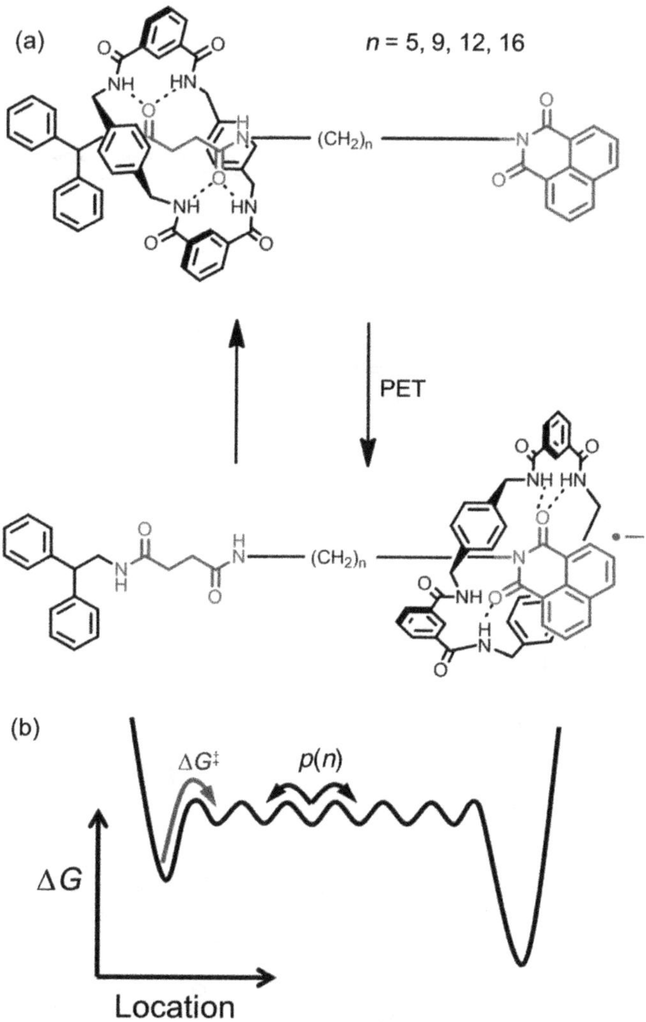

Figure 4.14 (a) The rate of translation of a macrocycle is derived from both the energy it takes to break the hydrogen bonds on the station on the left, $\Delta G^{\ddagger}$, as well as the probability, $p(n)$, of the macrocycle moving to the right across the alkyl linker. (b) Longer chains with more steps decrease the overall probability that the macrocycle will move right rather than returning to the initial station.

the rotaxane drives the macrocycle uphill in energy from the stronger binding site to the weaker binding site on the right-hand side. Here, the methyl–stilbene unit on the rod acts as a photoswitchable gate. When the stilbene is *cis* (state A), the gate is closed, and the macrocycle cannot pass. However, because the macrocycle is substituted with benzophenone, it can absorb light and transfer energy to the stilbene, causing its isomerization from *cis* to *trans*. The benzophenone acts like

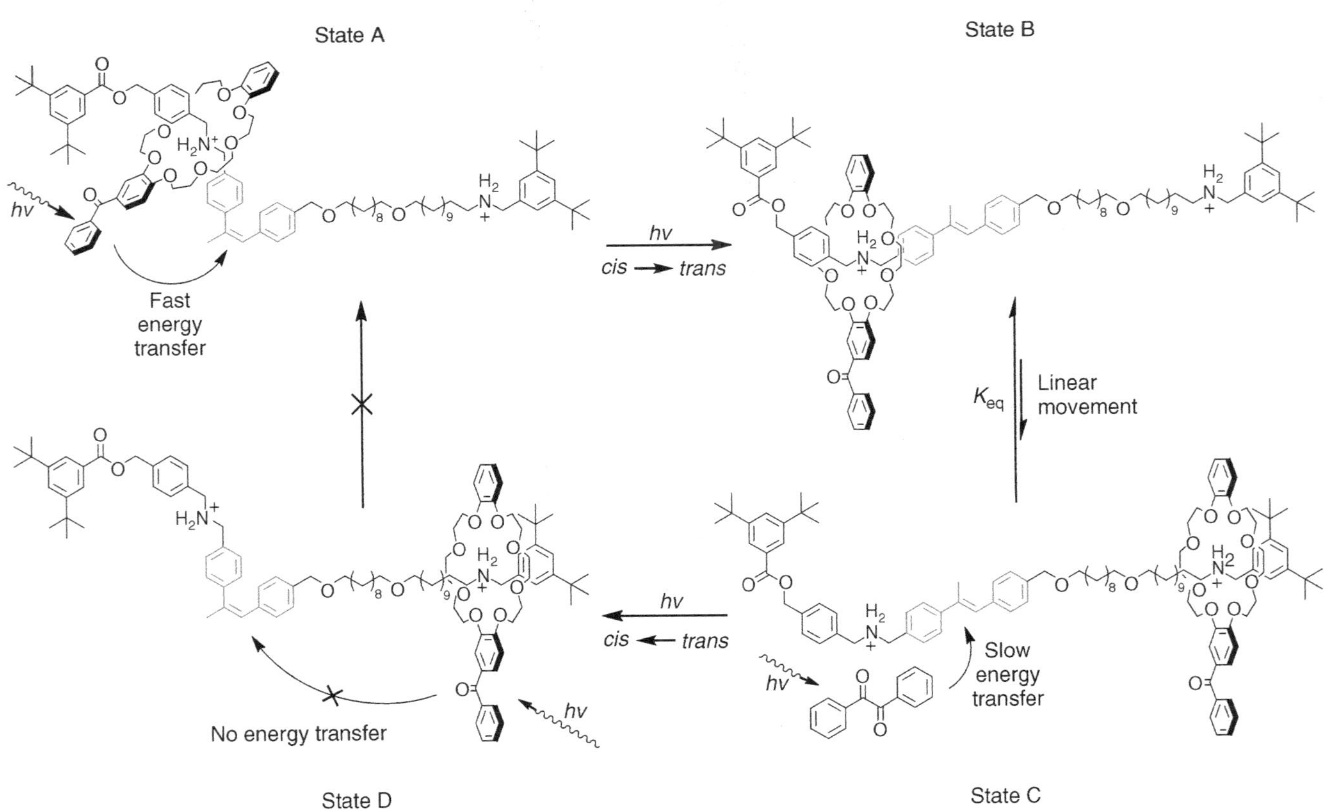

Figure 4.15 Information ratchet. The gate (*cis*-methylstilbene) can best be isomerized to *trans* in state A. Equilibration of the macrocycle is only possible in states B and C. Energy transfer from benzil to *trans*-methylstilbene closes the gate and prevents movement in state D. Even though the left station is preferred, the macrocycle populates the station to the right.

a key to open the gate (state B). This energy transfer is only possible when the macrocycle is nearby (i.e., at the station on the left). Consequently, the gate can be opened to the *trans* state and the macrocycle can pass from state B to state C. If, however, the macrocycle finds itself on the right-hand station, and the gate is closed (state D), the energy transfer is inefficient and the gate remains closed. Thus, information about the ring's location (left or right) is processed logically by the distance dependence of energy transfer. This logic represents the role normally played by Maxwell's demon. In order to close the gate and trap the ring at the right-hand and energetically uphill station, photoresponsive benzil is added into solution. Benzil can close the gate through a slower intermolecular energy charge transfer event that prevents the macrocycle from returning to the more stable station.

4.4.3 Bioinspiration in Rotaxanes

In all of the rotaxanes described herein, one can consider that bioinspiration motivated the first investigations into controllable stimuli-driven movements. Many other systems have been developed to exploit this capability. These studies also led to understanding the kinetics and mechanisms of motion. While there have been some key advances in this area as noted above, it is obvious that the field must continue to gain mastery over the sequence of events in a machine's operation to achieve the higher-level functions of biology. For instance, in the following cycle of motion—stimulate, move forward, stimulate, and move backward (see Figure 4.3)—it is clear that the length of time that the stimulation is held in place is actually determined by the time required for the movement. Not only has Nature got the timing of these events in sync but, as with all dynamic systems,[57] there is also an element of feedback control to achieve a responsive behavior. That very property is being explored under the guise of molecular robotics (vide infra).[15] Before turning to that topic, it is important to consider the extension of simple linear machines to muscle mimics, which also involve sliding movements.

4.4.4 Molecular Muscles as Length Changes

The simplest and broadest definition of a molecular muscle is one that displays a controllable length change. This action is displayed by the shortening and lengthening of skeletal muscles when they contract and relax. Yet not all muscle fibers are configured for large length changes; for example, pinnate muscle fibers have shorter contractions. Similarly, chemists have created multiple types of configurations to affect differing degrees of length changes.

The first molecular muscle deliberately created to mimic the contraction–extension cycles of motion in a sarcomere is based on a rotaxane dimer (Figure 4.16).[13] Movement is facilitated by the four-to-five expansion in the coordination number described previously for copper-based rotaxanes (Figure 4.12).[27, 49–51] Redox stimulation to the Cu(II) oxidation state was not able to cause the movement in this original design. Fortunately, it was possible to remove the Cu(I) ions and then add Zn(II) ions to cause the contraction movement. With Sauvage's leadership in

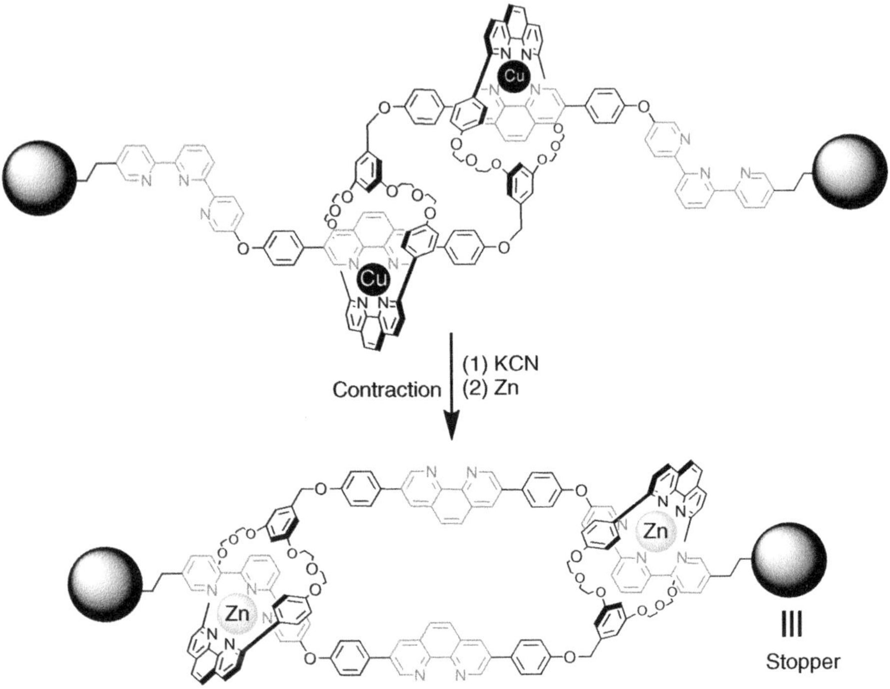

Figure 4.16 First molecular muscle: a metal-based daisy chain that contracts when Cu(I) is exchanged with Zn(II).

the synthesis and function of this complex structure, others have explored similar themes toward contractile changes.

The next molecular muscle proved to be redox-active in its reversible contraction and extension.[14] In this example (Figure 4.17), the simple two-station rotaxane described above (Figure 4.11) is dimerized together in a palindromic arrangement. This structure enables the two $CBPQT^{4+}$ rings to define an extended and contracted state depending on whether they are situated at the two TTF or two naphthalene stations. The molecular muscle was found to cycle through contraction–extension cycles in solution under redox control laying well the foundation to strap them down to cantilevers and to investigate if they could controllably bend the cantilever beams. The $CBPQT^{4+}$ rings were functionalized with disulfide-terminated linkers to enable the chemisorption of them to cantilevers that were top-coated with a thin gold layer (~20 nm). Being 500 μm long, 100 μm wide, and just 1 μm thin, the cantilevers were designed to bend. Being that large they were estimated to accommodate 6×10^9 of the molecular muscles on the gold surface. Although the muscles were randomly oriented, the intrinsic anisotropy of the cantilever with its 5 × 1 aspect ratio allowed bending along its length. Presumably, the molecule's contribution to beam bending depends on its projection onto the long axis of the cantilever. In this first study,[14] the cantilever was bent up and down through

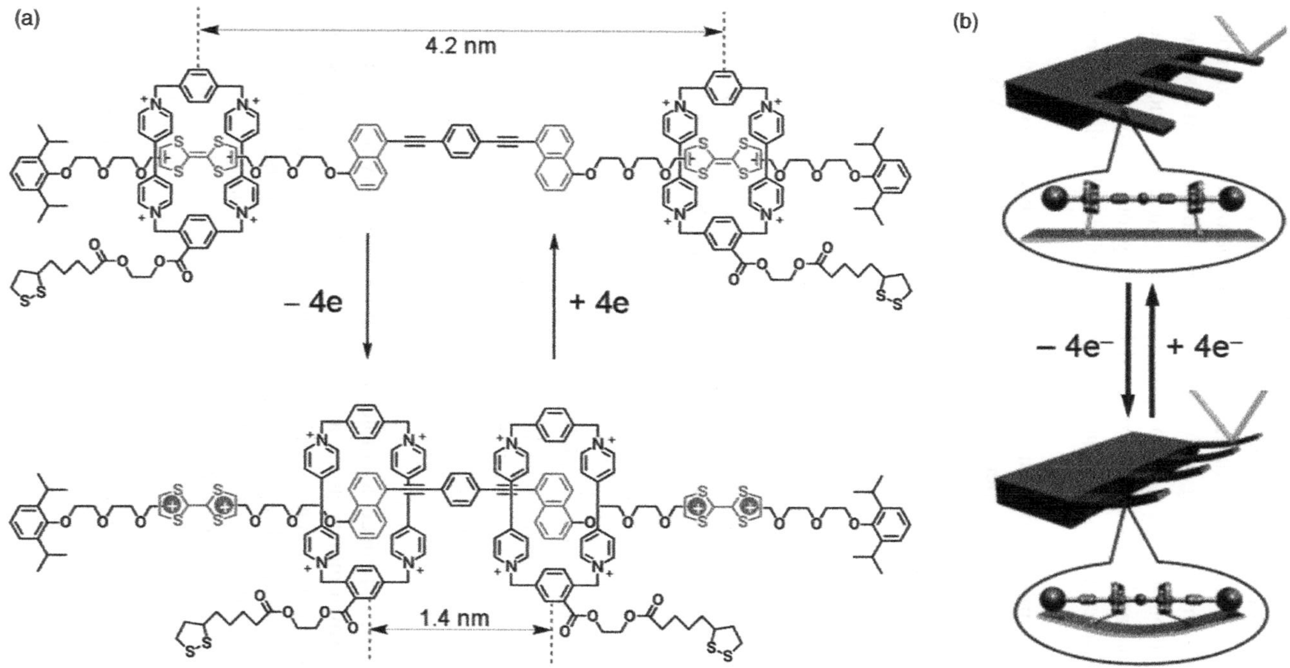

Figure 4.17 (a) Oxidation of the two TTF units causes the $CBPQT^{4+}$ macrocycles to move inward in this molecular muscle. This contraction can be reversed upon reduction. (b) When the rotaxanes are attached to a cantilever via the disulfide linker on the $CBPQT^{4+}$, the TTF oxidation-driven movements cause the cantilever to reversibly bend up and down. (Adapted with permission. Copyright the American Chemical Society: Ref. 14.)

35 cycles driven by the addition of an oxidant (bending up) followed by a reductant (bending down). This work was followed by a study using electrochemical stimulation[58] and both made use of control studies to confirm that the origin of the movement was the contractile motion of the molecular muscles. Use of the electrochemical apparatus to control[58] the cantilever provides one of the first examples of a large integrated system involving molecular machines. This system also resembles the way in which our muscles are composed of many molecular motors acting together and that their activities are coordinated by contiguous subsystems, where the nerves and capillaries for control and fueling are replaced by electrochemical apparatus.

These proof-of-principle demonstrations have motivated others to synthesize and test the contractile actuation of other classes of molecular muscles. A variety of published molecular machines take advantage of one of the foundational chemical insights of supramolecular chemistry, namely, the natural affinity of crown ethers for cations.[59] Different methods for generating cationic rotaxane stations exist, although protonation/deprotonation through pH modulation is currently the most common. A 2009 paper by Clark, Day, and Grubbs[60] (Figure 4.18) features a daisy chain molecule assembled around an ammonium cation. Under basic conditions, the crown ether rings migrate away from the ammonium cation, coming to rest on a biphenyl station. Reprotonation of the ammonia group returns the molecule to its original conformation. Once polymerized, this muscle could be induced to contract using an irreversible chemical reaction. A 2009 paper[61] by Stoddart and co-workers (Figure 4.18b) uses a similar approach, actuating the contraction/extension cycle of the daisy chain through the deprotonation/reprotonation of an ammonium cation. In this instance, however, replacing the biphenyl station with a bipyridinium dication creates a second station with secondary affinity for the macrocycle. This study also included the synthesis of a polymer stitched together using triazole click chemistry. The polymeric muscles[60, 61] suggest a future and exciting use for these materials in "functional nanomachinery."

Two related daisy chains from Coutrot (Figures 4.19a and 4.19b) share a similar starting point although examined as molecules rather than polymers. The initial coordination of the crown moiety about a substituted ammonium cation defines an extended state.[62] Upon deprotonation, the crown ether rings migrate to the adjacent methyl triazolium species, resulting in a contraction. In the second iteration of this molecule published in 2010,[63] a third station composed of distant pyridiniums was incorporated that were only accessible upon deprotonation of both the ammonium and triazolium species. This resulted in a "very contracted" coconformation of the daisy chain.

Another classical motif that has followed in the wake of earlier successes is a daisy chain that utilizes a cyclodextrin as the ring component. An early example was produced by Harada and co-workers (Figure 4.20a), who demonstrated actuation to mimic the contraction and extension of skeletal muscle[64] by controlling solvent polarity. Initially dissolved in dimethylsulfoxide, upon addition of water the muscle underwent a conformational change to conceal the hydrophobic alkyl chains

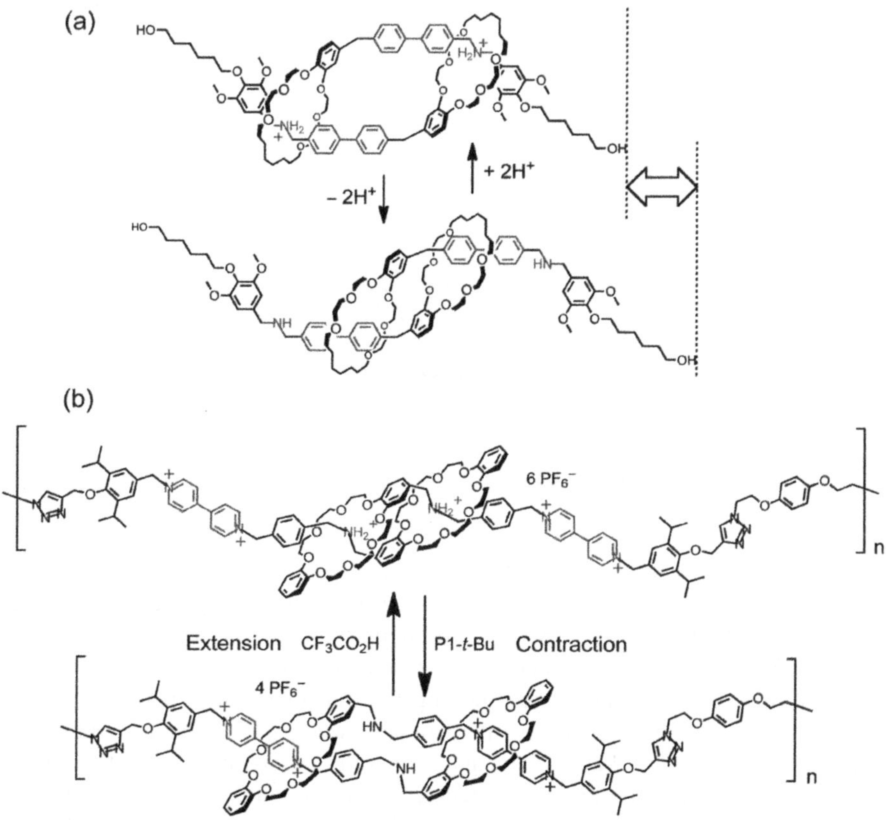

Figure 4.18 Deprotonation and reprotonation of the ammonium units alter the macrocycles' affinities, causing the lengthening and shortening of these daisy chain dimers as (a) single molecules and as (b) polymers.

from entropically disfavored interactions with the surrounding medium. Cyclodextrins also featured as the ring component in a photochemically actuated daisy chain designed by Easton and co-workers (Figure 4.20b).[65] This muscle utilized the well-established photoinduced conformational switching of stilbene to effect contractile motion. Interestingly, Easton was able to successfully generate an intermediate contracted state in the molecule, where one stilbene molecule had undergone transformation to the *cis* form while the other retained its *trans* conformation.

Other recent research has begun to explore unconventional means of producing length changes in linear molecules. A 2009 paper by Chuang et al.[66] made a variety of modifications to earlier precedents. First, they eschewed a conventional macrocycle in favor of a molecular cage that is structurally reminiscent of conjoined crown ether rings (Figure 4.21a). Second, unlike many rotaxanes, this molecule was capable of producing length changes without forming a daisy chain or other homodimeric structures. Instead, the molecular cage utilized in this design provided a suitably large cavity into which the dumbbell portion of the

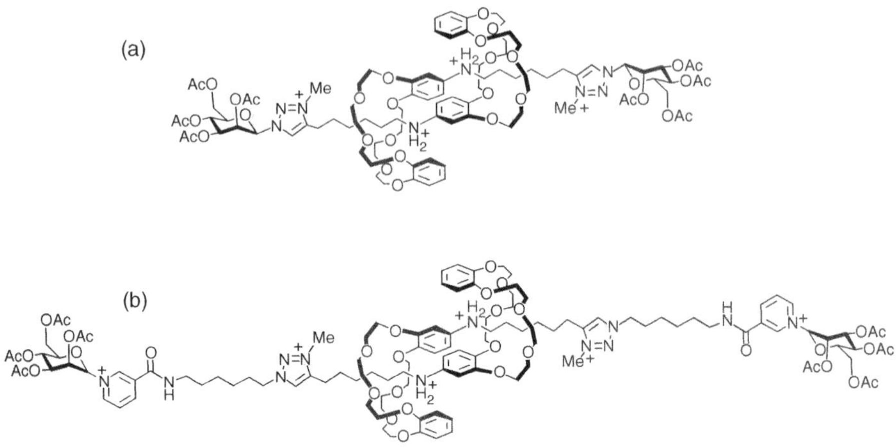

Figure 4.19 Two daisy chain dimers.

rotaxane could retract. While the affinity of the cage for the dumbbell was still governed by the familiar cation–crown interactions, the molecule demonstrated no pH-dependent switching response, even upon the addition of 10 equivalents of the base triethylamine. Consequently, the researchers sought to induce switching by modifying the environment surrounding the machine rather than the machine itself. Specifically, given that the interaction between the secondary ammonium cations and the crown ethers is dependant on solution counteranions, they added tetrabutylammonium fluoride to exchange hexafluorophosphate anions for F^- anions. Upon the addition of 4 equivalents of F^- to effect the exchange of all PF_6^- counterions, the hydrogen bonding between the crown ether and the ammonium cations is disrupted, favoring the formation of a stable interaction between the crown ethers and the distant pyridinum. Although a relatively slow switching process, it is worth note that it is an ion-specific machine, as the addition of Cl^--, Br^--, and I^--based tetrabutylammonium salts showed no actuation.

Another recent surprising molecular machine is a [1]rotaxane[67] (Figure 4.21b) reminiscent of the monomeric components of Sauvage's Cu(I)/Zn(II) exchanging muscle (Figure 4.16). Prior to this publication, investigations on [1]rotaxanes or "self-complexes"[68] were rare and never before featured such a molecule as a possible basis for reversible switching. While switching within pseudo[1]rotaxanes[69] was understood as accessible—namely, though the "end-in" method where the tail passes through the macrocycle like a thread going through the eye of a needle—it was decidedly ineffective for actuating motion as the tail had to be free to move between the free and entangled states. Anchoring both the macrocycle and the tail to a substrate could permit the molecule to exert physical influence on its surroundings, and this paper demonstrated an unexpected mechanism to do this. First synthesizing the "entangled" [1]rotaxane using copper templation followed by olefin metathesis, it was found that the molecule could undergo "capture and release" of the self-entangled morphology through the introduction and removal

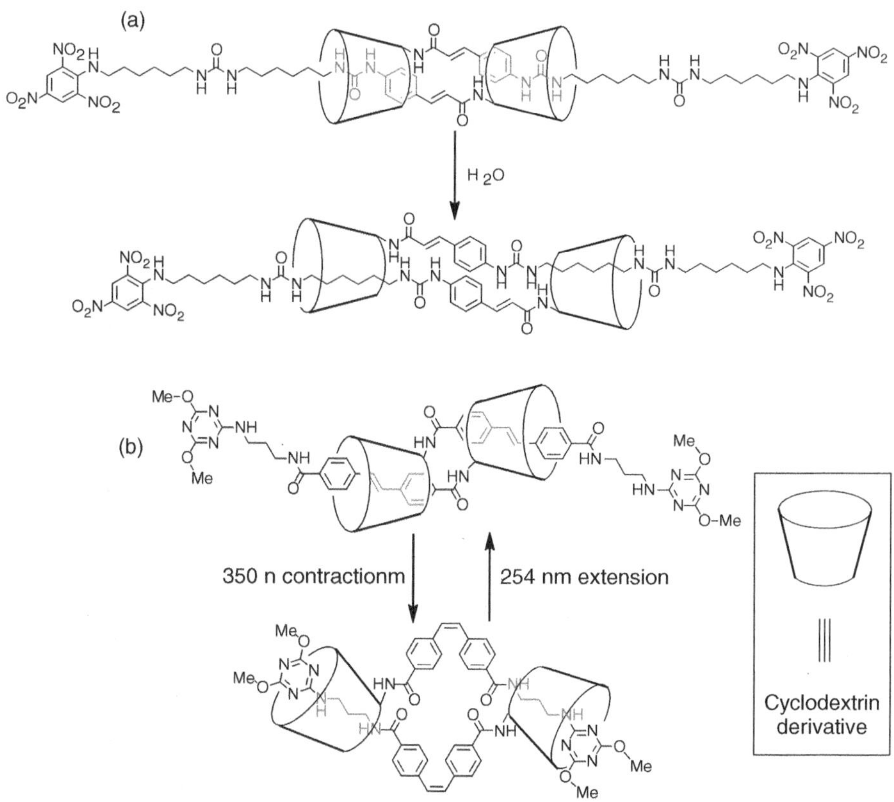

Figure 4.20 (a) Addition of water causes the contraction in this cyclodextrin-based dimer. (b) Isomerization of the stilbene causes the daisy chain to expand and contract.

of Cu(I). This process was made possible through the rotatability of the trisubstituted aryl ring in the macrocycle. Upon addition of Cu(I), the two phenanthrolines available in the molecule would proceed to coordinate directly to the copper ion, but rather than threading the macrocycle end-in (a process restricted by the stopper placed on the tail of the rotaxane), the tail would enter the annulus of the macrocycle as a loop, causing the aryl ring to flip through the center of the macrocycle, resulting in a molecule that resembles what we would expect had the molecule been formed from an "end-in" process.

There are many other departures from the rotaxane-based systems that help drive current molecular machine research. The dynamic contraction–extension process in a molecular spring shows the general trend toward the mastery of other aspects of bioinspiration. In one example, utilizing two different types of rotating modules—a cerium(IV) bis(porphyrinate) and a ferrocene—a molecular spring (Figure 4.22) was designed[70] that undergoes actuation through the cooperative binding of a small molecule, 1*R*, 2*R*-cyclohexane dicarboxylic acid (*RR*-CHDA). Upon adding *RR*-CHDA, the pyridine substituents of the bis(porphyrinate) rotators align to bind the

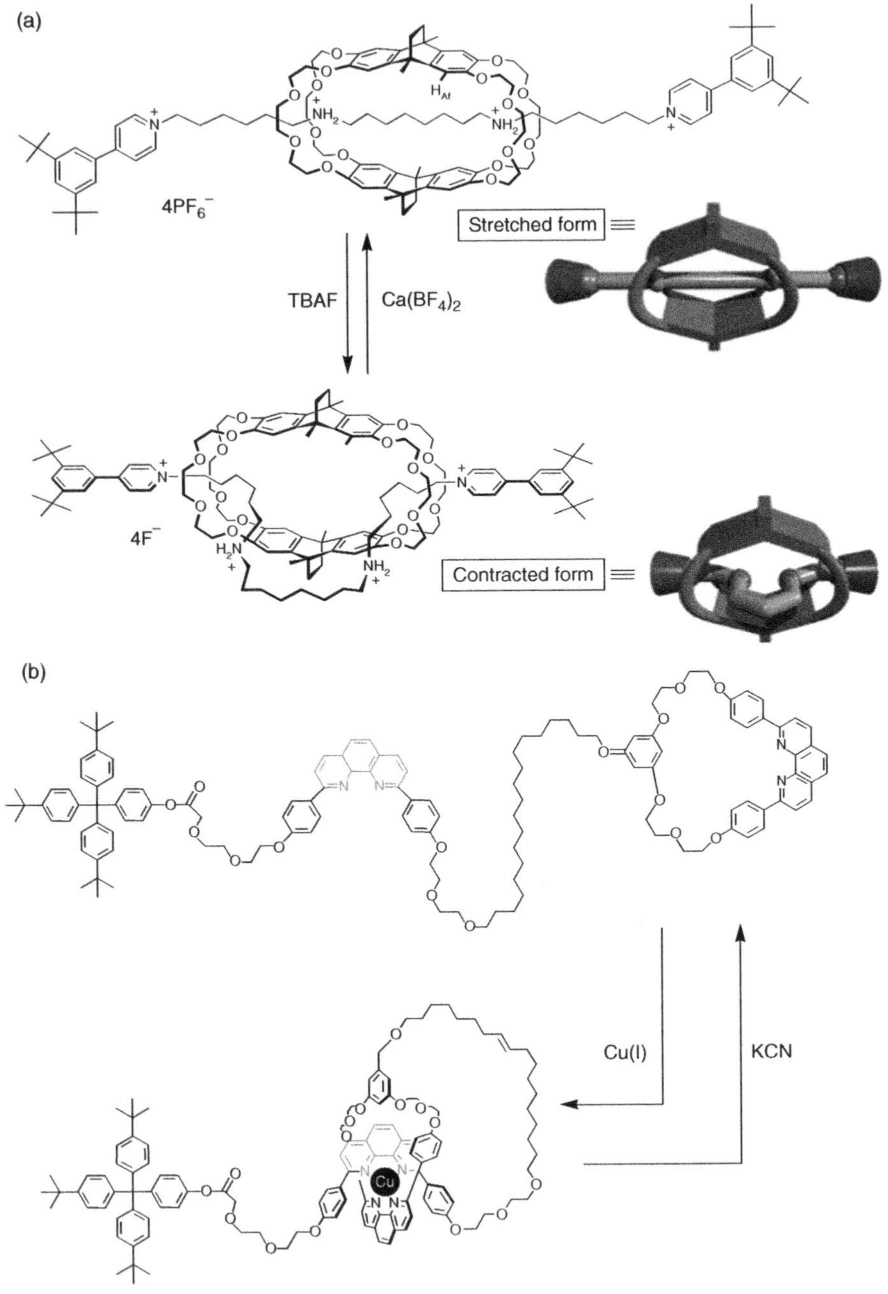

Figure 4.21 (a) An anion-sensitive molecular cage-based rotaxane demonstrating length changes. (Cartoons reproduced with permission. Copyright The American Chemical Society: Ref. 66.) (b) A self-entangling [1]rotaxane, which employs a "Sauvage-like" macrocycle.

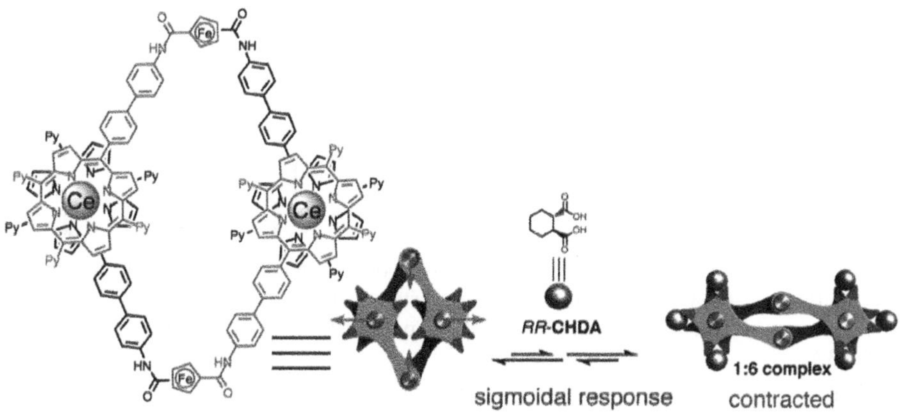

Figure 4.22 A molecular spring where the ferrocene and porphyrin groups act as hinges. Binding *RR*-CHDA to the pendant pyridine groups causes the molecule to isomerize. (Reproduced with permission. Copyright Elsevier: Ref. 70.)

six *RR*-CHDA guests resulting in the elongation of the spring. Interestingly, this molecule demonstrates homotropic allostery, in which the conformational change in the molecule does not follow a linear response upon the addition of the *RR*-CHDA guest. Instead, the molecule demonstrates a sigmoidal response, with maximum extension of the molecule occurring when the ratio of *RR*-CHDA to the molecular spring is 6:1. This work demonstrates not merely contractile actuation but the kind of cooperative tuning characteristic of biological machines, offering a glimpse into the creation of increasingly sophisticated biological facsimiles. New developments in stimuli-responsive foldamers are also contributing to the plethora of spring-like length changes.[71, 72]

Going beyond the use of single-molecule machines, functional crystals have also received attention as chemical entities capable of producing length changes. A recent paper[73] demonstrated substantial mechanical effects in response to photoinduced changes in the morphology of the component molecules within a diarylethylene crystal (Figure 4.23). Here, UV irradiation produces a persistent change in the shape of the crystal, brought on by the ring closing photoisomerization. Subsequent irradiation with visible light causes the crystal to return to its original shape. Amazingly, despite the fact that the photoisomerization only takes place at the surface of the crystal, the crystals are capable of lifting loads 600 times heavier than their weight. The molecules are remarkably resilient as well; while there are many crystals that undergo mechanical changes, they eventually fail due to extensive damage to the crystal structure. This system showed no damage to the crystal after 250 cycles.

4.4.4.1 Bioinspiration in Molecular Muscles While many molecular muscles clearly mimic the sarcomere, it is striking that some of them lack an obvious biological correlate. It is illustrative to consider, therefore, the relationship that these examples do have with biological machines. In all circumstances, we that see

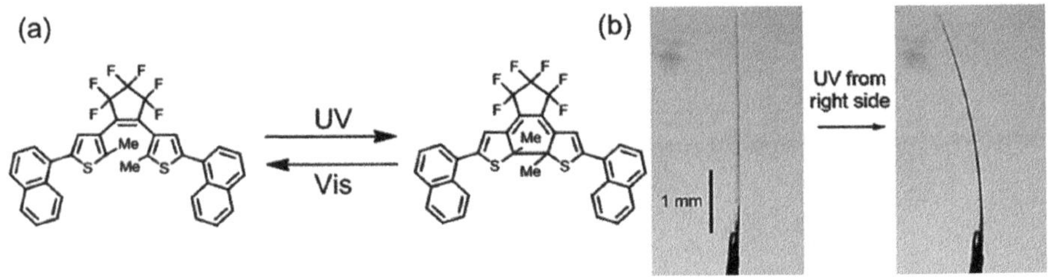

Figure 4.23 (a) The molecular unit of a crystal that changes conformation upon photoisomerization. (a) UV irradiation of these crystals can generate significant mechanical displacement. (Reproduced with permission. Copyright the American Chemical Society: Ref. 73.)

the stimuli-responsive modifications of some weak intermolecular forces produce observable changes in the net structure of the molecular machine.

It is also worthwhile to observe the extent to which integration is featured across the examples discussed here. While the architecture of the machinery or the mechanism driving the switching processes does vary, it is rare to see a machine that drastically departs from previous successes. Even in the fluorine counterion-sensitive machine[66] depicted in Figure 4.21a, the original aim of that machine's design was to utilize crown–cation affinities and the familiar ammonium protonation/deprotonation chemistry as reversible stimuli. These are features built in to a variety of synthetic machines. Similarly, the use and reuse of successful molecular components demonstrates a trend toward streamlining the process of integration. While the phenanthroline-based macrocycle employed by Sauvage and co-workers[49] in Figure 4.12 is a recurring element in a variety of molecular machines, it is revealing to consider that the creators of the [1]rotaxane[67] (Figure 4.21b) explicitly describe the macrocycle as "Sauvage-like." This usage has much in common with the way that experimental details collected together in the backmatter describe particulars of the make and model of an instrument. In short, the field is both structurally and functionally bioinspired. It has also begun to identify molecular components to serve as viable and interchangeable building blocks for formulating strategies during the design and creation of molecular muscles.

4.5 ROTARY MOTORS

In contrast to rotaxanes that act as linear molecular machines, several rotary machines[28] have been developed that exhibit unidirectional motion. Much like biological molecular motors, these artificial ones incorporate a form of a flashing ratchet[18] into the design in order to bias both the position and direction of motion. The challenges in synthesizing these systems is equally matched by their expected utility and has led them to being considered as a holy grail for molecular machines.

4.5.1 Interlocked Rotary Machines: Catenanes

Catenanes have often been staple sources of rotary motions.[27, 48] Leigh and co-workers synthesized a two-station unidirectional motor (Figure 4.24)[74] that operates on the same principle as the one described in Figure 4.13; that is, isomerization of the double bond in the larger ring changes the binding preference of the smaller macrocycle. In this system, the pathway of motion is controlled through the

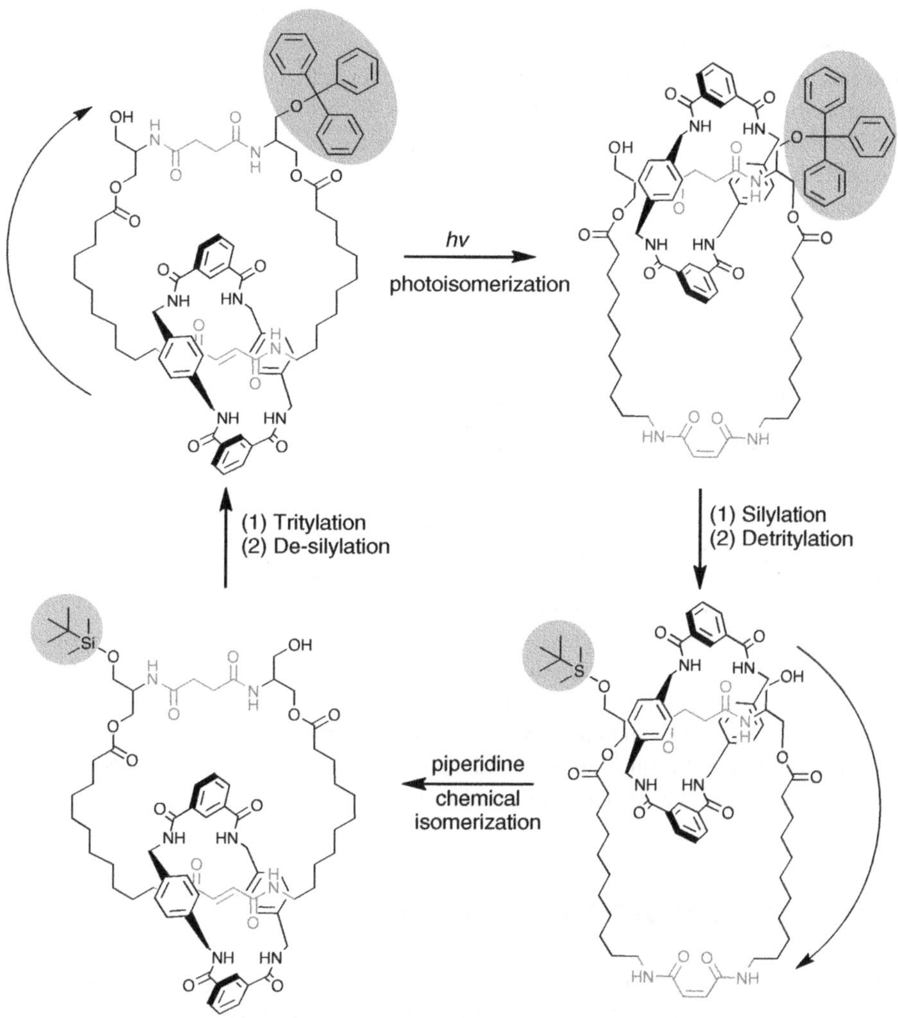

Figure 4.24 In this unidirectional motor, a smaller macrocycle traverses around the larger macrocycle of the catenane in a unidirectional manner. The power to drive this motion comes from flashing the relative energies of interaction between the smaller macrocycle and the two stations, and the directionality comes from selectively deprotecting one side of the larger macrocycle alternately with the other.

incorporation of alternative bulky protecting groups to emulate a flashing ratchet (Figure 4.10). Without the protecting groups, the smaller macrocycle can travel either clockwise or counterclockwise when triggered by light. Incorporation of one of the bulky stoppers prevents the macrocycle from passing over it, effectively blocking that particularly pathway. Unidirectional motion around the ring is achieved by first isomerizing the double bond, causing the smaller macrocycle to move clockwise around the larger ring because the counterclockwise pathway is blocked by the large trityl group. Switching the protecting groups silyl for trityl changes the path of movement, forcing the smaller macrocycle to continue moving clockwise after the double bond is reisomerized from *cis* to *trans*. Essentially, the thermodynamic and kinetic barriers can be switched in a stepwise fashion in a series of photochemical and chemical steps to achieve unidirectional motion. Verification of each individual state in the process of unidirectional motion was possible because kinetic barriers and thermodynamic wells could be changed independently, allowing each of the chemical species to be isolated. This feat stands as an amazing first proof-of-principle demonstration that overwhelms any criticisms that may be leveled at this system arising from the loss of efficiency introduced during the chemical steps in the unidirectional cycle of motion.

4.5.2 Unimolecular Rotating Machines

Feringa has pioneered many elegant examples of unidirectional motion in chiroptical switches achieved by sequentially applying light and then heat to the sterogenic isomerization of a restricted double bond.[75] Starting from the resting state, light at 365 nm causes the double bond to isomerize (Figure 4.25) in a *cis–trans* fashion, causing a clockwise movement in the rotor. Heat then allows for the phenyl rings to overcome their local rotational barrier and form the thermodynamic product of the switch state with another clockwise motion. These two steps cause a 180° unidirectional rotation by a large population of the sample. The process was repeated with 280 nm light and then heat to continue the motion in the same direction. This photochemical approach is matched by chemical and thermally stimulated examples of rotary motion in solution,[76] in the solid state,[77] and on surfaces.[78] This is certainly a fruitful area and the reader is directed to more in-depth reviews.[28, 77, 78]

4.6 MOVING LARGER SCALE OBJECTS

The harnessing and amplification of motion by many motor molecules is a hallmark of our muscles. Already discussed are the redox-active artificial molecular muscles (Figure 4.17)[14] and the photoactive crystals (Figure 4.23).[73] In addition, Feringa has embedded photodriven unidirectional motors into liquid crystals to rotate a micro-sized rod.[79] Such activities demonstrate how to bridge the nano-to-micro divide using soft materials. Another, and perhaps softer action, is to move fluids about.

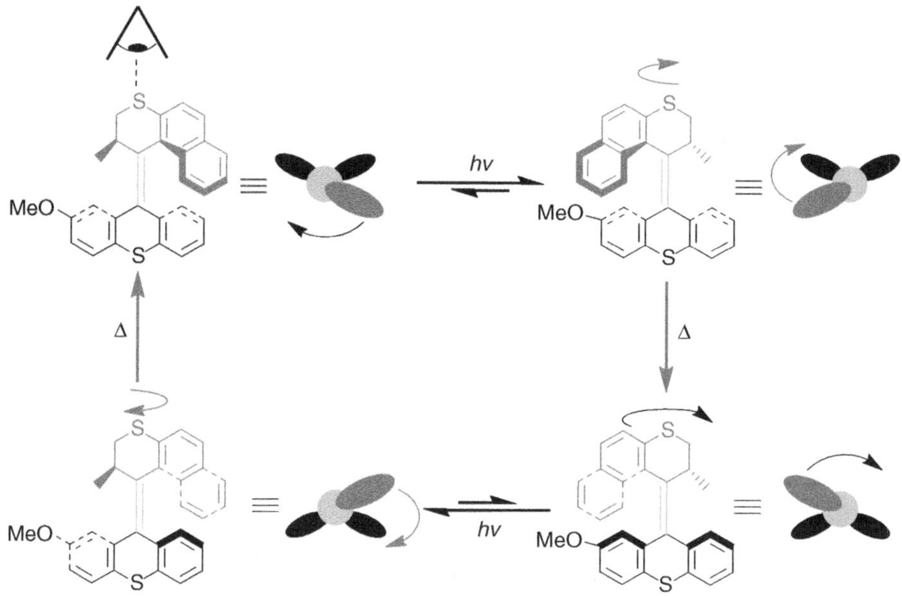

Figure 4.25 This photochemical chiroptical molecular motor is driven by two processes: one uses light to isomerize the double bond, and the other is a thermal equilibration allowing the *cis*-aromatics to overcome the steric barriers to local rotation. This overall process leads to unidirectional motion.

The photochemically driven [2]rotaxane[80] shown in Figure 4.26a switches with light according to the mechanism described in Figure 4.13. When the macrocycle moves, it can expose or conceal fluorinated alkanes, driving the molecule using fluorophobic/fluorophilic solution-interface interfacial interactions. When the macrocycle is around the station with the alkene, the fluorines on the left-hand station are exposed, making the molecule hydrophobic. When the macrocycle is around the fluorinated component, the [2]rotaxane exhibits hydrophilic properties. When the [2]rotaxane is attached to a surface patterned with fluorinated alkanes, light is able to drive the motion of a diiodomethane droplet uphill and downhill depending on the state of the molecular machine.

All of these systems fulfill the promise of replacing microsystems-based actuators with nanoscaled counterparts, yet there is still a long way to go. For instance, interfacing many molecules with a support material is nontrivial such that the rapid and high-throughput deployment and testing in engineered mechanical devices is itself a sizable undertaking. Once integrated, there is also the challenge of higher levels of organization, such as that displayed in muscle fibers. The polymer-based rotaxane dimers[61] take steps in that direction. Nonetheless, there remain outstanding challenges that will likely rely on multidisciplinary approaches—where different length scales and outcomes (chemical and mechanical) are brought together as integrated wholes.

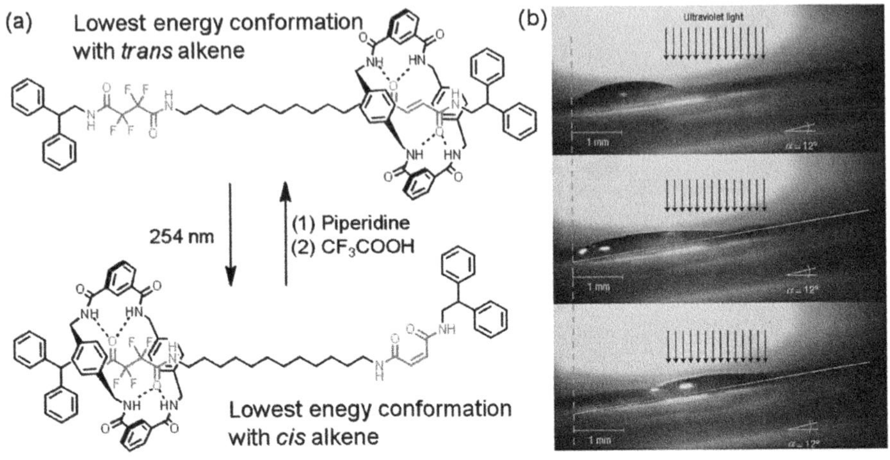

Figure 4.26 (a) Isomerization of the double bond in the right-hand station causes the macrocycle to move and cover the fluorinated groups of the rotaxane. (b) When these rotaxanes are patterned on a surface, the movement of the macrocycle by photoirradiation causes CH_2I_2 to move uphill. (Reproduced with permission. Copyright Nature Publishing Group: Ref. 80.)

4.7 WALKING MACHINES

Walking molecules like kinesin[4] and myosin V[41] present a unique case in the design of molecular machinery. By contrast to muscle, their primary actions are as single molecules. Beyond simply designing a system constrained to a single dimension, the machine must also proceed unidirectionally, rarely or never taking a step backward. Kinesin's importance as a model for this field is obvious, although feasible approaches to emulate its activity are less prevalent.

Arguably the most successful foray into this area was undertaken by Leigh and co-workers,[16] who designed a very simple walker (Figure 4.27) capable of proceeding down a track. The walker is comprised of two "feet," one capable of forming a hydrazone and another a disulfide. These feet have corresponding footholds along a molecular track, with the hydrazone forming at an aldehyde site and the disulfide at a thioether, interactions readily made and broken through dynamic covalent chemistry. The walker is first bound to a portion of the track. Present at an identical portion of the track is a disulfide-bonded "placeholder," which serves the role of introducing structural restriction and preventing unwanted backward steps. The walking process is induced by selectively making one bound foot labile through acid/base treatment, permitting it to rotate about the still "locked" foot, and step onto the next available station on the track. However, inducing the lability of the disulfide group also frees the placeholder, which provides no impetus to make a forward step, which could consequently result in a backward step during the subsequent lability of the hydrazone. The movement forward is thus encouraged by

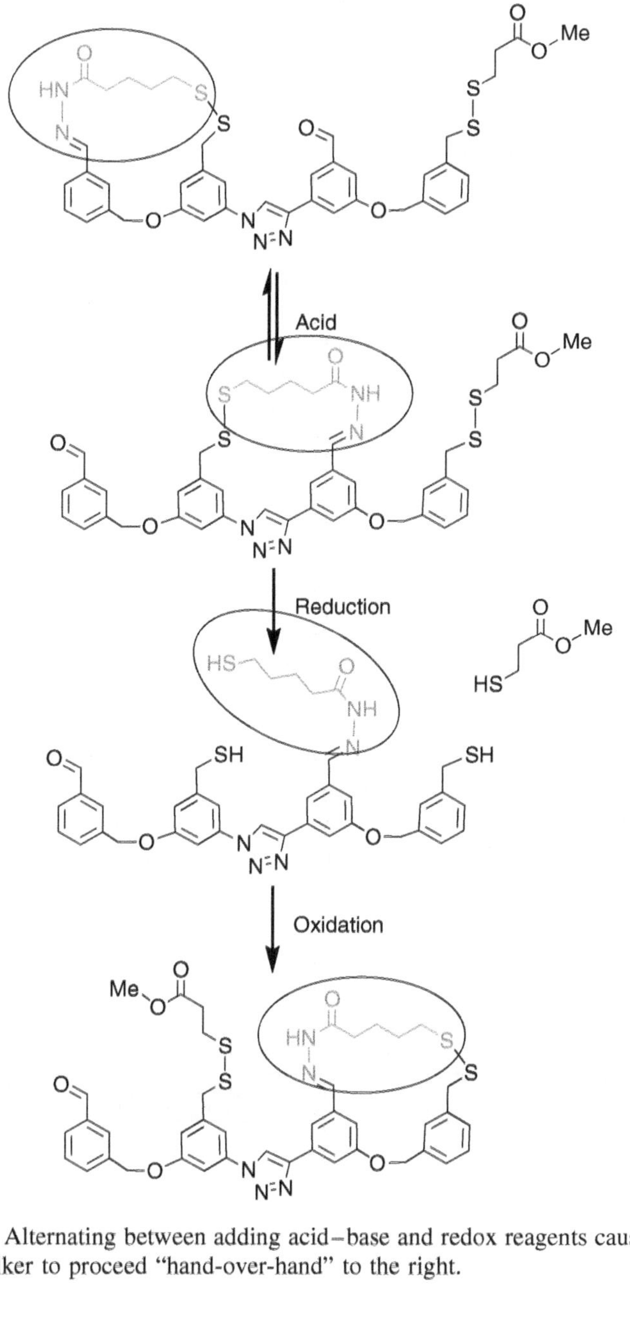

Figure 4.27 Alternating between adding acid–base and redox reagents causes the small molecular walker to proceed "hand-over-hand" to the right.

oxidizing the cleaved disulfides, producing a directional bias, driving movement down the track and replacing the placeholder "behind" the walker. In doing so, the molecule demonstrates a behavior that legitimately categorizes it as a Brownian rachet.

Prior to this work, the only synthetic unidirectional molecular walkers that existed were constructed from DNA,[15] a chemical chassis that has been subject to increasing sophistication. A DNA "spider" described in 2010 was shown to be capable of a variety of interactions with a DNA origami landscape (Figure 4.28), including commands to start, stop, and turn while proceeding down the track. The spider was created by attaching three legs comprised of molecular adaptations of the 8-17 DNA enzyme fastened to a streptavidin core. Walking is achieved by the incremental degradation of DNA footholds placed along the length of the origami track: when the spider's leg comes in contact with an intact DNA foothold, it hydrolyzes the molecule, allowing it to dissociate from the foothold and undergo a diffusional search for another foothold. Forward motion occurs on account of the difference in the length of time each leg occupies each foothold. Spider legs rest on intact DNA footholds for a discernably longer time than previously visited sites, leading the spider to move "forward" along the track.

These walking machines have the closest likeness to molecular robots, and, in fact, the DNA walker was described as such.[15] The elements of action, sensing, and logic that constitute a robot are themselves bioinspired from the sentient character of animals. What is surprising is that walkers like kinesin also act, sense, and proceed forward or backward according to cellular information as we would like robots to do. It is no wonder then that synthetic variants have emerged. It is now opportune to speculate that the variety of chemical motifs available in the molecular machinists' toolbox provide the power to imbibe autonomous walkers with many goals. Yet challenges remain, principally, to consider how to autonomously power their robotic actions using some form of catalysis analogous to ATP hydrolysis or otherwise.

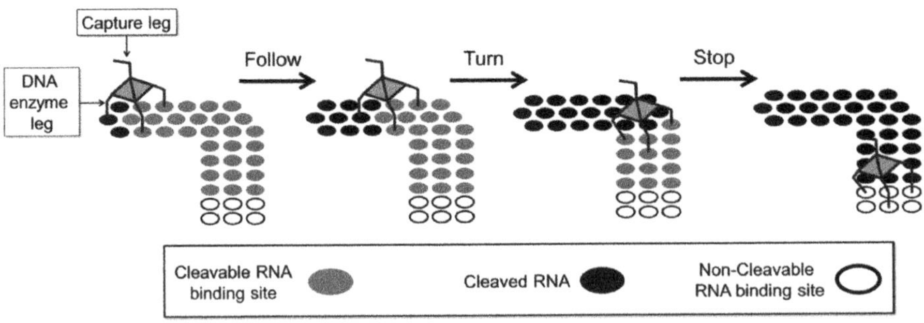

Figure 4.28 A DNA "spider" with a streptavidin core and DNase legs is programmed to move along a DNA origami track.

4.8 INGENIOUS MACHINES

Given the synthetic ingenuity and curiosity of the chemistry community, it is inevitable that after exploring different principles and components of "classical" synthetic molecular machines, new machines would begin to appear that explored functional space beyond previously conceived designs. What is remarkable about the systems described hereafter is their frequent use of chemical integration to capitalize on preexisting components.

4.8.1 Molecular Machines Inspired by Macroscopic Ones: Scissors and Elevators

Aida's "molecular scissors"[81] are an early example representative of this trend (Figure 4.29a). Combining a ferrocene unit, which acts as the pivot joining two scissor blades together, and an azobenzene unit to drive conformational changes between "open" and "closed" blades, the molecular scissors demonstrated the efficacy of transmitting conformational changes through rotatable joints. In another instance, the combination of previously explored machine morphologies was taken to a highly literal degree, where three rotaxanes were joined together to form a single "molecular elevator" (Figure 4.29b).[82] Here, a crown ether's interactions with ammonium and bipyridinium stations were arranged in a trifold symmetry, resulting in the formation of a crown ether "platform" capable of moving relative to the tripodal frame. The authors care to note an interesting point of diversion from macroscopic elevators: in this system, incremental deprotonation (multivalency) is observed, indicating that the elevator platform does not undergo a concerted movement. Instead, the individual crown ethers dip at noticeably different stages, giving rise to the capacity for multifunctional responses from the same stimuli.

4.8.2 Artificial Motility at the Nanoscale

While molecules like rotaxanes attempt to recapitulate the contraction–extension cycle of a sarcomere in miniature, other research has been done that attempts to

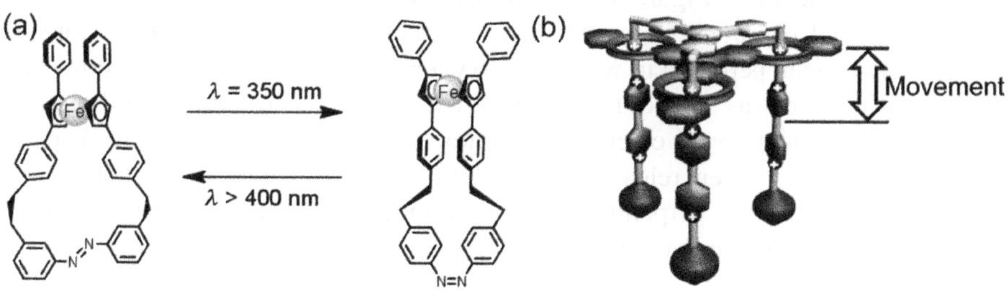

Figure 4.29 (a) Isomerization of the azobenzene group drives the rotation around the ferrocene group causing the "scissor blades" to open and close. (b) Acid–base stimulation causes the molecular elevator's platform to move up and down.

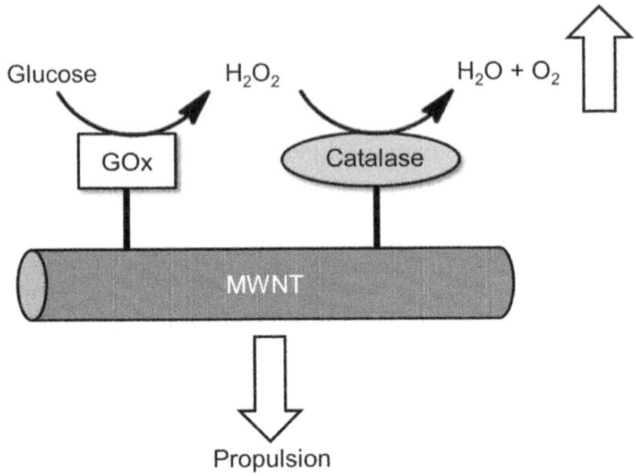

Figure 4.30 Catalysts bound to a multiwalled carbon nanotube (MWNT) propel the system when glucose is converted to H_2O and O_2.

replicate the chemotaxis of microorganisms. A 2008 example[83] proposes a system that combines materials taken from the biological and nanotech toolboxes in the form of enzymes anchored to multiwalled carbon nanotubes, producing a molecule that behaves much like a rocket (Figure 4.30). To generate propulsion in solution, the two enzymes glucose oxidase and catalase work cooperatively to break down glucose into oxygen gas at the surface of the nanotube. The formation of these bubbles acted as a propellant, sending the submicron-sized functionalized nanotubes jetting about as rapidly as 0.8 cm/s. Because of their size, the nanotubes were able to navigate without notable interference from Brownian motion. Instead, motion and direction were a function of the distribution of enzymes on the nanotube's surface, an observation suggesting that nonrandom navigation in solvent may be possible with the meticulous dispersal of enzymes along the surface of the nanotube. While autonomous movement guided by environmental information is the goal, a recent discovery[12] demonstrated the capacity for top–down control of "swimming" machines (Figure 4.2d). Similarly propelled by dioxygen formation, these directional nanorods featured a layer of platinum that catalyzed the conversion from hydrogen peroxide to water and oxygen. As would be expected, however, the path of the rod was subject to random fluctuations. To provide some direction, short nickel segments were electroplated onto the rods, allowing them to respond to magnetic fields. When released into solution with a nearby magnet, the rods would orient themselves perpendicularly to the magnetic field, giving operators the capacity to direct their movement.

4.8.3 Moving Molecules Across Surfaces

Other machines have attempted to capture "walking" and "rolling" movements along surfaces. In the instance of the "nanocar,"[84] molecules were designed to look like a car's chassis with fullerene wheels and proceeded along surfaces in a direction perpendicular to the axles of the car. The molecules also displayed a pivoting motion, giving them the capacity to move in two dimensions, where the authors had originally predicted only one. Similarly, the "nanohorse"[85] (Figure 4.31) was shown to undergo controlled diffusion along a surface at low temperatures, even showing a functional difference between molecules with different numbers of "legs."

4.9 USING SYNTHETIC BIOINSPIRED MACHINES IN BIOLOGY

Molecules have always had a natural role in biology on account of their size and the fact that they interact in the same way as *bio*chemicals. Synthetic molecular machines, as outlined in this chapter, have a direct connection to biomachines. Yet it is only recently that the synthetic variants are being exploited for possible applications in human health.

In an exemplary case,[86] the application by Stoddart and Zink of familiar molecular machines permitted the creation of a nanoparticle that acts a synthetic transport vesicle carrying molecular cargo, and potentially drugs (Figure 4.32). By utilizing mesoporous silica nanoparticles as the support structure for this mechanized nanoparticle, switchable machines—in this case, rotaxanes—can be attached to the surface to serve as valves at the openings of pores. The placement of these valves provides a mechanical barrier that can trap small molecules in the porous cavities of the nanoparticle, capable of transit, and ready to be released upon stimulation. Proof-of-principle experiments utilizing an acid/base sensitive valve assembly demonstrated the nanoparticles' ability to deliver an anticancer drug into living cells,[87] suggesting its eventual use as a tool to selectively target diseased tissue and release a payload of medication. Theoretically, this system can be functionalized to perform under any stimuli currently existing that effects switching, providing a very flexible model for improvement and specificity in design.

4.10 PERSPECTIVE

A great deal of human ingenuity originates from attempts to recreate the phenomena observed in the natural world using tools of human origin. The early saga of flight shows the result of this approach, which sometimes takes Nature far too literally. For instance, early flying machines of the ornithopter type[88] were doomed from the outset, lacking the modern understanding about how wings attached to a bird are flightworthy and the wings attached to a human are foolhardy. This approach entertained a simplified relationship between form and function, almost a kind of cargo cult[89] thinking. If we were to simply *look like* a bird, then we must necessarily be able to *fly like* one as well! While we still embrace the compelling relationship between form and function, experience has since taught us that it can be more subtle and deceptive than previously imagined. Flight is not merely the

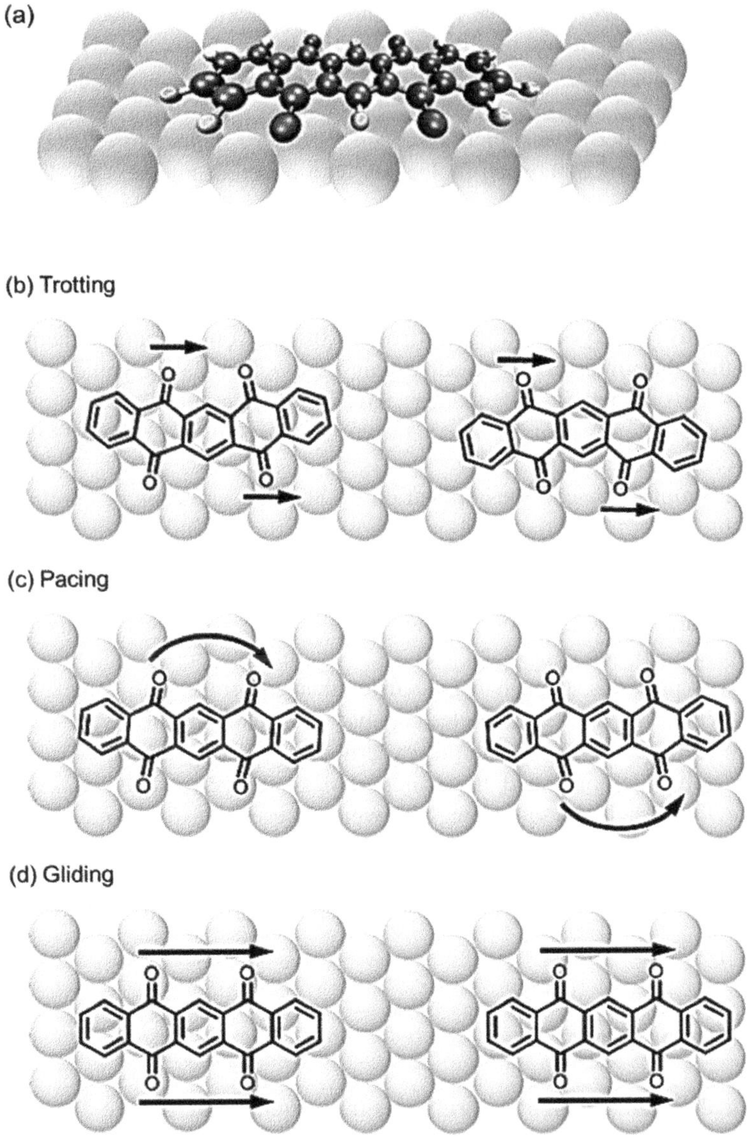

Figure 4.31 (a) Pentacenetetrone on a Cu(1,1,1) surface (b) trotting, (c) pacing, and (d) gliding across the copper surface.

product of a linear combination of physical components. Instead, it is an emergent phenomenon where those components take on an enhanced value when utilized together. This relationship captures the import of bioinspiration: *understanding the mechanisms* of the world around us makes the connection between the form we see and the function we desire.

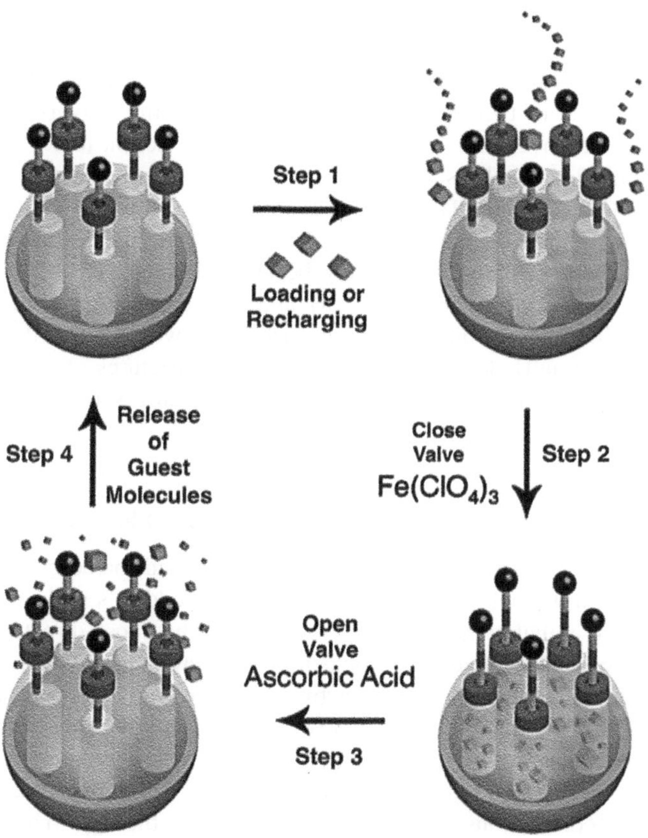

Figure 4.32 Porous silica nanoparticles can be opened and closed with rotaxanes, allowing for the loading and release of molecular cargo. (Reproduced with permission. Copyright the Royal Society of Chemistry: Ref. 86.)

A variation on this last notion of bioinspiration has a healthy life in our fertile cultural imagination—revisited in fiction and urban legend alike. The proposition has been made that the explosion in technological development over the past century or so came about when humanity reverse-engineered technology that was originally fabricated by advanced alien species. While absurd as an account of modern civilization, this sequence of events is somewhat analogous to chemistry's use of bioinspiration, which takes cues from Nature's mature "technology."

It is interesting to contrast the benefits inherent in this narrative to how computer engineers have always had to learn everything through trial and error. Although recent advances made toward understanding the human brain have provided inspiration to modify computers' processing power, historically there has never been any antecedent to the next generation computer save for the technological progress of the previous generation. Even then, they struggle with the same problem that plagues any field that must be cut from the whole cloth of the unknown: skirting a line between feasibility and failure without ever fully knowing where the boundary between the two lies.

It is the good fortune of molecular machinists[24] then, who are not only privileged by the bounty of their imagination but have an entire universe of molecular machines perfected over billions of years of natural selection to serve as a guide. Answers to questions such as "Can it be done?" or "How can it be done?" will do well to consider asking "Has Nature done it?" or "How has Nature done it?" Yet we are not limited by the confines of biology! Indeed, the necessity for the synthetic molecular machine is precisely because biology has limits and boundaries, for example, aqueous solutions, pH and temperature ranges, fuel sources, and molecular lifetimes. If biology is our teacher, we would do well to remember that we must eventually depart from the intellectual strictures of her classroom if we are to accomplish anything for ourselves. Given that assertion, to what extent have we done so?

4.10.1 Lessons and Departures from Biological Molecular Machines

Sliding machines encapsulate most of the length and breadth of conceptual exploration within the field of molecular machines. Stoddart's molecular muscle[14] (Figure 4.17) is arguably the molecule that bears the most direct resemblance to the structure of the sarcomere, replacing the stationary myosin core with a redox-active dumbbell, and the mobile actin scaffold with macrocycles anchored to a surface. Perhaps not coincidentally, use of this molecule was the first to achieve amplified movement at the macroscale. Nevertheless, it is scarcely a mimic; the resemblance is necessarily superficial, as the mode of action and mechanism used to activate the contractions are without precedent in biology. The majority of other molecular muscles become increasingly distant from the biological world, committing themselves to the daisy chain morphology[60–65] or even more exotic designs[66, 67, 73] such as the molecular spring.[70–72] Moreover, the stimuli that produce the length changes have explored virtually every energy source described by the theoretical switching cycle in Figure 4.3c with the notable exception of a reliable chemical, and catalytic, fuel source. As a result, rotaxane-based molecular muscles are becoming increasingly dissimilar in form to skeletal muscles even as they grow increasingly similar in performance. While this trend can fairly be expected to continue, it also serves to underscore the rich *variation* within the designs, and emphasizes the rotaxane's value as a flexible conceptual platform.

When considering walking machines (Section 4.7), Leigh's walker[16] is *kinesin* writ small. Designing a system capable of a hand-over-hand one-dimensional walk like kinesin was the expressed purpose of the molecule, and its success reveals the benefits of closely following biology's lead. The success comes from desiring to follow *kinesin's* alternating lability of the molecule's feet, and understanding the importance of having a complex (i.e., more than a simple on/off system running in parallel) catalytically powered cycle to encourage the unidirectional forward progress of the molecule. The DNA walker,[15] by contrast, illustrates a system premised upon the use of familiar biological principles (DNA complementarity) and then proceeds to follow a design that is truly innovative. Since necessity is the mother of invention, it is worth considering that the relatively unexplored challenge of unidirectional walking may drive the development of a plethora of solutions to the same problem.

One of the essential aspects of continued success within the field is keeping the practice of chemical integration alive and well. One of the principal insights of the Industrial Revolution was that of interchangeable parts. Realizing the benefits of mechanical standardization, engineers started to settle on common components to enable the rapid development and ease of maintenance of new technologies. It is encouraging then to see that this idea is not lost on modern chemists, who are clearly starting to embrace efforts to establish standard "parts" in the various classes of molecular machines. The recurring appearance of the crown/ammonium pair in pH-sensitive machines (Figures 4.15, 4.18, 4.19, 4.21a) or the "Sauvage-like" macrocycle (Figures 4.12, 4.16, 4.21b) in metal coordination-based machines displays a decided preference for generating a toolbox of ubiquitous components. The area of sliding machines (Section 4.4) with its many rotaxane-based motifs is exemplary of this idea. Walking machines (Section 4.7), however, are burgeoning and consequently much further from this ideal as they differ greatly from one design to another in their composition and mode of operation. These walkers have a promising future as they are apt to be the most capable of overcoming the interface problem, inherently designed to carry loads and interact with surfaces. Nonetheless, integration with support structures is a practice that is still somewhat rare, but represents a critical near-term benchmark for progress.

4.10.2 The Next Steps in Bioinspired Molecular Machinery

The successful exploration of increasing swathes of machine functionalities guided by the commonalities among biomachines described in Section 4.2.4 raises the question about the role of bioinspiration in the next generation of molecular machine technology. For instance, while controllable length changes like those seen in rotaxane-based molecular muscles have proved the efficacy of looking to Nature for direction (i.e., points 1, 5, and 6 in Section 4.2.4), it is no longer a monumental synthetic and conceptual leap to create some manner of molecular muscle; it is now a matter of patience and ingenuity. Can we now similarly use biology to help confront the next set of scientific hurdles, that is, the interface problem? Can these machines be automated, that is, capable of deriving their energy from catalytic subunits within the machine, and regulated by feedback from the surrounding environment?

Certainly intuition suggests the answer would be an emphatic yes! The field of origin for molecular machinery, namely, supramolecular chemistry, also proposes approaches to the design and manipulation of hierarchically organized systems.[90] Although biomachines are initially fabricated with covalent bonding (primary structure), the folding of proteins (secondary, tertiary) and their aggregation (quaternary) into larger networks is a process fundamentally driven by weak intramolecular forces. Studying Nature's methods in an attempt to learn the secret of creating organized muscle tissues from the rabble of myosin bundles may give us access to currently inconceivable functionalities. We must also remember the material lesson of modern flight and the successes of the molecular muscle.[14] That is, we must also consider "hard" yet flexible materials interfaced with the "soft" molecules as a complementary approach to access larger and ordered structures in ways not necessarily employed by Nature.

4.11 CONCLUSION

Herein, we have attempted to provide a picture of current research into molecular machines, with an emphasis on the way in which their design and operating principles are inseparably *inspired* by biological examples, *motivated* by engineering applications, and *realized* by chemistry's various synthetic tools. The form and function of each machine discussed can find its creative ancestry in each of these aspects. While we can consider the general features of biological machines as targets for future improvement, we are not bound to utilize all of them. After all, our goal is not to make synthetic machines into biological facsimilies, but rather to know that it is possible to install whatever biologically inspired functionality we need into the machines of our desire. This is, after all, the value of bioinspiration over biomimicry.

It is difficult to overstate the magnitude of the challenge inherent in trying to master, within the course of a few generations of scientific research, what has been an extremely long-lived "research program" by Nature. Yet it is also difficult to overstate the already-realized successes, the rapidity with which success continues to come, and the increasingly fine grasp of operating principles hitherto achieved by molecular machinists. Although attempting to set dates is pure speculation, it is not difficult to conceive of a day within the not too distant future, where machines of biological origin routinely share their molecular world with machines of synthetic origin.

REFERENCES

1. Kay, E. R.; Leigh, D. A.; Zerbetto, F. *Angew. Chem. Int. Ed.* **2007**, *46,* 72.
2. Balzani, V.; Credi, A.; Venturi, M. *Molecular Devices and Machines—A Journey into the Nano World;* Wiley-VCH Verlag GmbH & Co., KGaA, Weinheim, 2003.
3. Boyer, P. D. *Biochim. Biophys. Acta* **1993**, *1140,* 215.
4. Duncan, T. M.; Bulygin, V. V.; Hutcheon, M. L.; Cross, R. L. *Proc. Natl. Acad. Sci. U.S.A.* **1995**, *92,* 10964.
5. Yasuda, R.; Noji, H.; Kinosita, K.; Yoshida, M. *Cell* **1998**, *93,* 1117.
6. Yildiz, A.; Tomishige, M.; Vale, R. D.; Selvin, P. R. *Science* **2004**, *303,* 676.
7. Spudich, J. A. *Nature Rev.* **2001**, *2,* 387.
8. Feynman, R. P. *Eng. Sci.* **1960**, *23,* 22.

9. Leewenhoeck, A. *Philos. Trans*. **1684**, *14,* 568.
10. Harris, H. *Physiol. Rev*. **1954**. *34,* 529.
11. Ismagilov, R. F.; Schwartz, A.; Bowden, N.; Whitesides, G. M. *Proc. Natl. Acad. Sci. U.S.A.* **2002**, *41,* 652.
12. Kline, T.; Paxton, W.; Mallouk, T.; Sen, A. *Angew. Chem. Int. Ed*. **2005**, *44,* 744.
13. Jimenez, M. C.; Dietrich-Buchecker, C.; Sauvage, J.-P. *Angew. Chem. Int. Ed*. **2000**, *39,* 3284.
14. Liu, Y.; Flood, A. H.; Bonvallet, P. A.; Vignon, A. S.; Northrup, B.; Tseng, H.-R.; Jeppesen, J. O.; Huang, T. J.; Brough, B.; Baller, M.; Magonov, S.; Solares, S.; Goddard, W. A.; Ho, C.-M.; Stoddart, J. F. *J. Am. Chem. Soc*. **2005**, *127,* 9745.
15. Lund, K.; Manzo, A.; Dabby, N.; Michelotti, N.; Johnson-Buck, A.; Nangreave, J.; Taylor, S.; Pei, R.; Stojanovic, M.; Walter, N.; Winfree, E.; Yan, H. *Nature* **2010**, *465,* 206.
16. Delius, M.; Geertsema, E.; Leigh D. *Nature Chemistry* **2010**, *2,* 96.
17. Whitesides, G. M; Lipomi, D. J. *Faraday Discuss*. **2009**, *143,* 373.
18. Astumian, R. D. *Annu. Rev. Biophys*. **2011**, *40,* 289.
19. Lehn, J.-M. *Supramolecular Chemistry. Concepts and Perspectives*, VHC, Weinheim, Germany, 1995.
20. Steed, J. W.; Atwood, J. L. *Supramolecular Chemistry*, John Wiley & Sons, Chichester, UK, 2002.
21. Pederson, C. J. *J. Am. Chem. Soc*. **1967**, *89,* 7017.
22. Park, C. H.; Simmons, H. E. *J. Am. Chem. Soc*. **1968**, *90,* 2431.
23. Dietrich, B.; Lehn, J.-M.; Sauvage, J.-P. *Tetrahedron Lett*. **1969**, *34,* 2889.
24. Cram, D. J.; Cram, J. M. *Science* **1974**, *183,* 803.
25. Shinkai, S.; Nakaji, T.; Nishida, Y.; Ogawa, T.; Manabe, O. *J. Am. Chem. Soc*. **1980**, *102,* 5860.
26. Bissell, R. A.; Córdova, E.; Kaifer, A. E.; Stoddart, J. F. *Nature* **1994**, *369,* 133.
27. Dietrich-Buchecker, C. O.; Sauvage, J.-P.; Kern, J.-M. *J. Am. Chem. Soc*. **1984**, *106,* 3043.
28. Feringa, B. L. *J. Org. Chem*. **2007**, *72,* 6635.
29. Feringa, B. L. *Molecular Switches*, Wiley-VCH, Weinheim, Germany, 2001.
30. *Those Magnificent Men in Their Flying Machines, or How I Flew From London to Paris in 25 Hours 11 Minutes*. Annakin, K. 1965, 20th Century Fox.
31. Nicolaou, K.; Sorensen, E. *Classics in Total Synthesis: Targets, Strategies, Methods*, 1st ed., Wiley-VCH, Hoboken, NJ, 1996.
32. Hoffman, R. *The Same and Not the Same*, Columbia University Press, New York, 1997.
33. Boyle, M. M.; Smaldone, R. A.; Whalley, A. C.; Ambrogio, M. W.; Botros, Y. Y.; Stoddart, J. F., *Chem. Sci*. **2011**, *2,* 204.
34. van Leeuwen, P. W. N. M. *Supramolecular Catalysis*, Wiley, Hoboken, NJ, 2008.
35. Swiegers, G. F.; Huang, J.; Brimblecombe, R.; Chen, J.; Dismukes, G. C.; Mueller-Westerhoff, U. T.; Spiccia, L.; Wallace, G. G., *Chem. Eur. J*. **2009**, *15,* 4746.

36. Black, A. L.; Lenhardt, J. M.; Craig, S. L. *J. Mater. Chem*. **2011**, *21,* 1655.
37. Åström, K. J.; Murray, R. M. *Feedback Systems*, Princeton University Press, Princeton, NJ, 2008.
38. Carter, N. J.; Cross, R. A. *Nature* **2005**, *435,* 308.
39. Tajkhorshid, E.; Aksimentiev, A.; Balabin, I.; Gao, M.; Isralewitz, B.; Phillips, J. C.; Zhu, F.; Schluten, K. *Adv. Protein Chem*. **2003**, *66,* 195.
40. Noji, H.; Yasuda, R.; Yoshida, M.; Kinosita, K. *Nature* **1997**, *386,* 299.
41. Walker, M. L.; Burgess, S. A.; Sellers, J. R.; Wang, F.; Hammer, J. A.; Trinck, J.; Knight, P. J. *Nature* **2000**, *405,* 804.
42. Hess, H.; Clemmens, J.; Qin, D.; Howard, J.; Vogel, V. *Nano Lett*. **2001**, *1,* 235.
43. Toprak, E.; Yildiz, A.; Hoffman, M. T.; Rosenfeld, S. S.; Selvin, P. R. *Proc. Natl. Acad. Sci. U.S.A.* **2009**, *106,* 12712.
44. Howard, J. *Nature* **1997**, *389,* 561.
45. Feynman, R. P.; Leighton, R. B.; Sands, M. *The Feynman Lectures on Physics*, Vol. *1*. Addison Wesley, Boston, 2005.
46. Astumian, R. D. *Science* **1997**, *276,* 917.
47. Share, A. I.; Parimal, K.; Flood, A. H. *J. Am. Chem. Soc*. **2010**, *132,* 1665.
48. Asakawa, M.; Ashton, P. R.; Balzani, V.; Credi, A.; Hamers, C.; Mattersteig, G.; Montalti, M.; Shipway, A. N.; Spencer, N.; Stoddart, J. F.; Tolley, M. S.; Venturi, M.; White, A. J. P.; Williams, D. J. *Angew. Chem. Int. Ed*. **1998**, *37,* 333.
49. Durot, S.; Reviriego, F. Sauvage, J.-P.; *Dalton Trans*. **2010**, *39,* 10557.
50. Durola, F.; Lux, J.; Sauvage, J.-P. *Chem. Eur. J*. **2009**, *15,* 4124.
51. Periyasmy, G.; Sour, A.; Collin, J.-P.; Sauvage, J.-P.; Remacle, F. *J. Phys. Chem. B* 2009, *113,* 6219.
52. Altieri, A.; Bottari, G.; Dehez, F.; Leigh, D. A.; Wong, K. Y.; Zerbetto, Z. *Angew. Chem. Int. Ed*. **2003**, *42,* 2296.
53. Choi, J. W.; Flood, A. H.; Steuerman, D. W.; Nygaard, S.; Braunschweig, A.; Moonen, N. N. P.; Laursen, B. W.; Luo, Y.; DeIonno, E.; Peters, A. J.; Jeppesen, J. O.; Xu, K.; Stoddart, J. F.; Heath, J. R. *Chem. Eur. J*. **2006**, *12,* 261.
54. Nygaard, S.; Laursen, B. W.; Flood, A. H.; Hansen, C. N.; Jeppesen, J. O.; Stoddart, J. F. *Chem.Common*. **2006**, 144.
55. Serreli, V.; Lee, C.-F.; Kay, E. R.; Leigh, D. A. *Nature* **2007**, *445,* 523.
56. Panman, M. R.; Bodis, P.; Shaw, D. J.; Bakker, B. H.; Newton, A. C.; Kay, E. R.; Brouwer, A. M.; Buma, W. J.; Leigh, D. A.; Woutersen, S. *Science* **2010**, *328,* 1255.
57. Åström, K. J.; Murray, R. M. *Feedback Systems*, Princeton University Press, Princeton, NJ, 2008.
58. Juluri, B. K.; Kumar, A. S.; Liu, Y.; Ye, T.; Yang, Y.-W.; Flood, A. H.; Fang, L.; Stoddart, J. F.; Weiss, P. S.; Huang, T. J. *ACS Nano* **2009**, *3,* 291.
59. Ashton, P. R.; Chrystal, E. J. T.; Glink, P. T.; Menzer, S.; Schiavo, C.; Spencer, N.; Stoddart, J. F.; Tasker, P. A.; White, A. J. P.; Williams, D. J. *Chem. Eur. J*. **1996**, *2,* 709.
60. Clark, P. G.; Day, M. W.; Grubbs, R. H. *J. Am. Chem. Soc*. **2009**, *131,* 13631.

61. Fang, L.; Hmadeh, M.; Wu, J.; Olson, M.; Spruell, J.; Trabolsi, A.; Yang, Y.; Elhabiri, M.; Albrecht-Gary, A.; Stoddart, J. *J. Am. Chem. Soc.* **2009**, *131,* 7126.
62. Coutrot, F.; Romuald, C.; Busseron, E. *Org. Lett.* **2008** *10,* 3741.
63. Romuald, C.; Busseron, E.; Coutrot, F. *J. Org. Chem.* **2010**, *75,* 6516.
64. Tsukagoshi, S.; Miyawaki, A.; Takashima, Y.; Yamaguchi, H.; Harada, A. *Org. Lett.* **2007**, *9,* 1053.
65. Dawson, R.; Lincoln, S.; Easton, C. *Chem. Commun.* **2008**, *34,* 3980.
66. Chuang, C.; Li, W.; Lai, C.; Liu, Y.; Peng, S.; Chao, I.; Chiu, S. *Org Lett.* **2009**, *11,* 385.
67. Xue, Z.; Mayer, M. *J. Am. Chem. Soc.* **2010**, *132,* 3274.
68. Nygaard, S.; Liu, Y.; Stein, P. C.; Flood, A. H.; Jeppesen, J. O. *Adv. Funct. Mater.* **2007**, *17,* 751.
69. Liu, Y.; Muscovitz, R.; Flood, A. H.; Stoddart, J. F. *Chem. Eur. J.* **2005**, *11,* 369.
70. Ikeda, T.; Shinkai, S.; Sada, K.; Takeuchi, M. *Tetrahedron Lett.* **2009**, *50,* 2006.
71. Russew, M.-M.; Hecht, S. *Adv. Mater.* **2010**, *22,* 3348.
72. Hua, Y.; Flood, A. H. *J. Am. Chem. Soc.* **2010**, *132,* 12838.
73. Morimoto, M.; Irie, M. *J. Am. Chem. Soc.* **2010**, *132,* 14172.
74. Hernández, J. V.; Kay, E. R.; Leigh, D. A.; *Science* **2004**, *306,* 1532.
75. Koumura, N.; Geertsema, E. M.; van Gelder, M. B.; Meetsma, A.; Feringa, B. L. *J. Am. Chem. Soc.* **2002**, *124,* 5037.
76. Landge, S. M.; Aprahamian, I. *J. Am. Chem. Soc.* **2009**, *131,* 18269.
77. Khuong, T. A. V.; Nunez, J. E.; Godinez, C. E.; Garcia-Garibay, M. A. *Acc. Chem. Res.* **2006**, *39,* 413.
78. Kottas, G. S.; Clarke, L. I.; Horinek, D.; Michl, J. *Chem. Rev.* **2005**, *105,* 1281.
79. Vicario, J.; Katsonis, N.; Ramon, B. S.; Bastiaansen, C. W. M.; Broer, D. J.; Feringa, B. L. *Nature* **2006**, *440,* 163.
80. Berna, J.; Leigh, D. A.; Lubomska, M.; Mendoza, S. M.; Perez, E. M.; Rudolf, P.; Teobaldi, G.; Zerbetto, F. *Nature Mater.* **2005**, *4,* 704.
81. Muraoka, T.; Kinbara, K.; Kobayashi, Y.; Aida, T. *J. Am. Chem. Soc.* **2003**, *125,* 5612.
82. Badji, J.; Balzani, J.; Credi, A.; Silvi, S.; Stoddart, J. F. *Science* **2004**, *303,* 1845.
83. Pantarotto, D.; Browne, W.; Feringa, B. *Chem. Commun.* **2008**, 1533.
84. Shirai, Y.; Osgood, A.; Zhao, Y.; Kelly, K.; Tour, J. *Nano Lett.* **2005**, *5,* 2330.
85. Cheng, Z.; Chu, E.; Sun, D.; Kim, D.; Zhu, Y.; Luo, M.; Pawin, G.; Wong, K.; Kwon, K.; Carp, R.; Marsella, M.; Bartels, L. *J. Am. Chem. Soc.* **2010**, *132,* 13578.
86. Cotab, K.; Belowich, M.; Liong, M.; Ambrogio, M.; Lau, Y.; Khatib, H.; Zink, J.; Khashab, N.; Stoddart, J. F. *Nanoscale* **2009**, *1,* 16.
87. Meng, H.; Xue, M.; Xia, T.; Zhao, Y.-L.; Tamanoi, F.; Stoddart, J. F.; Nel, A. E.; Zink, J. I. *J. Am. Chem. Soc.* **2010**, *132,* 12690.
88. Richter, J. P., Ed. *The Notebooks of Leonardo da Vinci*, 1880.
89. Leighton, R., Ed. *Classic Feynman*, W. W. Norton & Company, New York, 2006.
90. Schneider, H. J.; Yatsimirsky A. *Principles and Methods in Supramolecular Chemistry*, John Wiley & Sons, Chichester, UK, 2000.

CHAPTER 5

Bioinspired Materials Chemistry I: Organic–Inorganic Nanocomposites

PILAR ARANDA, FRANCISCO M. FERNANDES, BERND WICKLEIN, and EDUARDO RUIZ-HITZKY

Materials Science Institute of Madrid, ICMM-CSIC, 28049 Madrid, Spain

JONATHAN P. HILL and KATSUHIKO ARIGA

World Premier International (WPI) Research Center for Materials Nanoarchitectonics (MANA), National Institute for Materials Science (NIMS), Tsukuba 305-0044, Japan

5.1 INTRODUCTION

Biological systems are a useful source of inspiration for designing and preparing organic–inorganic hybrid nanocomposites based on materials found in Nature. For example, nacre, ivory, and bone are naturally produced bionanocomposites whose remarkable properties are related to the hierarchical arrangement of their organic and inorganic components.[1–3] Apart from important advances made using biomimetic approaches involving carbonates and phosphates to prepare artificial nacre, bone, and other bionanocomposites,[3–9] siliceous (e.g., silica, silicates, and polysiloxanes) biosystems also represent a viable alternative for preparation of biohybrids since the chemistry of those compounds is extremely versatile, permitting formation of hierarchical superstructures, supramolecular composites, and other multifunctional bioinspired systems.[2, 10–15]

Mimetic approaches of the abovementioned processes have been applied to prepare artificial or synthetic bionanocomposites using bottom–up fabrication procedures that allow for controlled self-assembly of different types of silica-based building blocks and bioorganic species to achieve the required arrangements.[14] Consider, for instance, natural biomineralization mechanisms based on siliceous diatom formation. Here, paradigmatic approaches have been used in attempts to design and prepare new architectures by combining diverse silica sources and polymeric species.[16–20] In this way, complex silica structures that mimic diatom frameworks

Bioinspiration and Biomimicry in Chemistry: Reverse-Engineering Nature, First Edition.
Edited by Gerhard F. Swiegers.
© 2012 John Wiley & Sons, Inc. Published 2012 by John Wiley & Sons, Inc.

can be formed at room temperature from a silica source in the presence of silaffins and other proteins.[21] The soft conditions used for these biosilification processes allow incorporation of a variety of different biomolecules, such as enzymes[22] or cytochrome *c*,[23] with preservation of their biofunctionality.

To imitate the nanostructuring, porosity, and surface roughness of natural bone, any synthetic strategy requires generation of foam-like bionanocomposites bearing a suitable porosity and interconnectivity of pores. To this end, different approaches[24, 25] such as fiber bonding, phase separation, solvent casting/particle leaching, and gas foaming have been used while more recently freeze-drying approaches have been the most frequently employed.[26]

Alternatively, layer-by-layer (LbL) methodologies are easy to control techniques for preparing nanostructured materials based on the adsorption of charged species onto a substrate of opposite electrical charge[27] that have been demonstrated to be also very useful for assembly of inorganic and biological entities for developing nacre-like bionanocomposites.[28] In particular, formation of layered structures with inorganic and/or organic–inorganic composite nanostructures is one of the most useful approaches to construct bio-like hierarchical composites. In the final section of this chapter, preparation and functionalization of hierarchic nanocomposites by LbL assembly are introduced.

5.2 SILICATE-BASED BIONANOCOMPOSITES AS BIOINSPIRED SYSTEMS

One of the best examples of silicate-based bioinspired nanocomposites is artificial nacre. Nacre is a natural bionanocomposite built up by sequential assembly of inorganic layers of aragonite crystals ($< 0.5\ \mu m$) with organic thin films (<10 nm) mainly consisting of proteins that cement the carbonate layers.[29] Initial attempts to produce artificial nacre were by use of template-assisted thin film formation through surfactant self-assembly using tetraethoxysilane (TEOS) as silica precursor of the inorganic moiety and cetyltrimethylammonium chloride surfactant as the organic counterpart.[29] Studies by Schaffer and co-workers focused on an aragonite tablets model of abalone nacre growth and concluded that mineral bridge formation occurs through interlamellar organic sheets rather than heteroepitaxial nucleation.[30] Production by means of continuous self-assembly of organic (e.g., surfactant cetyltrimethylammonium bromide bilayer) and inorganic (silica-based compounds) counterparts gave rise to materials showing periodical and compositional changes that mimic nacre.[31]

The development of clay-based bionanocomposites opened the way to new approaches for the preparation of bioinspired materials for different applications.[1, 2, 13, 32, 33] In particular, chitosan–montmorillonite bionanocomposites first reported by Ruiz-Hitzky's group[34] illustrate the possibility to agglomerate at the nanometric scale alternative assemblies of inorganic (silicate) layers and charged layers of biopolymers (chitosan). Strong electrostatic interactions

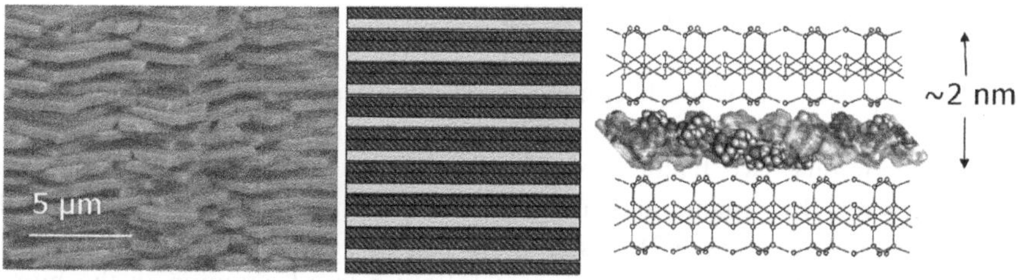

Figure 5.1 FE-SEM image of nacre natural bionanocomposite (left) and representations of synthetic clay-based bionanocomposite architectonics as biomimetic system at the nanometric scale (center and right).

between the negatively charged silicate sheets and the positive charges of chitosan are the predominant factor in the assembly giving rise to an alternation of inorganic–organic layers that mimics nacre at the nanometric scale (Figure 5.1). By controlling the amount of intercalated chitosan, the resulting materials exhibit tunable ion exchange capacities toward cations or anions, thus being of potential in the development of ion-sensing devices.[35] Large excesses of chitosan polysaccharide gives rise to exfoliated clay-nanocomposite materials where the layer stacking is lost.[36] Apart from chitosan, other biopolymers including polysaccharides, proteins, and nucleic acids have been assembled with different types of clay silicates as well as with other layered solids, including layered double hydroxides (LDHs), layered phosphates, and titanates.[1, 37]

By applying LbL techniques, Kotov and co-workers prepared highly ordered artificial nacre using Na–montmorillonite and polyelectrolytes such as poly(diallyldimethylammonium) chloride (PDDA) and poly(vinyl alcohol) (PVA) as starting components.[28, 38, 39] Other smectite clays such as saponite and stevensite[40] or synthetic Laponite[41] can be combined with PDDA and other polymers such as sodium polyacrylate, developing materials with excellent mechanical properties as well as heat resistance and gas barrier properties which make them of great interest for diverse applications. Natural or modified charged biopolymers such as chitosan[42] and sodium methylcellulosecarboxylate[40] have also been tested as the organic "cementing phase" of the clay-based artificial nacre bionanocomposites. In addition to good mechanical properties, the resulting materials can be conformed as films that exhibit elevated transparency to visible light, great thermal stability, and fire retardant ability. These aspects are of paramount importance for many applications but the high affinity to water molecules of those biopolymers is a limiting factor for preparation of stable and robust materials and must necessarily be improved for future applications. It should be noted that at the present time the development of silicate-based bionanocomposites mimicking nacre is still an emerging field of research. Advanced application developments are anticipated considering that biocompatibility, due to they being composed of natural polymers, leads to improved prospects for biomedical uses.

5.3 BIONANOCOMPOSITE FOAMS

Bionanocomposite foams are becoming more important in the field of bioinspired composite materials.[14, 43–45] Conceptually, these materials are of particular interest since they combine two features of the naturally occurring analogs. First, interactions between the organic and inorganic components are, as mentioned in the previous section (Section 5.2), inspired by self-assembly processes learnt from Nature. Second, bionanocomposite foams follow a trend common in the biomaterials field: that of minimizing the density of the material while maximizing its structural and functional properties (i.e., the formation of a porous structure). One of the first systematic descriptions regarding the relationship between structural performance and density in natural materials was by Ashby and co-workers,[46–48] whose work revealed the extraordinary mechanical properties of natural materials of very low density.

Porous materials, such as bionanocomposite foams, are usually characterized according to several structural features such as relative density, porosity, pore size, and type of pores or by their preparation procedures, although none of these characteristics is as important as the application envisaged for each material. Bone mimicking for scaffolding or direct replacement are recurring themes in the reported literature concerning macroporous bionanocomposites.[16, 49–52] Although the majority of such materials employ phosphates and carbonates as inorganic phase, siliceous materials such as silica and silicates have recently gained greater significance for tissue engineering applications.[14] Mimicking the architecture of bone is probably one of the most challenging tasks in materials science.[16] The association between collagen and hydroxyapatite in bone presents multiple levels of hierarchical structural organization resulting in two main types of material, cortical and cancellous bone.[53] Use of foamed bionanocomposites is aimed at mimicking the highest hierarchical order of cancellous (spongy zone) bone that can be characterized by open porosity, which allows the bone cellular communities to fully develop and migrate. Several strategies related to pore generation have been applied in this context to reproduce the architecture of bone. One of the most interesting routes when water-soluble polymers are used is freeze-drying because it permits fine-tuning of the pore typology by variation of the freezing conditions while guaranteeing the interconnectivity between pores. The possibilities for controlling the pore size and shape in bionanocomposite foams using freeze-drying techniques applied to aqueous mixtures of silicates and biopolymers have recently been disclosed by the porosity control observed in gelatin sepiolite foams.[54, 55] This technique has been applied in the preparation of chitosan–silicate hybrids with three-dimensional (3D) porous structures for tissue engineering scaffold applications.[56] The silicate component is formed in situ from biocompatible γ-glycidoxypropyltrimethoxysilane in the presence of chitosan and the mixture submitted to a freeze-drying process. Zhu and co-workers also reported the preparation of a bionanocomposite foam for bone scaffold applications based on freeze drying of a mixture of silk fibroin and silicate wollastonite.[57] The resulting three-dimensional foams prepared with 40 wt%

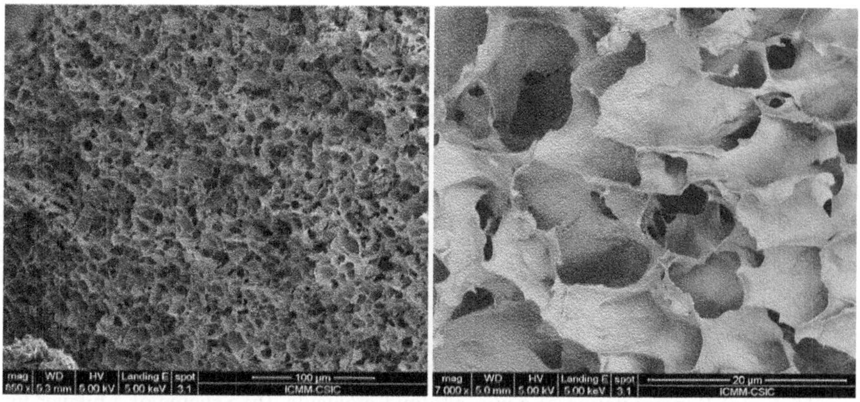

Figure 5.2 FE-SEM images of sepiolite-based bionanocomposite foams prepared by freeze-drying technique.

wollastonite presented enhanced fibroblast proliferation, cell adhesion, and elastic properties when compared with unmodified silk fibroin foams. Particulate leaching is another technique prone to produce bionanocomposites with controlled porosity. Other techniques for controlling the macroporosity in bionanocomposite materials are available including *rapid prototyping* or *solid freeform fabrication*[58]; however, these techniques have not been applied to silicate-based bionanocomposite foams. Figure 5.2 shows FE-SEM images of sepiolite-based bionanocomposite foams prepared by the freeze-drying technique.

Another interesting application of bionanocomposite foams is in the preparation of ultralightweight materials that combine excellent mechanical properties and low densities with biopolymers of natural origin.[43] Among these materials, layered silicate reinforced polylactide bionanocomposite foams are some of the most interesting due to their biodegradability.[59] These materials are prepared in a CO_2 pressure cell and subsequently foamed in a silicone oil bath and display a controlled closed cell morphology while the silicate layers are fully integrated in the pore walls.[60] Also, soy-based polyurethane foams filled with clays are of interest. Polyurethane materials usually rely on petroleum-based polyols as part of the polymerization foaming process. However, recent developments have led to the replacement of petroleum-based polyols with soy-based polyols.[61, 62] The inclusion of organically modified silicates in the soy-based polyurethane endows the thus prepared foams with thermomechanical performances superior to those of unfilled soy–polyurethane or even standard polyurethane foams.[63] This enhancement of the thermomechanical properties uses an eco-friendly polymer synthesis in the preparation of more effective material with numerous applications.

Surprisingly, although both bionanocomposite formulation and foam structuring techniques are well known, few examples have been reported. A broad field of research and applications can be envisaged for these materials including bone scaffolds, petro-based insulating materials (polyurethane and expanded polystyrene foams), and biocompatible containers for bioactive species and living organisms.

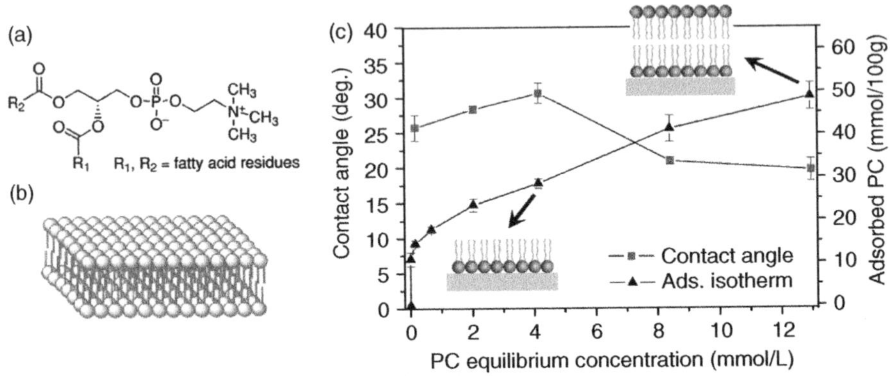

Figure 5.3 (a) Representation of a phospholipid molecule. (b) Scheme of a bilayer lipid membrane. (c) Phospholipid adsorption isotherm on sepiolite obtained from ethanol and water contact angle values.

5.4 BIOMIMETIC MEMBRANES

5.4.1 Phospholipid–Clay Membranes

Phospholipids are widely used in the preparation of biomimetic membranes.[64] The lipid phosphatidylcholine (PC) is the main constituent of cell membranes. It provides the structural framework and is crucial in cell metabolism and membrane signal tracking.[65] Two fatty acid chains are connected through a glycerol backbone to a zwitterionic head group leading to the amphiphilic nature of phosphatidylcholine (Figure 5.3a). The headgroup consists of an anionic phosphate group esterified with a cationic choline.[66]

As a result of its amphiphilicity, PC has the ability to form self-assembled structures such as micelles, tubes, liposomes, or supported artificial membranes[67] (Figure 5.3b). The self-assembly capacity of lipids has attracted considerable attention over the past decades from both fundamental and technological viewpoints and triggered the development of diverse lipidic systems.[68–70] For instance, freely suspended liposomes have gained attention as models for cell membranes in biochemical and physiological research,[64, 71] but also as biomimetic supports for pharmaceutical drug delivery,[72] gene delivery,[73] or immobilization of biological species such as membrane proteins.[74]

However, for many technological applications it is advantageous to immobilize the lipid membrane on a solid support. In the mid-1980s, the McConnell group[75, 76] introduced solid-supported lipid bilayers, which are well suited to serve as cell membrane mimics.[77–79] These initial works triggered the development of a variety of solid-supported membrane systems such as tethered lipid bilayers, polymer-cushioned lipid bilayers, supported vesicular layers, or hybrid bilayers.[80] The most common assembly methods are based on vesicle fusion[75] or LbL deposition using the Langmuir–Blodgett technique.[81]

The thus constructed supported lipid membranes find technological interest in biosensing and biocatalytic applications, for example, in the biomimetic association of bioactive species.[82–84] Another field of great importance is biocompatibilization of inorganic surfaces such as microfluid devices[85] and medical implants.[86]

The ionic head group makes PC prone to electrostatic interactions with silicate materials[87, 88] and silica,[89, 90] which provide a robust support and an "easy-to-handle" carrier system. Owing to their biointerface properties, lipid modified clay systems are promising materials for agricultural, clinical, and biotechnological applications. Examples may be sequestration of fungi produced mycotoxins[91, 92] or controlled herbicide release.[93]

Phospholipid molecules self-assemble on sepiolite clay in a controlled manner into mono- and bilayered structures, which allows tailoring of surface properties[92] (Figure 5.3c). Hydrogen bonding between PC headgroup moieties and silanol groups on the clay surface are the principal interaction mechanism.

On homoionic smectites, however, ion exchange is considered to be the controlling adsorption mechanism[94] providing the possibility to form intercalated lipid–clay compounds. Phospholipid adsorption from methanol on montmorillonite results in an expansion of the interlayer region by 4.2 nm that is compatible with a PC bilayer arrangement.[92] It could be demonstrated that beside the cation exchange mechanism water bridges are an important mechanism in the intercalation process.[92]

These clay–lipid biohybrids demonstrated good biocompatibility for the immobilization of enzymes and in their use as an active phase of a urea biosensor or cholesterol oxidase bioreactor.[95] In addition, these biomimetic membranes are currently being investigated as carrier for viral antigens (virus particles, hemagglutinin) in the preparation of influenza vaccines.[96] Preliminary results indicate that immobilization of these viral compounds on sepiolite-supported lipidic interfaces is advantageous in terms of biological activity, immunogenicity, and thermal stability as compared to aluminum hydroxide gel, which is frequently employed as commercial adjuvant in immunizations.

5.4.2 Polysaccharide–Clay Bionanocomposites as Support for Viruses

Certain bionanocomposite materials can provide a physicochemical environment that mimics the mucosal epithelial layer where some viruses naturally establish the first contact to initiate their inherent infection processes. The presence of negatively charged polysaccharides containing D-glucose units could favor the immobilization of viral particles of influenza. In this way, bionanocomposites built up by the assembly of xanthan gum and sepiolite are able to associate strongly with influenza viral particles, leading to very stable water dispersions suitable as intranasal or intramuscular vaccines which have been tested in mice.[97] Xanthan gum is a natural anionic polysaccharide consisting of a $\beta-(1\rightarrow 4)$-D-glucopyranose glucan backbone with side chains of D-mannopyranose derivative units and terminal mannose residues. The exposure of influenza virions (viral particles) to

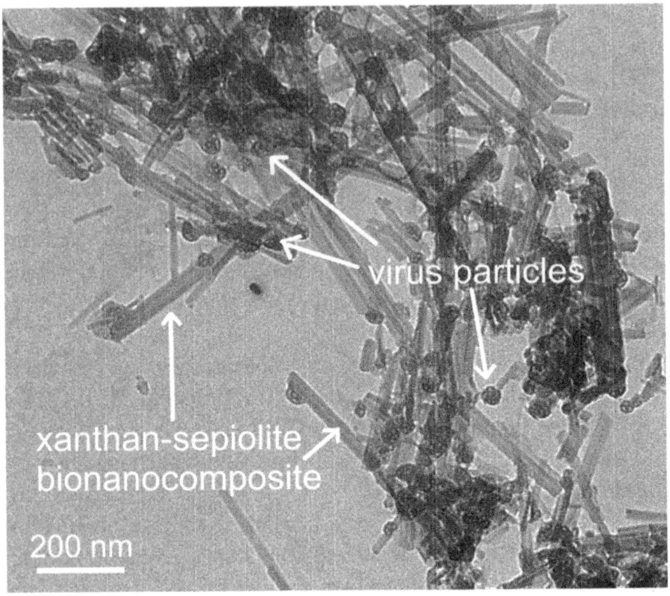

Figure 5.4 TEM image of influenza viral particles assembled with xanthan–sepiolite bionanocomposite. (Reproduced with permission. Copyright Wiley-VCH Verlag GmbH & Co. KGaA: Ref. 13.)

the xanthan–clay nanocomposites results in the assembly of both components leading to virus/biohybrid materials in which the virions are very homogeneously distributed on the bionanocomposites. Transmission electron microscopy (TEM) images of the immobilized viruses on the clay–bionanocomposite are shown in Figure 5.4 and reveal that the individual viral particles are well distributed within the solid bionanocomposite. Interestingly, the corresponding TEM images of the same pathogens not supported on the bionanocomposite show massive aggregates of viral particles.[13]

The integrity and bioactivity of the supported viral particles are preserved and allow novel biomedical applications of these biohybrids as an efficient and low-cost adjuvant for influenza vaccines and also potentially as biosensors for the rapid electrochemical detection of viral pathogens.[13, 97] In this way, experiments carried out in mice demonstrate that virus–clay bionanocomposites induce the formation of specific antibodies, thus providing effective protection against influenza virus. The most important feature is that these biohybrid materials provide the viruses with a physicochemical environment that may preserve native macromolecular conformation of the viral particle, thus maintaining its bioactivity. This new concept of biomimetic mucosa has been tested with the influenza virus and can be considered a model. However, it could be extended to other types of pathogens for developing new vaccines and biosensor devices.

5.5 HIERARCHICALLY LAYERED COMPOSITES

5.5.1 Layer-by-Layer Assembly of Composite-Cell Model

In many biological systems, structural hierarchies play important roles as seen in cell functions. Assembling nanocomposites into layered structures is one of the available methodologies for constructing bio-like hierarchical nanocomposites. Here, approaches to construct hierarchical layered nanocomposites with biomimetic functions based on layer-by-layer (LbL) assembly are introduced.

The LbL assembly[98–100] is mainly conducted through electrostatic interaction, although several modified methods have been proposed. The fundamental driving force of this technique is electrostatic interaction upon adsorption of charged materials on oppositely charged surfaces. An excess adsorption of the substances at a surface results in effective reversal of the surface charge through a number of ionic groups remaining at the film surface. Repeated alternating adsorption processes lead to an alternative change in the surface charge, resulting in a continuous assembly between positively and negatively charged materials. Film structure, including the number of layers and layering sequences, can freely be adjusted by controlling the adsorption process. In addition, this method can be used for many kinds of charged materials including polyelectrolytes, proteins, colloidal particles, and molecular assemblies. Therefore, this is a very powerful method for constructing bio-related layered composites. One of the most prominent advantages of the LbL assembly is the simplicity of the assembly procedure. The tools required for this procedure are just beakers and tweezers. This procedure can also be easily automated.

Katagiri and co-workers investigated the covalent linkage of a siloxane framework to a lipid bilayer vesicle and the sophisticated organization of multicellular systems. The resulting vesicles had a siloxane network covalently attached to the bilayer membrane surface and were named cerasome (ceramics + soma).[101, 102] Subjecting the cerasome structure to LbL techniques resulted in predesigned multicellular mimics. Figure 5.5 illustrates two modes of cerasome LbL assemblies reported by Katagiri and co-workers. LbL assemblies between cationic polyelectrolyte [poly(diallyldimethylammonium chloride), PDDA] and anionic vesicles have been investigated using a quartz crystal microbalance (QCM) that indicated successful LbL assembly according to the sensitive mass detection (based on QCM frequency shifts) of materials deposited on the surface.[103] Using both anionic and cationic cerasomes, direct LbL assembly of cerasome structures in the absence of polyelectrolyte counterions became possible.[104] TEM observation of the cerasomes revealed that cationic cerasomes are smaller in diameter (20–100 nm) than the anionic cerasomes (70–300 nm). The closely packed cerasome particles are like a stone pavement in both layers, as clearly confirmed by AFM observations of the surface of the assembled structures. This kind of assembly can be regarded as a multicellular mimic, and consequently it could be used as bioreactor or biosensor. Further functionalization of the cerasome surface using various biomolecules, such as enzymes and antibodies,

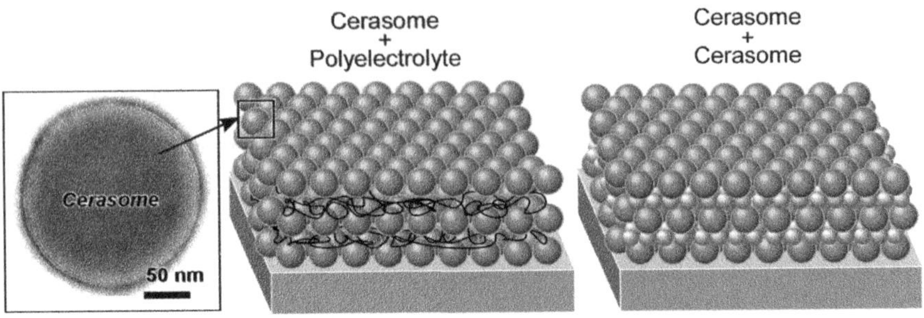

Figure 5.5 Cerasome and its LbL assemblies.

through covalent linkage indicates a great potential for creating various kinds of biomimetic silica nanohybrids.

5.5.2 Hierarchically Organized Nanocomposites for Sensor and Drug Delivery

Subjecting mesoporous materials to LbL technology can yield functional composites for sensor and drug delivery applications. Ariga and co-workers demonstrated preparation of a sensor device through LbL assembly of mesoporous carbon and polyelectrolytes.[105] Surface oxidation of carbon using ammonium persulfate enabled introduction of negative carboxylate groups onto mesoporous carbon, CMK-3, which was assembled with cationic polyelectrolytes in layer-by-layer mode (Figure 5.6). After a QCM sensor coated with an LbL film of mesoporous carbon was equilibrated in water, guests (tannic acids, catechin, and caffeine) were added to the water phase. Addition of tannic acid induced immediate and substantial reduction of frequency, indicating a strong adsorption of tannic acid onto the mesoporous carbon LbL film coated on the QCM. The frequency shifts upon adsorption of tannic acid to the mesoporous carbon LbL films greatly exceeds that for catechin and caffeine. The resulting sensitivity ratios of tannic acid to catechin or caffeine are 3.9 and 13.6, respectively. The selective adsorption

Figure 5.6 LbL assembly of mesoporous carbon CMK-3 on QCM sensor.

capacity for tannic acid likely originates from strong interactions between carbon and tannic acid in CMK-3 pores probably through $\pi-\pi$ interactions and hydrophobic effects. The adsorption quantities of tannic acid at equilibrium exhibited a sigmoidal profile at low concentrations, suggesting cooperative adsorption of tannic acid to CMK-3 pores. The highly cooperative adsorption might result from confinement effects during adsorption. The latter effect can be explained by enhanced guest–guest interaction within motion-restricted nanospaces.

A similar hierarchical structure with different function was also fabricated through the LbL assembly of mesoporous carbon capsules (Figure 5.7). [106] Tuning of the guest selectivity of the carbon capsule film became possible by impregnation of additional components, resulting in designable selectivity of volatile materials adsorption. In order to provide charges at the surface of mesoporous carbon capsules, the carbon capsules were coated with surfactant and dispersed in water. The surfactant-stabilized carbon capsules were then deposited alternately with counterionic polyelectrolytes, resulting in LbL films assembled on QCM electrodes. The thus-prepared films were exposed to vapors of various chemicals to check adsorption to the LbL film. Of the functional group-bearing guests, the carbon capsules have large affinities for aromatic guests such as aniline and pyridine. This selectivity could easily be tuned by impregnation with additional recognition components. Impregnation with lauric acid led to the greatest affinities for nonaromatic amines, where selectivity between nonaromatic amines and aromatic amines was completely reversed, probably reflecting the different basicity of these amines. On the other hand, impregnation of dodecylamine into the carbon capsule films resulted in a strong preference for acetic acid. The fabricated hierarchical LbL films will find widespread applications as sensors or filters because of their designable guest selectivity.

Ariga and co-workers further demonstrated use of molecularly layered graphene-sheet/ionic-liquid (GS-IL) composites on quartz crystal microbalances (QCM) for selective gas sensing.[107] This designed nanospace formed between sp^2-carbon nanosheets has a higher affinity for toxic aromatic hydrocarbons over their aliphatic analogs. Graphene oxide sheet (GOS) was prepared by oxidization of graphite under acidic conditions, followed by its reduction to graphene sheet (GS) in the presence

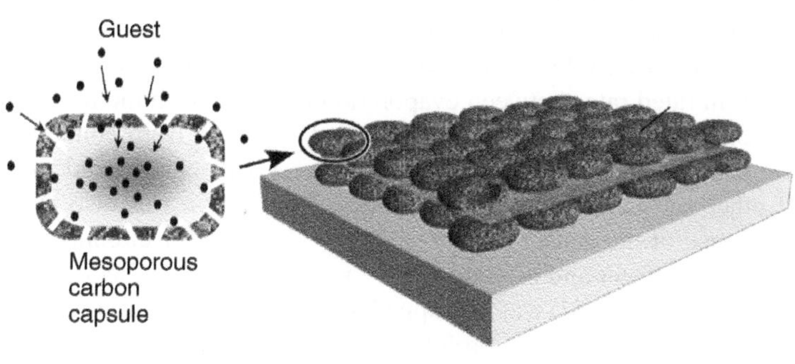

Figure 5.7 LbL assembly of mesoporous carbon capsules for gas sensing.

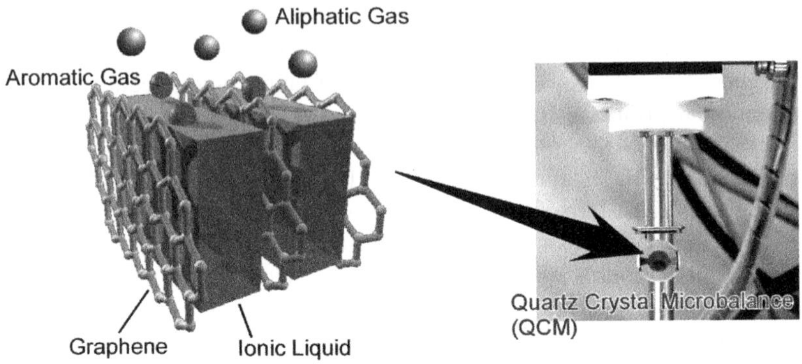

Figure 5.8 LbL assembly of graphene and ionic liquid for selective sensors for aromatic guest.

of ionic liquids in water. Composites of graphene sheet/ionic liquid (GS-IL) behave as charge-decorated nanosheets and were assembled alternately with poly(sodium styrenesulfonate) (PSS) by electrostatic LbL adsorption on appropriate solid supports to provide layered assemblies of GS-IL composite with PSS on the surface of a QCM resonator (Figure 5.8). Exposure of the composite films to various saturated vapors, after equilibration under an ambient atmosphere, caused an in situ decrease in frequency of QCM due to gas adsorption. These composite films have superior affinities for aromatic over aliphatic compounds. This behavior is a striking indication of the highly selective detection of aromatic guests within the well-defined π-electron-rich nanospace in the GS-IL films. The GS-IL films have a variety of potential practical applications including environment remediation through the capture of atmospheric CO_2. Adsorption of CO_2 vapors from a saturated sodium hydrocarbonate solution into the GS-IL films showed enhanced adsorption volume compared to the GS films without intercalated ionic liquids.

In another approach, mesoporous silica capsules were co-assembled with silica particles by the LbL technique with the aid of appropriate polyelectrolytes, resulting in mesoporous nanocompartment films.[108, 109] The mesoporous nanocompartment films possess special molecular encapsulation and release capabilities so that stimuli-free automodulated stepwise release of water or drug molecules was achieved through the mesopore channels of robust silica capsule containers embedded in the film. Stepwise release of water was reproducibly observed that originates in the nonequilibrated rates between evaporation of water from the mesopore channels to the exterior and the capillary penetration of water from container interior to the mesopore channels (Figure 5.9). This process was generalized to the evaporation of other substances such as fragrances. Application was also tested in the controlled release of the sunscreen UV-absorber (UV-S1) for circumvention of its rapid dissolution in water and prolongation of its prophylactic effect toward harmful ultraviolet radiation. UV-S1 was entrapped within the mesoporous nanocompartment films and was released in a prolonged stepwise mode. The nanocompartment

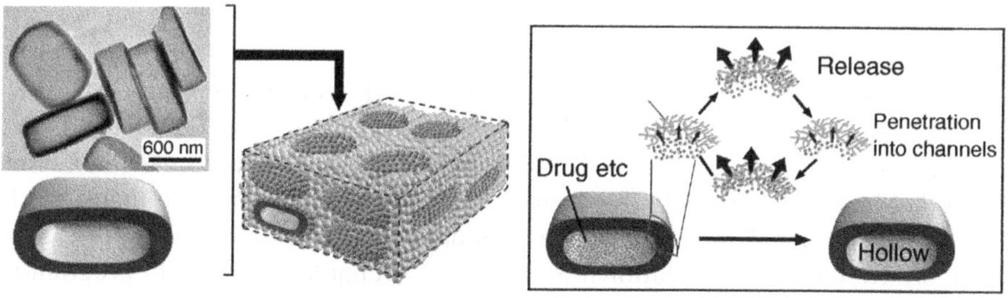

Figure 5.9 LbL films of mesoporous silica capsules with nanoparticles for drug delivery applications.

films developed in this research are promising materials for drug delivery since they allow gradual release of therapeutic agents with likely related improvements in their efficacy.

5.6 CONCLUSION

In summary, the strategies developed to prepare functional and structural materials from simple (and often natural) building blocks are strongly related to the assembly processes found in Nature. Although the materials referred to are, on occasion, not composed of biological entities, they reflect the change of paradigm that occurred in materials science by the introduction of concepts such as biomimetics[110] and self-assembly.[111]

This chapter focuses on some of the most relevant hybrid materials prepared using these biomimetic approaches, from bone-like materials to cerasome LbL assemblies. Most of the examples refer to silica- and silicate-based materials. However, these approaches may also be applied to other types of hybrid materials where natural architectures can serve to promote new ideas to assemble different building blocks, yielding functionality and preserving the natural structure of the mimicked motif. An illustrative example is that of cerasome LbL assemblies that replicate the architecture of nacre. These materials find application in drug delivery systems and sensors.[103]

This incipient field of research allows for developing complex systems that may eventually emulate or support biological functionality. Recent progress concerning these complex systems is the development of artificial mucosae based on the assembly between microfibrous clay particles and polysaccharides.[97]

As seen throughout this chapter, natural processes and architectures define the path to produce optimized materials, ready to cope with extreme service requirements, as in Nature. This field has witnessed several important contributions in the understanding of the underlying processes controlling natural self-assembly. Application of this knowledge has led to the preparation of bioinspired materials from different building block units. However, research in this field is still far from catching up with Nature's perfection and complexity.

ACKNOWLEDGMENTS

The authors are grateful for funding from CICYT (Spain) (project MAT2009-09960) and from CSIC-Academie Hassan II (project 2010 MA0003). FMF and BW acknowledge the Ministerio de Educación y Ciencia (Spain) and the Comunidad de Madrid, respectively, for their graduate fellowships. We also thank World Premier International Research Center Initiative (WPI Initiative), MEXT, Japan and Core Research for Evolutional Science and Technology (CREST) program of Japan Science and Technology Agency (JST), Japan.

REFERENCES

1. Ruiz-Hitzky, E.; Darder, M.; Aranda, P. *J. Mater. Chem.* **2005**, *15,* 3650.
2. Darder, M.; Aranda, P.; Ruiz-Hitzky, E. *Adv. Mater.* **2007**, *19,* 1309.
3. Fratzl, P.; Weinkamer, R. *Prog. Mater. Sci.* **2007**, *52,* 1263.
4. Ruiz-Hitzky, E.; Darder, M.; Aranda, P. In *Bio-inorganic Hybrid Nanomaterials* (Eds. Ruiz-Hitzky, E.; Ariga, K.; Lvov, Y.), Wiley-VCH, Weinheim, Germany, 2008.
5. Antonietti, M.; Fratzl, P. *Macromol. Chem. Phys.* **2010**, *211,* 166.
6. Zhao, F.; Yin, Y.J.; Lu, W. W.; Leong, J. C.; Zhang, W. J.; Zhang, J. Y.; Zhang, M. F.; Yao, K. D. *Biomaterials* **2002**, *23,* 3227.
7. Dorozhkin, S. V. *J. Mater. Sci.* **2009**, *44,* 2343.
8. Liu, X.; Smith, L. A.; Hu, J.; Ma, P. X. *Biomaterials* **2009**, *30,* 2252.
9. Bigi, A.; Torricelli, P.; Fini, M.; Bracci, B.; Panzavolta, S.; Sturba, L.; Giardino, R. *Int. J. Artif. Organs* **2004**, *27,* 664.
10. Livage, J.; Coradin, T.; Roux, C. In *Functional Hybrid Materials* (Eds. Gomez-Romero, P.; Sanchez, C.), Wiley-VCH, Weinheim, Germany, 2004, p. 387.
11. Coradin, T.; Allouche, J.; Boissiere, M.; Livage, J. *Curr. Nanosci.* **2006**, *2,* 219.
12. Avnir, D.; Coradin, T.; Lev, O.; Livage, J. *J. Mater. Chem.* **2006**, *16,* 1013.
13. Ruiz-Hitzky, E.; Darder, M.; Aranda, P.; Ariga, K. *Adv. Mater.* **2010**, *22,* 323.
14. Ruiz-Hitzky, E.; Aranda, P.; Darder, M.; Ogawa, M. *Chem. Soc. Rev.* **2011**, *40,* 801.
15. Sanchez, C.; Arribart, H.; Guille, M. *Nat. Mater.* **2005**, *4,* 277.
16. Vallet-Regi, M.; Colilla, M.; Gonzalez, B. *Chem. Soc. Rev.* **2011**, *40,* 596.
17. Gautier, C.; Lopez, P. J.; Livage, J.; Coradin, T. *J. Colloid Interface Sci.* **2007**, *309,* 44.
18. Allouche, J.; Boissiere, M.; Helary, C.; Livage, J.; Coradin, T. *J. Mater. Chem.* **2006**, *16,* 3120.
19. Watzke, H. J.; Dieschbourg, C. *Adv. Colloid Interface Sci.* **1994**, *50,* 1.
20. Shchipunov, Y.; Shipunova, N. *Colloid Surf. B: Biointerface* **2008**, *63,* 7.
21. Estroff, L. A.; Hamilton, A. D. *Chem. Mater.* **2001**, *13,* 3227.

22. Luckarift, H. R.; Spain, J. C.; Naik, R. R.; Stone, M. O. *Nat. Biotechnol*. **2004**, *22,* 211.

23. Vickery, J. L.; Thachepan, S.; Patil, A. J.; Mann, S. *Mol. Biosys*. **2009**, *5,* 744.

24. Thomas, V.; Dean, D. R.; Vohra, Y. K. *Curr. Nanosci*. **2006**, *2,* 155.

25. Rezwan, K.; Chen, Q. Z.; Blaker, J. J.; Boccaccini, A. R. *Biomaterials* **2006**, *27,* 3413.

26. Fernandes, F. M.; Darder, M.; Ruiz, A. I.; Aranda, P.; Ruiz-Hitzky, E. In *Nanocomposites with Biodegradable Polymers. Synthesis, Properties, and Future Perspectives* (Ed. Mittal, V.), Oxford University Press, New York, 2011.

27. Wang, Y.; Angelatos, A. S.; Caruso, F. *Chem. Mater*. **2008**, *20,* 848.

28. Tang, Z. Y.; Kotov, N. A.; Magonov, S.; Ozturk, B. *Nat. Mater*. **2003**, *2,* 413.

29. Aksay, I. A.; Trau, M.; Manne, S.; Honma, I.; Yao, N.; Zhou, L.; Fenter, P.; Eisenberger, P. M.; Gruner, S. M. *Science* **1996**, *273,* 892.

30. Schaffer, T. E.; IonescuZanetti, C.; Proksch, R.; Fritz, M.; Walters, D. A.; Almqvist, N.; Zaremba, C. M.; Belcher, A. M.; Smith, B. L.; Stucky, G. D.; Morse, D. E.; Hansma, P. K. *Chem. Mater*. **1997**, *9,* 1731.

31. Sellinger, A.; Weiss, P. M.; Nguyen, A.; Lu, Y. F.; Assink, R. A.; Gong, W. L.; Brinker, C. J. *Nature* **1998**, *394,* 256.

32. Ruiz-Hitzky, E.; Aranda, P.; Darder, M. Bionanocomposites. In *Kirk–Othmer Encyclopedia of Chemical Technology*, John Wiley & Sons, Hoboken, NJ, 2008, p. 1.

33. Ruiz-Hitzky, E.; Darder, M.; Aranda, P. In *Annual Review of Nano Research* (Eds. Cao, G.; Zhang, Q.; Brinker, C. J.), World Scientific Publishing, Singapore, 2010, Vol. 3, p. 149.

34. Darder, M.; Colilla, M.; Ruiz-Hitzky, E. *Chem. Mater*. **2003**, *15,* 3774.

35. Darder, M.; Colilla, M.; Ruiz-Hitzky, E. *Appl. Clay Sci*. **2005**, *28,* 199.

36. Wang, S. F.; Shen, L.; Tong, Y. J.; Chen, L.; Phang, I. Y.; Lim, P. Q.; Liu, T. X. *Polym. Degrad. Stabil*. **2005**, *90,* 123.

37. Darder, M.; Aranda, P.; Ruiz, A. I.; Fernandes, F. M.; Ruiz-Hitzky, E. *Mater. Sci. Technol*. **2008**, *24,* 1100.

38. Podsiadlo, P.; Paternel, S.; Rouillard, J. M.; Zhang, Z. F.; Lee, J.; Lee, J. W.; Gulari, L.; Kotov, N. A. *Langmuir* **2005**, *21,* 11915.

39. Podsiadlo, P.; Kaushik, A. K.; Shim, B. S.; Agarwal, A.; Tang, Z. Y.; Waas, A. M.; Arruda, E. M.; Kotov, N. A. *J. Phys. Chem. B* **2008**, *112,* 14359.

40. Ebina, T.; Mizukami, F. *Adv. Mater*. **2007**, *19,* 2450.

41. Vertlib, V.; Dietiker, M.; Plotze, M.; Yezek, L.; Spolenak, R.; Puzrin, A. M. *J. Mater. Res*. **2008**, *23,* 1026.

42. Yao, H. B.; Tan, Z. H.; Fang, H. Y.; Yu, S. H. *Angew. Chem. Int. Ed*. **2010**, *49,* 10127.

43. Ruiz-Hitzky, E.; Aranda, P.; Darder, M.; Fernandes, F. M.; Matos, R. **2010**, *WO* 2010081918, OEPM, CSIC, Spain.

44. Istrate, O. M.; Chen, B. *Soft Matter* **2011**, *7,* 1840.

45. Darder, M.; Aranda, P.; Ferrer, M. L.; Gutiérrez, M. C.; del Monte, R.; Ruiz-Hitzky, E. *Adv. Mater*. **2011**, *23,* 5262.

46. Ashby, M. F. *Acta Metall*. **1989**, *37,* 1273.

47. Ashby, M. F.; Gibson, L. J.; Wegst, U.; Olive, R. *Proc. Math. Phys. Sci*. **1995**, *450,* 123.

48. Gibson, L. J.; Ashby, M. F.; Karam, G. N.; Wegst, U.; Shercliff, H. R. *Proc. Math. Phys. Sci*. **1995**, *450,* 141.

49. van der Pol, U.; Mathieu, L.; Zeiter, S.; Bourban, P. E.; Zambelli, P. Y.; Pearce, S. G.; Bouré, L. P.; Pioletti, D. P. *Acta Biomater*. **2010**, *6,* 3755.

50. Hule, R. A.; Pochan, D. J. *MRS Bull*. **2007**, *32,* 354.

51. Manjubala, I.; Woesz, A.; Pilz, C.; Rumpler, M.; Fratzl-Zelman, N.; Roschger, P.; Stampfl, J.; Fratzl, P. *J. Mater. Sci. Mater. Med*. **2005**, *16,* 1111.

52. Taboas, J. M.; Maddox, R. D.; Krebsbach, P. H.; Hollister, S. J. *Biomaterials* **2003**, *24,* 181.

53. Rho, J.-Y.; Kuhn-Spearing, L.; Zioupos, P. *Med. Eng. Phys*. **1998**, *20,* 92.

54. Fernandes, F. M. PhD Thesis, Universidad Autónoma de Madrid, 2011.

55. Fernandes, F. M.; Ruiz-Hitzky, E. In preparation.

56. Shirosaki, Y.; Okayama, T.; Tsuru, K.; Hayakawa, S.; Osaka, A. *Chem. Eng. J*. **2008**, *137,* 122.

57. Zhu, H.; Shen, J.; Feng, X.; Zhang, H.; Guo, Y.; Chen, J. *Mater. Sci. Eng. C* **2010**, *30,* 132.

58. Salgado, A. J.; Coutinho, O. P.; Reis, R. L. *Macromol. Biosci*. **2004**, *4,* 743.

59. Ray, S. S.; Okamoto, M. *Macromol. Rapid Commun*. **2003**, *24,* 815.

60. Fujimoto, Y.; Ray, S. S.; Okamoto, M.; Ogami, A.; Yamada, K.; Ueda, K. *Macromol. Rapid Commun*. **2003**, *24,* 457.

61. Petrovic, Z. S.; Guo, A.; Zhang, W. *J. Polym. Sci. Polym. Chem*. **2000**, *38,* 4062.

62. Guo, A.; Zhang, W.; Petrovic, Z. S. *J. Mater. Sci*. **2006**, *41,* 4914.

63. Liang, H.-W.; Wang, L.; Chen, P.-Y.; Lin, H.-T.; Chen, L.-F.; He, D.; Yu, S.-H. *Adv. Mater*. **2010**, *22,* 4691.

64. Peetla, C.; Stine, A.; Labhasetwar, V. *Mol. Pharmaceutics* **2009**, *6,* 1264.

65. van Meer, G. *Nat. Rev. Mol. Cell Biol*. **2008**, *9,* 112.

66. Menger, F. M.; Chlebowski, M. E.; Galloway, A. L.; Lu, H.; Seredyuk, V. A.; Sorrells, J. L.; Zhang, H. L. *Langmuir* **2005**, *21,* 10336.

67. Zidovska, A.; Ewert, K. K.; Quispe, J.; Carragher, B.; Potter, C. S.; Safinya, C. R.; Nejat, D. In *Methods in Enzymology*, Academic Press, Waltham, MA, 2009, Vol. 465, p. 111.

68. Reimhult, E. *Biotechnol. Genet. Eng*. **2010**, *27,* 185.

69. Bally, M.; Bailey, K.; Sugihara, K.; Grieshaber, D.; Vörös, J.; Städler, B. *Small* **2010**, *6,* 2481.

70. Chemburu, S.; Fenton, K.; Lopez, G. P.; Zeineldin, R. *Molecules* **2010**, *15,* 1932.

71. Eytan, G. D. *BBA Rev. Biomembranes* **1982**, *694,* 185.

72. Puri, A.; Loomis, K.; Smith, B.; Lee, J. H.; Yavlovich, A.; Heldman, E.; Blumenthal, R. *Crit. Rev. Ther. Drug* **2009**, *26,* 523.

73. Zhang, S.; Zhao, Y.; Zhao, B.; Wang, B. *Bioconjugate Chem*. **2010**, *21,* 1003.

74. Steinem, C.; Janshoff, A. *Curr. Opin. Colloid Interface Sci*. **2010**, *15,* 479.

75. Brian, A. A.; McConnell, H. M. *Proc. Natl. Acad. Sci. Biol*. **1984**, *81,* 6159.

76. Tamm, L. K.; McConnell, H. M. *Biophys. J*. **1985**, *47,* 105.

77. McConnell, H. M.; Watts, T. H.; Weis, R. M.; Brian, A. A. *Biochim. Biophys. Acta* **1986**, *864,* 95.

78. Kam, L.; Boxer, S. G. *Langmuir* **2003**, *19,* 1624.

79. Sackmann, E. *Science* **1996**, *271,* 43.

80. Richter, R. P.; Bérat, R.; Brisson, A. R. *Langmuir* **2006**, *22,* 3497.

81. Petty, M. A. *Langmuir–Blodgett Films. An Introduction*, Cambridge University Press, New York, 1996.

82. Ramsden, J. J. *Biosens. Bioelectron*. **1998**, *13,* 593.

83. Nikolelis, D. P.; Hianik, T.; Nikoleli, G.-P. *Electroanalysis* **2010**, *22,* 2747.

84. Girard-Egrot, A. P.; Godoy, S.; Blum, L. J. *Adv. Colloid Interface Sci*. **2005**, *116,* 205.

85. Yang, T. L.; Jung, S. Y.; Mao, H. B.; Cremer, P. S. *Anal. Chem*. **2001**, *73,* 165.

86. Gong, M.; Yang, S.; Zhang, S. P.; Gong, Y. K. *Prog. Chem*. **2008**, *20,* 1628.

87. Rapuano, R.; Carmona-Ribeiro, A. M. *J. Colloid Interface Sci*. **1997**, *193,* 104.

88. Wiegart, L.; O'Flaherty, S. M.; Struth, B. *Langmuir* **2005**, *21,* 1695.

89. Anderson, T. H.; Min, Y. J.; Weirich, K. L.; Zeng, H. B.; Fygenson, D.; Israelachvili, J. N. *Langmuir* **2009**, *25,* 6997.

90. Bayerl, T. M. *Nature* **2004**, *427,* 105.

91. Wicklein, B.; Darder, M.; Aranda, P.; Ruiz-Hitzky, E. *Macla* **2008**, *9,* 257.

92. Wicklein, B.; Darder, M.; Aranda, P.; Ruiz-Hitzky, E. *Langmuir* **2010**, *26,* 5217.

93. Sanchez-Verdejo, T.; Undabeytia, T.; Nir, S.; Villaverde, J.; Maqueda, C.; Morillo, E. *J. Agric. Food Chem*. **2008**, *56,* 10192.

94. Ruiz-Hitzky, E. In *Functional Hybrid Materials* (Eds. Gómez-Romero, P.; Sanchez, C.) Wiley-VCH, Weinheim, Germany, 2004.

95. Wicklein, B.; Darder, M.; Aranda, P.; Ruiz-Hitzky, E. *ACS Appl. Mater. Interfaces* **2011**, *3*, 4339.

96. Wicklein, B.; Darder, M.; Aranda, P.; Martín del Burgo, M. Á.; Yuste, M.; del Real, G.; Ortín, J.; Ruiz-Hitzky, E. In preparation.

97. Ruiz-Hitzky, E.; Darder, M.; Aranda, P.; Martin del Burgo, M. Á.; del Real, G. *Adv. Mater*. **2009**, *21,* 4167.

98. Ariga, K.; Hill, P.; Ji, Q. *Phys. Chem. Chem. Phys*. **2007**, *9,* 2319.

99. Ariga, K.; Hill, J. P.; Ji, Q. *Macromol. Biosci*. **2008**, *8,* 981.

100. Ariga, K.; Hill, J. P.; Lee, M. V.; Vinu, A.; Charvet, R.; Acharya, S. *Sci. Technol. Adv. Mater*. **2008**, *9,* 014109.

101. Katagiri, K.; Ariga, K.; Kikuchi, J. *Chem. Lett*. **1999**, 661.

102. Katagiri, K.; Hashizume, M.; Ariga, K.; Terashima, T.; Kikuch, J. *Chem. Eur. J*. **2007**, *13,* 5272.

103. Katagiri, K.; Hamasaki, R.; Ariga, K.; Kikuchi, J. *Langmuir* **2002**, *18,* 6709.

104. Katagiri, K.; Hamasaki, R.; Ariga, K.; Kikuchi, J. *J. Am. Chem. Soc.* **2002**, *124,* 7892.

105. Ariga, K.; Vinu, A.; Ji, Q.; Ohmori, O.; Hill, J. P.; Acharya, S.; Koike, J.; Shiratori, S. *Angew. Chem. Int. Ed.* **2008**, *47,* 7254.

106. Ji, Q.; Yoon, S. B.; Hill, J. P.; Vinu, A.; Yu, J.-S.; Ariga, K. *J. Am. Chem. Soc.* **2009**, *131,* 4220.

107. Datta, K. K. R.; Subba Reddy, B. V.; Ariga, K.; Vinu, A. *Angew. Chem. Int. Ed.* **2010**, *49,* 5961.

108. Ji, Q.; Miyahara, M.; Hill, J. P.; Acharya, S.; Vinu, A.; Yoon, S. B.; Yu, J.-S.; Sakamoto, K.; Ariga, K. *J. Am. Chem. Soc.* **2008**, *130,* 2376.

109. Ji, Q.; Acharya, S.; Hill, J. P.; Vinu, A.; Yoon, S. B.; Yu, J.-S.; Sakamoto, K.; Ariga, K. *Adv. Funct. Mater.* **2009**, *19,* 1792.

110. Lowenstam, H. A. *Science* **1981**, *211,* 1126–1131.

111. Lehn, J.-M. *Angew. Chem. Int. Ed.* **1988**, *27,* 89–112.

CHAPTER 6

Bioinspired Materials Chemistry II: Biomineralization as Inspiration for Materials Chemistry

FABIO NUDELMAN and NICO A. J. M. SOMMERDIJK

Laboratory of Materials and Interface Chemistry and Soft Matter CryoTEM Unit, Eindhoven University of Technology, 5600 MB, Eindhoven, The Netherlands

6.1 INSPIRATION FROM NATURE

Living organisms are well known to exploit the material properties of amorphous and crystalline minerals in building a wide range of organic–inorganic hybrid materials for a variety of purposes, such as navigation, mechanical support, protection of the soft parts of the body, and optical photonic effects. The high level of control over the composition, structure, size, and morphology of biominerals results in materials of amazing complexity and fascinating properties that strongly contrast with those of geological minerals and often surpass those of synthetic analogs.[1a] It is no surprise, then, that biominerals have intrigued scientists for many decades and served as a source of inspiration in the development of materials with highly controllable and specialized properties. Indeed, by looking at examples from the biological world, one can see how organisms are capable of manipulating mineral formation as to produce materials that are tailor-made for their needs.

Many organisms, animals, and lower and higher plants form remarkable structures from amorphous silica. Diatoms and radiolarians are the most important biosilicifying organisms which have microskeletons and porous shells.[1b] Diatoms are unicellular algae that are abundant in fresh water and in marine environments, and most of the biosilica formation in the oceans is governed by these species. These algae have cell walls that are made of silica, with very complex morphologies on the nano- and microscale, thus representing the most remarkable example of nano- and microfabrication and patterning (Figure 6.1).[2] What is also interesting in this organism is that the silica structures are species specific, which points to genetic control over not only the formation of the walls themselves but also their

Bioinspiration and Biomimicry in Chemistry: Reverse-Engineering Nature, First Edition.
Edited by Gerhard F. Swiegers.
© 2012 John Wiley & Sons, Inc. Published 2012 by John Wiley & Sons, Inc.

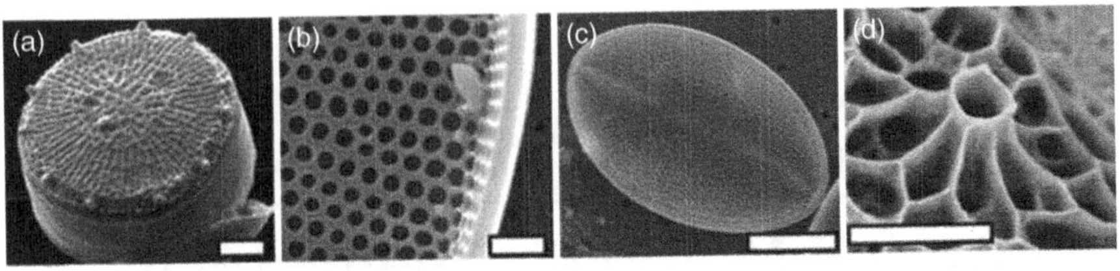

Figure 6.1 Diversity of silica structures from different species of diatoms. (a) *Thalassiora pseudonana*, scale bar: 1 μm. (b) *Coscinodiscus walensii*, sacle bar: 5 μm. (c) *Cocconeis species*, scale bar: 10 μm. (d) Rimoportula from *Thalassiosira weisflogii*, scale bar: 500 nm. (Adapted with permission. Copyright the American Chemical Society: Ref. 2.)

particular pattern. To date, synthesis of diatom-like silica structures has not yet been reproduced in the laboratory.

Bioinorganic iron oxides are formed by a variety of organisms and serve a broad range of functions, such as iron storage, sensing of magnetic fields, strengthing of the tissues, and hardening of teeth.[1c] Some of the well-known and fascinating forms of iron biomineral are the magnetic nanoparticles composed of magnetite or greigite that are found in magnetotactic bacteria (Figure 6.2).[3] The crystals are arranged into an intracellular chain of discrete crystals, where each crystal in the

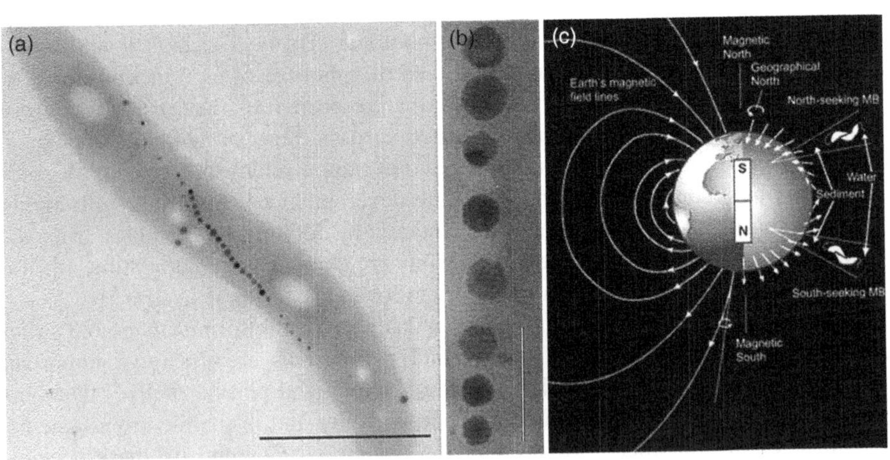

Figure 6.2 (a) Transmission electron microscopy image of a spirilium with a single chain of cuboctahedral magnetosomes. Scale bar, 1 μm. (b) Chain of magnetite crystals from a similar type of the magnetotatic bacteria shown in (a). Scale bar, 100 nm. (c) The intracellular magnetic dipoles of the magnetotatic bacteria allow the cells to align with the geomagnetic field lines while swimming. Due to the inclination of Earth's magnetic field (white arrows), north-seeking bacteria are present in the Northern Hemisphere and swim toward low oxygen concentrations. South-seeking bacteria in the Southern Hemisphere swim in the opposite direction to fulfill the same goal. (Adapted with permission. Copyright the American Chemical Society: Ref. 3.)

chain is located inside a specialized compartment called a magnetosome. It is inside this intracellular compartment that the magnetite crystals form and align in well-ordered chains. Typically, each crystal is 30–140 nm in size, within the single magnetic domain size range where the particles are highly efficient as a permanent magnetic carrier. It has been statistically demonstrated that magnetosomes have crystal size distributions that are narrow, asymmetrical, and negatively skewed with sharp cutoffs toward larger size and with shape factor consistent for a given strain.[3] Thus, the chain of crystals enables the organism to align itself along the Earth's magnetic field, functioning as a navigational device for the alignment along chemical gradients in aquatic habitats.

Calcium carbonate-based biominerals are the most abundant biogenic minerals found in Nature. They are especially present in fresh water and marine organisms, such as sea urchins, sea shells, sponges, and crustaceans but also form gravity sensors in marine and land animals.[1a] Calcium carbonate ($CaCO_3$) can occur in the form of three anhydrous crystalline polymorphs—vaterite, aragonite, and calcite; and three hydrated forms—amorphous calcium carbonate (ACC), calcium carbonate monohydrate, and calcium carbonate hexahydrate. Of these polymorphs, calcite and aragonite are the most thermodynamically stable forms,[4] comprising the majority of calcium carbonate biominerals.

Various sea shell types contain both *calcite* and *aragonite* as a hard part of the mollusk (Figure 6.3). Generally, the outer prismatic layer of the shell consists of calcite, and the inner part, nacre, is in the form of plate-like aragonite crystals.[5] It is noteworthy, however, that inorganically formed calcite cleaves easily along the {104} plane and therefore is not very suitable material for protection of the soft parts of the organism.[6] Yet, mollusks are not alone in employing calcite. Sea urchins, for example, have spicules that are composed of single crystals of calcite that are several millimeters long and exhibit very smooth surfaces, not corresponding to the well-defined rhombohedral morphology of calcite crystals.[6] In order to enhance the mechanical properties of the calcite crystals and to decrease their brittleness, proteins are occluded inside the crystal, preferentially in the planes parallel to the c-axis, causing dislocations on the planes that are oblique to the [104] cleavage planes.[6–8] The result is an efficient crack deviation mechanism, such that the spines cleave conchoidally, as if they were composed of glassy materials. This design strategy essentially introduces anisotropic fracture behavior into a material that is still highly anisotropic at the atomic level. Thus, mimicking this approach of occluding polymers inside crystals may well have applications in the field of materials fabrication. Furthermore, by using amorphous calcium carbonate as a precursor to calcite, organisms are able to shape the crystals into nearly any desirable morphology.[9] As for the aragonitic nacreous layer of sea shells, much of its mechanical strength derives from its superstructure, where plate-shaped aragonite crystals ~500 nm thick are arranged into parallel layers that are separated by a sheet of organic matrix.[10–13] This arrangement and combination of organic–inorganic materials makes the nacre 3000 times tougher than pure inorganic aragonite.

Calcite can also be employed by animals for other purposes besides structural support and protection. For example, brittlestars use single calcite crystals not

Figure 6.3 (a) Scanning electron microscopy image of a cross section of the shell of *Atrina rigida*, showing the nacreous (white star) and prismatic (white circle) layers. Scale bar: 10 μm. (b) Scanning electron micrograph of a fracture surface of the cross section of the nacreous layer. Scale bar: 1 μm. (c) Scanning electron micrograph of the surface of the prismatic layer. Scale bar: 50 μm. (Adapted with permission. Copyright the Royal Society of Chemistry: Ref. 14.)

only for skeletal construction but also for specialized photosensory organs.[15] The labyrinthic calcitic skeleton has a regular array of spherical microstructures that have a characteristic double-lens design. These microlenses are optical elements that guide and focus the light inside the tissue. The lens array senses the light from a specific direction; it is generated in order to reduce the spherical aberration and birefringence. Thus, these animals show photosensitivity from a largely light-indifferent behavior to pronounced color change and rapid escape behavior. They can sense predators at a distance by detecting their shadows and quickly escape into dark crevices.

Calcium phosphate minerals are known to exist in various compositions, and the most well-known examples are found in vertebrate bone and teeth. Bone has unique mechanical properties, defined by its chemical composition and structural organization (Figure 6.4).[16] It is a nanocomposite composed primarily of collagen type I fibrous matrix that is a scaffold and template, within which carbonated apatite crystals are embedded.[17, 18] The major content of the noncollagenous organic part consists of highly acidic proteins, which are important to control apatite formation inside the collagen.[19] One of the most interesting characteristics of bone is its hierarchical structure, going from the nanometer to the macroscopic scale (Figure 6.5).[16] As such, while all types of bone have the same building block (the mineralized collagen fibril), arrays of fibrils can be organized in different patterns, generating a structural diversity that is optimized to functional need. Typical examples are the woven bone, where fibrils are loosely packed and poorly oriented; the rotated plywood structure that is common to lamellar bone; arrays of parallel fibers, found in mineralized tendons; and radial fibril arrays, as found in dentin.[16] All these different arrangements will lead to structures with different mechanical properties.

When looking at the above examples of the level of sophistication found in biominerals in terms of their adaptation to function, it is no wonder that many efforts in the fields of chemistry, physics, and materials science have been made in order to mimic these inspiring structures and their properties. However, controlling the structure and the morphology of these organic–inorganic composite materials is

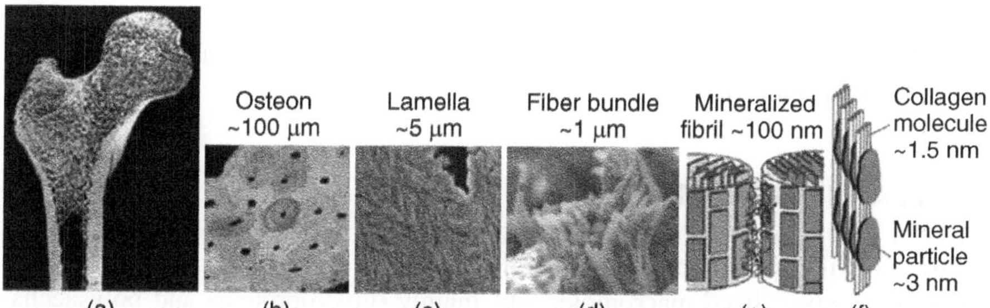

Figure 6.4 Schematic representation of the hierarchical structure of a human femur. Interfaces at many scales contribute to the extraordinary toughness of bone. (a) Cross section of a human femur. (b) This image depicts the osteons, which are cylindrical structures surrounding blood vessels in compact bone. (c) At this level compact bone consists of lamellae, which can have different architectures according to the type of bone. (d) The lamellae are built by collagen fibrils aligned parallel to each other. (e) Mineralized collagen fibril, which consists of collagen type I and crystals of hydroxyapatite. (f) The crystals of hydroxyapatite are located inside the fibrils, closely associated to the collagen molecules and oriented with their *c*-axis along the long axis of the fibril. (Adapted with permission. Copyright Annual Reviews: Ref. 20.)

Figure 6.5 (a) Optical microscopy image under polarized light of a patterned crystalline $CaCO_3$ film prepared in the presence of polyacrylic acid. (b) Higher magnification of (a). (c)–(e) Scanning electron microscopy images of $CaCO_3$ films grown in the presence of DNA on a poly(caprolactone) scaffold. Panels (d) and (e) show higher magnification of (c), highlighting the ability of the inorganic coating to follow the contours of the scaffold. (Panels (a) and (b) reproduced with permission. Copyright Wiley-VCH Verlag GmbH & Co. KGaA: Ref. 47. Panels (c)–(e) adapted with permission. Copyright the Royal Society of Chemistry: Ref. 48.)

still a challenge and requires a profound fundamental knowledge of the mechanisms involved in the biogenic processes.

6.2 LEARNING FROM NATURE

Control over the formation of biominerals occurs at several levels. Most important is the use of specialized macromolecules, mainly (glyco)proteins and polysaccharides, that are assembled into a three-dimensional (3D) organic matrix framework[1a] and provide the microenvironment where mineral deposition occurs.[21] Some of these macromolecules are also occluded inside the mineral phase,[6, 8] where they presumably exert direct control over crystal growth, polymorph type, morphology, and material properties.[22–24] Thus, the organic macromolecules form an intimate mix with the mineral phase at all different hierarchical levels, from the nanometer to the millimeter scale. In addition, the structure–function relationship between the different components also plays a significant role,[6] first in that it is crucial for the proper templating of mineral formation, and second, because it determines the properties of the material as a whole, whether they are mechanical, optical, magnetic, and soon. Therefore, these are the concepts that need to be investigated and applied to the synthesis of materials.

It is very common that, when compared to their geological counterparts, biogenic crystals possess different morphologies and growth behavior and nucleate from different faces. These differences are clearly due to the directing effect that the organic phase exerts over the incipient mineral and are among the most intriguing characteristics of biominerals. Generally, acidic macromolecules such as polysaccharides and (glyco)proteins, which are rich in aspartatic acid or glutamic acid residues and/or phosphate moieties, are involved in the control over crystal formation, in particular, crystal nucleation.[1a] Therefore, studies aiming to mimic these templates and to understand how they induce oriented crystal nucleation have focused on surfaces having acidic functionalities, such as carboxylic, phosphate, and sulfate groups. The first synthetic system to investigate crystal nucleation on a synthetic surface was performed using polyaspartic acid adsorbed on a sulfonated polystyrene film as a scaffold for calcium carbonate nucleation.[22] In this study, oriented nucleation of calcite was obtained, demonstrating the importance of the ordered arrangement of the functional groups and the cooperativity between the carboxylates and sulfates in templating calcite nucleation. Subsequently, two other biomimetic systems were developed: Langmuir monolayers of fatty acids on aqueous subphases[25] and self-assembled monolayers (SAMs) on solid substrates.[26] Studies on Langmuir monolayers were pioneered by Mann et al.,[25] who demonstrated the controlled crystallization of calcium carbonate under monolayers of stearic acid. Further studies on SAMs, which predominantly used functionalized long chain thiols on gold and silver surfaces, have shown that in addition to the nature of the head group (i.e., COO^-, $-OH$, $-SO_3^-$ and $-PO_3^{2-}$), the organization and orientation of the thiol chains is important to effectively nucleate calcium carbonate crystals with a high degree of orientational specificity.[27–29] These studies show that the stereochemical and geometrical match between the functional groups in the organic template and the ions in the organic phase dictates the orientation of the crystal.

A further subject of interest in material science concerns the question of how to precisely control the morphology of a given mineral. Organisms do this through two main routes. The first involves the use of water-soluble, generally acidic, macromolecules, which interact with specific faces of a crystal.[23, 30, 31] These biomolecules may also select a polymorph type either by inhibiting the formation of the most stable polymorph or by promoting the development of the less stable forms. The second route is the growth of the mineral within a confined space with a predefined shape that acts as a mold for the incipient crystal.[1a, 32, 33] In order to understand how organisms make use of soluble additives to control crystal morphology and transpose this knowledge to the synthesis of artificial materials, it is necessary first to look at the composition of such biomolecules. Using biochemical and molecular biology tools, many of such proteins have been purified and sequenced, most notably from mollusk shells,[34] bone apatite,[19] and magnetite in magnetotactic bacteria.[3] Several of these proteins were used as additives during crystal growth and provided valuable mechanistic insights into the control over mineral formation. For example, the Mms6 protein involved from magnetotactic bacteria was shown to form in vitro superparamagnetic cuboidal crystals 20–30 nm in size, while the particles synthesized without the protein were not homogeneous in size or shape.[35–37] This protein clearly has a role in the formation of uniform, monodisperse magnetite crystals, and a next step would be to produce synthetic polymers that mimic the effect of Mms6.[38] For calcium carbonate, synthetic polymers such as polyaspartic acid and polyacrylic acid, among others, have successfully been used to tune the polymorph type and morphology of crystals.[39] Most interesting is the formation of chiral morphologies in calcium carbonate through the interaction with chiral molecules,[40] and the formation of hierarchical structures using low molecular weight and polymeric additives.[41, 42]

Although most biomimetic systems deal with only one or very few organic components, when trying to mimic Nature, one must always keep in mind that the organic matrix constituents that control the formation of biominerals generally do not function in isolation. Rather, the three-dimensional assembly of the biomolecules into a framework is crucial for proper control over mineralization and over the properties of the material. Therefore, an understanding of the structure–function relationship of the organic matrix–mineral composite cannot be neglected. The importance of understanding how the matrix components function together in mineral formation was well represented in the work of Falini et al.[24] The authors assembled in vitro the major organic components of the aragonitic nacreous layer of mollusk shells: silk, purified from the cocoon of a silk worm; β-chitin from the pen of a squid; and acidic proteins extracted from both the aragonitic and calcitic layer of the shell. The adsorption on the silk–chitin scaffold of proteins extracted from the calcitic layer resulted in the formation of calcite, while adsorption of proteins extracted from the aragonitic layer resulted in the formation of aragonite crystals. In both cases crystallization inside the chitin scaffold occurred only when the acidic macromolecules were present. Furthermore, in the absence of silk or when chitin was substituted by a polystyrene scaffold, the acidic proteins from the aragonitic layer lost their ability to induce aragonite formation. While this example deals with a structure–function relationship in terms of understanding the formation of biominerals, this aspect is also crucial when the interest is to mimic the properties of a biomaterial. Most notable examples are hierarchical materials such as bone, teeth, and the skeleton of the glass sponge *Euplectela*.[43] Their mechanical properties, and hence their function, are highly dependent on the assembly of the basic building blocks, from the nanometer to the macroscopic level.

6.3 APPLYING LESSONS FROM NATURE: SYNTHESIS OF BIOMIMETIC AND BIOINSPIRED MATERIALS

So far we have discussed the major principles behind formation of biogenic minerals in Nature, and how one can learn from the biological system. In the present section, we discuss how we can apply what we learned from Nature to the synthesis of bioinspired materials with tunable morphologies and properties. Given the complexity of biominerals, it is very important to keep in mind which aspect of the biomaterial (i.e., morphology, mechanical, optical, or magnetic properties) are to be reproduced, and in what length scale.

6.3.1 Biomimetic Bone Materials

Biologically inspired materials have a great potential in the fields of regenerative medicine and biomedical engineering.[44] In this case, most approaches focus on reproducing the overall properties of the biomineral in order to restore its function in the body. One biological tissue that has been the subject of research for biomimetic replacement materials is bone. Bone is a tissue that provides structural support for our bodies and has unique mechanical properties that arise from its hierarchical structure and may vary according to the function that a particular bone performs at a particular location in the body.[16] Although bone is capable of self-repair, this capability is limited to small defects and further decreases with age and is affected by diseases. In case of severe traumas, the tissue needs to be replaced using artificial materials in order to restore its function. Thus, there is great interest in developing bioinspired materials that possess osteoinductive properties, being capable of inducing bone regeneration and eventually being resorbed by the organism and replaced by bone, or that can directly be used as replacement material.

Biogenic calcium carbonate, as found in coral skeletons and sea urchin spines, was found to be a promising material for bone replacement and regeneration, since it can easily be resorbed by osteoclasts and be replaced by native bone.[44, 45] Furthermore, nacre was shown to have osteoconductive properties, meaning that it stimulates the activity of osteoblasts and induces the formation of new bone. Therefore, synthetic, bioinspired organic–inorganic composites are emerging as new materials for bone regeneration and offer more possibilities to tune biocompatibility, biodegradability, and mechanical properties.[44, 46] Based on the osteoconductive properties of biogenic calcium carbonate, synthetic calcium carbonate has been investigated as a material with potential application in bone regeneration. Indeed, it has been shown that thin films of crystalline $CaCO_3$ can be used as substrates for cell culture, being capable of supporting rat bone marrow stromal cell attachment and differentiation into osteoblast- and osteoclast-like cells.[47] These films were prepared on a patterned block copolymer film consisting of alternating of mineralized and nonmineralized regions (Figures 6.5a and 6.5b). Furthermore, polyacrylic acid was used to stabilize an amorphous $CaCO_3$ precursor phase, which is important in that the amorphous material can easily be molded into any shape before it crystallizes. Thus, it is possible to generate different patterns of $CaCO_3$ in a variety of substrates. While these results were obtained on two-dimensional (2D) substrates, a follow-up investigation was performed, using the same methodology, on three-dimensional (3D) polymer scaffolds (Figures 6.5c–e).[48] The jump from a 2D to a 3D scaffold is an important step since biological tissues have a complex 3D architecture that is crucial not only for its mechanical properties, as is the case of bone, but also for providing spatial and organizational cues toward morphogenesis. Bone cells are sensitive to the physical properties of their environment, such that the composition, topography, and roughness are key determinants in osteogenesis.[49, 50] The deposition of calcium carbonate on 3D scaffolds allows the construction of hybrid ceramic scaffolds with controllable architecture, porosity, and mechanical properties, which allows their application in bone tissue engineering.

It is evident that an ideal biomimetic scaffold should be as close to the biological tissue as possible, in terms of composition, structure, and properties. Although designing and producing such a scaffold is still a challenge, a large step was made a few years ago, when it was demonstrated that collagen mineralization could be mimicked in vitro by substituting the noncollagenous proteins by a synthetic polymer, polyaspartic acid (Figure 6.6a).[51] For the first time, intrafibrillar mineralization of collagen was achieved under artificial conditions, where the apatite crystals formed inside the collagen fibrils had the same morphology and orientation as in bone (Figures 6.6c and 6.6d).[51, 52] These findings open new possibilities in developing bone-replacement scaffolds composed entirely of mineralized collagen, with optimal osteoconductive properties.

In a similar approach, collagen–apatite composites were prepared using a neutralization reaction, where $Ca(OH)_2$ was directly mixed with a phosphoric acid solution containing disassembled collagen.[53] This reaction yielded calcium phosphate in the form of apatite crystals, which nucleated in close association to the collagen, while its assembly into fibrils occurred, triggered by the increase in pH. Based on this methodology, a three-layered scaffold was constructed, consisting of a layer of mineralized collagen, mimicking subchondral bone; an intermediate layer also of mineralized collagen, however, with lower mineral content, simulating the tidemark layer, which separates hyaline cartilage from subchondral bone; and a layer of hyaluronic acid–collagen, reproducing the cartilage.[54] Tissue culture tests were done, where it was shown that articular chondrocytes loaded into the three-layered scaffold yielded cartilaginous tissue formation selectively in the cartilage-mimicking layer. Additionally, the scaffold was seeded with stromal cells and implanted in mice, resulting in bone formation within the layer of mineralized collagen and loose connective tissue in the cartilaginous layer. Thus, even though one could argue that this scaffold mimicks the biological tissue in composition but not in structure (i.e., the apatite crystals are most likely outside the collagen fibrils and not as well organized and oriented as in bone), it still has very good potential as an implant material with osteoconductive properties. This methodology for collagen mineralization was further developed to allow the incorporation of magnetite nanoparticles during the stage of apatite nucleation and collagen assembly, with the aim of improving the stability, biocompatibility, and mechanical properties of the scaffold.[55] Indeed, the authors observed that the stiffness of the composite material when under compression increased, even with a higher collagen/apatite ratio; its cytotoxicity decreased, and it was able to support cell growth. Last, the magnetization properties can be exploited for further applications.

In addition to being able to induce osteogenesis and eventually be replaced by bone, a scaffold also mimics the hierarchical structure and mechanical properties of the biological tissue. Although we still lack the knowledge to produce such materials, an interesting approach recently developed is to modify wood templates

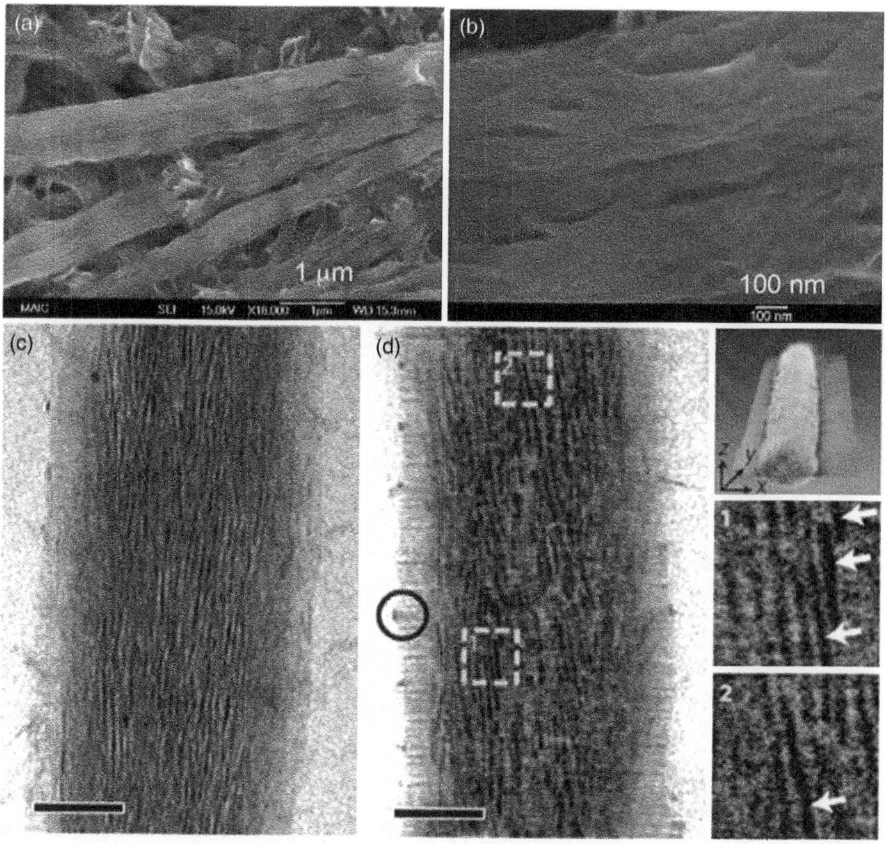

Figure 6.6 (a) Scanning electron microscopy image of collagen fibrils mineralized with hydroxyapatite, using polyaspartic acid as a directing agent. (b) Higher magnification of (a). (c) Cryo-transmission electron microscopy image of a collagen fibril mineralized with hydroxyapatite, using polyaspartic acid as a directing agent. (d) Slice from a section of the three-dimensional reconstruction of a mineralized collagen fibril, after cryo-electron tomography. Crystals are viewed edge on (insets 1 and 2, white arrows). Note how the long axis of the crystals is aligned parallel to the long axis of the fibril. Black circle: amorphous calcium phosphate infiltrating into the fibril. Scale bars: 100 nm. (Panels (a) and (b) adapted with permission. Copyright Elsevier: Ref. 51. Panels (c) and (d), reproduced with permission. Copyright Macmillan Publishers Ltd, Nature Materials, www.nature.com/nmat: Ref. 52.)

to obtain an organic–inorganic composite scaffold containing a three-dimensional morphology and hierarchical architecture (Figure 6.7).[56] The development of such a scaffold was achieved by using a multistep process, involving (1) pyrolysis of the wood specimens to decompose mainly the cellulose, hemicelluloses, and lignin and produce a carbon template; (2) carburization to transform the carbon template into calcium carbide; (3) oxidation of the calcium carbide template to yield calcium oxide; (4) carbonation to convert the calcium oxide into calcium

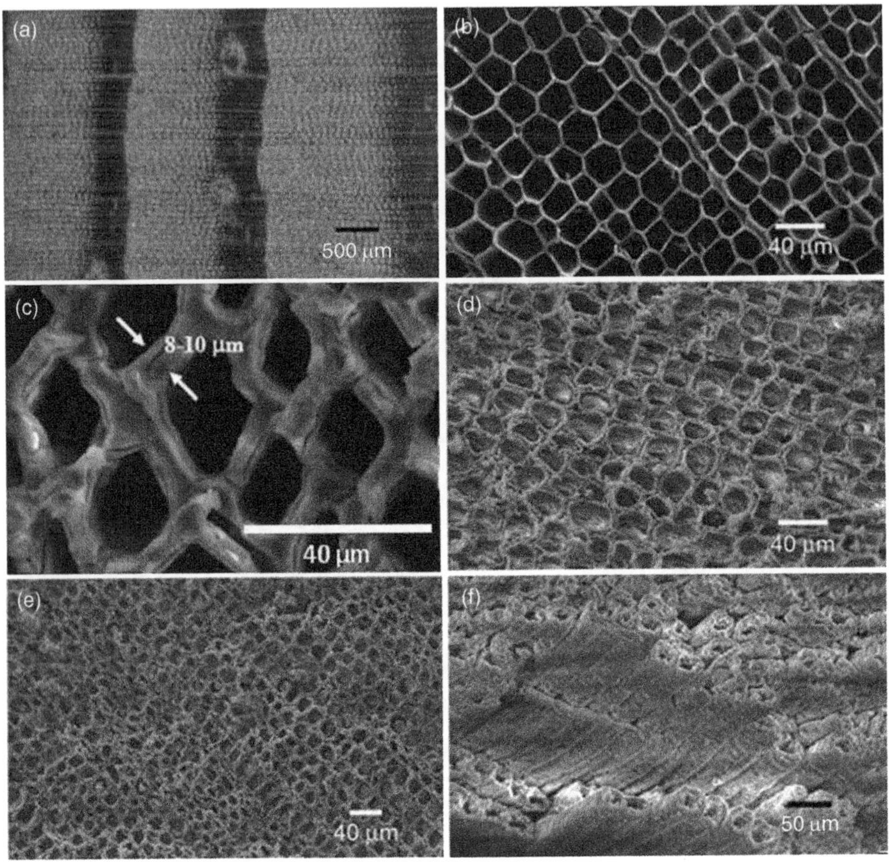

Figure 6.7 Scanning electron micrographs depicting the different stages in converting pine wood into an organic–inorganic composite of native pine wood. (a) Native pine wood. (b) Wood after pyrolysis. (c) Calcium carbide obtained from pyrolyzed pine wood. (d) Pine wood-derived calcium oxide. (e) Pine wood-derived calcium carbonate. (f) Pine wood-derived hydroxyapatite. Note how the natural wood microstructure is preserved throughout the procedure, culminating with parallel fastened hydroxyapatite microtubes. (Adapted with permission. Copyright the Royal Society of Chemistry: Ref. 56.)

carbonate; and (5) phosphatization to transform the calcium carbonate template into hydroxyapatite. After this process, the obtained biomaterial preserved the structure and morphology of the original wood template. The hierarchical structure of this scaffold, combined with the hydroxyapatite constituting phase, is very valuable. First, it ensures a multilevel organized morphology characterized by unidirectional oriented pore structures that is necessary for cell-in growth and reorganization and provides the necessary space for vascularization. Second, the fascicular matrix may be able to satisfy biomechanical requirements, providing the mechanical properties that are required for the tissue. This process is highly versatile and can be adapted to a number of different organic structures found in Nature, allowing the production of scaffolds with different three-dimensional architectures.

Ultimately, an ideal biomimetic scaffold should be able to reproduce the composition, three-dimensional structure, and overall properties of a biological tissue and thus be able to restore its function. However, since biological tissues, and bone in particular, have quite complex architectures that are directly correlated to their function and overall properties, producing such a scaffold is still a challenge. Nevertheless, a biomimetic approach, namely, to finding new approaches to understand and mimic the way that mineralized biological tissues are designed and formed, has provided significant advances in the design of bioinspired materials that have great potential for tissue engineering.

6.3.2 Semiconductors, Nanoparticles, and Nanowires

At the nanoscopic scale, mimicking the ability of organisms to tune the size and morphology of crystals is relevant to a number of synthetic materials. One example is of semiconductor materials, which have unique optical, electrical, and optoelectrical properties that can effectively be controlled by tuning the size, composition, and crystal structures of the nanocrystals.[57] Indeed, over the last decade, methodologies have been developed to use organic templates for the molecular manipulation of semiconducting microstructures, such as CdS and CdSe.[57] For example, organic surfactants have been used to produce II–VI semiconductor nanocrystals that were highly monodisperse and regular in shape.[58–63] In this case the organic template ensured not only the formation of nanocrystallites that were homogeneous in size and morphology, but also the surfactant formed molecular monolayers around the nanocrystals, preventing the formation of disordered structures. The result was the self-organization of the quantum dots in superlattices that formed two- and three-dimensional networks. Langmuir monolayers have also been used as templates for semiconductor growth, resulting in nanocrystals with different morphologies, such as rods, triangles, or a continuous network (Figures 6.8a and 6.8b).[64] Another approach is to produce a polycrystalline semiconducting continuum with periodic nanometer-sized cavities, directly templated by assemblies of organic molecules.[65, 66] The nanometer-sized cavities are an interesting feature in semiconducting materials because, for instance, their presence could produce a periodic array of antidots that could modify the electronic properties of the material.[67–69] A further possibility is to use the cavities to selectively adsorb, transport, or transform molecules diffusing through the cavities according to the electronic and photonic properties of the semiconductor.[57] There are also reports where the self-assembly properties of large molecules and subsequently their supramolecular structure were exploited to directly template the formation of the semiconductor following the morphology of the template. In this case, CdS nanoribbons could be produced, which were composed of polycrystalline domains of 4–8 nm (Figures 6.8c and 6.8d).[70]

More recently, core–shell CdSe/ZnS nanocrystals were synthesized using bifunctional peptides composed of two different domains, one containing a

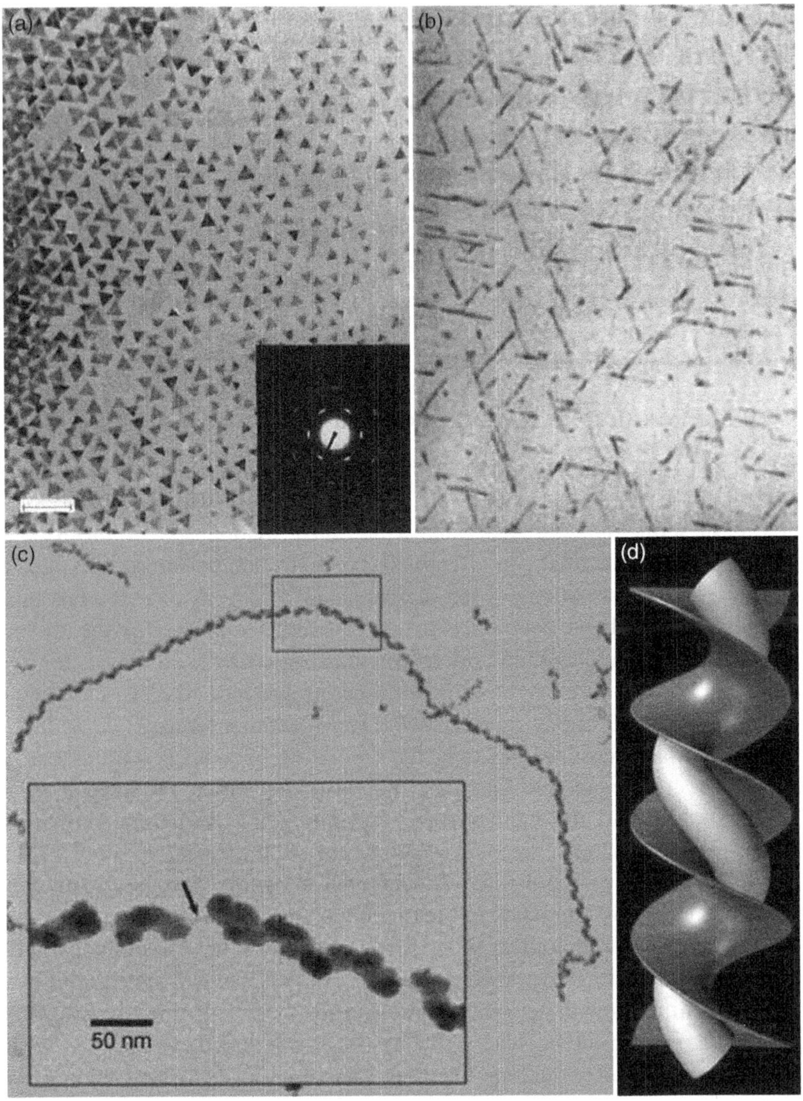

Figure 6.8 (a) Transmission electron microscopy image of a film of particulate PbS formed by the infusion of H_2S to a monolayer of arachidic acid. Scale bar: 200 nm. Inset: Electron diffraction of a PbS domain. (b) Transmission electron microscopy image of a film of particulate PbSe formed by the infusion of H_2Se to a monolayer of arachidic acid. (c) Transmission electron microscopy image of CdS helixes precipitated in a suspension of dendron rondcoil nanoribbons[71] in ethyl methacrylate. (d) Schematic representation of a possible templating mechanism, in which a coiled CdS helix (in light gray) is produced from a twisted helical template through growth along one face of the template (in dark gray). (Panels (a) and (b) adapted with permission. Copyright Wiley-VCH Verlag GmbH & Co. KGaA: Ref. 64. Panels (c) and (d) adapted with permission. Copyright Wiley-VCH Verlag GmbH & Co. KGaA: Ref. 70.)

CdSe-binding domain and the second comprised of a ZnS-binding domain.[71] By using this method, it is possible to modulate the thickness of the shell using another peptide capable of controlling the growth of the ZnS shell. In addition, since it is the peptide sequence that guides the formation of the nanocrystals, it can be selectively fine-tuned to control and direct the formation of other inorganic materials and structures. Recently, new methods to produce nanoshells were developed using enzymes as a template and nanoreactor for the reaction. In this example, ZnO nanoshells were produced using *urease* as a catalytic template.[72] This enzyme has an overall negative charge and thus can electrostatically interact with zinc precursors, and the ammonia generated by the enzyme through the hydrolysis of urea increases the pH, which is suitable for the formation and growth of ZnO. The procedure used allows the synthesis of semiconductors at room temperature, under mild conditions, and can be extended to other materials such as ZrO_2, SnO_2, Ga_2O_3, WO_3, IrO_2, NiO, and TiO_2. Furthermore, since the size of the nanoshells is determined by the size of the enzyme core, their diameter can be tuned by employing isoenzymes of different molecular weights. Viruses have also been exploited as scaffolds for semiconductors. Peptides that control the size, composition, and phase during the nucleation of nanoparticles were expressed on the capsid of the virus and served as templates for the formation of ZnS and CdS nanocrystals 3–5 nm in size that were in close contact and preferentially aligned.[73] Upon annealing, the removal of the organic template allowed the polycrystalline assemblies to form single crystal nanowires of high aspect ratio, being several hundreds of nanometers in length and only about 20 nm in width (Figure 6.9). By changing the substrate-specific peptide, nanowires of ferromagnetic FePt and CoPt were also produced, highlighting the versatility of this system.

Biomimetic systems have also been applied to the synthesis of silver nanoparticles and nanowires. Drawing inspiration from the biosynthesis of silver nanoparticles by bacteria,[74] Naik et al.[75] have used a phage display library to select peptides capable of precipitating flat silver crystals, 60–150 nm in size and about 15–18 nm in thickness. By patterning the adsorption of the peptides on a substrate, they could also form ordered arrays of nanoparticles. However, although the peptides could induce nucleation of crystals, they could not precisely control their morphology. Silver nanowires could also be produced, using amyloid-based polypeptides.[76] These polypeptides self-assembled into hollow nanotubes that were a few micrometers in length and could subsequently be filled with silver nanoparticles (Figure 6.10). Upon reduction of the silver with citric acid, the amyloid nanotubes served as molds for casting the metal. After degradation of the polypeptide chains with proteases, discrete nanowires with high persistence length were obtained. Contrary to what we have discussed so far, where organic macromolecules directly control the nucleation and growth of nanoparticles, the key function of the amyloid fibers is to template the morphology of the silver deposits simply by providing a casting mold. Therefore, controlling the self-assembly of the polypeptide chain into higher structures is the crucial step in order to template the deposition of the metal into desired morphologies.

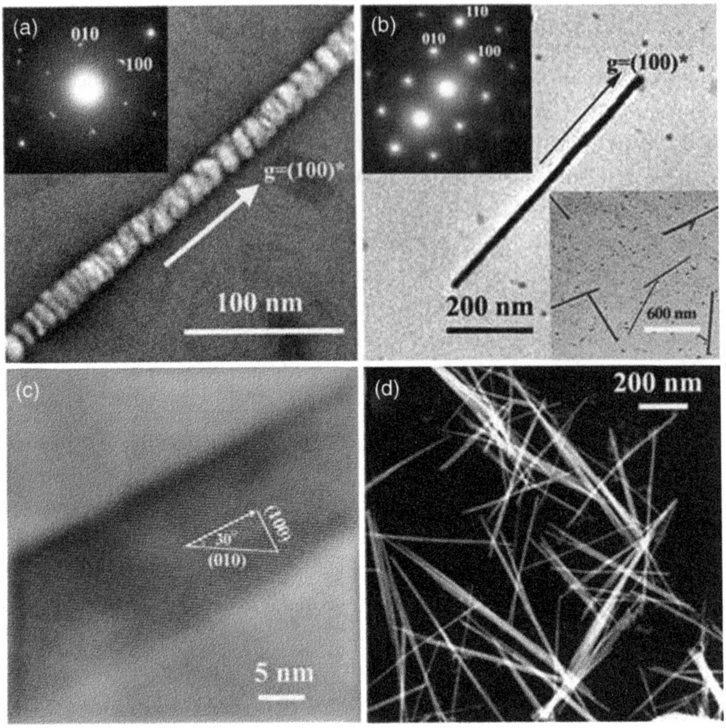

Figure 6.9 Electron microscopy of both the pre- and postannealed ZnS and CdS viral nanowires. (a) Dark-field diffraction-contrast imaging of the preannealed ZnS system using the (100) reflection, showing the crystallographic ordering of the nucleated nanocrystals, in which contrast stems from satisfying the (100) Bragg diffraction condition. Inset: Electron diffraction pattern of the polycrystalline preannealed wire showing the wurtzite crystal structure and the single-crystal type [001] zone axis pattern, suggesting a strong [001] zone axis preferred orientation of the nanocrystals on the viral template. g = (100)* denotes the reciprocal vector of (100) crystal planes, which is perpendicular to the (100) planes and has a length inversely proportional to the interplanar spacing of the (100) planes. (b) Bright-field TEM image of an individual ZnS single-crystal nanowire formed after annealing. Inset: (Upper left) Electron diffraction pattern along the [001] zone axis shows a single crystal wurtzite structure of the annealed ZnS nanowire. Inset: (Lower right) Low-magnification TEM image showing the monodisperse, isolated single-crystal nanowires. (c) HRTEM of a ZnS single-crystal nanowire showing a lattice image that continually extends the length of the wire, confirming the single-crystal nature of the annealed nanowire. The measured lattice spacing of 0.33 nm corresponds to the (010) planes in wurtzite ZnS crystals. (d) HAADF-STEM image of single-crystal ZnS nanowires. (e) HAADF-STEM image of CdS single-crystal nanowires. (f) A HRTEM lattice image of an individual CdS nanowire. (Adapted with permission. Copyright American Association for the Advancement of Science, AAAS: Ref. 73.)

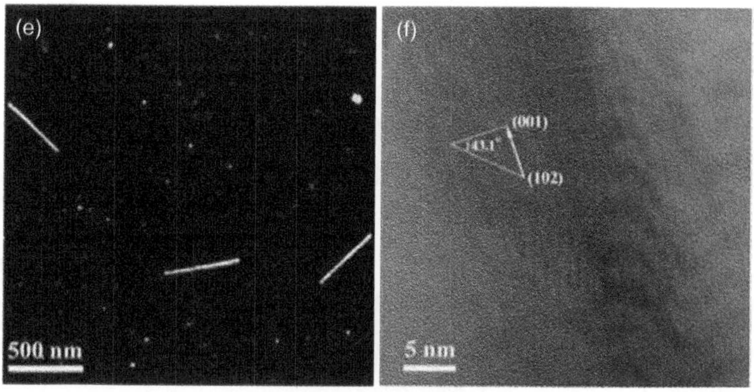

Figure 6.9 (*Continued*)

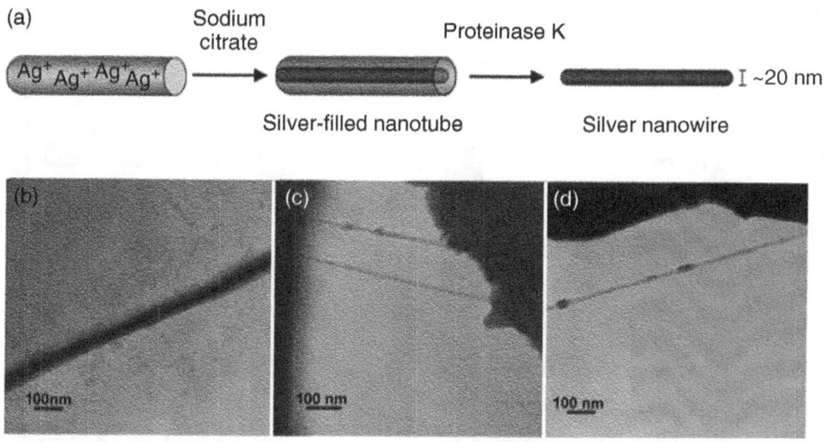

Figure 6.10 (a) Schematic representation of the casting of silver nanowires using the peptide nanotubes as templates. (b) Transmission electron micrograph of peptide tubes filled with silver nanowires. (c) and (d) Transmission electron micrographs of the silver nanowires after digestion of the peptide tubes with proteinase K. (Adapted with permission. Copyright American Association for the Advancement of Science, AAAS: Ref. 76.)

The potential for such methodology for the construction of functional nanometer-scale electronic devices has already been demonstrated. Braun et al.[77] used DNA strands to connect two gold electrodes, which was followed by selectively depositing silver on the DNA molecules through Ag^+/Na^+ ion exchange and formation of complexes between the Ag ions and the DNA bases. The silver ions were then reduced to form nanometer-sized metallic silver aggregates, bound to the DNA backbone, forming a conductive metal wire connecting both electrodes (Figure 6.11).

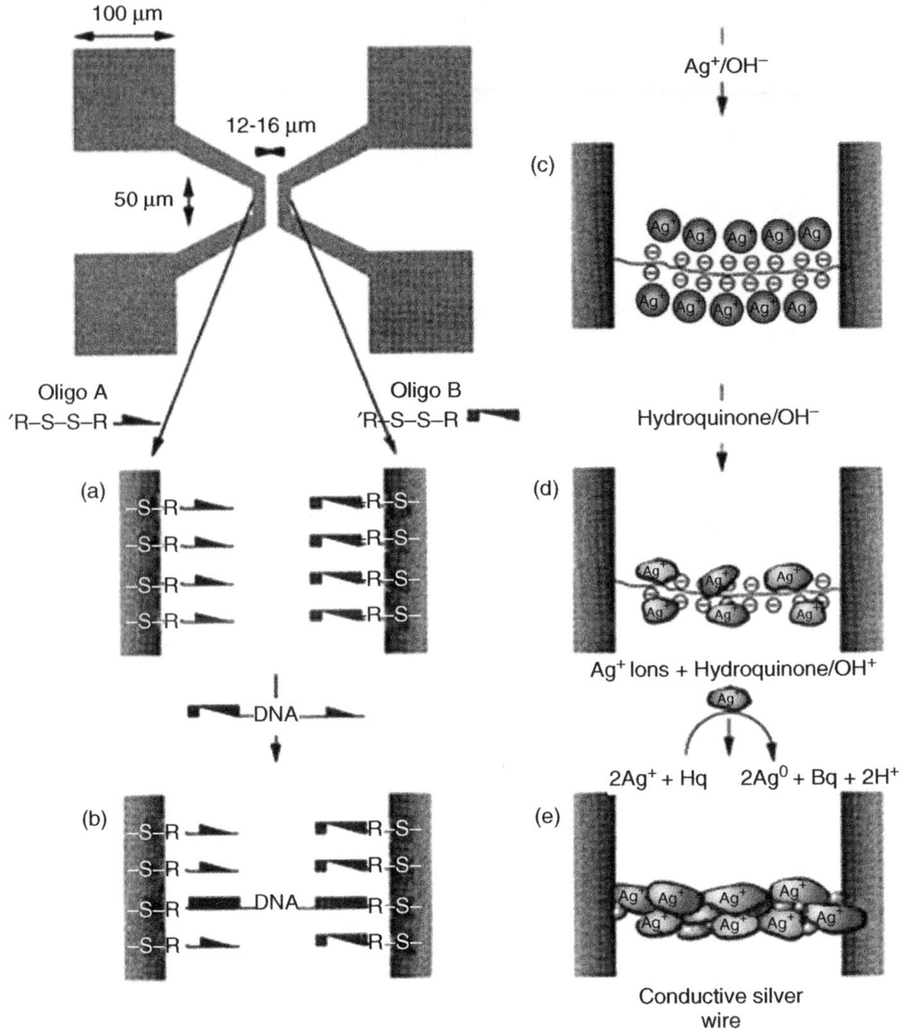

Figure 6.11 Schematic representation of the construction of a silver wire connecting two gold electrodes. (a) Oligonucleotides with two different sequences are attached to the electrodes. (b) A DNA bridge connects both electrodes. (c) The DNA bridge is loaded with silver ions. (d) Metallic silver aggregates and binds to the DNA backbone. (e) Silver wire fully formed on the DNA substrate. (Adapted with permission. Copyright © 1998 Macmillan Publishers Ltd, Nature, www.nature.com: Ref. 77.)

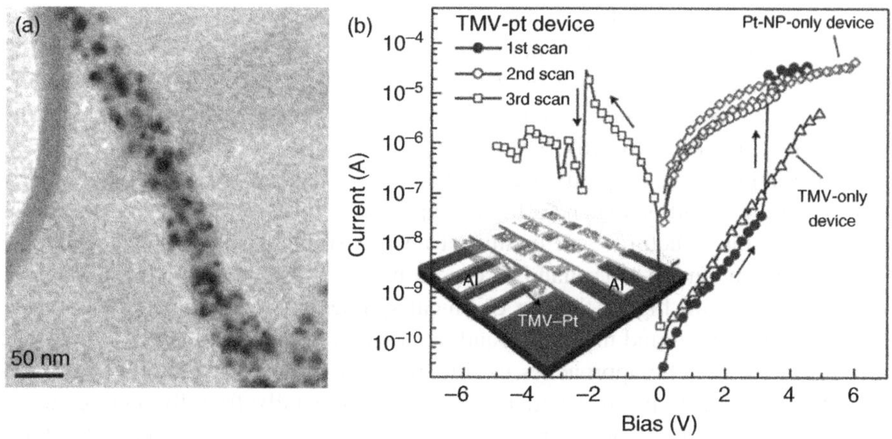

Figure 6.12 (a) Transmission electron micrograph of a tobacco-mosaic virus (TMV) impregnated with Pt nanoparticles. (b) Current–voltage (I-V) characteristics of the TMP–Pt composite. Filled circles represent the first bias scan and show that the device switches to the ON state at 3.1 V and stabilizes (empty circles). A reverse scan (squares) shows that the device switches back to the OFF state at 22.4 V. No conductance switching was observed for TMV-only (triangles) and Pt nanoparticles-only (diamonds) devices. Inset: Schematic representation of the device structure, with an active layer of dispersed TMV–Pt nanowires. (Adapted with permission. Copyright Macmillan Publishers Ltd, Nature Nanotechnology, www.nature.com/nnano: Ref. 79.)

Further works exploited this idea, for example, depositing the DNA strands in ordered arrays that were used as a substrate to form parallel one-dimensional and two-dimensional crossed-metallic nanowire arrays of Pd.[78] In these cases, however, the DNA not only templates the morphology of the nanowires but its interaction with the metal ions during the metal deposition process also assists in controlling the formation of the nanowires. Also in the field of nanoelectronics, tobacco-mosaic viruses were used as substrates for the incorporation of Pt, yielding nanowires with remarkable nonvolatile memory properties that are the result of the combination of the platinum nanoparticles and the virus (Figure 6.12).[79] Here, the virus serves not just as a scaffold for the organization of the nanoparticles but also plays an integral role in the process as a charge donor and in stabilizing the charges and creating a repeatable memory effect.

6.3.3 Biomimetic Strategies for Silica-Based Materials

Another class of materials that have attracted interest in bioinspried design is silica. Silica and silica-based materials are widely used in industrial and technological applications.[80] A few examples of such applications include (1) zeolites, which are products of silicalites or aluminosilicates that are applied as thermostable catalysts in chemical reactions; (2) molecular sieves in separation, purification, and ion exchange, including (3) as carriers of detergents in washing powders, (4) absorbents (e.g., in refining, cleaning, and clarification), (5) abrasives (e.g., in toothpaste, cleaners), or as (6) fillers and whiteners (e.g., in paints, food, cosmetics, plastics).[81] Thus, there is great demand for producing improved silica structures with specific properties, such as mechanical strength, pore volume, pore-size distribution, specific surface area, or surface reactivity.

Industrial processes currently used to synthesize silica particles generally employ harsh reaction conditions such as high temperatures, and in most cases the reproducibility of the structures and quality at an industrial scale is limited. Biogenic silica formation, on the other hand, occurs at mild conditions with a level of reproducibility that is far higher than in chemical syntheses. Remarkable examples are the silica structures found in diatoms and in the glass sponge *Euplectela*.[43, 82] Thus, developing biomimetic approaches to silica formation may provide efficient routes to synthesize industrial silica under mild, environmentally friendly and economically attractive reaction conditions, with highly reproducible, tailor-made structures and properties.

Formation of biogenic silica in diatoms is a process controlled by small polycationic proteins named silafins, which also contain several phosphorylated amino acids, in this case serine.[83–85] These proteins self-assemble and induce precipitation of silica and it is thought that they relate to the surfactants or templates used in the preparation of mesoporous silica.[85] Based on these findings, Naik et al.[86] investigated the function of a 19 amino acids long peptide derived from silaffin-1 A in silica precipitation. The authors of this study showed that particles with different morphologies, ranging from spheres to organized and complex fibrils, could be obtained through the manipulation of the environment and the use of mechanical force. In a different study, poly-L-lysine was used as a template, and fiber-like and ladder-shaped silica morphologies with periodic voids could be obtained, depending on whether the reaction mixture was flowing through a tube or stirred during the reaction, respectively.[87] Furthermore, it was shown that biopolymers rich in amine-containing peptides can be used as gelating agents of silica oligomers in the case of silicic acid, and as flocculating agents in the case of silica solutions.[88, 89] These observations provide that in vitro biocatalysis may successfully be employed to tailor-make silica structures. Another promising approach, which may provide a cheap and versatile substitute for silafins, polyamino acids, and other biomolecules involved in silica precipitation, is polyethylene glycol (PEG).[81, 90] It has been shown that PEG chains of different lengths, as well as different PEG/silica ratios, can be used to tune the pore dimensions, going from less than 2 nm to 20 nm.

One aspect that is very important in the synthesis of silica is to be able to control and produce structures with well-defined morphologies. With this aim, a number of amine-containing organogels that self-assemble into fiber-like structures have been used successfully to produce fibrillar structures of silica.[91, 92] Examples include hollow silica fibers with linear, helical, and multilayered morphologies[93, 94]; silica fibrils with a double-stranded helical structure[95]; and silica nanotubes with adjustable meso- or macroscale inner diameters.[96] One particularly interesting example is the synthesis of chiral silica nanotubes, synthesized using an organogel from a dimeric azobenzene-appended cholesterol derivative, offering potential for

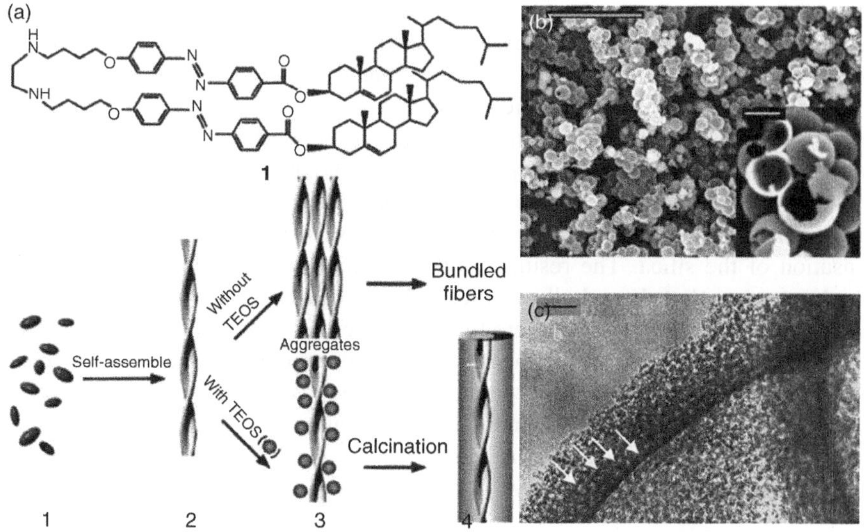

Figure 6.13 (a) Schematic representation of the creation of the helically structured silica from the organogel based on azobenzene-appended cholesterol derivative: **1**, gelator; **2**, incipient organogel fiber; **3**, silica adsorption (lower) and aggregation of organogel fiber (upper); and **4**, structure of the silica formed after calcinations. (Reproduced with permission. Copyright the American Chemical Society: Ref. 98.) (b) Scanning electron micrograph of silica particles prepared from PEO–PPO–PEO–based emulsions, showing the monodispersity of the particles. Scale bar: 10 μm. Inset: High magnification, showing that the spheres are hollow. Scale bar: 1 μm. (c) High-resolution transmission electron micrograph, demonstrating the multilamellar structure of the shell (arrows) of the hollow silica spheres. Scale bar: 33 nm. (Adapted with permission. Copyright Wiley-VCH Verlag GmbH & Co. KGaA: Ref. 99.)

applications in chiral separation and catalysis (Figure 6.13a).[97] Although in all these cases different molecules were used, it is important to note that the presence of hydrogen-bonding sites, such as amino groups, is indispensable for a successful transcription of the organic template into the silica structure.

Hierarchical structures of silica and, in particular, mesoporous silica with hollow morphologies have attracted a lot of attention because they combine low density with thermal and mechanical stability, making interesting materials for insulators, catalysts, sorbents, and containers for chemically active or vulnerable agents (e.g., in controlled drug delivery).[80] Sun and co-workers showed that an EO_{76}–PO_{29}–EO_{76} triblock copolymer-based emulsion in combination with the inexpensive sodium silicate solution can be used to generate hollow silica spheres[98] with well-defined multilamellar structure and high monodispersity (Figure 6.13b).[99] By performing the reaction at room temperature, temperature-sensitive or volatile chemicals can be enclosed inside the structure, making it suitable for drug delivery, for instance. The reaction can also be performed at 80 °C, in which case thermally stable materials with well-defined wall structures are obtained.

Additionally, hollow spheres with ultralarge mesoporous structures were produced using EO_{39}–BO_{47}–EO_{39}-based reverse emulsions, where the silica wall of the product consists of uniform and hexagonally ordered mesopores with pore sizes of up to 50 nm, with applications in the storage, release, and transport of biomolecules.[100] Another interesting approach was to apply sonication to a mixture of PEO-polymer Tergitol $C_{15}(EO)_{12}$ and silicon alkoxide in order to form a stable emulsion.[101] The ultrasonication created cavitation bubbles that became trapped in the solution and frozen by the hydrolysis of silicon alkoxide and condensation of the silica. The result was the retention of nitrogen within the voids, which suggests that these hollow silica spheres can be used to store or stabilize volatile compounds. Last, Wong et al.[102] employed block-copolypeptides that were designed with specific recognition sites for gold nanoparticles and silica, resulting in hollow spheres with amorphous walls composed of two distinct layers of silica–gold nanoparticles. This method is quite interesting in that it represents an approach to produce silica with a hierarchical organization of nanoparticles into multidimensional composite arrays.

6.4 CONCLUSION

Biomimetic approaches for the design of materials is very promising. Biomineralization principles learned from Nature, such as supramolecular template synthesis, template-directed crystal growth, phase separation, and self-assembly, offer great potential as strategies for tailoring the structure and function of materials from the nanometer to the macrometer scale. However, the current knowledge in generating such functional materials for industrial and technological applications in a cost-effective way is still very limited and much research is necessary. As our understanding of the mechanisms of formation and of the structure–function relationship of biomaterials expands, so will our capability of applying these principles in the fields of material science, chemistry, and biomedical engineering to produce materials with tunable functions, morphologies, and properties.

ACKNOWLEDGMENTS

The authors thank the Netherlands Science Foundation, NOW, for financial support.

REFERENCES

1. Lowenstam, H. A.; Weiner, S. *On Biomineralization*, Oxford University Press, New York, 1989: (a) p. 3–48; (b) p. 50; (c) p. 197.
2. Hildebrand, M. *Chem. Rev*. **2008**, *108,* 4855.
3. Faivre, D.; Schuler, D. *Chem. Rev*. **2008**, *108,* 4875.
4. Jamieson, J. C. *J. Chem. Phys*. **1953**, *21,* 1385.

5. Carter, J. G. *Skeletal Biomineralization: Patterns, Processes and Evolutionary Trends*, Van Nostrand Reinhold, New York, 1990.
6. Weiner, S.; Addadi, L. *J. Mater. Chem*. **1997**, *7,* 689.
7. Addadi, L.; Weiner, S. *Angew. Chem. Int. Ed. Engl*. **1992**, *31,* 153.
8. Berman, A.; Hanson, J.; Leiserowitz, L.; Koetzle, T. F.; Weiner, S.; Addadi, L. *Science* **1993**, *259,* 776.
9. Raz, S.; Hamilton, P. C.; Wilt, F. H.; Weiner, S.; Addadi, L. *Adv. Funct. Mater*. **2003**, *13,* 480.
10. Gregoire, C. *J. Biophys. Biochem. Cytol*. **1957**, *3,* 797.
11. Gregoire, C. In *Chemical Zoology* (Ed. Florkin, B. T. S.), Academic Press, New York, 1972, pp. 45–102.
12. Wada, K. *Bull. Nat. Pearl Res. Lab*. **1961**, *7,* 703.
13. Addadi, L.; Joester, D.; Nudelman, F.; Weiner, S. *Chem. Eur. J*. **2006**, *12,* 981.
14. Nudelman, F.; Chen, H. H.; Goldberg, H. A.; Weiner, S.; Addadi, L. *Faraday Discuss*. **2007**, *136,* 9.
15. Aizenberg, J.; Tkachenko, A.; Weiner, S.; Addadi, L.; Hendler, G. *Nature* **2001**, *412,* 819.
16. Weiner, S.; Wagner, H. D. *Annu. Rev. Mater. Sci*. **1998**, *28,* 271.
17. Hulmes, D. J. S.; Wess, T. J.; Prockop, D. J.; Fratzl, P. *Biophys. J*. **1995**, *68,* 1661.
18. Traub, W.; Arad, T.; Weiner, S. *Proc. Natl. Acad. Sci. U.S.A.* **1989**, *86,* 9822.
19. George, A.; Veis, A. *Chem. Rev*. **2008**, *108,* 4670.
20. Dunlop, J. W. C.; Fratzl, P. *Annu. Rev. Mater. Res*. **2010**, *40,* 1.
21. Weiner, S.; Traub, W.; Lowenstam, H. A. In *Biomineralization and Biological Metal Accumulation* (Eds. Westbroek, P.; de Jong, E. W.), Reidel Publishing Company, Dordrecht, 1983, pp. 205–224.
22. Addadi, L.; Moradian, J.; Shay, E.; Maroudas, N. G.; Weiner, S. *Proc. Natl. Acad. Sci. U.S.A.* **1987**, *84,* 2732.
23. Addadi, L.; Weiner, S. *Proc. Natl. Acad. Sci. U.S.A.* **1985**, *82,* 4110.
24. Falini, G.; Albeck, S.; Weiner, S.; Addadi, L. *Science* **1996**, *271,* 67.
25. Mann, S.; Heywood, B. R.; Rajam, S.; Birchall, J. D. *Nature* **1988**, *334,* 692.
26. Tremel, W.; Kuther, J.; Balz, M.; Loges, N.; Wolf, S. E. In *Handbook of Biomineralization: Biomimetic and Bioinspired Chemistry* (Ed. Baeuerlein, E.), Wiley-VCH, Weinheim, Germany, 2007, p. 209.
27. Aizenberg, J.; Black, A. J.; Whitesides, G. H. *J. Am. Chem. Soc*. **1999**, *121,* 4500–4509.
28. Aizenberg, J.; Black, A. J.; Whitesides, G. M. *Nature* **1998**, *394,* 868.
29. Aizenberg, J.; Black, A. J.; Whitesides, G. M. *Nature* **1999**, *398,* 495.
30. Albeck, S.; Aizenberg, J.; Addadi, L.; Weiner, S. *J. Am. Chem. Soc*. **1993**, *115,* 11691.
31. Albeck, S.; Weiner, S.; Addadi, L. *Chem. Eur. J*. **1996**, *2,* 278.
32. Beniash, E.; Addadi, L.; Weiner, S. *J. Struct. Biol*. **1999**, *125,* 50.
33. Wada, K. *Bull. Nat. Pearl Res. Lab*. **1968**, *13,* 1561.

34. Marin, F.; Luquet, G. *C. R. Palevol* **2004**, *3,* 469.

35. Arakaki, A.; Webb, J.; Matsunaga, T. *J. Biol. Chem.* **2003**, *278,* 8745.

36. Tanaka, M.; Mazuyama, E.; Arakaki, A.; Matsunaga, T. *J. Biol. Chem.* **2011**, *286,* 6386.

37. Prozorov, T.; Mallapragada, S. K.; Narasimhan, B.; Wang, L. J.; Palo, P.; Nilsen-Hamilton, M.; Williams, T. J.; Bazylinski, D. A.; Prozorov, R.; Canfield, P. C. *Adv. Funct. Mater.* **2007**, *17,* 951.

38. Arakaki, A.; Masuda, F.; Amemiya, Y.; Tanaka, T.; Matsunaga, T. *J. Colloid Interface Sci.* **2010**, *343,* 65.

39. Sommerdijk, N. A. J. M.; de With, G. *Chem. Rev.* **2008**, *108,* 4499.

40. Orme, C. A.; Noy, A.; Wierzbicki, A.; McBride, M. T.; Grantham, M.; Teng, H. H.; Dove, P. M.; DeYoreo, J. J. *Nature* **2001**, *411,* 775.

41. Chen, S. F.; Yu, S. H.; Wang, T. X.; Jiang, J.; Colfen, H.; Hu, B.; Yu, B. *Adv. Mater.* **2005**, *17,* 1461.

42. Mukkamala, S. B.; Powell, A. K. *Chem. Commun.* **2004**, 918.

43. Aizenberg, J.; Weaver, J. C.; Thanawala, M. S.; Sundar, V. C.; Morse, D. E.; Fratzl, P. *Science* **2005**, *309,* 275.

44. Green, D.; Walsh, D.; Mann, S.; Oreffo, R. O. C. *Bone* **2002**, *30,* 810.

45. Westbroek, P.; Marin, F. *Nature* **1998**, *392,* 861.

46. Green, D.; Walsh, D.; Yang, X. B.; Mann, S.; Oreffo, R. O. C. *J. Mater. Chem.* **2004**, *14,* 2206.

47. Popescu, D. C.; van Leeuwen, E. N. M.; Rossi, N. A. A.; Holder, S. J.; Jansen, J. A.; Sommerdijk, N. A. J. M. *Angew. Chem. Int. Ed.* **2006**, *45,* 1762.

48. Sommerdijk, N. A. J. M.; van Leeuwen, E. N. M.; Vos, M. R. J.; Jansen, J. A. *Cryst. Eng. Comm.* **2007**, *9,* 1209.

49. Boyan, B. D.; Hummert, T. W.; Dean, D. D.; Schwartz, Z. *Biomaterials* **1996**, *17,* 137.

50. Schwartz, Z.; Lohmann, C. H.; Sisk, M.; Cochran, D. L.; Sylvia, V. L.; Simpson, J.; Dean, D. D.; Boyan, B. D. *Biomaterials* **2001**, *22,* 731.

51. Olszta, M. J.; Cheng, X. G.; Jee, S. S.; Kumar, R.; Kim, Y. Y.; Kaufman, M. J.; Douglas, E. P.; Gower, L. B. *Mater. Sci. Eng. R* **2007**, *58,* 77.

52. Nudelman, F.; Pieterse, K.; George, A.; Bomans, P. H. H.; Friedrich, H.; Brylka, L. J.; Hilbers, P. A. J.; de With, G.; Sommerdijk, N. A. J. M. *Nat. Mater.* **2010**, *9,* 1004.

53. Tampieri, A.; Celotti, G.; Landi, E.; Sandri, M.; Roveri, N.; Falini, G. *J. Biomed. Mater. Res. A* **2003**, *67A,* 618.

54. Tampieri, A.; Sandri, M.; Landi, E.; Pressato, D.; Francioli, S.; Quarto, R.; Martin, I. *Biomaterials* **2008**, *29,* 3539.

55. Tampieri, A.; Landi, E.; Valentini, F.; Sandri, M.; D'Alessandro, T.; Dediu, V.; Marcacci, M. *Nanotechnology* **2011**, *22,* 015104.

56. Tampieri, A.; Sprio, S.; Ruffini, A.; Celotti, G.; Lesci, I. G.; Roveri, N. *J. Mater. Chem.* **2009**, *19,* 4973.

57. Stupp, S. I.; Braun, P. V. *Science* **1997**, *277,* 1242.

58. Herron, N.; Calabrese, J. C.; Farneth, W. E.; Wang, Y. *Science* **1993**, *259,* 1426.
59. Murray, C. B.; Kagan, C. R.; Bawendi, M. G. *Science* **1995**, *270,* 1335.
60. Murray, C. B.; Norris, D. J.; Bawendi, M. G. *J. Am. Chem. Soc.* **1993**, *115,* 8706.
61. Peng, X. G.; Wilson, T. E.; Alivisatos, A. P.; Schultz, P. G. *Angew. Chem. Int. Ed. Engl.* **1997**, *36,* 145.
62. Vossmeyer, T.; Reck, G.; Katsikas, L.; Haupt, E. T. K.; Schulz, B.; Weller, H. *Science* **1995**, *267,* 1476.
63. Donners, J. J. J. M.; Hoogenboom, R.; Schenning, A. P. H. J.; van Hal, P. A.; Nolte, R. J. M.; Meijer, E. W.; Sommerdijk, N. A. J. M. *Langmuir* **2002**, *18,* 2571.
64. Fendler, J. H.; Meldrum, F. C. *Adv. Mater.* **1995**, *7,* 607.
65. Braun, P. V.; Osenar, P.; Stupp, S. I. *Nature* **1996**, *380,* 325.
66. Tohver, V.; Braun, P. V.; Pralle, M. U.; Stupp, S. I. *Chem. Mater.* **1997**, *9,* 1495.
67. Fleischmann, R.; Geisel, T.; Ketzmerick, R. *Phys. Rev. Lett.* **1992**, *68,* 1367.
68. Kang, W.; Stormer, H. L.; Pfeiffer, L. N.; Baldwin, K. W.; West, K. W. *Phys. Rev. Lett.* **1993**, *71,* 3850.
69. Weiss, D.; Roukes, M. L.; Menschig, A.; Grambow, P.; Vonklitzing, K.; Weimann, G. *Phys. Rev. Lett.* **1991**, *66,* 2790.
70. Sone, E. D.; Zubarev, E. R.; Stupp, S. I. *Angew. Chem. Int. Ed.* **2002**, *41,* 1705.
71. Singh, S.; Bozhilov, K.; Mulchandani, A.; Myung, N.; Chen, W. *Chem. Commun.* **2010**, *46,* 1473.
72. de la Rica, R.; Matsui, H. *Angew. Chem. Int. Ed.* **2008**, *47,* 5415.
73. Mao, C. B.; Solis, D. J.; Reiss, B. D.; Kottmann, S. T.; Sweeney, R. Y.; Hayhurst, A.; Georgiou, G.; Iverson, B.; Belcher, A. M. *Science* **2004**, *303,* 213.
74. Klaus, T.; Joerger, R.; Olsson, E.; Granqvist, C. G. *Proc. Natl. Acad. Sci. U.S.A.* **1999**, *96,* 13611.
75. Naik, R. R.; Stringer, S. J.; Agarwal, G.; Jones, S. E.; Stone, M. O. *Nat. Mater.* **2002**, *1,* 169.
76. Reches, M.; Gazit, E. *Science* **2003**, *300,* 625.
77. Braun, E.; Eichen, Y.; Sivan, U.; Ben-Yoseph, G. *Nature* **1998**, *391,* 775.
78. Deng, Z. X.; Mao, C. D. *Nano Lett.* **2003**, *3,* 1545.
79. Tseng, R. J.; Tsai, C. L.; Ma, L. P.; Ouyang, J. Y. *Nat. Nanotechnol.* **2006**, *1,* 72.
80. Sun, Q. Y.; Vrieling, E. G.; van Santen, R. A.; Sommerdijk, N. A. J. M. *Curr. Opin. Solid State Mater. Sci.* **2004**, *8,* 111.
81. Vrieling, E. G.; Sun, Q. Y.; Beelen, T. P. M.; Hazelaar, S.; Gieskes, W. W. C.; van Santen, R. A.; Sommerdijk, N. A. J. M. *J. Nanosci. Nanotechnol.* **2005**, *5,* 68.
82. Sundar, V. C.; Yablon, A. D.; Grazul, J. L.; Ilan, M.; Aizenberg, J. *Nature* **2003**, *424,* 899.
83. Kroger, N.; Deutzmann, R.; Sumper, M. *Science* **1999**, *286,* 1129.
84. Kroger, N.; Deutzmann, R.; Sumper, M. *J. Biol. Chem.* **2001**, *276,* 26066.
85. Kroger, N.; Lorenz, S.; Brunner, E.; Sumper, M. *Science* **2002**, *298,* 584.
86. Naik, R. R.; Whitlock, P. W.; Rodriguez, F.; Brott, L. L.; Glawe, D. D.; Clarson, S. J.; Stone, M. O. *Chem. Commun.* **2003**, 238.

87. Patwardhan, S. V.; Mukherjee, N.; Steinitz-Kannan, M.; Clarson, S. J. *Chem. Commun*. **2003**, 1122.

88. Coradin, T.; Durupthy, O.; Livage, J. *Langmuir* **2002**, *18,* 2331.

89. Coradin, T.; Roux, C.; Livage, J. *J. Mater. Chem*. **2002**, *12,* 1242.

90. Sun, Q. Y.; Beelen, T. P. M.; van Santen, R. A.; Hazelaar, S.; Vrieling, E. G.; Gieskes, W. W. C. *J. Phys. Chem. B* **2002**, *106,* 11539.

91. Gronwald, O.; Snip, E.; Shinkai, S. *Curr. Opin. Colloid Interface Sci*. **2002**, *7,* 148.

92. Soler-illia, G. J. D.; Sanchez, C.; Lebeau, B.; Patarin, J. *Chem. Rev*. **2002**, *102,* 4093.

93. Prins, L. J.; Reinhoudt, D. N.; Timmerman, P. *Angew. Chem. Int. Ed*. **2001**, *40,* 2382.

94. van Esch, J. H.; Feringa, B. L. *Angew. Chem. Int. Ed*. **2000**, *39,* 2263.

95. Sugiyasu, K.; Tamaru, S.; Takeuchi, M.; Berthier, D.; Huc, I.; Oda, R.; Shinkai, S. *Chem. Commun*. **2002**, 1212.

96. Jung, J. H.; Shinkai, S.; Shimizu, T. *Nano Lett*. **2002**, *2,* 17.

97. Jung, J. H.; Shinkai, S.; Shimizu, T. *Chem. Mater*. **2003**, *15,* 2141.

98. Botterhuis, N. E.; Sun, Q. Y.; Magusin, P. C. M. M.; van Santen, R. A.; Sommerdijk, N. A. J. M. *Chem. Eur. J*. **2006**, *12,* 1448.

99. Sun, Q. Y.; Kooyman, P. J.; Grossmann, J. G.; Bomans, P. H. H.; Frederik, P. M.; Magusin, P. C. M. M.; Beelen, T. P. M.; van Santen, R. A.; Sommerdijk, N. A. J. M. *Adv. Mater*. **2003**, *15,* 1097.

100. Yu, C. Z.; Tian, B. H.; Fan, J.; Stucky, G. D.; Zhao, D. Y. *Chem. Lett*. **2002**, 62.

101. Prouzet, E.; Cot, F.; Boissiere, C.; Kooyman, P. J.; Larbot, A. *J. Mater. Chem*. **2002**, *12,* 1553.

102. Wong, M. S.; Cha, J. N.; Choi, K. S.; Deming, T. J.; Stucky, G. D. *Nano Lett*. **2002**, *2,* 583.

CHAPTER 7

Bioinspired Catalysis

GERHARD F. SWIEGERS, JUN CHEN, and PAWEL WAGNER

Intelligent Polymer Research Institute and ARC Centre of Excellence for Electromaterials Science, University of Wollongong, Wollongong, NSW 2522, Australia

7.1 INTRODUCTION

Catalysts are species that accelerate chemical reactions without themselves being consumed in the process. The most efficient catalysts by far are the catalysts of biology, known as *enzymes*. The maintenance and creation of life in all its many and varied forms on Earth depends on the ability of enzymes to speed chemical transformations in biochemical systems. To this end, enzymes often display truly amazing vigour, specificity, and reliability. The fact that life itself depends on the action of enzymes testifies to their remarkable power.

To illustrate enzymatic capacities, it is worth considering that a single example of the enzyme carbonic anhydrase has the capacity to convert 600,000 CO_2 molecules per second in our muscles into H_2CO_3 in our blood.[1, 2] It can do this repeatedly, without fail, at body temperature (37 °C) in the extraordinarily mixed reactant feedstock that is a biological fluid, and at a CO_2 partial pressure of ≤ 1 atmosphere.[1, 2] Moreover, it selectively transforms CO_2 in the presence of a wide variety of other possible reagents without becoming deactivated.

By comparison, modern industrial catalysts are rudimentary in their operation. For example, the economically important Haber–Bosch process for the production of ammonia from nitrogen and hydrogen typically requires temperatures of 500 °C with the reagent gases compressed to 1000 atmospheres. Despite these extremes, ammonia is generated in only 15–25% yield. The catalyst, a heterogeneous iron and oxide mix, must be replaced periodically because it is poisoned by even miniscule impurities in the feedstocks.

How do enzymes achieve such feats? More pertinently, how can we replicate them?

Because of the remarkable versatility and efficiency of enzymes, understanding and applying Nature's catalytic principles in nonbiological systems is exceedingly

Bioinspiration and Biomimicry in Chemistry: Reverse-Engineering Nature, First Edition.
Edited by Gerhard F. Swiegers.
© 2012 John Wiley & Sons, Inc. Published 2012 by John Wiley & Sons, Inc.

important. Ronald Breslow of Columbia University outlined it as one of the Holy Grails of chemistry in a landmark 1995 scientific publication in the journal *Accounts of Chemical Research*.[3] He coined the term *biomimetic chemistry*, which is defined as[3] "imitating the style of enzyme catalyzed processes in an effort to achieve some of the advantages, which Nature has realized by the use of enzymes."

There is no doubt that biomimetic chemistry holds substantial promise. A true understanding of, and implementation of, Nature's catalytic principles offers the prospect of transforming industry in innumerable ways. For example, by avoiding the extreme temperatures and pressures required in many modern-day catalytic processes, catalysts that are truly biomimetic could allow for small-scale, localized production of industrial chemicals. Such an elimination of the need for economies of scale would likely create major changes in the production of chemical feedstocks. Farmers may, for example, be able to distribute on their fields catalysts that directly convert atmospheric nitrogen to ammonia. The industrial production of fertilizer would then be shifted from large-scale centralized facilities to small-scale, local processes. Many other industries could be similarly transformed by an understanding and implementation of Nature's catalytic principles.

The problem with this scenario, however, is that we do not yet fully understand how enzymes work. This is proved by the fact that we have not been able to harness Nature's catalytic principles to create new, highly efficient catalyst systems with capacities similar to those of enzymes. If we did truly understand enzymes, we would, undoubtedly, have developed such catalysts.

This leads us to a secondary and perhaps a more important problem. A powerful antireductionist sentiment seems to prevail among some sections of the enzymology community, who consider that the unique features and complexity of biological catalysis makes it, effectively, impossible to implement Nature's catalytic principles in human-made systems. This approach is, in many ways, a more significant problem because it suppresses and subjugates innovation in a profoundly disempowering manner. One can never achieve something you truly believe to be impossible. It arguably also closes the mind to the possibility that Nature can teach us something which lies beyond the existing paradigms.

The fact of the matter is that humankind has a very highly tuned capacity for innovation. It is not inconceivable that this capacity for innovation may ultimately prove to be more than a match for Nature in the realm of catalysis; that is, we may come up with unexpected and convenient ways to harness Nature's catalytic principles. Moreover, there is always something more to learn. But the first step to acquiring deeper understanding is to acknowledge that Nature has something more to teach.

Biological systems are certainly and admittedly extraordinarily complicated. This is especially true of enzymatic catalysis, which is characterized by remarkable chemical dynamism, feedback loops, and all manner of complexity, both subtle and extreme. The fundamental, underlying principles by which enzymes operate are therefore well concealed. They are clearly not amenable to ready elucidation using the physical and spectroscopic techniques that we currently have available. Unearthing the fundamental processes that govern enzymic action and adapting them to human use is therefore an extremely challenging endeavor.

The same was true of an earlier example of a biological phenomenon that challenged human ingenuity: heavier-than-air flight. Until the Wright brothers, nobody understood that birds exploit a rather simple principle to fly—the law of the aerofoil. Thus, while the action of a bird flapping its wings is bewilderingly complicated and incompletely understood to this day, beneath that complexity lay a fairly straightforward underlying principle. That principle, once it was understood, could be innovatively adapted and applied to build aircraft. The genius of the Wright brothers was not to clarify in detail the action of birds during flight. Rather, it was to build the first wind tunnel and thereby correctly identify and formulate the law of the aerofoil as the key phenomenon underlying flight by heavier-than-air objects. This is what made the first flight at Kitty Hawk possible.

The challenge in respect of understanding and mimicking enzymes may be similar. We need to devise systems and experiments that elucidate and clarify, in crisp relief, the key underlying principle(s) of enzymatic catalysis. We then need to innovatively implement that understanding in a nonbiological setting.

The historic example of heavier-than-air flight illustrates another feature of biomimetic chemistry. Every winged aircraft and putative aircraft ever built comprises nothing less than a test of the builder's understanding of the underlying principle by which birds fly, namely, the law of the aerofoil. In achieving powered flight at Kitty Hawk, the Wright brothers not only solved the problem of mimicking flight by birds but also demonstrated in an entirely unequivocal way that they fully understood the principle that underlies it.

In the same way, by developing catalysts that are bioinspired or biomimetic, we are not only seeking to harness Nature's catalytic principles, we are also testing our own understanding thereof. We can only ever be sure that we truly understand Nature's catalytic principles if we can design, prepare, and demonstrate catalysts whose feats are comparable to enzymes. The development of bioinspired or biomimetic catalysis therefore encapsulates both an aim and a test.

In this chapter, we consider biomimetic chemistry in this light. We start by briefly discussing how catalysts operate in general and then how we have historically understood enzymes to operate. Thereafter, we examine and consider a range of representative examples of "bioninspired/biomimetic catalysts" that illustrate the key theories of enzymatic catalysis. Finally, we step back and consider how one may implement those principles that seem to be critical in enzymes. We ask whether there is an underlying principle and, if so, how we can harness it to develop new, highly active, highly selective nonbiological catalytic systems. In the last section of the chapter, we remark briefly on how close (or far) we appear to be to a full understanding and implementation of Nature's catalytic principles.

7.2 A GENERAL DESCRIPTION OF THE OPERATION OF CATALYSTS

Figure 7.1 schematically depicts a general example of a catalytic process in which two reactants, A and B, are transformed into products. The left-hand side of Figure 7.1a shows the two reactants binding reversibly to the catalyst (the open circle) to form a catalyst–reactant intermediate. The catalyst may be a molecule dissolved in liquid solution (in which case it is termed a *homogeneous catalyst*)[4] or it may be a solid state material whose surface is catalytically active (in which case it is termed a *heterogeneous catalyst*).[4] Enzymes are widely considered to be examples of homogeneous catalysts,[4] albeit natural ones not synthetic ones. In enzymes, the catalyst–reactant intermediate is called the *Michaelis complex*.

Once the reactants are bound, the catalyst will typically engineer "collisions" between them (to use the terminology of *collision theory*), to thereby form a short-lived, high-energy *transition state*. The square bracket in the middle of Figure 7.1a depicts the transition state. The catalyst may mediate such a reactant collision in several ways. In the case of a molecular homogeneous catalyst dissolved in open solution, the collision may be created by a conformational

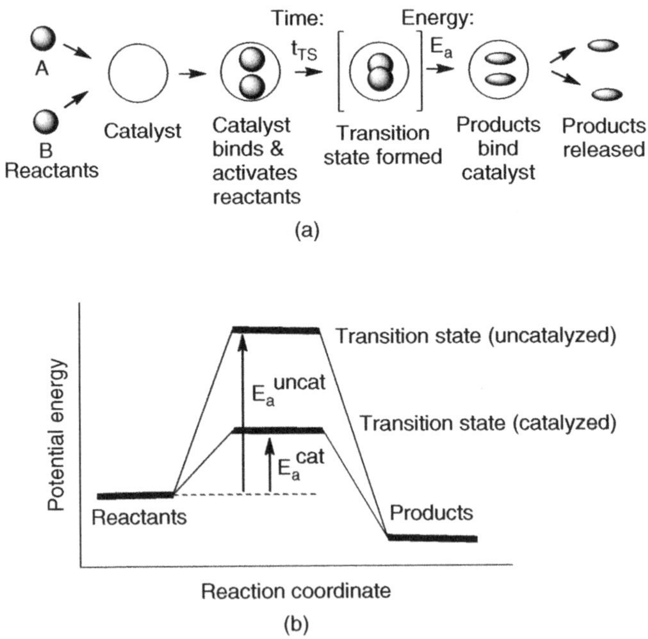

Figure 7.1 Schematic depiction of (a) a catalyzed, reactive encounter between two molecules, A and B, leading to the formation of products, and (b) the energetic profile followed during the catalyzed reaction, showing the minimum threshold energy (the activation energy, E_a^{cat}) needed for reaction, relative to the same threshold for the uncatalyzed reaction (E_a^{uncat}). (Adapted with permission. Copyright John Wiley & Sons: Ref 21b.)

change within the structural framework of the catalyst. Alternatively, it may be generated by a molecular rearrangement about a particular atom within the catalyst. In a solid state heterogeneous catalyst, the collision may be created by movement of bound reactants from surface atom to surface atom over the catalyst surface. Alternatively, reactants bound within the crystal lattice of the catalyst may collide due to oscillatory thermal motion.

In all cases, collisions of this type occur with a particular *collision frequency*, meaning that there is a certain average time, t_{TS}, between each collision and the next. This time, t_{TS} (units: seconds), is the reciprocal of the collision frequency (units: s^{-1}).

Once a transition state is formed in a collision, it must overcome the minimum threshold energy required for reaction; namely, the *activation energy*, E_a. Figure 7.1b schematically depicts the activation energy that must be overcome relative to the corresponding activation energy in an uncatalyzed process. If the transition state has sufficient energy to overcome this barrier, the reactants are transformed into products, which are initially bound to the catalyst and then released, thereby regenerating the catalyst for another cycle. If the collision is insufficiently energetic, then the catalyst–reactant intermediate is reformed and no products are created.

What, then, is the principle on which catalysts accelerate chemical reactions? As depicted in Figure 7.1b, catalysts have the general property that they often drastically diminish the activation energy, E_a, that must be overcome to bring about reaction. They do this, at least in part, by binding to the transition state and stabilizing it. In an uncatalyzed reaction, no such stabilization is available, meaning that higher energies must be achieved in the transition state in order for it to form products. Because of this lowering of the energy barrier, the intermediacy of a catalyst allows reactions to proceed at a faster rate than is the case when a catalyst is not present.

(Explanatory comment: Readers kindly note that for convenience and continuity in argument, we continue to use the terms *collision* and *collision frequency* in the remainder of this chapter to describe contact between reactant functionalities in liquid phase homogeneous catalysts. We acknowledge that collision is a gas phase term and the liquid phase equivalent is, more correctly, termed an encounter.)

7.3 A BRIEF HISTORY OF OUR UNDERSTANDING OF THE OPERATION OF ENZYMES

Numerous theories have been proposed to explain the catalytic power of enzymes over the last 100 years. At least 21 distinct hypotheses of enzymatic catalysis were listed in a 1989 publication.[5] Space precludes a description of all of these theories; however, we consider here the critical ones that have led to the current conceptualization of enzymatic catalysis as it is applied in bioinspired/biomimetic studies. Several of these historically key theories have been the subject of intensive investigation in this respect.

7.3.1 Early Proposals: Lock-and-Key Theory, Strain Theory, and Induced Fit Theory

In 1894, Emil Fischer came to the insight that the specificity of glycolytic enzymes indicates that they must have a particular shape into which the substrate fits exactly.[1] He described the process of enzyme–substrate interactions as being similar to a key fitting a lock.[1] The substrate only bound and was transformed if its shape was complementary to the docking site presented by the active site of the enzyme. In effect, he was proposing that molecular recognition played a key role in the process of enzymatic catalysis. The resulting theory came to be known as the *lock-and-key* theory.

In the 1930s, J. R. Haldane proposed a modification of the lock-and-key theory.[6] He suggested that enzymes and their substrates do not bind in perfect complementarity to each other. Rather, they bind somewhat imperfectly, so that the substrate becomes structurally strained upon binding. Haldane stated: "Using Fischer's lock and key simile, the key does not fit the lock perfectly, but exercises a certain strain upon it."[6] This strain acts to accelerate the catalytic transformation of the substrate by the enzyme.

In 1958, Daniel Koshland proposed a further adaption along this line of reasoning.[7] He postulated that enzyme active sites were, in fact, relatively flexible. What mattered, according to Koshland, was not that the "key" (the substrate) distorted to become precisely complementary to the "lock" (the enzyme active site), but rather that the two could be induced to structurally match each other.[7] In other words, if the active site could distort to precisely accommodate the substrate, then strong enzyme–substrate binding would occur. The resulting molecular recognition could potentially explain the specificities that characterize enzymatic catalysis. Fersht later suggested that whereas the substrate was strained, the enzyme was "stressed."[6]

7.3.2 The Critical Role of Molecular Recognition in Enzymatic Catalysis: Pauling's Concept of Transition State Complementarity

In 1946, Linus Pauling made an important advance in the understanding of enzymes. He noted that many enzyme active sites appeared to be structurally complementary to the *optimum* transition state of the reaction that they catalyzed.[8] He suggested that this structural complementarity likely caused the enzyme to form a transition state that was close to ideal and also to bind that transition state significantly more strongly than it did the substrate. The effect was to diminish the overall energy barrier (the activation energy, E_a) that must be overcome in the reaction.[9] A smaller energy barrier necessarily means that the reaction will proceed more rapidly, thereby explaining the remarkable rate accelerations that are engineered by enzymes. Enzymes have been shown to accelerate the rates of chemical reactions by up to 10^8-fold relative to the equivalent uncatalyzed reaction.[9]

Pauling's proposal has found strong supporting evidence over the years. For example, studies have shown that, while substrate binding is generally strong in enzymes ($1/K_M \approx 10^{4\pm3}$ M^{-1}),[1] transition state binding is substantially stronger by several orders of magnitude ($1/K_M \approx 10^{16\pm4}$ M^{-1}).[10]

Thus, it appears that, by dint of the structure of their active site, enzymes may exert a powerful stabilizing influence on the transition state of the reaction. Not only can this explain the rate accelerations observed for enzymes,[10] but tight binding of the transition state could also explain why specific substrates are cleaved more rapidly than nonspecific ones; that is, it could explain the specificity of enzymatic catalysis.[10]

While there is good evidence for Pauling's theory, the one problem it has encountered is that its insights have proved difficult to reduce to practice in a nonbiological setting. That is, powerful synthetic catalysts capable of mimicking the feats of enzymes have not been developed using Pauling's proposal.

In other words, while studies of enzymatic systems have shown that molecular recognition of the substrate and, more particularly, of the transition state plays a critical role in enzymatic catalysis, attempts to design nonbiological molecular catalysts that exploit this effect have not, to date, yielded new, practical, highly active and specific nonbiological catalysts. This is disturbing because it suggests an inconsistency between theory and practice.

There are a range of possible reasons for the fact that Pauling's theory has not led to new, powerful synthetic catalysts capable of emulating the feats of enzymes. We list below several possibilities in this regard. We critically evaluate and consider these possibilities later in this chapter.

1. *A Very High Degree of Structural Complementarity to the Transition State May Be Needed.* It is possible that the synthetic systems that have been studied to date are not complementary enough to their reactants and transition state to achieve enzyme-like accelerations and specificity. That is, the molecular recognition of their reactants and, more especially, of the transition state by the synthetic catalysts may have been too weak to yield an effect as powerful as that displayed by enzymes. In that case, efforts need to be focused on developing catalysts whose structures more closely complement the transition states of the reactions that they catalyze.

2. *Pauling's Theory May Be Incomplete.* It is conceivable that other factors not described by Pauling's conceptualization may be critical in enzymes. Such factors may potentially be even more critical than Pauling's insights. The absence of these features in the synthetic systems that have been studied to date would then have blocked a true mimicry of the feats of enzymes. If this is the case, then we need to focus on identifying those missing factors and including them in future bioinspired/biomimetic catalysts.

3. *Our Understanding of Molecular Recognition May Be Incomplete.* It is not inconceivable that there may be a fundamental misconception in our understanding of the process of molecular recognition itself. The synthetic systems that have been studied to date will then have been incorrectly designed. In that case, we need to fully define molecular recognition and design bioinspired/biomimetic catalysts that properly harness it.

7.3.3 The Critical Role of Approach Trajectories in Enzymatic Catalysis: Orbital Steering, Near Attack Conformers, the Proximity Effect, and Entropy Traps

One consequence of Pauling's theory was the realization that if enzyme active sites complement their optimum transition state, then they must also control the way in which substrate functional groups approach each other or disengage from each other during reaction. In fact, they must limit this approach or disengagement to trajectories that are close to ideal since the optimum transition state represents the best arrangement for reaction.

This realization spawned a range of theories seeking to capture the concept of optimized approach trajectories. One of the earliest and most prominent in this respect was the theory of *orbital steering*, which was proposed by Koshland in 1971.[11] According to this conceptualization, the rate accelerations displayed by enzymes can be explained by a severe angular dependence in the interaction of the relevant orbitals on the participating substrate functionalities. In effect, Koshland proposed that a perfectly linear approach of the interacting orbitals creates a much lower energy barrier for reaction than is the case if the orbitals are even slightly displaced from linear. That is, even miniscule optimizations in the reactant approach trajectories may have a very substantial effect on the energy barrier that must be traversed in practice.

While orbital steering generated much controversy at the time and was later effectively disavowed by its author,[12] its central thesis remains a perceived general wisdom in the field. This thesis states, in the language of collision theory,[13] that enzymes optimize the way that bound reactants approach each other immediately prior to collision and thereby create thermodynamically "ideal" or "near-ideal" collisions.

Several other theories effectively also advanced this concept, including, for example, Bruice's concept of near attack conformers.[14] Bruice further proposed the notion of a *proximity effect* in enzymes,[14] which recognized that the first step in achieving optimum collisions involves localizing the reactants in close proximity to each other and thereby constraining the way in which they may approach each other. Page, Jencks, and others have proposed that enzymes harness entropy in this respect to accelerate their reactions; that is, enzymes were suggested to be *entropy traps*.[15]

7.3.4 The Critical Role of Conformational Motion in Enzymatic Catalysis: Coupled Protein Motions

The concept of an "ideal" collision requires motion in the catalyst that mediates such a collision. It was perhaps for this reason that molecular motions were proposed to be critical to the catalytic process in enzymes. Starting in the 1970s, several biochemists started examining the link between conformational motion in enzymes and their catalytic properties. Evidence has since been collected for the existence of a network of coupled protein motions that facilitate enzymatic catalysis.[16]

This network appears to be comprised of fast, equilibrium thermal motions that contribute to slower conformational changes that are involved in the reaction.[16]

For example, recent NMR studies on enzymes like *cyclophilin A* and *triosephosphate isomerase* have detected regular, repeated conformational fluctuations in their active sites that occur on a time scale which correlates closely with the measured rate of substrate turnover.[17] The unambiguous implication of these studies is that the conformational oscillations are intimately connected to the catalytic process. However, the way in which they are connected and the fundamental nature of this connection were not clarified by the experiments.

Some investigators have consequently suggested that the conformational fluctuations may mediate "gating" and solvent exclusion of the enzyme active site, in which polypeptide loops or domains close and open over the catalytic site.[18] During the times they are open, substrates and solvent molecules are proposed to have access to the active site. When closed, substrates bound within the active site are predicted to be transformed chemically, with many forward and backward transits of the reaction possible.[18]

The latter proposition could potentially explain the fact that many enzymes appear capable of catalyzing both the forward and the reverse of their reactions, a property known as *bidirectional catalysis*.[18] Bidirectional catalysis is critical to maintaining homeostasis in biological entities, including in humans.[1] Except for certain acid, base, and other catalyzed reactions in synthetic organic chemistry, bidirectional catalysis is relatively unknown outside biology.[18c]

7.3.5 Enzymes as Molecular Machines: Dynamic Mechanical Devices and the Entatic State

In the early 1990s, Bob Williams at Oxford University put together the concepts of molecular recognition (via structural complementarity), optimized approach trajectories, and the role of regular, repeated conformational motion in his description of enzymes as dynamic mechanical devices.[19] In effect, Williams recognized that mechanical devices also display all of the above elements, albeit at a macro- not at a molecular scale. In fact, machines are the only contrivance known to humans that incorporate all of the elements mentioned above.

To illustrate this statement, consider Figure 7.2, which depicts the interdigitating cogs of a machine. As can be seen, the cogs employ structural complementarity to interlock their teeth and thereby constrain their trajectories and motions to pathways that are optimum. Their motions are driven by a regular, repeated mechanical impulse, which is qualitatively identical to the mechanical impulse imparted by regular, repeated conformational flexing on the molecular scale.

In the same way, Williams suggested that protein motions in enzymes guide the substrate through a limited set of movements that are comparable to the way in which a glove is guided, via structural complementarity, onto the fingers of a hand.[19] The underlying concept was that the conformational fluctuations constrain the way in which bound substrates approach each other. They do so in a manner analogous to the way in which the components in mechanical devices

Figure 7.2 A mechanical system. Note that the cogs employ structural complementarity to interlock their teeth and thereby constrain their trajectories and motions to pathways that are optimum. Their motions are driven by a regular, repeated mechanical impulse, which is qualitatively identical to the mechanical impulse imparted by regular, repeated conformational flexing on the molecular scale. The resulting cooperative synchronization of actions yields an outcome that is more than the simple sum of the parts (synergy). With such synchronization, mechanical devices can perform astounding feats. Without it, they are mere collections of machined parts. The same may be true of enzymes. (Reproduced with permission. Copyright John Wiley & Sons: Ref. 21e.)

are constrained by their structure to interact with each other only in an optimum fashion.[19]

Williams further (and separately) proposed that, like a machine, relatively unfavorable, energetically uphill actions may be driven at certain points in the system by the combined, synchronized motions. Thus, coupled protein motions in an enzyme may drive the formation at the active site of an energetically relatively unfavorable, high-energy arrangement—the entatic state.[20] Such an action would be qualitatively identical to the way in which, for example, a bottle-making machine uses the cumulative, synchronized mechanical impulse of the system to force plastic starting materials into the shape of a bottle.[21] In other words, coupled protein motions in an enzyme could have sufficient momentum to drive substrates into relatively unfavored chemical transformations and thereby generate new products.

While not noted by Williams, one may also mention that, in analogy with the bottle-making machine mentioned above, such transformations would occur within a structure that complements the desired outcome[21d]—as noted and predicted by Pauling.

7.3.6 The Fundamental Origin of Machine-like Actions: Mechanical Catalysis

Swiegers subsequently elaborated upon the proposals of Williams and explained their fundamental origin in his book *Mechanical Catalysis*.[21] He noted that the concept of a thermodynamically "ideal" collision between reactant functionalities necessarily required a new view of the catalytic process.

As mentioned earlier, the currently accepted view is that catalysts bring about rate accelerations by lowering the energy barrier that must be overcome in the reaction, namely, the activation energy, E_a.[13] The more this barrier is decreased, the larger the resulting rate acceleration. In a typical catalytic reaction there are many reactant collisions of which only a few lead to reaction. The role of the catalyst is to increase the proportion of those collisions that are successful. In other words, catalysts are believed to act by improving the thermodynamics of reactant collision.[13]

However, in the situation where the catalyst constrains the reactants to optimum approach trajectories, *every* reactant collision is thermodynamically "perfect" or "near perfect". The catalytic rate cannot then depend on the *activation energy* of the collision because there will no longer be unsuccessful collisions. Rather, it will depend on the catalyst-mediated collision frequency; that is, on how often the catalyst mediates collisions between the bound reactants.[21] In other words, the catalytic rate will then be governed by the *mechanics* of reactant collision, not by its *thermodynamics*.[21] The role of the catalyst is then to facilitate as many reactant collisions as possible per unit time. To do this, the catalyst needs to be engaged in rapid, repeated motion leading to reactant collisions. In molecular species the motion will typically be coupled conformational motions, like the protein motions of enzymes. However, in lattice-based, solid state catalysts it may, theoretically, also be oscillatory, thermal motion.

In effect, there are, generally, two critical steps in catalytic processes. As depicted in Figure 7.1a, the first step involves the formation and processing of the catalyst–reactant intermediate, leading to catalyst-mediated reactant collision. This step is characterized by a certain collision frequency ($1/t_{TS}$), which controls the rate at which collisions mechanically occur between the bound reactants. The collision frequency will generally depend on the rate of repeated motion in the catalyst and the extent to which it is synchronized with reactant binding. Since frequency (units: s^{-1}) is, as noted earlier, essentially a measure of time, this step can be said to be time dependent.

The second step, also depicted in Figure 7.1a, is a thermodynamic step, involving the need to overcome the activation energy, E_a. This barrier has units of kJ/mol or kcal/mol, so that it may be considered to be an energy-dependent step.

The important point to be made is that in any process involving a sequence of steps, the overall rate and pathway will depend on the slowest and least likely of these steps. Thus, whichever of the above steps is the slowest and least likely will control not only the rate but also the character of the catalytic process.

In typical nonbiological catalyzed reactions, the slowest step is the energy-dependent step of overcoming the activation energy. In such a system there will therefore be many catalyst-mediated collisions, of which only a proportion will lead to products. The catalyst accelerates the reaction by increasing the proportion of successful collisions.

However, in the case where the catalyst mediates only "ideal" or "near-ideal" collisions, virtually every collision leads to products. In that situation, the slowest and least likely step is the time-dependent step of catalyst-mediated reactant collision. The action of the catalyst, including its overall rate, is then dependent on and governed by the temporal and spatial fluctuations involved in creating collisions. That is, the action of the catalyst is dominated by the mechanics of reactant encounter, not the thermodynamics thereof.

Catalysts of this type may consequently be termed *mechanical catalysts*.[21] This term is also apt to the analogy with mechanical devices mentioned in the previous section. In fact, it explains why catalysts controlled by their collision frequency are like machines. The key features of such catalysts include their ability to (1) repeatedly and reliably carry bound reactants into reactive collisions along optimum trajectories (achievable through structural complementarity with the transition state), and (2) harness the mechanical impulse of regular, repeated conformational motion to generate products at a high rate. This mode of action is very different and quite distinct from that employed by conventional (energy-dependent) catalysts.

In summary, according to this conceptualization,[21] there are two distinct realms of catalysis: thermodynamic (or energy-dependent) and mechanical (or time-dependent) catalysis. Catalysts in the former realm are governed by the need to overcome the activation energy. In the latter realm, however, the most demanding step involves creating collisions.

A time-dependence mode of action could potentially explain numerous aspects of enzymatic catalysis. For example, as noted earlier, it could explain why enzymatic action is often associated with vigorous, repeated conformational motion in proteins.[16, 17] Because the catalytic rate depends on the collision frequency, it is governed by how often the enzyme mediates reactant collisions. This, in turn, depends on the rate of conformational flexing in the active site, since conformational change is how most enzymes create collisions between bound reactants.

Enzymic action is therefore postulated to resemble the molecular/ion pumps of biology, which also rely on and harness rapid, repeated conformational motion. The only difference is that, whereas molecular/ion pumps bring about substrate transport across membranes, enzymes are proposed to bring about substrate transformation.

Several other general properties of enzymatic catalysis are also potentially explicable if they employ a mechanical mode of action. For example, the ubiquity of *Michaelis–Menten* kinetics in enzymes can be rationalized by the fact that the reaction rate in a mechanical catalyst is determined by the formation and processing of the rapidly equilibrating reactant–catalyst intermediate, which goes on to form the transition state (see Figure 7.1a). As noted earlier, in enzymes this intermediate is called the *Michaelis complex*. By contrast, the reaction rate in conventional catalysts is governed by the activation energy, E_a.

In later sections of this chapter we discuss several other aspects of enzymes that can be explained by these concepts.

Finally, we should note that there are several compelling precedents for the concept of catalysis governed by the collision frequency. The first of these comprises the fact that uncatalyzed reactions controlled by their collision frequency are unusual but well known in chemistry.[21c, 22] They are termed diffusion-controlled reactions. An example of such a reaction is the acid–base reaction ($H^+ + OH^- \rightarrow H_2O$; $k \approx 10^{-10}$ s), which is limited only by the rate at which the species diffuse to each other. This reaction is certainly not controlled by its activation energy.

While common in uncatalyzed processes, catalyzed reactions controlled by their collision frequency have, with only one obscure exception,[21g, 23] never been recognized as distinct, nor studied at all.[21c, 22] There can be no justification for this since catalyzed reactions have lower activation energies in general than uncatalyzed ones and must therefore be more likely to be governed by their collision frequency.[21c, 22]

We should note at this juncture that the term "diffusion controlled" is also used in enzymology. However, it designates enzymes that display an extremely high overall catalytic rate that is governed only by the rate at which the substrates bind the enzyme.[24] That is, this term indicates a dependence on the rate of diffusion by the reactants to the catalyst, not a dependence on the catalyst-mediated collision frequency. No description has hitherto existed to describe homogeneous catalysts, including enzymes, that are governed by their catalyst-mediated collision frequency.

The exception referred to above provides the second compelling precedent. A 1950's study of *volcano plots* in various heterogeneous metal catalysts established—unequivocally—the existence of two distinct realms of catalysis, governed by their activation energies and collision frequencies respectively.[21g, 23] If such realms exist in heterogeneous catalysis, then they must necessarily also exist in other classes of catalyst, including nonbiological homogeneous catalysts and enzyme catalysts.

7.4 REPRESENTATIVE STUDIES OF BIOINSPIRED/BIOMIMETIC CATALYSTS

In order to evaluate the conceptualizations described above and verify that they truly trend toward a fuller understanding of enzymes, it is necessary to test them against reality. To do this we need to compare the characteristics of putative bioinspired or biomimetic catalysts with those of enzymes as a group. The better the match, the more confident one may reasonably feel in this respect.

In this section we briefly list several distinctive properties of enzymes as a class of catalyst. We then examine bioinspired/biomimetic models that display one or more of these properties. The properties may be incidental to the remarkable feats of enzymes, or they may be critical. They may also be an outcome of another property. Regardless of their origin, it is only by recognizing their existence, rationalizing their origin, and understanding their interrelatedness that we can hope to expose the underlying principle(s) that govern enzymatic catalysis.

7.4.1 Important General Characteristics of Enzymes as a Class of Catalyst

Some of the general properties of enymes that could serve as useful comparators for evaluating bioinspired/biomimetic catalysts include the following:

- Structural complementarity of the active site with the reaction transition state.[8] As noted earlier, the active site of most enzymes appears to structurally complement the reaction transition state. The existence of a catalytic effect in most ezymes is typically extremely sensitive to this feature. For example, minor changes in the sequence of amino acid residues in enzyme active sites often diminish this complementarity, resulting in a complete and total, nonlinear loss of the catalytic properties.[9, 10b] Even modifications to remote residues may dramatically influence enzymatic rates because they affect the structural complementarity to the transition state.

- An apparent association with vigorous, repeated conformational motion for the catalytic effect. Anything that slows the rapid, regular, and repeated protein flexing usually dramatically diminishes the overall reaction rate.[16, 17]
- *Michaelis-Menten* kinetics. Such kinetics are characterized by a reaction rate controlled by the rapidly equilibrating catalyst–reactant intermediate. This intermediate is observed in the rate expression.[1, 21d] Because enzymes routinely display Michaelis-Menten kinetics, enzymological kinetics has been formulated exclusively in terms of this type of behavior.
- Remarkable activity and specificity. Enzymes routinely achieve turnover frequencies in the range 1–10,000 s^{-1}.[1] Moreover, they often operate with near-total selectivity in the extraordinarily mixed reagent streams of biology; despite the choice of possible substrates, they typically select and transform only one substrate. They do this without getting poisoned and at remarkably mild temperatures and pressures.

The most critical and desirable property of enzymes is the latter one—their often awe-inspiring activity and specificity. This property undoubtedly offers the key measure of success in seeking to understand and emulate Nature's catalytic principles. If we can develop catalysts that match the activities and specificities of enzymes, we can truly say that we know how they work and can harness the principles that they employ. This particular characteristic therefore offers the target and benchmark against which we can evaluate our efforts.

7.4.2 Bioinspired/Biomimetic Catalysts that Illustrate the Critical Importance of Reactant Approach Trajectories

A very large body of work has examined, or sought to examine, catalytic species that act by constraining their bound reactants to approach each other along increasingly more optimum trajectories immediately prior to collision. The majority of this work has involved bimetallic catalysts with tethers between the metal ion catalytic groups that act to limit and constrain the approach trajectories that are available to the bound reactants.[25] To illustrate the principle that has been studied, we discuss a representative example in this respect.

7.4.2.1 Monomeric Chromium Epoxidation Catalysts Figure 7.3 depicts a dichromiun epoxidation catalyst system studied by Jacobsen and co-workers.[26] In these species, the Cr ions in **1a–c** or **2** bind to an epoxide or an azide (N_3) as depicted at the bottom of the figure.[26] In so doing, they activate these reactants, meaning that they polarize the electronic arrangement of the reactants to thereby make them conducive to reaction when they are brought into contact with each other.[22] The polarization may involve the metal ions withdrawing electron density

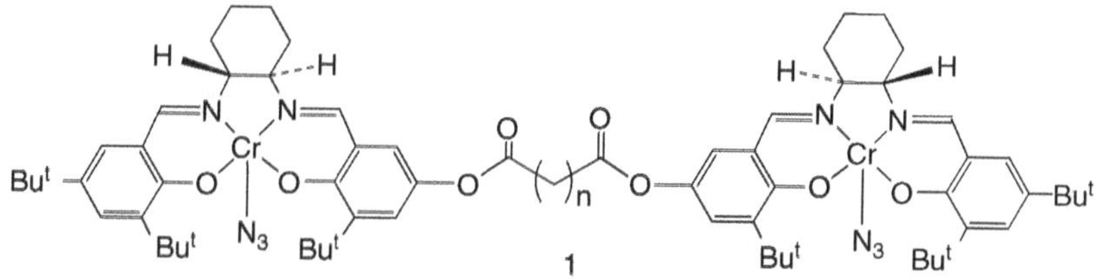

1

a: n = 5 **b**: n = 2 **c**: n = 10

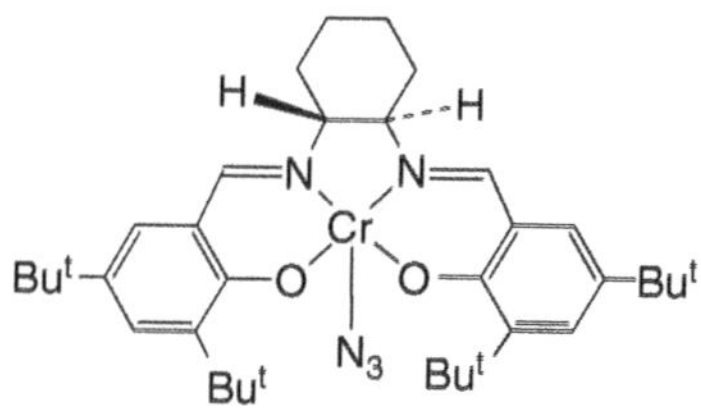

2

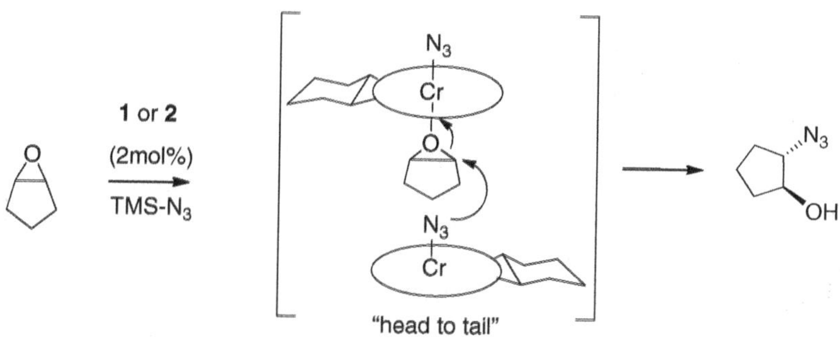

Intramolecular reaction rate constant (k_{intra}): **1a:** $42.9 \times 10^{-2}\ min^{-1}$
1b: $4.4 \times 10^{-2}\ min^{-1}$
1c: $3.8 \times 10^{-2}\ min^{-1}$

Figure 7.3 Bioinspired bimetallic catalysts that illustrate the critical role of reactant trajectory. Catalytic ring-opening reaction of an epoxide. The reaction involves two catalytically functioning Cr groups.[37] (Reproduced with permission. Copyright John Wiley & Sons: Ref. 21i.)

from the reactant bonding orbitals or donating electron density into the antibonding orbitals of the reactants. Figure 7.4 illustrates how a catalyst may activate its bound reactants.[21j] The extent of the activation depends entirely on the nature of the catalytic group. A transition metal catalytic group such as a Cr ion will typically induce a stronger polarization (activation) than, say, an amino acid residue catalytic group in an enzyme.

In the next step of the process, the catalyst brings the polarized reactants into physical contact with each other ("collision"). In the case of the monomeric catalyst **2**, collisions between the Cr-bound azide and epoxide reactants occur in open solution between free, monomeric species. The collisions therefore occur

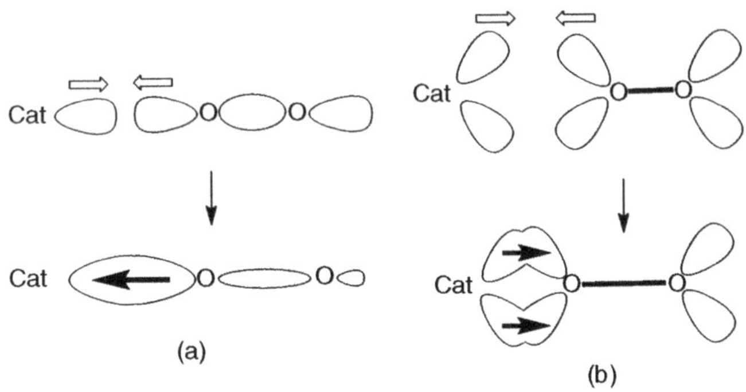

Figure 7.4 Reactant activation by a catalyst. Schematic illustrating a catalyst (CAT) activating a representative reactant (O_2) by increasing the O–O bond in length in order to facilitate its cleavage. The activation is achieved by (a) withdrawing electron density from its σ-bonding molecular orbital and/or (b) increasing electron density in its π^*-antibonding orbital. (Reproduced with permission. Copyright John Wiley & Sons: Ref. 21j.)

with entirely random orientations. Many of these orientations will be highly inopportune for reaction in that the wrong parts of the respective functional groups will be brought into contact with each other. In those cases, the activation provided by the Cr-ion catalytic groups will be insufficient to bring about reaction. In other words, in many of the collisions that occur between the bound azides and epoxides on **2**, the reactants will not be polarized enough to react and form products. The catalyst will then not have provided sufficient energy to facilitate the reaction.

In a much smaller proportion of the collisions, however, the reactants will happen to be well oriented and disposed for reaction. In these cases, the activation provided by the catalyst will be sufficient to bring about reaction. That is, in these collisions, the reactants will be polarized enough to react with each other and the catalyst will have provided sufficient energy to facilitate reaction.

In the monomeric catalytic system involving **2**, catalysis therefore involves many collisions, of which only a small fraction are successful. This fraction is, nevertheless, greater than it would have been in the equivalent uncatalyzed reaction. In that case, a still smaller proportion of the collisions between the reactants would have been successful, because the reactants were not made more reactive due to polarization imparted by a catalyst.

The reaction mediated by **2** in Figure 7.3 is an example of a classical energy-dependent catalytic process, where the catalyst acts to make the reactants more reactive, thereby decreasing the threshold energy barrier (the activation energy, E_a) that must be overcome.[21f] The more opportunely the catalyst polarizes the reactants, the greater the proportion of successful collisions and the faster the overall catalytic rate.

One way that a catalytic species may help overcome the activation barrier is by polarizing (activating) the reactants more intensely and thereby increasing their overall reactivity.[21f] This can potentially be achieved for example, by, using a suitable metal ion in higher oxidation state. A tactic of this type amounts, in effect, to a brute force approach involving an application of greater energy. The stronger and more intense the activation provided by the catalytic groups, the wider will be the range of approach trajectories and pathways that will lead to reaction upon collision, and the faster will be the overall catalytic rate.

7.4.2.2 Tethered Dichromium Epoxidation Catalysts Another, more subtle way to accelerate the overall catalytic rate is to design the catalyst so as to ensure that a greater proportion of the reactant collisions occur in orientations and dispositions that are suitable for reaction. In other words, instead of applying more energy to increase the range of orientations that will result in reaction, one can limit the ways in which the reactants are able to approach each other to those that are most suited for reaction. If successful, this should have the effect of increasing the proportion of collisions that lead to reaction and generate products.

The dichromium catalytic species **1a–c** in Figure 7.3 employ this approach. The presence of a tether between the two Cr-ion catalytic groups limits the ways in which the activated, bound azides and epoxides can approach each other immediately prior to collision. Thus, species **1a–c** differ from **2** in having the catalytic groups physically connected to each other, which means that the reactants become localized in close proximity to each other. They also differ among themselves in the length of the tethers that join the two Cr-ion catalytic groups in **1a, 1b**, and **1c**.

Listed at the bottom of Figure 7.3 are the comparable intramolecular reaction rate constants for **1a–c**.[4] As can be seen, these constants differ significantly. Thus, **1a** has the highest rate constant, indicating that it constrains the approach trajectories and orientations of its bound, activated epoxide and azide reactants to the best effect. For **1b**, the tether is clearly too short to achieve the same extent of optimization. In **1c** the tether is too long.

The only way to rationalize these rate constants is to recognize that there must be an "ideal" set of orientations and trajectories for reactant collision leading to product formation. If it was simply a case of creating more collisions, then **1b** would have the highest reaction rate, because the short tether would see it create the greatest number of intramolecular collisions per unit time. The fact that **1a** is, instead, the most reactive can only be explained by it having the greatest population of "ideal" trajectories and orientations prior to collision. In other words, a set of key, important conformational interconversions that bring about successful reactant collisions must be more populated in **1a** than in **1b**. These conformational changes must drive the attached and activated reactants along particular, suitable approach pathways into collisions that lead to reaction.

While **1c** likely has the same set of key conformational interconversions available, these are clearly less populated than in **1a**. This may be, in part, because the longer tether diminishes the frequency with which the reactant-bound Cr groups approach each other.

Thus, the representative example of **1a–c** and **2** confirms and demonstrates the critical role played by the approach trajectories and pathways followed by the bound reactants at collision. It also illustrates the very important role of the conformational interconversions in molecular catalysts that control these parameters.

7.4.2.3 The Likely Origin Within Enzymes of Structural Complementarity to the Optimum Transition State In summary, there are two ways to bring about increased catalytic rates. One is a brute force approach that involves the application of ever greater amounts of reactant activation by the catalyst; that is, to improve the thermodynamics of reactant collision.

The other approach is to engineer a subtle dance that progressively constrains the bound reactants to collisions along approach trajectories that are most likely to yield products. That is, to iteratively improve the *mechanics* of reactant collision.

Taken to its ultimate conclusion, the latter approach will necessarily lead to catalysts where virtually every collision is successful. When that occurs, the catalyst will, moreover, necessarily be repeatedly and rapidly flexing about a structure that complements the optimum transition state of the reaction. It will do this because that is the only way for the catalyst to ensure that every collision is successful.

The structural complementarity to the optimum transition state that is observed in enzymes likely has its origin in this effect. That is, enzymes appear to have evolved to take full advantage of the latter approach and advance it to its ultimate embodiment.

7.4.3 Bioinspired/Biomimetic Catalysts that Demonstrate the Importance and Limitations of Molecular Recognition

One of the most widely cited features of enzymes is their ability to harness molecular recognition. By this is meant that they have a high affinity for substrates whose structures are complementary to their active sites.[1] As noted earlier, enzymes also have an even higher affinity for their transition states.[10] Given the important role that molecular recognition plays, several researchers have developed bioinspired catalysts that seek to replicate and examine this feature of enzymatic catalysis.

7.4.3.1 Mn Porphyrin Oxidation Catalysts Figure 7.5 depicts a catalytic system in this respect that was developed by Breslow and co-workers.[27] A Mn porphyrin catalyst **3** was laterally (*trans*-) appended with two β-cyclodextrins. When treated with a linear olefin substrate **4** of length appropriate to the cyclodextrin–cyclodextrin distance in **3**, the long-chain tails of the olefin should be included in the two, opposing cyclodextrin cavities, with the substrate C=C double bond then held atop the Mn ion. Thus, molecular recognition in this system should see oxidative epoxidation strongly favored at this bond. The transition state shown in square brackets at the bottom of Figure 7.5 should be selectively generated, thereby yielding the epoxidation product depicted on the right of the figure.

By contrast, a control porphyrin **5**, which was *cis*-appended with cyclodextrin groups, could reasonably be expected to display little or, at least, substantially less molecular recognition because the substrate cannot be included in both porphyrins simultaneously. Thus, the desired transition state should not be selectively generated, yielding a less selective formation of the epoxidation product on the right of the figure.

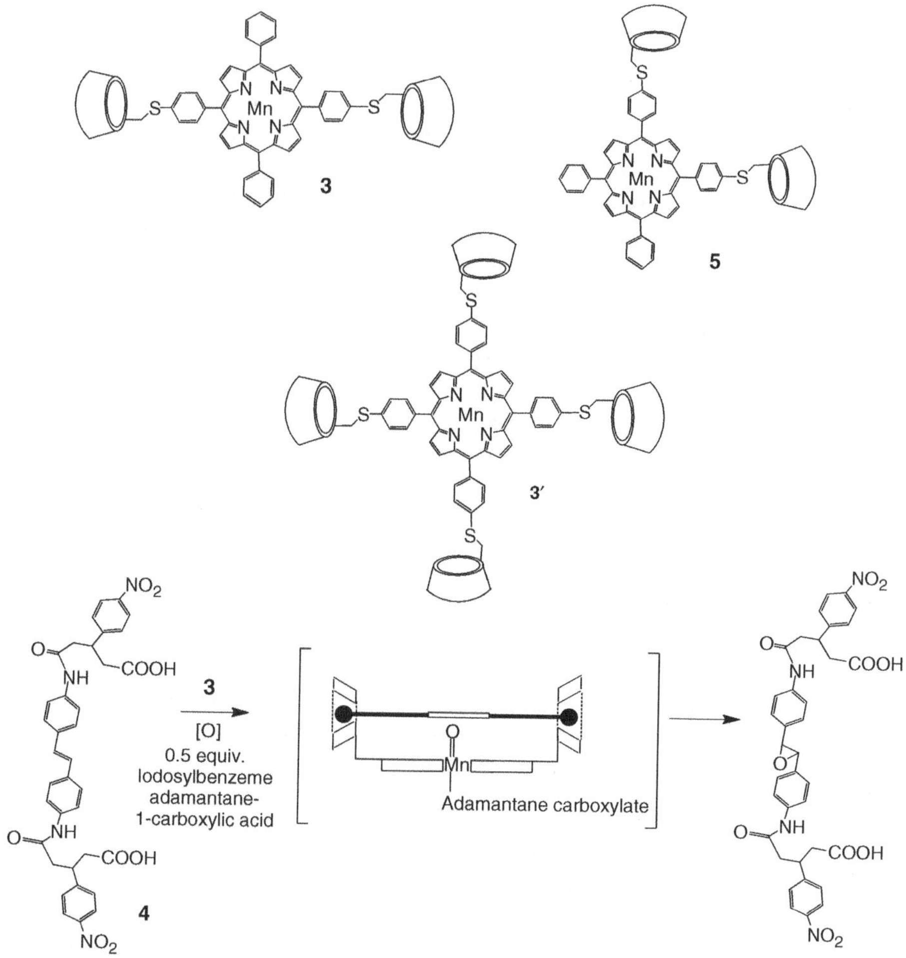

Figure 7.5 Biomimetic epoxidation catalyst based on molecular recognition. Mn porphyrin **3**, which contains two laterally appended cyclodextrins, oxidizes substrate **4** with high, but not perfect selectivity. Porphyrin **5**, which contains angularly appended cyclodextrin groups, shows little selectivity under comparable conditions. Porphyrin **3′** achieves the highest selectivity. (Adapted with permission. Copyright John Wiley & Sons: Ref. 21k.)

To test these propositions and the role of molecular recognition in oxidative catalysis, porphyrins **3** and **5** were converted to their Mn(III) complexes, which were used as catalysts for the oxidation of substrate **4** at approximately 1 mM with 0.5 equiv of iodosylbenzene.[27] 1-Adamantanecarboxylic acid was also added to block the bottom site on the Mn porphyrins. Authentic samples of the product epoxides were prepared and competitive oxidations were performed with substrate mixtures. The conversion was carried to only 20%, and the product mixture was analyzed by ^{1}H NMR.

The results of this study showed that **3** facilitated the desired oxidation of **4** at a rate which was 49 times faster than that achieved with a similar but more weakly binding and longer substrate having carboxylate termini. By contrast, **5** accelerated the desired oxidation of **4** at a rate only 4 times faster than that achieved with the longer, carboxylate-terminated substrate.[27] Thus, there was an approximate 12-fold acceleration in the selective catalytic oxidation of **4** by **3** relative to **5**. This can only have been because of the expected molecular recognition discussed above.

While the selectivity achieved by **3** is certainly impressive, it was not the highest selectivity observed. This was obtained by porphyrin **3′**, which facilitated the desired oxidation of **4** at a rate that was 50 times faster than that achieved with the longer, carboxylate-terminated substrate. Thus, **3** was at least 2% unselective in its catalytic performance relative to **3′** (i.e., 50 times faster compared to 49 times faster). While that may sound very little, enzymes routinely achieve 100% selectivity in entioselective catalytic processes that are far more demanding.

The point of this exercise is to demonstrate a limitation of molecular recognition, or at least our conceptualization thereof, in catalytic processes. Catalyst **3** generates its selectivity in a purely thermodynamic way. That is, by including the substrate, the cyclodextrin groups are favored to stabilize and populate the desired transition state shown near the bottom of Figure 7.5. However, this does not guarantee the formation of this particular transition state, as indicated by the observed 2% nonselectivity.

Thus, while **3** undoubtedly has a structure that is complementary to its desired transition state, it does not achieve optimum selectivity. This is fundamentally because it does not display a feature common to enzymes known as functional convergence. In the next section we discuss this feature.

7.4.3.2 Functional Convergence The term *synergy* refers to the situation where the components of a system interact in such a way that the overall properties of the system exceed the simple sum of the properties of its individual elements.[28] For example, a soccer team may comprise of a collection of players, each of whose individual skills are relatively mediocre. The team will then display *synergy* if they play far better as a unit than would be expected from the simple sum of their individual capabilities.

There are various ways in which such synergy can be created; these have been examined in detail in the field of complex systems science.[29] One way is through *functional complementarity*.[29] This class of synergy arises when the different elements of the system each perform complementary tasks. When the tasks are combined, the system displays amplified overall properties. For example, if each of the members of the soccer team alluded to above is good at a particular, different task, then the team as a whole will enjoy capabilities that combine the best abilities of each of the players.

Another way in which synergy can be created is called *functional convergence*.[29] This is a extraordinarily powerful form of synergy in which the system is so contrived that its elements act in unison with each other to thereby amplify the overall system properties. That is, the elements act to mutually augment and magnify their respective capabilities. The elements also rely entirely on each other in this respect.

The archetypal functionally convergent system is a mechanical device of the type shown in Figure 7.2, where the individual cogs combine synchronously to generate an overall capability that far surpasses the sum of the properties of the individual cogs.[21e] However, if just one cog is left out or placed incorrectly, the entire system fails and the machine cannot operate at all. What is left is a mere collection of machined parts with no operational capacity whatsoever.

A soccer team displaying functional convergence would similarly be considered "a machine". The players would always be at the right place at the right time and acting in mutually interdependent unison with each other. Such a team would generate remarkable overall outcomes. But, if one player was left out, or if one player could not be depended upon, then the team would fail entirely.

The distinguishing feature of convergence is a group dynamic in which apparently innocuous system components work cooperatively to all succeed together or all fail together. When they succeed they do so in an awe-inspiring way.

Many enzymes appear to display functional convergence of this type. They combine and exploit simple amino acids, which are not known to display catalytic properties outside biology,[26] to bring about awe-inspiring feats. In effect, enzymes turn these innocuous species into truly potent catalysts. They appear to do this by ensuring that the amino acids act in a coordinated, mutually reliant manner. For this reason, apparently minor changes to the sequence or shape of the enzyme protein, even at a distance from the active site, may lead to total inactivation.[9, 10b]

In effect, enzymes rely on coordinated, simultaneous actions by each of the catalytic groups to bring about a catalytic effect. Without such synchronicity, catalysis becomes impossible.

7.4.3.3 Selectivity in Mn Porphyrin Oxidation Catalysts Illustrates the Likely Origin of Enzymatic Selectivity The fundamental reason that **3** displays at least 2% nonselectivity in its catalytic action is because it is not functionally convergent in its action.[21l] In other words, it does not depend on coordinated, synchronized actions by the Mn and cyclodextrin catalytic groups for its catalytic effect. The Mn=O oxo functionality can facilitate oxidation independent of, and without the involvement of, the cyclodextrins.

In essence, **3** has a structure that complements the desired transition state, but it does not have a way of ensuring that the substrate is correctly and simultaneously attached to all of the catalytic groups at the precise moment of reaction.[21l] The substrate may therefore be attached to only one, or even none, of the cyclodextrins at the instant of reaction. On such occasions, a nonselective product is generated.

The only way that **3** could ensure simultaneous attachment of all of the catalytic groups at the precise moment of reaction would be if each of the catalytic groups was constrained to act only when all of the other catalytic groups acted too, in unison. This is only possible in a functionally convergent system. (In a later section in this chapter we describe the action of a representative nonbiological catalyst that operates in a functionally convergent manner.)

The point of this discussion is that simple complementarity to the optimum transition state is, on its own, insufficient to create a totally selective catalytic effect. Something more is needed. That something involves a method for ensuring *concurrent actions* during the catalysis.

In fact, **3** employs functional complementarity in its catalytic action. The presence and positions of the cyclodextrin and Mn groups complement each other and thereby create the selectivity that is observed. This selectivity arises from the fact that the reaction pathway in which **4** is simultaneously included in both of the cyclodextrins of **3** during the transition state has a substantially lower overall energy than the competing pathways, where **4** is not included in one or both of the cyclodextrins. This is quite different from the requirement for concurrent actions that exists in functional convergence.

7.4.3.4 Enzymatic Selectivities Likely Derive from Functional Convergence As mentioned earlier, the selectivity that arises out of functional complementarity derives from the thermodynamics of the system. However, thermodynamics seldom allows for total selectivity of the type displayed by many enzymes. Thus, a thermodynamic system will always distribute itself over the energetically most-favored pathways. In a chemical reaction, the thermodynamically most stable pathway will be followed in greatest proportion. However, other products will also be generated in proportion to their pathway stabilities relative to the most-favored product. In other words, selectivity is not created by the absolute stability of the product or of its reaction pathway, but rather by its stability relative to other, competing products and pathways.

To generate the near-total specificities that are common in enzymes, one would need an extraordinarily powerful and extreme relative thermodynamic driving force. That is, the selected pathway would have to be many orders of magnitude more stable than any potential competing pathway. It is an open question as to whether this is possible, even in the near-perfectly shaped active sites found in enzymes. The problem is that substrates bind dynamically not statically to enzymes. That is, enzyme–substrate binding typically involves a multitude of binding contacts between the enzyme and the substrate, each of which is weak and dynamic. These contacts usually comprise hydrogen bonding, ion-pairing, and *van der Waals* binding. They constantly form and release. Thus, at any one instant there will be a multiplicity of incompletely bound substrates present. Some of these will be missing only a single van der Waals contact. Each, nevertheless, offers an alternative, competing reaction pathway.

Enzymologists would argue that enzymes bind their substrates strongly overall ($1/K_M = 10^{4\pm3}\ M^{-1}$) and their transition states still more strongly ($1/K_M \approx 10^{16\pm4}\ M^{-1}$) so that there is a powerful driving force for selective reaction.[10] However, because the binding is not static, the values quoted above are averages, which combine the binding strength of multiple, partly bound substrates at any one instant. Under these circumstances, it is not clear how any one pathway could achieve the required, very large thermodynamic stabilization that would be needed relative to every other pathway.

By contrast, if enzymes acted in a functionally convergent manner where reaction was only possible if all of the binding contacts were simultaneously intact in a mutually dependent manner, then it is perfectly reasonable to conceive of total catalytic selectivity. All of the partially bound substrates present at any one instant in time would simply be unable to undergo catalytic transformation. Only the fully bound substrate would react.

7.4.3.5 An Improved Conceptualization of Molecular Recognition?

The above discussion leads us to the suggestion that there may be a misconception in our understanding of the fundamental origin of molecular recognition in biology. We currently view molecular recognition in a static manner when, in fact, it is highly dynamic. There may be a time component to molecular recognition, at least in biological systems. The fundamental origin of molecular recognition may not relate directly to the average strength of enzyme–substrate or enzyme–transition state binding as it is currently determined. This may be only a proxy to a deeper, underlying phenomenon involving the length and regularity of the time periods during which the substrate or the transition state is fully bound at all contact points to the enzyme. Such time periods are known in catalysis science as the residence time.[21d] The longer and more regular the residence time, the better the molecular recognition may be. The binding strength that we currently use to quantify molecular recognition may reflect this more fundamental effect.

7.4.4 Bioinspired/Biomimetic Catalysts that Operate Like a Mechanical Device

As noted earlier, several studies in the field of heterogeneous catalysis have used volcano plots to show that a range of metal catalysts may be governed by their collision frequency.[21g, 22] In all of these cases, the step of collision becomes rate determining only when the individual catalyst–reactant binding contacts are weak and dynamic.[21g, 22] In other words, when the reactant functionalities bind and release the catalyst rapidly and dynamically at each point of contact, then they are often attached only briefly so that there is little time for them to collide with each other. Achieving collision then becomes the key impediment that must be overcome during catalysis.

Enzyme–substrate binding contacts are similarly generally weak and dynamic, comprising hydrogen bonding, ion-pairing, or van der Waals contacts. This factor could be the key reason that enzymes may generally be governed by their collision frequencies.

To find nonbiological molecular, homogeneous catalysts with a mechanical action, it makes sense to look at systems in which the reactants dynamically interact with the catalyst. Out of these one may reasonably expect to find catalysts governed by their collision frequency. A review of the literature in this respect indicated that species of this type were extremely rare; only a few were identified. These were later documented in a book[21c] and a review article.[22]

In the following section we discuss the operation of a representative example of such a catalyst and compare its action to that of a typical enzyme.

7.4.4.1 Hydrogen Reduction Catalysis by Ferrocenophane Catalysts

The [1.1]ferrocenophane catalyst **6** facilitates the conversion of protons (H^+) to hydrogen (H_2).[30, 31] It is believed to do so using the mechanism depicted in Figure 7.6.[22]

Ferrocene is only very weakly basic ($pK_{a1} - 6.5, pK_{a2} - 7.1$),[32] so that proton binding to each Fe ion in **6** is highly dynamic; the protons bind and release the ferrocene Fe ions rapidly and repeatedly. During the infrequent periods of ferrocene–H^+ binding, the protons also participate in a rapid, dynamic equilibrium involving Fe–H and agostic Cp-ring C–H isomers.[31, 33]

During the time that the protons are bound exclusively to the ferrocene Fe ions, density functional theory (DFT) calculations indicate that their positive charge is largely delocalized onto the Fe, with the attached H atom effectively being highly reactive atomic hydrogen ($H^{\bullet}$).[31–33] DFT also indicates that this $^{+}Fe–H^{\bullet}$ state is of higher energy and its lifetime is exceedingly short; too short to be measured at ambient temperature.[33] As such, it must be considered to be the activated form of the bound reactant. Recently, **7** could be observed directly by ^{1}H NMR at −122 °C.[33] This confirmed that both iron atoms became protonated and that the exchange processes were too rapid to be resolved by NMR. This was further confirmed by the fact that **1** progressively loses its activity in solution as it is cooled. At −122 °C, Fe–proton binding is still rapid and dynamic, but conformational flexing is halted on the NMR time scale.[33]

The effect of this dynamic and transient proton binding is to make the creation of collisions between two such highly reactive atomic hydrogen atoms ($H^{\bullet}$) during conformational flexing the key impediment to catalysis. When such collisions occur, as shown in $[7]_{TSC}$ (Figure 7.6), they must involve an exceedingly low activation energy, E_a, for H_2 formation. But they do occur and, in fact, they occur often. [1.1]Ferrocenophane is a remarkably active and long-lived homogeneous catalyst at room temperature. Individual molecules of **6** readily turn over more than 1 million H_2 molecules without significant loss of activity.[34a] When bound

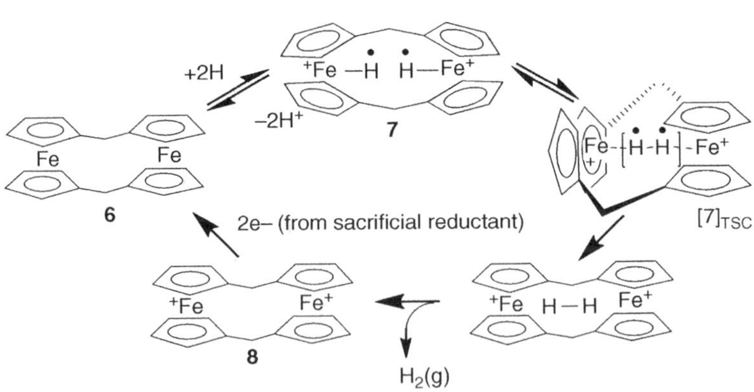

Figure 7.6 Hydrogen reduction catalysis by [1.1]ferrocenophane, **6**. (Adapted with permission. Copyright Wiley-VCH: Ref. 22.)

to polystyrene and coated on a *p*-type silicon photocathode, **6** was found to turn over approximately 5 molecules of H_2 per second for at least 5 days of continuous operation.[34b] Such remarkable durability can only be due to exceptional selectivity for protons, which prevents the formation of nonfunctional intermediates and avoids deactivation. Production of such vigor and high fidelity is typical of a machine—in this case, a "molecular machine".[22]

7.4.4.2 The Catalytic Action of the Ferrocenophanes Of significant interest is the catalytic action of **6**. Two highly dynamic processes must be synchronized in order to achieve a catalytic effect by **6**: (1) dynamic proton binding at each ferrocene Fe ion and (2) conformational flexing of **6**.

Collisions only occur when both ferrocene Fe ions bear an $H^{\bullet}$ at the precise instant that conformational flexing brings the two ferrocene Fe ions into their closest proximity to each other. But each of these events occurs only very briefly and infrequently. Moreover, they occur independently of each other. The creation of a catalytic effect therefore comes down to the problem of synchronizing these events so that they have a high likelihood of occuring simultaneously.[21d, 22] How the catalyst does that is of great significance.

It turns out that **6** is only catalytically active if it rapidly and regularly flexes about a structure that complements the optimum transition state of the reaction (shown in [7]TSC) (Figure 7.6). Changes to the character of the flexing or the rapidity of the flexing induce drastic declines in, and even completely halt, catalysis.

Thus, for example, **9**, which is identical to **6** except for the presence of a single methyl group on each bridging C atom, catalyzes hydrogen reduction much more slowly than does **6** (Figure 7.7).[34] The addition in **10** of two methyl groups on

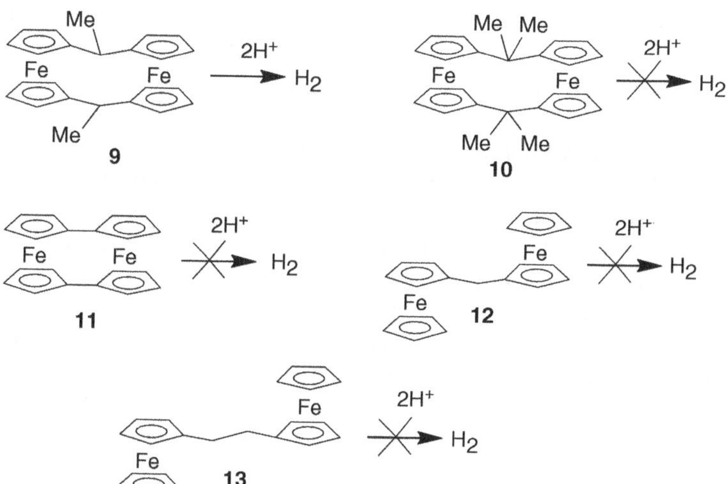

Figure 7.7 Hydrogen reduction catalysis by other ferrocenophanes. (Adapted with permission. Copyright Wiley-VCH: Ref. 22.)

each bridging C atom suppresses the catalytic effect completely, at least at room temperature.[34] Both **9** and **10** flex in the same way as **6** but they do so more slowly; **9** flexes slower than **6** but faster than **10**. Thus, the rate of flexing controls the catalytic process and it does so in a nonlinear way. The faster the flexing, the more the catalytic activity is accelerated. However, even moderate declines in the flexing rate destroy the catalytic effect entirely.

Nonlinear behavior of this type can only arise out of the extreme brevity of $H^{\bullet}$ formation at each Fe centre. In effect, there is only a very limited and short time to sequester and exploit this highly reactive intermediate. The more rapidly the catalyst oscillates about the optimum transition state structure, the more likely it is to trap and harness the $H^{\bullet}$ at each Fe centre.

The same is true if the character of the flexing is altered. Thus, for example, derivatives **11–13** are all entirely inactive as hydrogen reduction catalysts (Figure 7.7).[34] This arises because they do not flex about a structure that complements the optimum transition state of the reaction. As such, they have no chance of capturing and exploiting the short-lived $H^{\bullet}$ formed at each Fe centre.

7.4.4.3 The Catalytic Action of the Ferrocenophanes Is Machine-like

In the above example, it is clear that the action of **6** is governed by the spatial and temporal fluctuations that lead to collisions, not by the need to overcome the activation energy. In other words, it is a mechanical catalyst. This terminology is apt because its action displays numerous characteristics of a mechanical device.

This can be seen in the fact that catalysis by **6** is driven by a mechanical impulse, namely, conformational flexing.[21d] The speed and efficiency of this flexing is critical to creating the overall effect. When the processes of catalyst flexing and binding are synchronized, protons are dynamically bound, activated, and carried—repeatedly and rapidly—along unchanging, optimum pathways into reactive collisions with each other. These collisions occur within a structure that complements the desired outcome, namely, the optimum transition state. The H_2 products are then dynamically ejected and new reactants taken up for the next cycle.

This *binding*, *transformation*, and *ejection* process is comparable to a mechanical device.[21d] For example, to use an earlier-mentioned analogy, a machine making plastic bottles dynamically takes in polymer starting materials and then mechanically forces them into the shape of a bottle. It does so within a mold that complements the shape of a bottle. The bottle is then dynamically ejected and a new cycle starts. If the interactions between the machine and the polymer starting materials (or bottle products) are not dynamic, then the machine will jam and become inoperable in a wholly nonlinear manner. It will not progressively slow down in a linear fashion.

Like all mechanical devices, the components and motions of **6** must be cooperatively synchronized in order to generate a catalytic effect. That is, the catalytic groups must act in a coordinated or, more correctly, a convergent manner for a catalytic effect to be realized.[21l, 35] In other words, their actions must converge in a mutually reinforcing way.[21l, 35] If the necessary synchronicity is not achieved or if it is only partly achieved, then the overall effect will be curtailed in a drastic and nonlinear way. Synchronization is achieved in **6** by constraining the catalyst to rapid, repeated conformational flexing about a structure which complements the transition state of the reaction.[21d, 22]

Characteristics of this type are seen in machines of all kinds. A car engine, for example, depends entirely on synchronicity in its components. Many engines employ "timing belts" to ensure that the timing of the spark and the movement of the pistons is synchronized with those of the valves. If the timing belt breaks, the engine becomes inoperable. The gear box and differential are needed to modify and transmit the synchronicity of the engine to the wheels.

It is the fine balance and timing of dynamic uptake, transformation, and release that gives machines their unique properties. In the same way, it is the fine balance and timing of dynamic uptake, transformation, and release of reactant H^+ and product H_2 that gives **6** its catalytic properties.

As mentioned earlier, a mechanical mode of operation of this type is already known to exist in biology, in the molecular/ion pumps that can be found in cell membranes. These pumps employ a regular, repeated mechanical action to facilitate *transportation* of substates into and out of the cell. It is not inconceivable that enzymes use a qualitatively similar action to facilitate transformation of substrates.

7.4.4.4 Similarities in the Catalytic Action of the Ferrocenophanes and Those of Enzymes in General In the [1.1]ferrocenophanes we therefore have an example of a catalytic process that is dependent on, and governed by, conformational dynamics, much like enzymes seem to be governed by their conformational dynamics. Moreover, the individual binding contacts between the reactants and the catalyst are weak and dynamic—again, like enzymes. The catalytic effect is dependent, furthermore, on the catalyst rapidly and repeatedly flexing about a structure that complements the optimum transition state for the reaction. This is also comparable to enzymes, whose active sites are known to structurally complement the optimum transition state. Finally, even apparently minor changes to the structure or flexing of **6** generate drastic, nonlinear declines in the overall catalytic rate; that is, there is an enormous disparity between the activity of **6** in hydrogen reduction catalysis and that of even closely related members of the ferrocene family of compounds. In the same way, enzymes are known to display catalytic rates that are many orders of magnitude larger than even closely related proteins (up to 10^8-fold greater).[9]

It is further worth noting that ferrocene Fe ions, which are the catalytic groups in this case, are not known to act catalytically outside this one example. This is undoubtedly because of the extremely dynamic and transient way in which they bind and activate protons. In **6**, conformational dynamism turns these unconventional catalytic groups into potent catalysts. In the same way, the amino acid residues of enzymes are not known to act as catalytic groups outside biology. Of all of the available monomers and oligomers of amino acids, only artificial L-prolines have been reported to be catalytically active in open solution outside enzymology.[36] This is, undoubtedly, because of the dynamism and brevity of the binding contacts that they use to bind and activate substrates. Yet, in enzymes, these apparently unconventional catalytic groups are tranformed into extraordinarily powerful catalysts.

There are other important similarities to enzymes. In the single kinetic study that has been carried out on catalysis by **6**, the data conformed to a Lineweaver–Burke plot (with K_m for **6** ≈ 0.2 μM).[32] Lineweaver–Burke plots are routinely used in enzymology to determine the key parameters of Michaelis–Menten kinetics.[37] Thus, catalysis by **6** displays Michaelis–Menten kinetics and its overall rate is controlled by the rapidly equilibrating catalyst–reactant intermediate **7**. Enzymes similarly display Michaelis-Menten kinetics and are governed by their rapidly equilibrating Michaelis complexes.

The activation energy of H_2 formation during collisions by two $H^{\bullet}$ species in **6** is undoubtedly low. The actual energy requirement arises almost solely out of the energy involved in flexing by **6**; this energy is provided by the thermal background (i.e., the local temperature). In the same way, enzymes are known to generally display the lowest activation energies of any class of catalyst,[37] although this may be misleading if most of the activation energy derives from the energy involved in conformational flexing, not from the thermodynamics of the collision itself.

Finally, and perhaps most significantly of all, **6** is an extraordinarily active and durable catalyst. It is also highly selective for proton binding and hydrogen generation. Enzymes are similarly highly active and selective. Indeed, the earlier mentioned turnover frequency that was measured for **6** of 5 H_2 molecules s^{-1} $catalyst^{-1}$, falls within the range that is considered to be characteristic of enzymes (1–10,000 s^{-1} $enzyme^{-1}$).[1]

An ability to cooperatively sequester and exploit reactive intermediates that are very short-lived—too short-lived to be harnessed in other classes of catalyst—may explain why enzymes are able to catalyze numerous reactions that cannot presently be catalyzed by nonbiological catalysts.[1]

7.5 THE RELATIONSHIP BETWEEN ENZYMATIC CATALYSIS AND NONBIOLOGICAL HOMOGENEOUS AND HETEROGENEOUS CATALYSIS

For many years, catalysis science has sought to combine its subdisciplines of enzymatic, homogeneous, and heterogeneous catalysis into a single, unified theory of catalysis.[38–40] This has, to date, been hindered by the fact that each subdiscipline has its own set of fundamental principles and conceptualizations. These appear—at least superficially—to be unrelated to those of the other fields. Thus, it has not proved possible to find a common thread that can connect the different fields and serve as the basis on which they may be unified.

However, the existence of two distinct realms of catalytic action—thermodynamic and mechanical actions—potentially explains the physical and conceptual distinctions that exist between homogeneous, heterogeneous, and enzymatic catalysis.

Thus, enzymes can be understood to differ from other classes of catalysts in that they typically employ optimized approach trajectories, which lead to "ideal" or "near-ideal" reactant collisions.[21d] Collisions of this type are uncommon, but not unknown, in nonbiological homogeneous catalysts.[22] In other words, the distinctive kinetic, binding, and other effects that characterize enzymes can now be understood to derive from their mechanical mode of action.[21d, 22]

By contrast, nonbiological homogeneous catalysts have typically not been designed to constrain their reactants to optimum approach trajectories. For this reason they typically employ a thermodynamic mode of action.[21f] Their general properties can be understood to derive therefrom. Thus, their slowest, rate-determining step normally involves the proportion of reactant collisions that are successful. It does not involve the step leading up to reactant collision. The rate expressions of such catalysts will therefore not contain a rapidly equilibrating catalyst–reactant intermediate of the type observed in Michaelis-Menten kinetics.[21f]

In the case of nonbiological heterogeneous catalysts, volcano plots have previously shown that both thermodynamic and mechanical modes of action may be harnessed in, for example, metal catalysts.[21g, 23] This is because a large variety of different catalytic sites are typically present on the surfaces of heterogeneous catalysts, including face, edge, defect, and other sites. The surface of a heterogeneous catalyst can, in fact, be considered to be a combinatorial experiment, which may contain sites that facilitate "perfect" or "near-perfect" collisions under the applied conditions.[21h] If such sites are present and the applied conditions are suitable, then a mechanical mode of action will likely dominate the catalytic process. If not, then a thermodynamic action will. Other factors, such as dynamic reactant binding, may also induce a mechanical action.

The combinatorial experiment that occurs on the surface of a heterogeneous catalyst arguably mirrors the combinatorial experiment that is biological catalysis.[21h] That experiment has led, over many eons of evolution, to enzymes, which are catalysts with a mechanical mode of action.

Homogeneous, heterogeneous, and enzymatic catalysis, which appear so different and which have had to be formulated in such dissimilar terms, can therefore be related back to a single fundamental mode of catalytic action, which is exploited in different ways in the different classes of catalyst.

7.6 SELECTED HIGH-PERFORMANCE NONBIOLOGICAL CATALYSTS THAT EXPLOIT NATURE'S CATALYTIC PRINCIPLES

The question that we now address is: How can we use innovation to practically implement Nature's catalytic principles—at least as they are currently understood—in human-made, nonbiological systems? The problem in this respect appears to be similar to that of designing a machine but at a molecular not a macroscopic scale. The scale makes it drastically more difficult since we have far less information about the nanoscopic chemical world than we do of the macroscopic physical world. We do not, for example, know nearly as much about the shapes and structures of optimum transition states as we do of the complementarity needed in machine tools. Nor do we know as much about the dynamism that will be needed in the interactions between the catalyst and the potential reactants. How, then, are we to design from first principles, a catalyst that exploits dynamic binding contacts and synchronous conformational flexing to create a molecular machine?

Two prospective starting points to this problem have been examined in recent years.[21c] Both have generated interesting and promising new catalytic systems. The two approaches involve:

1. Developing or adapting models of enzymes to facilitate a mechanical mode of action.[21c, 22] This approach relies on the proposition that, if enzyme active sites are complementary to their transition state, then molecular models with similar shapes and constituents should also be. Provided that these models display the requisite conformational and binding dynamism, they should act as catalysts employing a mechanical mode of action.
2. A combinatorial technique involving the creation of statistical proximity catalysts.[21c, 35] In this approach, monomers containing prospective catalytic groups are very highly concentrated within limited volumes in the expectation that some small, but statistically significant proportion of them will happen to thereby become optimally disposed to facilitate a mechanical mode of catalysis. The idea of using monomeric catalytic groups is to avoid the complex synthetic procedures often needed in approach 1 above, and thereby obtain inexpensive, practically useful catalysts.

In the following sections, we briefly consider representative examples of these approaches.

7.6.1 Adapting Model Species of Enzymes to Facilitate Machine-like Catalysis

7.6.1.1 Water Oxidation Catalysis by a Molecular Cubane After an extensive search,[21c, 22] a model complex of an enzyme active site was identified that appeared to facilitate—via a mechanical process—the same reaction as the enzyme. In 1997, Dismukes and Ruettinger reported the molecular cubane $Mn_4O_4L_6$, **14** [L = $(p\text{-MeO–}C_6H_4)_2PO_2^-$] (Figure 7.8), which was a model complex of the photosystem II water oxidizing complex (PSII-WOC).[41, 42] The PSII-WOC is the only known species capable of catalyzing sustained, highly efficient, light-driven water oxidation. Figure 7.9 depicts a model of the PSII-WOC based on X-ray crystal structure data.[43–45] As can be seen, the core of the active site appears very similar to that of **14**.

In later studies, it was shown that **14** yielded dioxygen when illuminated with UV light ($\lambda = 350$ nm) in the gas phase, with accompanying ejection of a phosphinate ligand L to form the cationic species $Mn_4O_4L_5^+$ **15**.[42] The remarkable aspect of this reaction was that the quantum efficiency of O_2 release approached

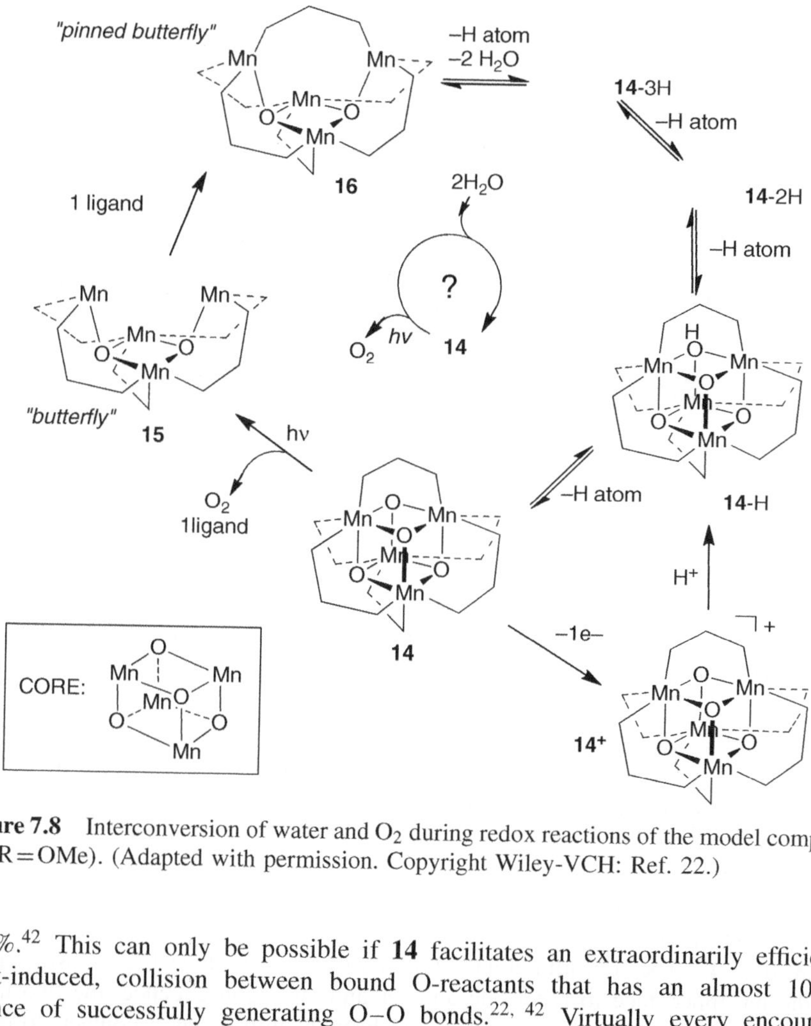

Figure 7.8 Interconversion of water and O_2 during redox reactions of the model complex **14** (R=OMe). (Adapted with permission. Copyright Wiley-VCH: Ref. 22.)

100%.[42] This can only be possible if **14** facilitates an extraordinarily efficient, light-induced, collision between bound O-reactants that has an almost 100% chance of successfully generating O–O bonds.[22, 42] Virtually every encounter between reactant O-atoms that is brought about by the catalyst, must therefore lead to O–O bond formation. That is, the step of O–O bond formation, which is known to be the most critical and difficult of the steps in water oxidation catalysis, must be extremely efficient in **14**.[22, 42] This raised the possibility that **14** eliminated O_2 by a mechanical process having a very low activation energy.

Separate studies revealed that, in solution, **14** is capable of reversibly abstracting $4\ H^+ + 4\ e^-$ via proton-coupled electron transfer (PCET) processes and releasing $2\ H_2O$ molecules, to form a "pinned butterfly" species $Mn_4O_2L_6$ **16** that is, effectively, **15**, with one extra bridging ligand L.[41, 42] Thus, a pathway potentially existed from **15** back to **14** via **16**. This pathway (shown at the top left and right of Figure 7.8) involved **15** taking up a sixth ligand L to yield **16**, which may then take up two water molecules with accompanying release of $4\ H^+$ and $4\ e^-$ by PCET reactions to generate **14**.

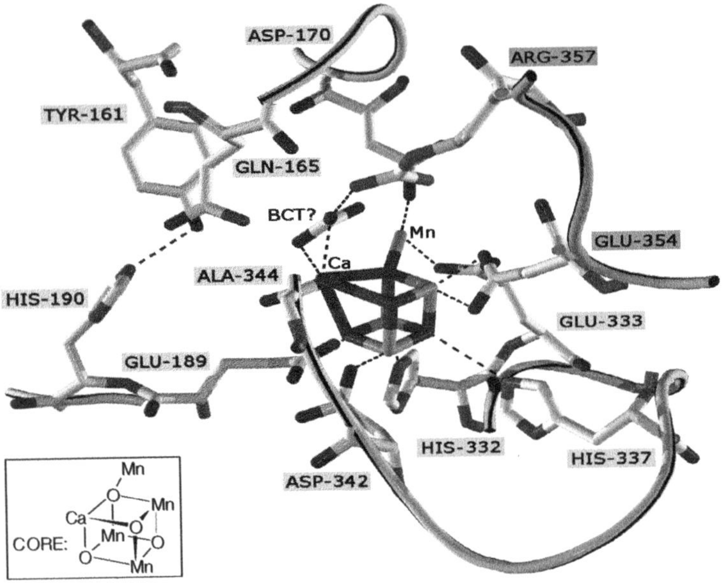

Figure 7.9 Model of the photosystem II water oxidizing complex (PSII-WOC) showing a proposed model of the core active site. (Reproduced with permission. Copyright Wiley-VCH: Ref. 22.)

In other words, a possible overall photocatalytic cycle existed in which **14** photogenerates O_2 and is then recycled by taking up 2 H_2O and releasing $4H^+ + 4e^-$ (shown as the framed question mark in the middle of Figure 7.8).

On the basis of this work and associated theoretical calculations,[46] a mechanism for the water oxidation step was proposed; this is depicted in Figure 7.10.[22] A critical first step in this mechanism is believed to be the photoinduced loss of one of the bridging phosphinate ligands L to form the uncapped open-faced cubane structure $Mn_4O_4L_5^+$ **17**. The conformational flexibility of **17** then allows the two Mn centers on the uncapped face to repel one another and, as a consequence, the two tri-μ-oxo bridging atoms on the same face are brought into reactive contact, forming, first, the peroxo (O_2^{2-}) (**18**) and then the superoxo (O_2^-) species (**19**). Subsequent release of dioxygen results in the "open butterfly" complex $Mn_4O_2L_5^+$ (**15**).

According to a DFT study by Musaev and co-workers,[46] the key step of O–O bond formation (**17** to **18** in Figure 7.10) proceeds via a rate-determining barrier of approximately 120 kJ mol^{-1}. This barrier was entirely due to the energy involved in conformational flexing, not to the thermodynamics of the O–O collision. The earlier mentioned quantum efficiency of O_2 release indicates that this step occurs with near 100% efficiency in the gas phase. The calculated barriers of the remaining steps were found to be insignificant.

Based on these facts, **14** was studied as a potential mechanical catalyst of photocatalytic water oxidation. To immobilize it, **14** was ion-exchanged as **14**$^+$, into a thin layer of Nafion deposited on a conducting surface. When the resulting

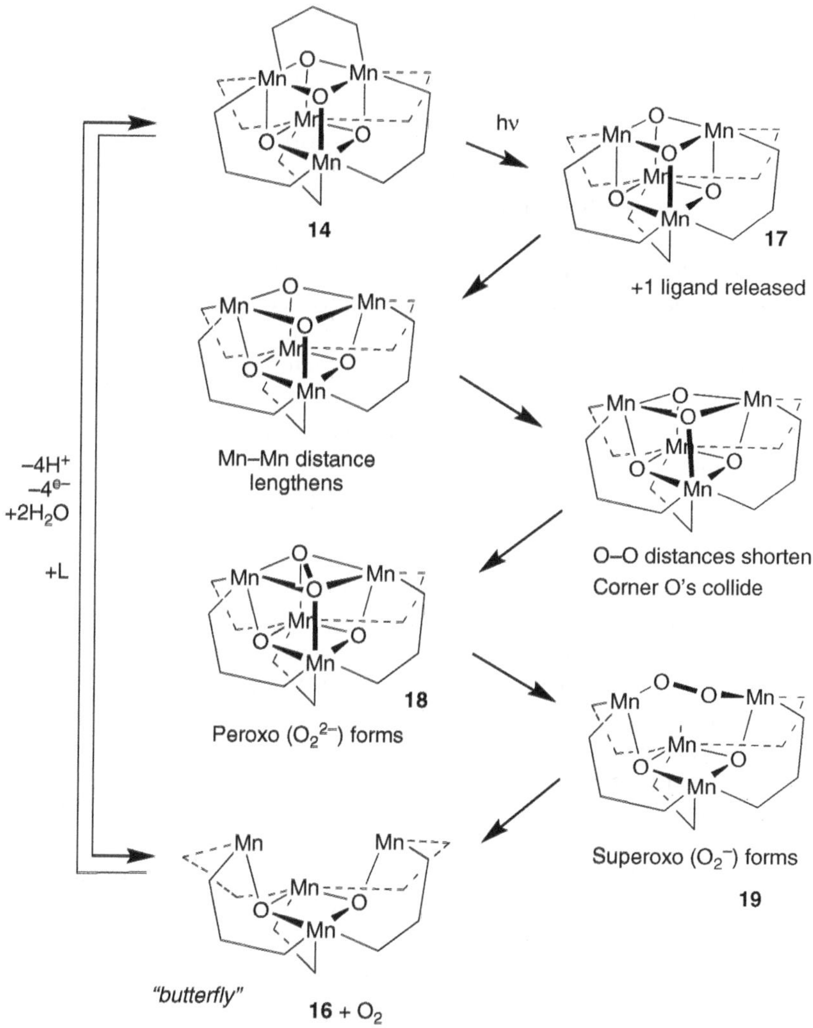

Figure 7.10 Simulation of O_2 formation and release by **14** under illumination (λ = 350 nm) in the gas phase, according to Dismukes and colleagues,[42] and later confirmed in DFT calculations by Musaev and co-workers.[46] Cluster charges have been excluded for clarity. (Adapted with permission. Copyright Wiley-VCH: Ref. 22.)

14$^+$/Nafion layer was then biased at 1.20 V (vs. SHE) and illuminated with sunlight, it rapidly and readily catalyzed the conversion of H_2O into O_2, with accompanying H_2 formation at the other electrode.[47] The catalytic effect, which was substantial and sustained, occurred at a voltage similar to that employed by the PSII-WOC, which is 1.25 V (vs. SHE).

While **14**$^+$/Nafion has no light-harvesting system and can therefore not be nearly as vigorous a catalyst as the PSII-WOC, it was nonetheless found to be one of the most active and durable synthetic, homogeneous water oxidation photocatalysts.[47–50]

In subsequent work, **14**$^+$/Nafion was adapted to operate in a free-standing, water-splitting dye-sensitized solar cell (DSSC).[51] The photoanode of the DSSC was prepared by attaching a Ru(II) dye, $[Ru^{II}(bipy)_2(bipy(COOH)_2)]$ (bipy = 2, 2′-bipyridyl), onto a titania-coated, conductive FTO glass electrode, followed by overcoating with a thin Nafion membrane into which **14**$^+$ was introduced by ion exchange.[51] Upon illumination with visible light while immersed in water and connected to a Pt counterelectrode, the photoanode spontaneously oxidized H_2O to release O_2 gas and protons to solution. An electrical current simultaneously flowed through the external circuit, to reduce protons in the solution to H_2 at the other electrode. Thus, the cell spontaneously split water into O_2 and H_2 upon illumination with light; no other energy source was needed.

Despite being far from optimized, the cell generated a potential of 0.6 V in air, was active under illumination of up to $\lambda < 625$ nm, and achieved an IPCE of 1.7% at $\lambda = 450$ nm.[51] There was no need for an externally applied voltage to drive the system because the oxidized Ru^{3+} form of the photoanode dye that is created by absorption of a photon, and the reduced form of the cubane catalyst that is created by O_2 release, mutually cycle each other during turnover.[51]

This photoanode arguably represents a functional analog of the PSII-WOC.[50] It contains all of the same elements that are present in the PSII-WOC.[50]

The **14**$^+$/Nafion catalyst has also been shown to operate cleanly in seawater, catalyzing only water oxidation with no formation of chlorine.[52] This effect is comparable to the ability of the PSII-WOC in aquatic organisms to catalyze, exclusively, water oxidation in seawater at its known 1.25 V driving potential (vs. SHE). It stands in contrast to commercial water electrolyzers, which generate chlorine gas at their anodes when filled with seawater.

ESR and other studies suggested a catalytic mechanism similar to that shown in Figures 7.8 and 7.10.[48] However, recent XAFS work suggests that a MnO_2 mineral, *birnessite*, may also be formed in the Nafion layer under suitable conditions and this may also be catalytically active.[53]

7.6.1.2 An Apparent Structural Convergence in Enzymatic, Homogeneous, and Heterogeneous Catalysts of Water Oxidation

Most recently, it has been shown that several newly discovered, active Mn or Co molecular or solid state water oxidation catalysts employ a *cubane* core that is structurally almost identical to the catalytic core present in the PSII-WOC.[54] That is, not only is the structural motif of the catalytic core of these species a cubane, but the actual physical dimensions of the cubane core also closely match those of the Mn_3O_4 unit, which is capped by the Ca ion in the PSII-WOC. Included among these catalysts are (1) Nocera's Co-phosphate,[55, 56] (2) Frei and earlier workers' Co_3O_4 spinel (B-site),[57] (3) MnO_2 birnessite,[53] (4) Hill's recently discovered Co-polyoxotungstate,[58] (5) Dismuke's λ-MnO_2 spinel (B-site),[59] as well as (6) Dismuke's molecular cubane **14**.[41, 42, 47–52] A Co cubane with a cubical structure that is active in water oxidation catalysis has recently also been reported.[60]

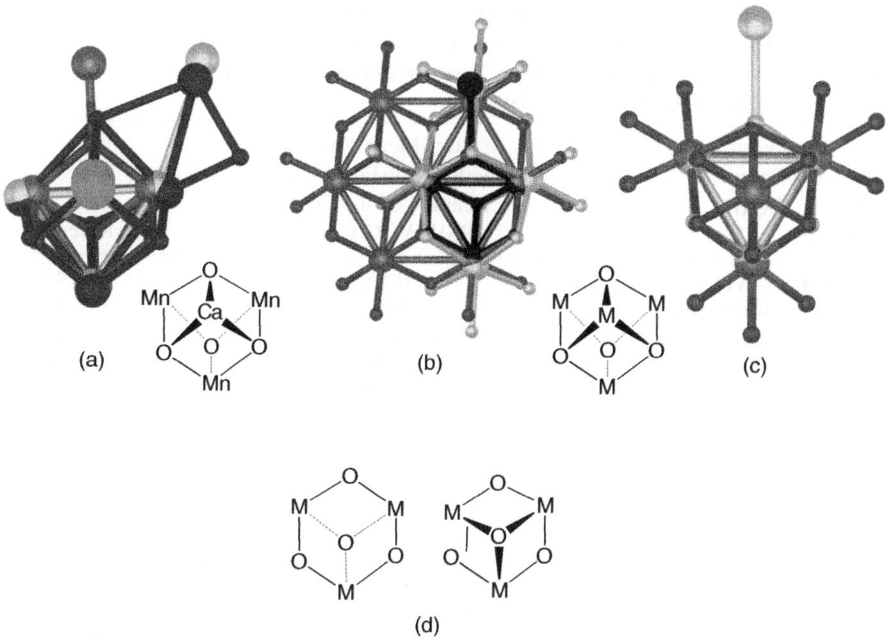

Figure 7.11 (a) Dimensionally accurate superimposition[54] of the $CaMn_4O_4$ core of the photosystem II water oxidizing complex (PSII-WOC) from the *London, Berlin*, and *Osaka* single-crystal X-ray structures.[43–45] The darkest depiction is that of the *Osaka* structure, which was resolved to the highest resolution of 1.9 Å.[45] As can be seen, the core structures in these X-ray structures coincide to all intents and purposes. (b) Dimensionally accurate superimposition[54] of the X-ray-derived structure of the Co-phosphate water oxidation catalyst (large structure),[55, 56] the *London* single crystal X-ray structure of the $CaMn_4O_4$ core of the PSII-WOC (dark small structure),[43] and the XRD structure of the Co_4O_4 core of the Hill's Co-polyoxotungstate catalyst (light small structure).[58] The Ca ion in the PSII-WOC has been excluded for clarity (M = Mn or Co). As can be seen, the cores of these catalysts structurally coincide to all intents and purposes. (c) Dimensionally accurate superimposition[54] of the single-crystal X-ray structure of the $CaMn_3O_4$ core of the *London* structure of the PSII-WOC (light structure)[43] and the single-crystal X-ray structure of the B-site of λ-Mn_sO_4 spinel (dark structure).[59] The Ca ion in the PSII-WOC has been excluded for clarity (M = Mn or Co). As can be seen, the core structures of these catalysts coincide to all intents and purposes. (d) Structural formations at the surface of the Co_3O_4 and λ-Mn_sO_4 spinels that derive from the B-site (M = Mn or Co).[54, 57, 59] Similar structures will exist at the surface of MnO_2 birnessite and Co-phosphate.[53, 55, 56] (All images adapted with permission. Copyright the Royal Society of Chemistry: Ref. 54.)

Figure 7.11 illustrates the extent of these similarities. Figure 7.11a superimposes on each other the $CaMn_4O_4$ core of the PSII-WOC as it has been elucidated in the three most detailed single-crystal X-ray crystal structures of this enzyme. These are (1) the *London* structure by Barber, Iwata, and colleagues, which was resolved to 3.5 Å,[43] (2) the *Berlin* structure by Loll, Kern, and co-workers, which was resolved to 2.9 Å,[44] and (3) the *Osaka* structure by Umena, Kawakami, Shen,

TABLE 7.1 Root Mean Square (RMS) Goodness-of-Fit Measure Relative to the Osaka PSII-WOC Core of Positional Overlays for Three Metals and Their Bridging O, [54] as calculated with Mercury (CSD v 2.4, 2010, CCDC 2001–2010)

High-Resolution, Single-Crystal X-Ray or EXAFS Structure	Root Mean Square "Goodness-of-Fit" Measure[a] Relative to the Osaka PSII-WOC Core[54]	References
London PSII-WOC core	0.255	43
Berlin PSII-WOC core	0.0757	44
Osaka PSII-WOC core	0 (perfect fit)	45
Co-phosphate	0.262	55
Co_3O_4 spinel, B-site	0.235	54, 57
Cubane **14**	0.219	42
Birnessite MnO_2	0.245	53, 54
λ- Mn_2O_4 spinel, B-site	0.237	59
Co-polyoxotungstate	0.296	58

[a] A value of zero indicates a perfect fit. For **14**, Mn_3 and Mn_4 have been used. For the Berlin structure, only the metals have been used.
Source: Macrae, C. F.; Bruno, I. J.; Chisholm, J. A.; Edgington, P. R.; McCabe, P.; Pidcock, E.; Rodriguez-Monge, L.; Taylor, R.; van de Streek, J.; Wood, P. A. *J. Appl. Crystallogr*. **2008**, *41*, 466.

and Kamiya, which was resolved to 1.9 Å.[45] The Osaka structure offers the most detailed picture of the PSII-WOC core. As can be seen in Figure 7.11a, in all of these X-ray structures, the $CaMn_3O_4$ core is essentially identical. The only significant difference is in the location of the fourth, outlying, "dangler" Mn ion.

The lack of evolutionary and structural diversity in the PSII-WOC core cubane arrangement implies that combinatorial biosynthesis in Nature has yielded only this one catalytic structure capable of facilitating sustained water oxidation catalysis. The question arises as to why that is the case.

Figure 7.11b superimposes the PSII-WOC core as revealed in the London structure[43] upon Nocera's Co-phosphate catalyst[55] and the core of Hill's Co-polyoxotungstate catalyst.[58] As can be seen, there is a close structural match. In fact, the core M_3O_4 (M = Co, Mn) structures are virtually identical.

Figure 7.11c superimposes the PSII-WOC core as seen in the London structure,[43] upon the B-site of Dismukes's λ-Mn_2O_4 spinel catalyst.[59] Once again, there is a close structural match.

In fact, as shown by the root mean square (RMS) "goodness-of-fit" comparison in Table 7.1,[54] the active site structures of all of the above catalysts closely match the Mn_3O_4 core in the PSII-WOC. As can be seen in this table, a greater variation exists between the London, Berlin, and Osaka structures of the PSII-WOC than does between the Osaka structure and most of the synthetic catalysts discussed above.[54]

Thus, human efforts to develop new water oxidation catalysts appear to have inadvertently converged on a cubane structure that is not only qualitatively but also quantitatively identical, in large measure, to the cubane active site of the PSII-WOC.[54] These apparent commonalities in an otherwise disparate and unconnected range of homogeneous, heterogeneous, and enzymatic catalysts are remarkable. Human studies appear to be confirming the findings of combinatorial biosynthesis regarding the utility of the cubane structure in water oxidation catalysis.

The commonality may also extend to the actions of these catalysts. For example, several of these catalysts appear to undergo spontaneous, *disassembly* and *reassembly* of the cubane, at an open face (or surface) during catalysis. Figure 7.11d depicts the open face structures that must exist in the Co_3O_4 and λ-Mn_2O_4 spinel catalysts.[54] Similar open faces are present in the Co-phosphate, MnO_2 birnessite, and molecular cubane **14** (immediately prior to O_2 formation—see Figure 7.10).[54] Given that Hill's Co-polyoxotungstate self-assembles, a library of species exists in solution; this likely includes an open-faced arrangement. We should note here too that Hill's catalyst achieves the highest recorded turnover frequency of any abiological catalyst for catalytic water oxidation: $>5\ s^{-1}$ at pH 8.[58] This falls within the range of turnover frequencies typically achieved by enzymes: $1-10{,}000\ s^{-1}$.[1]

The only reasonable explanation for the fact that all of these heterogeneous, homogeneous, and enzymatic catalysts employ so similar a structure is that the cubane arrangement is needed to constrain the reactant O's to a single, optimum approach trajectory during collision. The pathway involved may be the same as that shown in going from **17** to **18** in Figure 7.10. The only difference would be that, whereas the two tri-μ-oxo species are brought into reactive contact across the open face of **14** by the conformational flexing of the cubane structure (Figure 7.10), in the solid state catalysts, they may be brought into collision across the open face (or surface) by oscillatory thermal motion.[54] This proposal is supported by Dismukes's recent finding that $LiMn_2O_4$ spinel becomes active (as λ-Mn_2O_4) only when the Li^+ ion is removed to impart freedom of motion to the bridging cubical B-site O atoms.[59]

These findings have potentially significant implications in biology and biochemistry, as well as in nonbiological homogeneous and heterogeneous catalysis. They also strongly buttress the strategy of mimicking enzymes by copying the structures of their active sites while ensuring dynamism in catalyst binding and flexing.

7.6.2 Statistical Proximity Catalysts

A second approach to exploiting Nature's catalytic principles involves trying to develop combinatorial catalysts where prospective, dynamically binding catalytic groups are drastically concentrated in monomeric form within a small volume.[21c, 35] The idea is that some small proportion of these monomers may thereby become ideally disposed for catalysis involving a mechanical mode of action. Provided that the reactive intermediates formed on the monomers are too short-lived to be sequestered and exploited in any way other than a mechanical interaction, only the product deriving from such cooperative catalysis should be obtained. In this way it may be possible to circumvent the need for complex structures in biomimetic catalysts and create inexpensive and practical new catalysts.

A first "statistical proximity" catalyst of this type was based on **6** and described in 2004.[61] Extreme concentration of monomeric ferrocene sulfonate ($FcSO_3$) within a thin layer of polypyrrole (PPy) yielded.a composite material that was found to be an extraordinarily powerful hydrogen reduction catalyst. At the test voltage of −0.44 V versus Ag/AgCl, it was approximately sevenfold more active than an equivalent area of platinum metal in generating hydrogen gas from 1 M acid.[61] Platinum is the industry-best catalyst for this reaction. PPy-$FcSO_3$ was at least as active as **6**; however, it was far simpler and less time consuming to make. By contrast, an equivalent control layer of polypyrrole containing *p*-toluene sulfonate was several orders of magnitude less active. Thus, the simple act of drastically concentrating prospective monomeric catalytic groups within a thin layer yielded a new catalyst that far exceeded the capacity of the industry standard catalyst, platinum. This is precisely what would be expected from a successful catalyst that makes use of Nature's catalytic principles.

Later work suggested that the polypyrrole matrix may participate in and also act as a statistical proximity catalyst in its own right. By manipulating the morphology of this layer, high rates of hydrogen reduction catalysis could be achieved.[21m] A powerful, all-polymer hydrogen-generating catalyst was subsequently developed by incorporating polyethyleneglycol (PEG) into the polypyrrole.[62] The resulting PPy-PEG catalyst was at least as active as an equivalent area of platinum, but orders of magnitude less expensive. A range of new, all-polymer catalysts has since been developed by expanding on this approach.[63] Chapter 11 (Section 11.5) describes various polymer and gel-type enzyme-like catalysts that have been obtained by imprinting techniques; all display Michaelis-Menten kinetics.

Another type of statistical proximity catalyst was reported in 2007.[35] This species was based on the cofacial Co diporphyrins **20–21** that are known to be excellent four-electron catalysts of dioxygen (O_2) reduction to water (H_2O) (Figure 7.12). The equivalent monomeric Co porphyrins, like **22**, are, by contrast, exclusively catalysts of the much slower two-electron reduction of O_2 to H_2O_2. However, when monomer **22** was drastically concentrated within a layer of polypyrrole (PPy) that was densely deposited from the vapor phase, the resulting composite PPy−**22** was found to be an excellent catalyst of four-electron reduction of O_2 to H_2O.[35] Thus, the same functionality that exists in the synthetically challenging bimetallic complexes **20–21** was created by concentrating the readily available and inexpensive monomer **22** into a thin layer.

Significantly, the electrochemical response of PPy−**22** was not consistent with a uniformly distributed catalyst. Rather, it was indicative of a microarray electrode in which there were a very few, highly dispersed but highly active catalytic sites present.[35] This concurred with the expectation that a small, but statistically significant, proportion of the monomers happened to find themselves optimally disposed within PPy−**22**. These sites facilitated four-electron reduction of O_2 to H_2O.

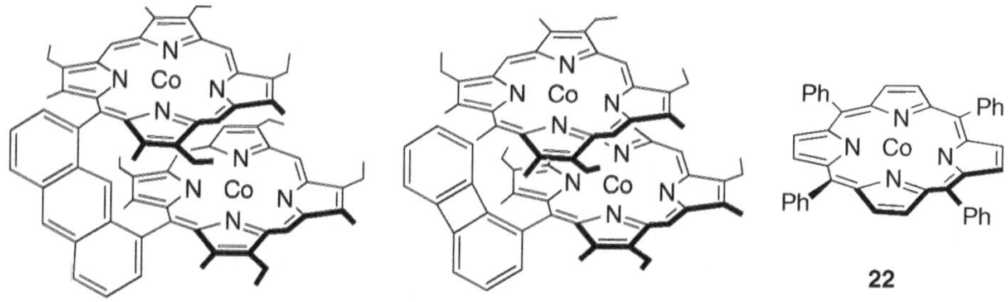

20 **21** **22**

$$O_2 \xrightarrow{\textbf{20} \text{ or } \textbf{21}} H_2O \text{ (4e}^- \text{ process)}$$

$$O_2 \xrightarrow{\textbf{22}} H_2O_2 \text{ (2e}^- \text{ process)}$$

$$O_2 \xrightarrow{\textbf{22} \text{ in Polypyrrole}} H_2O \text{ (4e}^- \text{ process)}$$

Figure 7.12 Cofacial Co diporphyrins **20–21** catalyze the four-electron reduction of O_2 to H_2O. The corresponding monomers, like **22**, catalyze exclusively two-electron reduction to H_2O_2. However, when **22** is immobilized in high concentrations within a thin layer of vapor-phase deposited polypyrrole, then the resulting composite catalyzes the four-electron reduction of O_2 to H_2O. (Adapted with permission. Copyright the Royal Society of Chemistry; Ref. 35.)

7.7 CONCLUSION: THE PROSPECTS FOR HARNESSING NATURE'S CATALYTIC PRINCIPLES

Despite many decades of investigation, Nature's principles of catalysis are still the subject of intense study and discussion. Significant progress, however, has been made. Powerful "machine-like" catalysts and simple-to-prepare "statistical proximity" catalysts have been discovered based on the current understanding of how enzymes operate. A wide range of nonbiological homogeneous and heterogeneous catalysts with active sites that are in large measure physically identical to those of the relevant enzyme have also been prepared and studied. A range of vigorous all-polymer catalysts is being developed.

The capacities of the above catalysts exceed, in several cases, those of long-established industry-best and industry-standard catalysts. For example, the statistical proximity catalyst PPy–$FcSO_3$ is significantly more active than the industry-best catalyst for hydrogen reduction, platinum, while the all-polymer PPy–PEG catalyst at least matches it.[61–63]

Certain of the new catalysts also achieve turnover that is sustained over extended periods of operation, with turnover frequencies that fall within the general range observed for enzymes. For example, the machine-like catalyst **6** turns over 5 H_2 s^{-1} for at least 5 days of continuous operation with no observable deactivation.[31] It is just as active at the end of the 5 days as it was at the beginning. Its turnover number during that time exceeds 1,000,000 per catalyst molecule.[38] Hill's Co-polyoxotungstate water oxidation catalyst also achieves an enzyme-like turnover frequency of >5 s^{-1} at pH 8.[58]

Several of the new catalysts, furthermore, open new vistas in fields that are critically important to humankind but extraordinarily challenging. For example, solar water splitting is achieved without need of an external energy input by $\mathbf{14}^+$/Nafion in a dye-sensitized solar cell.[51] The ability of $\mathbf{14}^+$/Nafion to selectively split seawater without generating toxic Cl_2 gas must also rank as an important advance.[55] Industry standard water oxidation catalysts like Pt and commercial electrolyzers cannot achieve this feat.

These developments are cumulatively extremely promising and bode well for the future. While it is still an open question as to whether the current understanding of enzymes is complete, we have, at the very least, discovered a range of new and powerful catalysts. These catalysts are a significant advance on what was previously possible; they set a new benchmark for synthetic catalysis in a variety of reactions. This is precisely what would be expected if the current understanding of Nature's catalytic principles has at least some measure of validity and is trending toward a fuller understanding.

The challenge now is to continue the quest and to harness the knowledge that we have available in the development of still more powerful synthetic catalyst systems. Whatever the pitfalls, bioinspired catalysis is clearly a field with enormous potential that is eminently worth pursuing.

REFERENCES

1. Stryer, L. *Biochemistry*, 3rd ed., W. H. Freeman and Company, New York, 1988, pp. 177–200.
2. Kuchel, P. W.; Ralston, G. B. *Theory and Problems of Biochemistry*, Schaum's Outline Series, McGraw Hill, New York, 1988, p. 225.
3. (a) Breslow, R. *Acc. Chem. Res*. **1995**, *28,* 146 and references therein; (b) Breslow, R. *Chem. Soc. Rev*. **1972**, *1,* 553; (c) See also Breslow, R.; Dong, S. D. *Chem. Rev*. **1998**, *98,* 1997.
4. Masel, R. I. *Chemical Kinetics and Catalysis*, Wiley-Interscience, Hoboken, NJ, 2001, and references therein. For a description of homogeneous catalysis in general, see: (a) Stern, E. W. In *Transition Metals In Homogenous Catalysis* (Ed. Schrauzer, G. N.), Marcel-Dekker, New York, 1971, Chapter 4 and references therein; (b) Parshall, G. W.; Ittel, S. D. *Homogeneous Catalysis: The Applications and Chemistry of Catalysis by Soluble Transition Metal Complexes*, Wiley, Hoboken, NJ, 1992, and references therein.
5. Page, M. I. In *Enzyme Mechanisms* (Eds. Page, M. I.; Williams, A.), Royal Society of Chemistry, London, 1989, Chapter 1 and references therein.

6. Haldane, J. B. S. *Enzymes*, Longmans, Green and Co., Harlow, UK, 1930, p. 182. See also: Fersht, A. *Enzyme Structure and Mechanism*, 2nd ed., W. H. Freeman and Company, New York, 1977, p. 331, and references therein.
7. Koshland, D. E. *Proc. Natl. Acad. Sci. U.S.A.* **1958**, *44,* 98.
8. (a) Pauling, L. *Chem. Eng. News* **1946**, *24,* 1375; (b) Pauling, L. *Nature* **1948**, *161,* 707.
9. (a) Menger, F. M. *Acc. Chem. Rev*. **1993**, *26,* 206, and references therein; (b) Menger, F. M. *Acc. Chem. Rev*. **1985**, *18,* 128, and references therein; (c) Menger, F. M. *Pure Appl. Chem*. **2005**, *77,* 1873.
10. (a) Williams, D. H.; Stephens, E.; Zhou, M. *Chem. Commun*. **2003**, 1973, and references therein. (b) Another demonstration of the importance of structural complementarity to the optimum transition state is the fact that even minor modifications to the protein sequence can dramatically diminish enzymatic activity and specificity; see, for example: Lipscombe, W. N. In *Structural and Functional Aspects of Enzymatic Catalysis* (Eds. Eggerer, H.; Huber, R.), Springer-Verlag, Berlin, 1981, p. 17, and references therein.
11. Storm, D. R.; Koshland, E. E. *Proc. Natl. Acad. Sci. U.S.A.* **1970**, *66,* 445. See also: Storm, D. R.; Koshland, D. E. *J. Am. Chem. Soc.* **1972**, *94,* 5805; Dafforn, A.; Koshland, D. E. *Proc. Natl. Acad. Sci. U.S.A.* **1971**, *68,* 2463.
12. Bruice, T. C.; Brown, A.; Harris, D. O. *Proc. Natl. Acad. Sci. U.S.A.* **1971**, *68,* 658.
13. Atkins, P. W. *Physical Chemistry*, Oxford University Press, New York, 1978, pp. 52–782.
14. Near-attack conformers: (a) Bruice, T. C. *Acc. Chem. Res*. **2002**, *35,* 139, and references therein. See also: (b) Bruice, T. C.; Lightstone, F. C. *Acc. Chem. Res*. **1999**, *32,* 127; (c) Lau, E.; Bruice, T. C. *J. Am. Chem. Soc*. 2000, *122,* 7165; (d) Bruice, T. C.; Benkovic, S. J. *Biochemistry* **2000**, *39,* 6267. Proximity effect: Bruice, T. C. *Annu. Rev. Biochem*. **1976**, *45,* 331.
15. See, for example: Page, M. L.; Jencks, W. P. *Proc. Natl. Acad. Sci. U.S.A.* **1971**, *68,* 1678. The role of entropy was first raised in Westheimer, F. H. *Adv. Enzymol*. **1962**, *24,* 441.
16. (a) Tousignant, A.; Pelletier, J. N. *Chem. Biol*. **2004**, *11,* 1037; (b) Hammes-Schiffer, S. *Biochem*. **2002**, *41,* 13335; (c) Hammes-Schiffer, S.; Benkovic, S. J. *Annu. Rev. Biochem*. **2006**, *75,* 519; (d) Benkovic, S. J.; Hammes-Schiffer, S. *Science* **2003**, *301,* 1196; (e) Boehr, D. D.; McElheny, D.; Dyson, H. J.; Wright, P. E. *Science* **2006**, *313,* 1638; (f) Vendruscolo, M.; Dobson, C. *Science* **2006**, *313,* 1586.
17. (a) Eisenmesser, E. Z.; Bosco, D. A.; Akke, M.; Kern, D. *Science* **2002**, *295,* 1520; (b) Eisenmesser, E. Z.; Millet, O.; Labeikovsky, W.; Korzhnev, D. M.; Wolf-Watz, M.; Bosco, D. A.; Skalicky, J. J.; Kay, L. E.; Kern, D. *Nature* **2005**, *438,* 117.
18. (a) Workshop on future directions of catalysis science, *Catal. Lett*. **2001**, *76,* 111, and references therein; (b) Jogl, G.; Rozovsky, S.; McDermott, A. E.; Tong, L. *Proc. Natl. Acad. Sci. U.S.A.* **2003**, *100,* 50, and references therein. Bidirectional catalysis is also observed in various acid-catalyzed and other reactions in, for example, synthetic organic chemistry: see, for example, (c) Gugger, P. A.; Hockless D. C. R.; Swiegers, G. F.; Wild, S. B. *Inorg. Chem*. **1994**, *33,* 5671; Hockless, D. C. R.; Lindoy, L. F.; Swiegers, G. F.; Wild, S. B. *Perkin Trans. 1* **1998**, *1,* 117.

19. Williams, R. J. P. *Trends in Biochemical Science* **1993**, *18,* 115, and references therein.

20. Vallee, B. L.; Williams, R. J. P. *Proc. Natl. Acad. Sci. U.S.A.* **1968**, *59,* 498.

21. (a) Swiegers, G. F. (Ed.). In *Mechanical Catalysis; Methods of Heterogeneous, Homogeneous, and Enzymatic Catalysis*, John Wiley & Sons, Hoboken, NJ, 2008; see also in this work: (b) p. 70; (c) Chapter 10; (d) Chapter 5; (e) p. 15; (f) Chapter 6; (g) Chapter 4 and references therein; (h) Chapter 9; (i) p. 142; (j) p. 71; (k) p. 199; (l) Chapter 8; (m) Chapter 12; (n) Chapter 11.

22. Swiegers, G. F.; Huang, J.; Brimblecombe, R.; Chen, J.; Dismukes, G. C.; Mueller-Westerhoff, U. T; Spiccia, L.; Wallace, G. G. *Chem. Eur. J*. **2009**, *15,* 4746.

23. (a) Balandin, A. A. *Adv. Catal*. **1958**, *10,* 120, and references therein; (b) Bond, G. C. *Heterogeneous Catalysis. Principles and Applications*, 2nd ed. Oxford Science Publications, Oxford, UK, 1987, pp. 62–64, and references therein; (c) Bond, G. C.; *Catalysis by Metals*, Academic Press, London, 1962, pp. 476–478; (d) Fahrenfort, J.; van Reyen, L. L.; Sachtler, W. M. H. In *The Mechanism of Heterogeneous Catalysis, Proceedings of the Symposium held in Amsterdam, 12–13 November 1959* (Ed. De Boer, J. H.), Elsevier, Amsterdam, 1960, pp. 23–48, and references therein; (e) Sachtler, W. M. H. *Z. Phys. Chem*. **1960**, *26,* 16; (f) Fahrenfort, J.; van Riegen, L. L.; Sachtler, W. H. M. *Z. Electrochemistry* **1960**, *64,* 216.

24. Kinetic "perfection" in an enzymatic system is described in: Knowles, J. R.; Albery, W. J. *Acc. Chem. Res*. **1977**, *10,* 105.

25. See, for example, Kirby, A. J. *Angew. Chem. Int. Ed. Engl*. **1996**, *35,* 707, and references therein.

26. Konsler, R. G.; Karl, J.; Jacobsen, E. N. *J. Am. Chem. Soc*. **1998**, *120,* 10780.

27. Breslow, R.; Zhang, X.; Xu, R.; Maletic, M.; Merger, R. *J. Am. Chem. Soc*. **1996**, *118,* 11678.

28. Karlin, S.; Taylor, H. E. *A First Course in Stochastic Processes*, 2nd ed., Elsevier Science, New York, 1975.

29. (a) Gleick, J. *Chaos: Making a New Science*, Penguin, London, 1988; (b) Alligood, K. T.; Sauer, T. D.; Yorke, J. A. *Chaos. An Introduction to Dynamical Systems*, Springer-Verlag, New York, 2000; (c) Strogatz, S. H. *Nonlinear Dynamics and Chaos: With Applications to Physics, Biology, Chemistry and Engineering*, Addison Wesley, Reading, MA, 1994; (d) Katok, A.; Hasselblatt, B. *Introduction to the Modern Theory of Dynamical Systems*, Cambridge University Press, Cambridge, 1996; (e) Stewart, I. *Does God Play Dice? The New Mathematics of Chaos*, Penguin, London, 1997; (f) *Chaos and Fractals: New Frontiers of Science* (Eds. Peitgen, H-O; Juergens, H.; Saupe, D.) Springer-Verlag, New York, 1992.

30. Bitterwolf, T. E.; Ling, A. C. *J. Organomet. Chem*. **1973**, *57,* C15.

31. Mueller-Westerhoff, U. T. *Angew. Chem. Int. Ed. Engl*. **1986**, *25,* 702, and references therein. See also: Mueller-Westerhoff, U. T.; Haas, T. J.; Swiegers, G. F.; Leipert, T. K. *J. Organomet. Chem*. **1994**, *472,* 229.

32. Hillman, M.; Michalle, S.; Felberg, S. *Organometallics* **1985**, *4,* 1258. Figure 4 in this paper is effectively a Lineweaver–Burke plot.

33. (a) Karlsson, A.; Broo, A.; Ahlberg, P. *Can. J. Chem*. **1999**, *77,* 628; (b) Karlsson, A.; Himersson, G.; Ahlberg, P. *J. Phys. Org. Chem*. **1997**, *10,* 590.

34. (a) 10 mg of **6** will dissolve at least 5 g of lead as sacrificial reductant during H_2 generation; see Mueller-Westerhoff, U. T. *Angew. Chem. Int. Ed. Engl*. **1986**, *25,* 702, and references therein; (b) a coating containing 2 M [**6**] over an area of 0.16 cm^2 and a response thickness estimated to be 1 μm generates a saturation current of 37 mA during catalysis; see Mueller-Westerhoff, U. T.; Nazzal, A. *J. Am. Chem. Soc*. **1984**, *106,* 5381.

35. Chen, J.; Zhang, W.; Officer, D.; Swiegers, G. F.; Wallace, G. G. *Chem. Commun*. **2007**, 3353.

36. (a) Cobb, A. J. A.; Shaw, D. M.; Longbottom, D. A.; Gold, J.; Ley, S. V. *Org. Biomol. Chem*. **2005**, *3,* 84, and references therein; (b) Gröger, H.; Wilken, J. *Angew. Chem. Int. Ed. Engl*. **2001**, *40,* 529; (c) Kelly, D. R.; Roberts, S. M. *Chem. Commun*. **2004**, 2018; (d) Kelley, D. R.; Meek, A.; Roverts, S. M. *Chem. Commun*. **2004**, 2021; (e) Cordova, A.; Zou, W.; Ibrahem, I.; Reyes, E.; Engqvist, M.; Liao, W.-W. *Chem. Commun*. **2005**, 3586; (f) Davies, S. G.; Sheppard, R. L.; Smith, A. D.; Thomson, J. E. *Chem. Commun*. **2005**, 3802.

37. See, for example, for *catalase*: (a) E_a7 kcal mol^{-1}: Lehninger, A. L. *Biochemistry*, Worth Publishing, New York, 1975, p. 189; (b) E_a 2.5 kcal mol^{-1}: Brown-Peterson, N. J.; Salin, M. L. *J. Bacteriol*. **1993**, 4197. Some other examples of low energy barriers in enzymatic catalysis: (c) Kolawole, A. O.; Sanni, D. M.; Ajele, J. O. *J. Plant. Sci*. **2007**, *2,* 425; Kolawole, A. O.; Sanni, D. M.; Ajele, J. O. *J. Plant. Sci*. **2006**, *1,* 273; (d) Mirjafar, P.; Aghari, K.; Mahinpey, N. *Ind. Eng. Chem. Res*. **2007**, *46,* 921.

38. Thomas, J. M.; Williams, R. J. P. *Philos. Trans. R. Soc. A* **2005**, *363,* 765.

39. Corma, A. *Catal. Rev*. **2004**, 46, 369.

40. Moss, D. W. *Enzymes*, Oliver and Boyd, London, 1968, p. 68, and references therein.

41. Ruettinger, W.; Campana, C.; Dismukes, G. C. *J. Am. Chem. Soc*. **1997**, *119,* 6670.

42. (a) Ruettinger, W.; Yagi, M.; Wolf, K.; Bernasek, S.; Dismukes, G. C. *J. Am. Chem. Soc*. **2000**, *122,* 10353; (b) Yagi, M.; Wolf, K. V.; Baesjou, P. J.; Bernasek, S. L.; Dismukes, G. C. *Angew. Chem. Int. Ed. Engl*. **2001**, *40,* 2925.

43. Ferreira, K. N.; Iverson, T. M.; Maghlaoui, K.; Barber, J.; Iwata, S. *Science* **2004**, *303,* 1831.

44. (a) Loll, B.; Kern, J.; Saenger, W.; Zouni, A.; Biesiadka, J. *Nature* **2005**, *438,* 1040; (b) Yano, J.; Kern, J.; Irrgang, K.-D.; Latimer, M. J.; Bergmann, U.; Glatzel, P.; Pushkar, Y.; Biesiadka, J.; Loll, B.; Sauer, K.; Messinger, J.; Zouni, A.; Yachandra, V. K. *Proc. Natl. Acad. Sci. U.S.A.* **2005**, *102,* 12047.

45. Umena, Y.; Kawakami, K.; Shen, J.-R.; Kamiya, N. *Nature* **2011**, *473,* 55.

46. Kuznetsov, A. E.; Geletii, Y. V.; Hill, C. L.; Musaev, D. G. *J. Phys. Chem. A* **2010**, *114,* 11417.

47. Brimblecombe, R.; Swiegers, G. F.; Dismukes, G. C.; Spiccia, L. *Angew. Chem. Int. Ed. Engl*. **2008**, *120,* 7445.

48. Brimblecombe, R.; Swiegers, G. F.; Dismukes, G. C.; Spiccia, L. *Inorg. Chem*. **2009**, *48,* 7269.

49. Brimblecombe, R.; Bond, A. M.; Dismukes, G. C.; Swiegers, G. F.; Spiccia, L. *Phys. Chem. Chem. Phys*. **2009**, *11,* 6441.

50. Dismukes, G. C.; Brimblecombe, R.; Felton, G. A. N.; Pryadun, R. S.; Sheats, J. E.; Spiccia, L.; Swiegers, G. F. *Acc. Chem. Res*. **2009**, *42,* 1935.

51. Brimblecombe, R.; Koo, A.; Dismukes, G. C.; Swiegers, G. F.; Spiccia, L. *J. Am. Chem. Soc.* **2010**, *132,* 2892.

52. Brimblecombe, R.; Chen, J.; Wagner, P.; Buchhorn, T.; Dismukes, G. C.; Spiccia, L.; Swiegers, G. F. *J. Mol. Catal. A* **2011**, *338,* 1.

53. Hocking, R. K.; Brimblecombe, R.; Chang, L.-Y.; Singh, A.; Cheah, M. H.; Glover, C.; Casey, W. H.; Spiccia, L. *Nature Chem.* **2011**, *3,* 461.

54. Swiegers, G. F.; Clegg, J. K.; Stranger, R. *Chem. Sci.*, **2011**, *2,* 2254.

55. Kanan, M. W.; Nocera, D. G. *Science* **2008**, *321,* 1072.

56. (a) Lutterman, D. A.; Surendranath, Y.; Nocera, D. G. *J. Am. Chem. Soc.* **2009**, *131,* 3838; (b) McAlpin, G.; Surendranath, Y.; Dinca, M.; Stich, T. A.; Stoian, S. A.; Casey, W. H.; Nocera, D. G.; Britt, R. D. *J. Am. Chem. Soc.* **2010**, *132,* 6882; (c) Kanan, M. W.; Yano, J.; Surendranath, Y.; Dinca, M.; Yachandra, V.; Nocera, D. G. *J. Am. Chem. Soc.* **2010**, *132,* 13892, and references therein.

57. Jiao, F.; Frei, H. *Angew. Chem. Int. Ed.* **2009**, *48,* 1841, and references therein.

58. Yin, Q.; Tan, J. M.; Besson, C.; Geletti, Y. V.; Musaev, D.; Kuznetsov, A. E.; Luo, Z.; Hardcastle, K. I.; Hill, C. L. *Science* **2010**, *328,* 342. Note added in proof: since publication of this work, a new record has been achieved for turnover frequency in water-oxidation catalysis, see: Duan, L.; Bozoglian, F.; Mandal, S.; Stewart, B.; Privalov, T.; Llobet, A.; Sun, L. *Nature Chem.* **2012** advance article (DOI: 10.1038/NCHEM.1301)".

59. Robinson, D. M.; Go, Y. B.; Greenblatt, M.; Dismukes, G. C. *J. Am. Chem. Soc.* **2010**, *132,* 11467.

60. A molecular Co-cubane water oxidation catalyst is described in: McCool, N.; Robinson, D. M.; Sheats, J. E.; Dismukes, G. C. *J. Am. Chem. Soc.* **2011**, *133,* 11446.

61. Chen, J.; Swiegers, G. F.; Too, C. O.; Wallace, G. G. *Chem. Commun.* **2004**, 308.

62. Winther-Jensen, B.; Fraser, K.; Ong, C.; Forsyth, M.; MacFarlane, D. R. *Adv. Mater.* **2010**, *22,* 1727.

63. Winther-Jensen, B.; MacFarlane, D. R. *Energy Environ. Sci.* **2011**, *4,* 2790.

CHAPTER 8

Biomimetic Amphiphiles and Vesicles

SABINE HIMMELEIN and BART JAN RAVOO

Organic Chemistry Institute and Graduate School of Chemistry, Westfälische Wilhelms-Universität Münster, Corrensstrasse 40, 48149 Münster, Germany

8.1 INTRODUCTION

Without exaggeration, it could be stated that a bilayer of phospholipid molecules is all that separates "internal" from "external" in living organisms. Biological membranes play a key role in the protection of the integrity, stability, and shape of a cell as well as its interaction and communication with its environment. Moreover, crucial processes such as recognition and adhesion of cells, membrane fusion, signal transduction, and enzyme catalysis occur at the membrane surface. In addition, even the simplest of cells contain numerous subcompartments, each of which are secluded by a separate bilayer membrane. Several key functions of cell membranes are highlighted in Figure 8.1. It was shown by Bangham and Horne in 1964 that phospholipid bilayer membranes can easily be formed in vitro[1] and it was reported by Kunitake and Okahata in 1977 that the formation of bilayers is not restricted to biological phospholipids.[2] Vesicles (Lat. *vesicula* = small bubble) have been an important topic in both chemistry and the life sciences ever since. On the one hand, vesicles are of interest as highly dynamic supramolecular structures that mimic the remarkable properties of biological membranes. On the other hand, vesicles are of interest as self-assembled responsive capsules that may be applied in drug delivery, as nanoreactors and nanosensors, or in the design of soft materials.

This chapter focuses on the use of abiological amphiphiles to generate biomimetic vesicles. The chapter opens with a general introduction to the use of synthetic amphiphiles as building blocks for biomimetic vesicles. The major part of the chapter is organized according to a number of sophisticated functions that are typical of biological cell membranes. First, we discuss biomimetic vesicles that display molecular-recognition-induced adhesion and fusion of membranes. Second, we describe vesicles that are senstitive to stimuli-responsive shape control. Third, we focus on vesicles that are capable of transmembrane signaling

Bioinspiration and Biomimicry in Chemistry: Reverse-Engineering Nature, First Edition.
Edited by Gerhard F. Swiegers.
© 2012 John Wiley & Sons, Inc. Published 2012 by John Wiley & Sons, Inc.

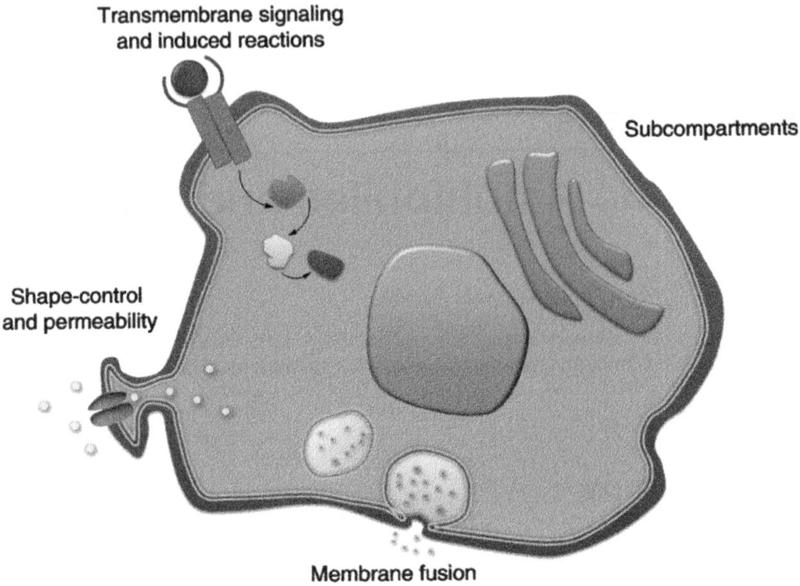

Figure 8.1 Schematic representation of key functions of a biological cell membrane.

and vesicles that can be used to compartmentalize chemical reactions. Finally, we address the issue of formation of subcompartments within a vesicle. The chapter closes with a brief outlook.

This chapter is intended to provide insights into the fascinating supramolecular chemistry of biomimetic vesicles by highlighting a selected number of recent publications. A comprehensive review of the literature is not provided. Langmuir–Blodgett films and supported bilayers—which are also versatile biomimetic membranes—are not covered in this chapter.

8.2 SYNTHETIC AMPHIPHILES AS BUILDING BLOCKS FOR BIOMIMETIC VESICLES

Vesicles are dynamic supramolecular structures that consist of a molecular layer that encapsulates a small amount of solvent. Vesicles are predominantly formed in aqueous solution. The term "liposome" is generally reserved for vesicles composed of natural phospholipids, while the term "vesicle" includes vesicles composed of synthetic amphiphiles, phospholipids, or any other components. "Polymersomes" are vesicles composed of polymers.

Bilayer vesicles are closely related to liposomes and biological membranes. Most molecules that form bilayer vesicles in water are amphiphilic: they have a hydrophobic and a hydrophilic part. The hydrophilic part ("head group") of the molecule interacts favorably with the surrounding water, while the hydrophobic part ("tail") minimizes its exposure to water. Hence, the amphiphiles arrange in a bilayer and the formation of vesicles is driven primarily by hydrophobic interaction. Typically, the head group is polar and/or charged and contains phosphate, sulfate, ammonium, amino acid, peptide, carbohydrate, or oligo(ethylene glycol) groups. Typically, the "tail" is apolar and uncharged. The tail is usually composed of long hydrocarbon chains, which may be saturated or unsaturated, linear, cyclic or branched, aromatic or aliphatic, or fluorinated. In accordance with the concept of the packing parameter,[3] the amphiphile must have an approximately cylindrical shape, so that the molecules arrange into a bilayer, which may be slightly curved so that it can close into a spherical vesicle. It should be noted that the packing parameter cannot be defined exclusively on geometric considerations: attractive and repulsive interactions of head groups should also be taken into account.

With the advent of polymersomes and vesicles of other "nonconventional" (i.e., not phospholipid-like) amphiphiles, numerous examples of *monolayer vesicles* in water have been reported. Typically, monolayer vesicles are prepared from small molecules with a hydrophobic core and two hydrophilic head groups (bolaform amphiphiles, see below) or from triblock copolymers with two hydrophilic terminal blocks. The molecule must have a cylindrical or rectangular shape, so that it can arrange into a monolayer.

Irrespective of their composition, it is useful to differentiate between small unilamellar vesicles (SUVs, <100 nm), large unilamellar vesicles (LUVs, 100–1000 nm), giant unilamellar vesicles (GUVs, >1 μm), and multilamellar vesicles (MLVs) (Figure 8.2). SUVs, LUVs, and GUVs have a unilamellar membrane composed of a single molecular bilayer (or monolayer). SUVs and LUVs are the most widely studied types of vesicles. GUVs are of particular interest as biomimetic membranes since their size is comparable to living cells.[4] MLVs have an onion-like structure and consist of many concentric bilayer (or monolayer) membranes. It can easily be calculated that the smallest SUVs (ca. 50 nm) of small amphiphiles contain about ten thousand molecules, whereas LUVs contain about a hundred thousand molecules, and GUVs and MLVs contain many millions of molecules. The number of molecules in one vesicle can only be

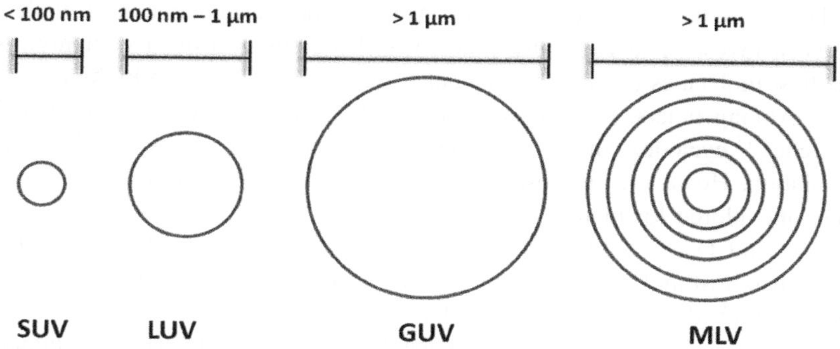

Figure 8.2 Small unilamellar vesicles (SUVs), large unilamellar vesicles (LUVs), giant unilamellar vesicles (GUVs), and multilamellar vesicles (MLVs).

given as an approximate average, because it is difficult to prepare vesicles of an exactly defined size (see Section 8.4).

The first report on bilayer vesicles formed from synthetic amphiphiles dates from 1977, when Kunitake and Okahata described the formation of vesicles from di-*n*-dodecyl dimethyl ammonium bromide in aqueous solution.[2] In the 1980s, it was shown by many groups that a wide range of amphiphilic molecules can form vesicles in water. In a sense, these amphiphiles are all very similar to phospholipids: they generally have two hydrophobic tails and a hydrophilic head group, so that the molecule has a cylindrical shape and packs efficiently into a bilayer sheet, which curves and closes into a vesicle. On the other hand, the structural variety of synthetic amphiphiles provides vesicles with a range of functions that clearly surpass the properties of liposomes. Among others, synthetic vesicles can be made light sensitive, vesicles can be made pH sensitive, and vesicles can be polymerized. The pioneering work on synthetic vesicles is summarized in reviews by Kunitake,[5] Ringsdorf et al.,[6] and Engberts et al.[7]

Small amphiphiles must not necessarily have a phospholipid-like structure with two tails and one head group. For example, *bolaform* (or bipolar) amphiphiles are amphiphilic molecules that contain two head groups separated by an extended hydrophobic chain. Bolaform amphiphiles form monolayer vesicles, in which each amphiphile extends across the monolayer membrane, exposing both head groups to water and sheltering the hydrophobic chain from water.[8] However, these types of vesicle-forming amphiphiles are in fact also inspired from Nature: many extremophilic bacteria have membranes that contain a high percentage of bolaform amphiphiles.[9] Monolayer membranes of bolaform amphiphiles are much more robust than bilayer membranes and contribute to the stability of extremophiles in acidic, alkaline, and hot environments.

The innovative design of small amphiphiles continues to give rise to unconventional vesicles. Recently, there has been increasing interest in synthetic amphiphiles equipped with a recognition unit.[10] In many cases, such amphiphiles can form vesicles by themselves. Alternatively, they can be mixed with conventional amphiphiles or phospholipids. In this way, it is possible to functionalize vesicles with complementary recognition motifs that can guide the selective adhesion and even fusion of bilayer vesicles (see Section 8.3). Amphiphilic macrocyclic host molecules have been investigated for many years. Vesicles composed of or containing synthetic host molecules are versatile model systems for receptors in biological membranes. Among others, it is known that amphiphilic crown ethers,[11] cryptands,[12] calixarenes,[13] cyclodextrins,[14] and cucurbiturils[15] can form bilayer vesicles in aqueous solution. However, the host–guest chemistry of such host vesicles remained largely unexplored for many years.

Ravoo and Darcy prepared bilayer vesicles composed entirely of amphiphilic cyclodextrin host molecules.[14] Such vesicles have a membrane that displays a high density of embedded host molecules that bind hydrophobic guest molecules. The characteristic size-selective inclusion behavior of the cyclodextrins is maintained, even when the host molecules are embedded in a hydrophobic membrane. For example, adamantane carboxylate binds preferentially to β-cyclodextrin vesicles

($K_a = 7000\ M^{-1}$), weaker to γ-cyclodextrin vesicles ($K_a = 3000\ M^{-1}$), and very poorly to α-cyclodextrin vesicles ($K_a < 100\ M^{-}1$).[16]

Cucurbiturils are another class of host molecules that have been assembled into bilayer vesicles in water. Kim and co-workers synthesized an amphiphilic cucurbit[*6*]uril that forms vesicles and forms host–guest complexes at the vesicle surface.[15] It is possible to decorate the surface of the host vesicles with guest molecules. Exposure of cucurbituril vesicles to a fluorescent spermidine derivative leads to fluorescent vesicles. Exposure of the cucurbituril vesicles to α-mannose substituted spermidine leads to vesicles coated with α-mannose, which bind specifically to the lectin concanavalin A (ConA). ConA does not bind when the vesicles are coated with a galactose spermidine conjugate. These experiments demonstrate how synthetic host membranes can interact with proteins via multivalent interactions mediated by carbohydrates.

A major innovation in the area of vesicles was triggered by Eisenberg and co-workers, who demonstrated in 1995 that very large amphiphilic molecules can form vesicles.[17] In a pioneering *Science* report, it was shown that poly(styrene)-*block*-poly(acrylic acid) can form bilayer vesicles in water. These polymersomes were prepared by slow addition of water to a DMF solution of the block copolymer, followed by dialysis to remove the remaining DMF. The hydrophobic poly(styrene) forms the interior of the bilayer membrane, while the hydrophilic poly(acrylic acid) is exposed to water. It has been shown since that many block copolymers can form vesicles. Important advantages of polymersomes include their high kinetic stability and their very low membrane permeability (which increases with the length of the hydrophobic block).

Although block copolymers that merely contain a hydrophobic block connected to a hydrophilic block can still be considered rather straightforward high molecular weight analogs of conventional small amphiphiles, the field of polymersomes has benefited tremendously from the design of more complex block copolymer architectures using new polymerization methods (such as atom transfer radical polymerization, ATRP) and highly efficient conjugation protocols (such as click chemistry). For example, ABA block copolymers can form monolayer vesicles. In fact, ABA block copolymers are the macromolecular equivalent of bolaform amphiphiles. ABC block copolymers can form bilayer vesicles if A and B (but not C), or B and C (but not A), are of similar polarity, but they can also form monolayer vesicles if A and C (but not B) are of similar polarity. Figure 8.3 outlines the most important block copolymer architectures for polymersomes. Block copolymers can also contain biopolymer segments, such as polypeptides or polysaccharides conjugated to synthetic segments ("biohybrid copolymers").[18, 19] Biohybrid copolymers are important biomimetic amphiphiles, since they combine the biological activity of membrane-embedded biomacromolecules with the ability to self-organize in kinetically stable vesicles and microcapsules. The blossoming field of polymersomes has been the subject of several reviews.[20, 21]

The versatility of polymersomes was significantly advanced by Nolte and co-workers, who expanded the scope of block copolymers from large to giant biohybrid amphiphiles.[22] The key innovation in their work is the conjugation

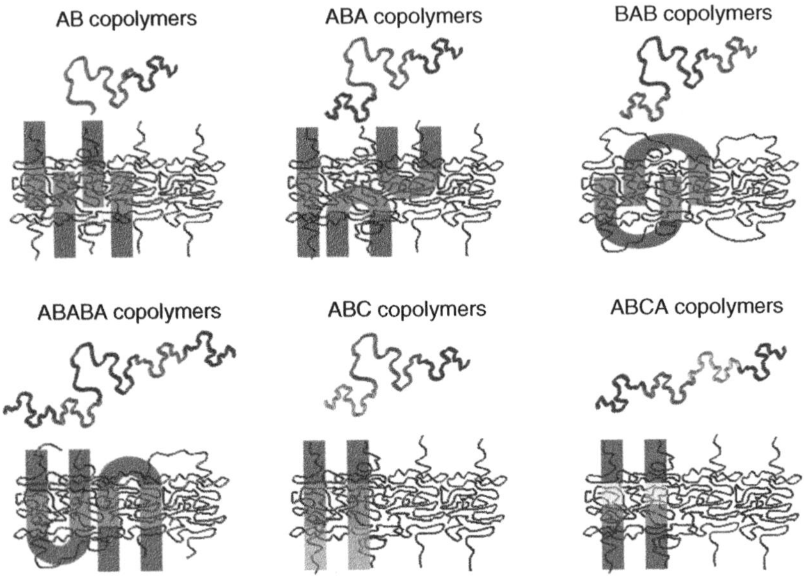

Figure 8.3 Block copolymers for polymersomes. (Reproduced with permission. Copyright Royal Society of Chemistry: Ref. 21.)

of very large hydrophilic proteins to hydrophobic synthetic polymers. These biohybrid block copolymers differ from other protein–polymer conjugates in the sense that the protein to polymer ratio is predefined and the position of the conjugation site is precisely known. In a particularly elegant experiment, giant biohybrid amphiphiles self-assembled by cofactor reconstitution of poly(styrene) modified heme and apo-horseradish peroxidase (HRP) as well as apomyoglobin. The biohybrid amphiphiles were obtained by adding a THF solution of the heme cofactor-appended poly(styrene) to an aqueous solution of the apoenzyme. TEM revealed the formation of LUVs with diameters of 80–400 nm. The activity of the HRP and myoglobin enzymes is retained in the polymersomes.

Vesicle-forming amphiphiles must not be held together by covalent interactions exclusively: it is easily conceivable that an amphiphile is formed by noncovalent interaction of two (or more) components. Hence, although the individual components cannot form vesicles, vesicles self-assemble upon mixing of the components in the appropriate molar ratio. A remarkable example of a ternary complex that self-assembles into vesicles was reported by Kim and co-workers.[23] It was shown that vesicles are formed spontaneously in a mixture of cucurbit[*6*]uril, *n*-alkyl viologen, and dihydroxynaphthalene (Figure 8.4). Viologen and dihydroxynaphthalene form a stable charge transfer complex in the cavity of the cucurbituril host. The ternary complex is amphiphilic due to the presence of the long alkyl chain on the viologen. If the alkyl substituent is *n*-dodecyl, SUVs are formed; if the alkyl substituent is *n*-hexadecyl, LUVs are formed. The vesicles can be imaged by scanning

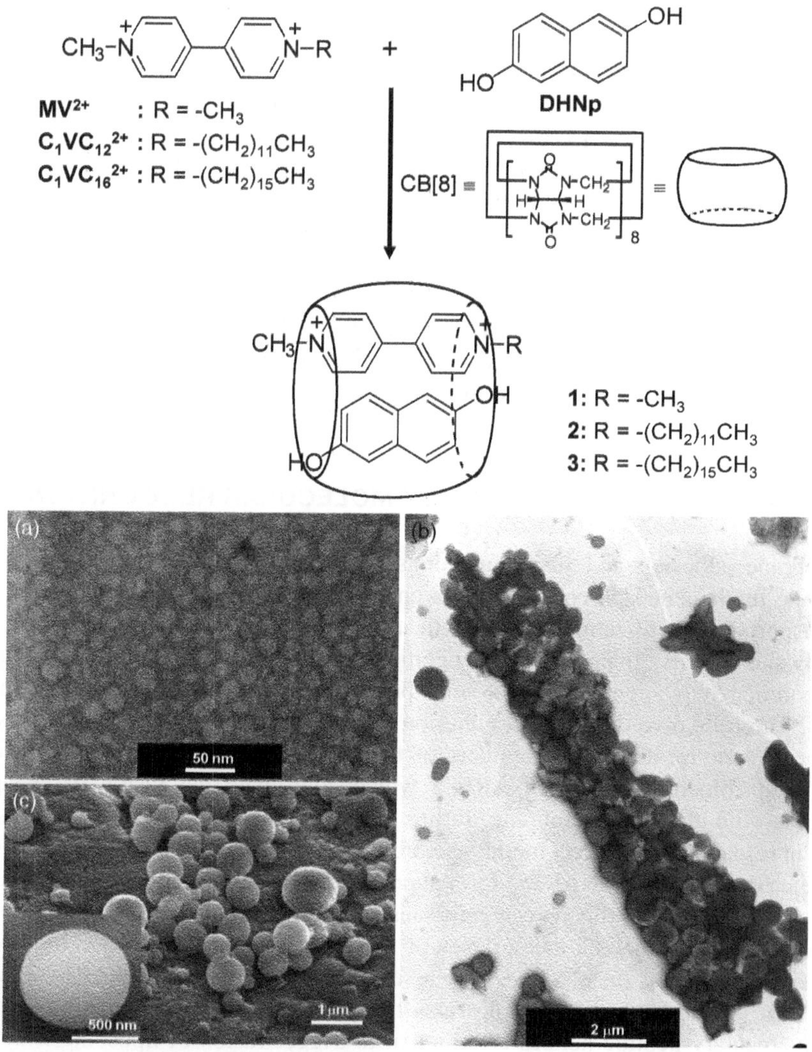

Figure 8.4 Vesicles formed of a charge transfer complex of *n*-alkyl viologen and dihydroxynaphtalene in the cavity of cucurbit[6]uril. (a)-(c) TEM and SEM images of complex **2** and **3**. (Reproduced with permission. Copyright Wiley-VCH: Ref. 23.)

electron microscopy (SEM), demonstrating the robustness of the supramolecular structure.

In recent years, it has become clear that vesicles can also be assembled from building blocks that are in no way reminiscent of phospholipids. In fact, the building block must not even be amphiphilic. For example, an Israeli–Indian team reported the formation of all-peptide vesicles in 2007.[24] It was shown by TEM, SEM, and fluorescence microscopy that a trimer of ditryptophan assembles into vesicles in a mixture of water and methanol. The peptide forms a π-stacked network across the surface of the vesicle, similar to the cytoskeleton in biological cells. Araki and co-workers reported the formation of highly stable LUVs composed of two-dimensional hydrogen-bonded sheet structures of guanosine.[25] The building block for these LUVs is a guanosine substituted with a phenylsilyl unit and an oligo(ethylene glycol unit). This molecule is not an amphiphile! Nevertheless, it assembles into vesicles. It is shown that the membrane of the vesicles is stabilized by a two-dimensional hydrogen-bonding network of guanosine, while the oligo(ethylene glycol) unit is exposed to water. The formation of polymersomes can also be induced by electrostatic interaction or "electrostatic self-assembly." Among others, it was shown that LUVs are obtained when a polycation is mixed with a polyanion-*block*-poly(ethylene glycol).[26] The vesicle size can be controlled by changing the polymer concentration.

8.3 VESICLE FUSION INDUCED BY MOLECULAR RECOGNITION

Membrane adhesion and fusion reactions are vital for cell function since they mediate fundamental processes such as endocytosis and exocytosis, synaptic neurotransmission, fertilization, cell growth, and viral infection. In Nature, fusion is highly site specific, given that the recognition of two different organelle membranes is coordinated by noncovalent protein–protein or protein–carbohydrate interactions. These involve diverse cell adhesion molecules (CAMs) present in opposing bilayers. Among others the members of the soluble *N*-ethylmaleimide-sensitive-factor attachment proteins (SNAREs), which coordinate the fusion of synaptic vesicles with the cell membrane during neurotransmission, are in the focus of current research.[27–30] Target membrane proteins (t-SNAREs) and secretory vesicle-associated proteins (v-SNAREs) form supramolecular helix bundle complexes in a zipper-like fashion. In this way, membranes are brought into close contact, allowing calcium bridging between the surfaces of the opponent membranes and lipid rearrangements causing fusion. Recently, a semisynthetic "minimal SNARE system" has been proposed, which demonstrates that the interaction of membrane-bound complementary peptide helices is sufficient to trigger fusion of liposomes.[31]

Complete membrane fusion requires two different organelles to merge their membranes and to mix the aqueous compartments encapsulated by the membranes. Generally this occurs in two steps (Figure 8.5): (1) two vesicles cluster into an intervesicular complex due to site-specific recognition; and (2) the lipid bilayers are close enough to interact, resulting in the fusion of the two vesicles. According to the stalk hypothesis, fusion proceeds by an ordered sequence of steps. First two vesicles form close contacts. In this "docking" stage the contacting proximal monolayers ("cis") have fused, while the distal monolayers ("trans") remain intact. The curved intermediate is referred to as the "stalk." Subsequently a hemifusion intermediate is formed from the interaction of the trans monolayers, which results in the formation of a fusion pore. Finally, opening of the fusion pore allows the mixing of the inner compartments.[27, 28] In this section we discuss a selection

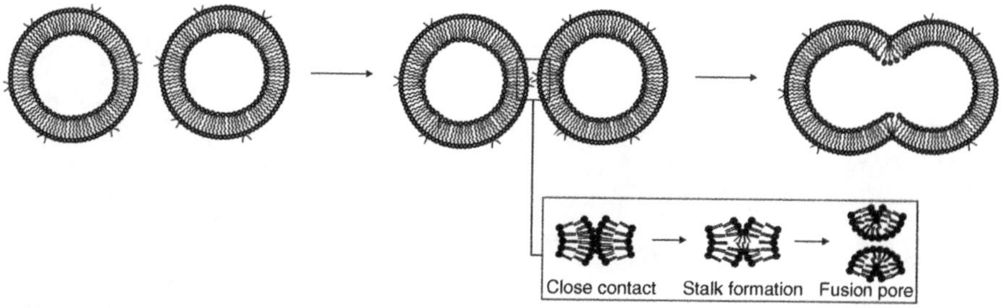

Figure 8.5 General mechanism of membrane fusion. Two vesicles cluster into an intervesicular complex due to site-specific recognition and the lipid bilayers form a close contact of the outer bilayer leaflets. The vesicles merge upon stalk formation and opening of a fusion pore.

of appealing examples where chemists designed membrane-bound receptors that function as CAM mimics in order to induce vesicle adhesion or fusion. We note that membrane model systems have also been the topic of a recent tutorial review.[32]

An early example for the interaction of vesicles induced by hydrogen bonding was described by Zasadzinski and co-workers.[33] Unilamellar vesicles of dilauroylphosphatidylcholine (DLCP) and biotin conjugated dipalmitoylphosphatidylethanolamine (DPPE–biotin) were prepared in order to obtain SUVs with an average of 80 biotin ligands per vesicle exposed on the outer monolayer. Addition of streptavidin at a ligand–receptor ratio of about 7:1 caused the liposomes to aggregate immediately. Streptavidin is able to bind four biotin molecules and hereby tethers the liposomes together. Interestingly, cryogenic electron microscopy (cryo-TEM) showed that the vesicles remained intact during aggregation and no fusion was induced. This is remarkable since biotin and streptavidin form a very strong supramolecular complex ($K_a \sim 10^{14}$–$10^{15} M^{-1}$) held together by multiple hydrogen bonds. Moreover the ligand–receptor bonds could be reversed by addition of soluble biotin as a competitor, leading to redispersion of the vesicles.

In analogy to Zasadzinski's pioneering work, Constable and co-workers presented the first example of metal ion directed aggregation of phosphatidylcholine (PC) liposomes directed by specific metal coordination.[34] The introduction of terpyridine functionalized phospholipids into PC liposomes led to vesicles equipped with an average of 970 ligand molecules. By addition of Fe^{2+} into the modified vesicle solution the formation of a 1:2 metal–ligand complex was observed by its characteristic purple color. Additionally, titration of Fe_2SO_4 resulted in an increasing hydrodynamic radius until the precipitation of a purple material. Cryo-TEM imaging showed that the majority of vesicles are aggregated into clusters at a concentration of Fe^{2+}:terpy-ligand of 4.5:1. The addition of an excess of EDTA reversed the Fe^{2+}-induced aggregation. Because of the reversibility of the aggregation process it can be concluded that no fusion into GUVs occurred.

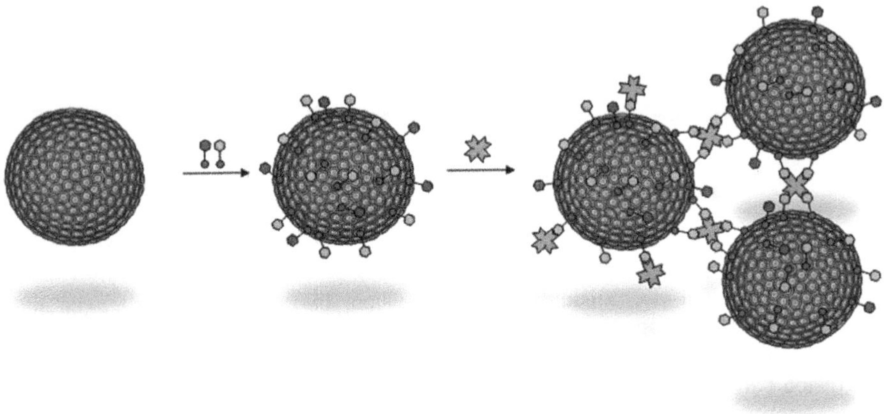

Figure 8.6 Lectin concanavalin A (ConA) induces the aggregation of cyclodextrin vesicles in the presence of bifunctional maltose–adamantane conjugate. Lectin peanut agglutinin (PNA, not shown) induces the aggregation of cyclodextrin vesicles in the presence of lactose–adamantane conjugate. The rate and extent of aggregation require a critical surface density of carbohydrate.

Not only vesicles composed of phospholipids but also unilamellar bilayer vesicles of amphiphilic β-cyclodextrins were shown to reversibly aggregate by a biomimetic interaction, namely, specific carbohydrate–lectin recognition.[35] Cyclodextrin vesicles consist of a membrane with a high density of embedded host molecules that bind hydrophobic guests.[14] An artificial glycocalix was formed by decorating cyclodextrin vesicles with maltose and lactose by host–guest interactions. Agglutination of cyclodextrin vesicles was induced by specific binding of maltose to ConA and lactose to peanut agglutinin (PNA), respectively (Figure 8.6). The lectins ConA and PNA form tetramers at neutral pH and each lectin is able to bind four sugar molecules, in a similar fashion as the streptavidin–biotin system described above. Dynamic light scattering (DLS) showed an increase of the average particle size in the vesicle solution from about 100 nm to 320 nm. Cryo-TEM revealed that the vesicles remain intact and only deform slightly to establish extended areas of contact between membranes during agglutination; no fusion of vesicles could be observed. This is in accordance with the biological role of lectins, as they mediate the adhesion, not the fusion, of membranes. Moreover, the clustering was reversed by the addition of competitive inhibitors, such as D-glucose or β-cyclodextrin.

Metal-ion recognition can not only induce aggregation of vesicles but also leads to complete fusion as reported by Lehn and co-workers.[36] Synthetic ligands bearing a bipyridine head group for metal coordination were synthesized and incorporated into phospholipid LUVs. Addition of the transition metal ions Ni^{2+} or Co^{2+} led to the formation of intravesicular 1:1 and 1:2 metal–ligand complexes, followed by lipid exchange with formation of intervesicular complexes. Strong adhesion with multilayer formation resulted spontaneously in fusion pore opening and MLVs were

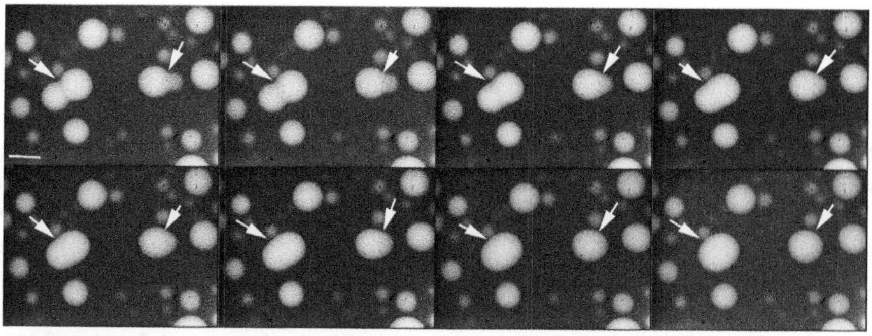

Figure 8.7 Formation of giant vesicles by fusion of LUVs equipped with an amphiphilic dipyridine ligand and filled with rhodamine sulfonate (50 μM) in the presence of $NiCl_2$ (0.1 μM) observed by fluorescence microscopy. The time between the first panel (upper left) and the last one (lower right) was 7 s. Scale bar: 10 μm. (Reproduced with permission. Copyright National Academy of Sciences, U.S.A.: Ref. 36.)

produced. In order to follow the fusion process by fluorescence microscopy in real time, vesicles were filled with rhodamine sulfonate (Figure 8.7). Within 10 s after the addition of $NiCl_2$, many fusion events could be observed. No leakage of the fluorescent solution into the external medium was detected. The authors showed further that the formation of intervesicular complexes is only possible if a sufficiently long poly(ethylene glycol) (PEG) spacer between the vesicle surface and the bipyridine group is present. No fusion could be observed for ligands containing only four PEG units. Ten PEG groups were required to induce fusion and ligands holding fourteen PEG entities led to the formation of the largest MLVs. Additionally, it was observed that a minimal surface concentration of bipyridine ligands was required (3 mol %) for stabilizing the adhesion state. The authors expect that by increasing the length of the PEG spacer as well as the concentration of the ligands adhesion and fusion processes should be enhanced.

In 2005, Webb and co-workers took one step further toward biomimetic membrane adhesion and fusion. They presented a system in which multiple metal–ligand interactions were used to create networks of vesicles.[37] Furthermore, they showed that the composition of the lipid bilayer tilts the balance between stable vesicle adhesion and membrane fusion. Vesicle adhesion and fusion were induced by multivalent coordination of Cu(iminodiacetate) (IDA) complexes with poly-L-histidine, which functions as supramolecular glue. The metal-binding group was embedded into the membrane with a pyrene group, allowing direct visualization of changes in the lipid distribution via fluorescence spectroscopy. LUVs consisting of distearoylphosphatidylcholine (DSCP) or egg yolk phosphatidylcholine (EYPC) and 5 mol % of amphiphilic IDA ligand only showed monomer emission of the pyrene unit, which indicates a homogeneous distribution of the ligand within the membrane. The titration of Cu^{2+} to a solution of DSPC or EYPC vesicles decorated with IDA ligand led to a decreasing pyrene fluorescence caused by Cu^{2+}-induced quenching. Titration experiments revealed the formation of ML and ML_2 complexes with an average binding constant $K_a = 4 \times 10^7\ M^{-1}$. Cu^{2+} titration into a solution of DSPC–IDA vesicles followed by titration of poly-L-histidine resulted in a strong increase of turbidity. A $(His)_2Cu(IDA)$ complex was formed which led to the aggregation of vesicles without fusion. Interestingly, the titration of Cu^{2+} into a solution of EYPC–IDA vesicles showed a strong increase in turbidity caused by the formation of very large vesicles by fusion without leakage. Addition of poly-L-histidine showed no further effect on turbidity. These observations are attributed to the higher fluidity of the EYPC bilayer membranes because it contains a large amount of unsaturated phospholipids. This allows the IDA amphiphiles to migrate to the vesicle adhesion interface to form intervesicular $Cu(IDA)_2$ complexes. When the multivalent binding is close to a 1:2 ratio of metal to ligand, the unsaturated phospholipids at the contact area promote the formation of a stalk, initiating irreversible membrane fusion. In contrast DSPC bilayers lack fusogenic unsaturated lipids and vesicle adhesion therefore is irreversible.

In 2006 Webb and co-workers presented an even more elaborate system which mimics the formation of biological lipid rafts.[38] The localization of CAMs within phase separated domains or lipid rafts is essential in cell–cell adhesion processes. Membrane proteins such as cadherins accumulate at cell–cell contact sites and act as adhesion-activated signaling receptors.[39] In order to create lipid rafts, the authors introduced a fluorinated pyrene anchor attached to the Cu(IDA) complex and a histidine ligand functionalized with an alkyl chain was used as complementary binding partner. Fluorine–fluorine interactions enabled the synthetic Cu(IDA) lipids to phase separate even at low membrane concentrations (1% mol/mol) and caused the receptor to cluster. It should be noted that the relatively weak interaction between Cu(IDA)–histidine bonds ($K_a \sim 10^3\ M^{-1}$) mimics natural adhesive interactions like selectin–sialyl Lewis X ($K_a \sim 10^4\ M^{-1}$).[40] The ratio of excimer to monomer emission intensity of the pyrene groups reflects the extent of lipid phase separation of the Cu(IDA) ligand in the vesicle membrane and allows visualization of the vesicles by fluorescence microscopy (Figure 8.8). Vesicles containing histidine ligand were doped with a red fluorescent rhodamine dye, which contrasts with the blue fluorescence of the pyrene moiety. Mixing vesicles of dimyristoylphosphatidylcholine (DMPC) and cholesterol containing 5 mol % of Cu(IDA) complex with vesicles containing 5 mol % of histidine ligand led to a strong increase of turbidity caused by the formation of vesicle aggregates. No vesicle fusion occurred. The formation of vesicle aggregates was shown to be dependent on the degree of phase separation of the Cu(IDA) complex. Without cholesterol and in the case of a nonfluorinated Cu(IDA) ligand no phase separation and no adhesion were observed. Thus, adhesion of the vesicles is based on receptor clustering in the membrane and multivalent metal–ligand coordination in the intervesicular contact area, just like the adhesion of biological membranes depends on the clustering of CAMs.

As shown for metal–ligand coordination, the specific recognition of complemetary pairs by hydrogen bonds has also been reported to promote fusion. Lehn and co-workers incorporated the well-known barbituric acid (BAR)–2,4,6-triaminopyrimidine (TAP) pair into EPC liposomes by conjugation to an amphiphilic molecule.[41] LUVs of 100 nm diameter and 10% of BAR or TAP were

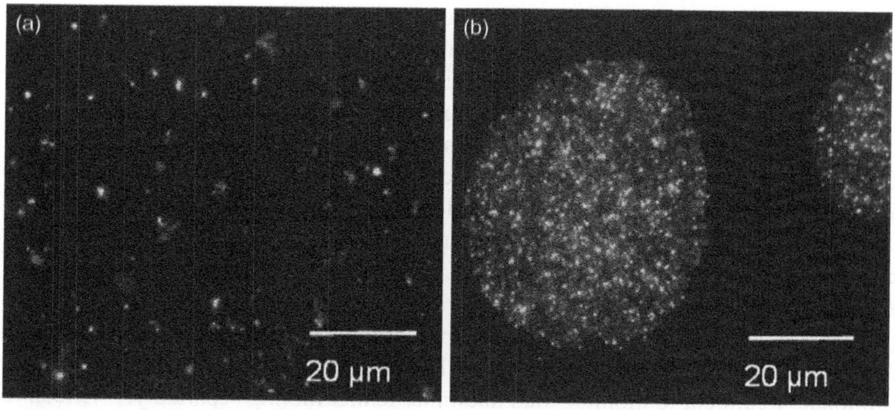

Figure 8.8 Fluorescence micrographs of vesicles containing 5 mol % of amphiphilic histidine (red) mixed with vesicles containing 5 mol % of Cu(IDA) (blue) in (a) DMPC and (b) DMPC/cholesterol. (Reproduced with permission. Copyright American Chemical Society: Ref. 38.) (See insert for color representation of this figure.)

prepared by extrusion. It could be shown via a FRET assay using amphiphilic acceptor and donor dyes[42] that there is an exchange of lipids between the BAR and the TAP vesicles as a result of vesicle aggregation and fusion. The complementary surface charge of the (negatively charged) TAP and (positively charged) BAR vesicles is critical to the intervesicular interaction. The aggregation of TAP and BAR vesicles leads to fusion into GUVs.

In 2009 Kashiwada and co-workers presented a stimulating example for biomimetic membrane fusion based on hydrogen bond interactions.[43] In Nature the recognition of simple as well as complex carbohydrates plays an important role in many cellular processes. Enveloped viruses, for example, recognize the site of sialic acid on the membranes of the host cells via fusion proteins, which then insert into the target membrane.[28] The authors constructed for the first time a carbohydrate-selective membrane fusion system by incorporating boronic acid derivatives into EPC vesicles. Previous studies had shown that the boronic acid functional group forms reversible covalent linkages with diols, making it a common recognition moiety for carbohydrate sensing.[44] On this basis a selective fusion system toward a phosphatidylinositol (PI)-containing vesicle membrane was generated. The system uses the boronic acid and *cis*-diol structure of the inositol-glycocalix as surface bound CAMs. Vesicles with an average diameter of 100 nm and 5 mol % of boronic acid ligand (with a stearic acid moiety and a PEG5 spacer) or PI were prepared. DLS experiments showed an increase of vesicle size in the mixture of the complementary molecular recognition partners to around 150 nm. This corresponds to the fusion of one donor vesicle (ca. 130 nm) with one target vesicle (ca. 90 nm) rather than aggregation. Vesicle fusion was further monitored by a FRET assay verifying total and inner leaflet mixing.[42] Addition of competitor molecules (free myoinositol) and pH-dependent experiments confirmed that the molecular recognition is essential for inducing fusion in this system. Based on their results the authors propose a fusion mechanism involving (1) formation of intervesicular complexes through molecular recognition, (2) attachment of donor with target vesicles, (3) joining and mixing of donor and target membranes, and (4) preceding pore formation and intermixing of vesicle contents.

Inspired by SNARE proteins Höök and co-workers presented a system where fusion of phospholipid vesicles is induced by hybridization of complementary DNA strands (Figure 8.9).[45] Cholesterol modified DNA strands were incooperated into 100 nm LUVs. The length and sequence of the DNA strands were designed such that hybridization occurs in a zipper-like fashion, similar to the formation of helix bundles in SNAREs,[27, 28] and thereby forces vesicles with complementary DNA into close contact. Lipid mixing was probed by conventional NBD-PS/rhodamine-DHPE dequenching FRET assay in conjugation with sodium dithionate treatment to monitor total and inner leaflet mixing separately.[42] Effective fusion was demonstrated for four model membranes composed of different lipid classes in the presence of complementary DNA strands, whereas the presence of noncomplementary DNA did not lead to fusion. DNA-induced fusion furthermore was shown to be sensitive to the presence of lipids such as 1,2-dioleyl-*sn*-glycero-3-phospho-etanolamine (DOPE) and cholesterol, which are known to promote the transport of DNA-liposomes across membranes. Therefore, the authors suggested that a stalk intermediate and raft structures may be formed during fusion. In a further study Höök and co-workers investigated how parameters such as length of the DNA strands, anchoring strategy, and number of DNA strands affected the

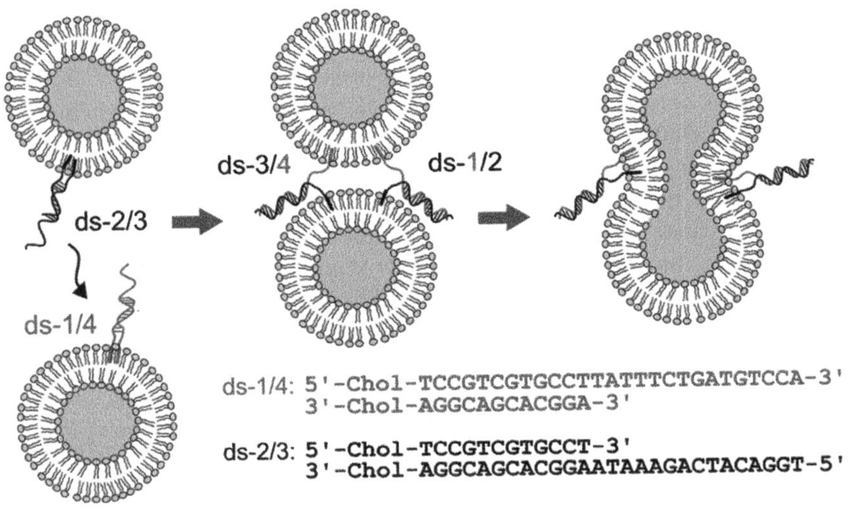

Figure 8.9 Vesicle fusion mediated by DNA hybridization. Initially, vesicles are modified with the double cholesterol terminated DNA strands ds-1/4 and ds-2/3 (left). As ds-1/4 and ds-2/3 encounter each other, they hybridize in a zipper-like fashion (middle). In this way the bilayers are brought into close contact, which eventually enables opening of the fusion pore (right). (Reproduced with permission. Copyright American Chemical Society: Ref. 45.)

DNA's ability to induce membrane fusion.[46] It was found that anchoring DNA with two cholesterol groups is essential for fusion, whereas single cholesterol anchoring caused rapid dissociation of the DNA-bridged vesicle complex. The length and coverage of DNA only slightly affected fusion. It was concluded that the DNA and lipid rearrangements that take place at the vesicle–vesicle contact zone were limiting for fusion.

Another biologically inspired fusiogenic system based on multiple hydrogen bonding was reported by Bong and co-workers.[47, 48] In this study selective vesicle fusion is driven by small molecule recognition between vancomycin glycopeptides and its native target the D-Ala-D-Ala dipeptide. Vancomycin is an antibiotic that binds to D-Ala-D-Ala peptides via five hydrogen bonds (Figure 8.10) and thereby inhibits the transpeptidation step of peptidoglycan synthesis in the cell wall of bacteria.[49] Membrane display of the dipeptide was accomplished by coupling with a phospholipid anchor and vancomycin derivatives were conjugated to a magainin II peptide anchor. Mangainin II is an antimicrobial peptide from frog skin and was used to anchor vancomycin because it has the ability to insert selectively into the hydrophobic matrix of negatively charged membranes. In doing so, it destabilizes membranes in a concentration-dependent manner. The noncovalent binding of the peptide to the membrane disrupts the hydrophobic packing of the lipids and triggers lipid mixing and fusion. Both receptors were separately embedded into EPC liposomes without inducing an increase in the size of the liposomes. However, mixing vesicles that expose complementary binding partners led to a strong increase in size monitored by DLS. Fusion of total and inner leaflet was verified with a donor and acceptor dye FRET assay in conjugation with sodium dithionate treatment.[42] The results indicate a fusion process in which surface binding initiates a highly aggregated state where fusion occurs rapidly and slows as the lipid binding partners increasingly occupy the same membrane. Moreover, it was shown that relatively nonperturbative POPE lipid anchor instead of the magainin anchor resulted in liposome aggregation without fusion. Also, the addition of gel-phase lipids to the EPC membrane significantly increased the membrane fusion rate and decreased leakage. The authors suggest that membrane subdomains may be formed by lipid mismatch, resulting in clusters of lipids and fusogens in the bilayer, enhancing binding and function.

In summary, it is evident from a number of examples that specific molecular recognition between complementary recognition units can induce selective adhesion and fusion of vesicles, much like CAMs mediate the adhesion and fusion of biological membranes. Clearly, the multivalent noncovalent interactions can overcome the surface repulsion that normally keeps membranes away from each other. However, to the best of our knowledge, there is no example yet of a specific fusion of polymersomes, which are generally much more stable than vesicles composed of small amphiphiles. In this respect, it is clear that biological membranes are just thick and tight enough to pose a significant barrier, yet thin and flexible enough to be prone to fusion. It is interesting to note that many researchers have chosen a reductionist approach toward mimicking protein–carbohydrate-induced adhesion as well as protein-induced fusion of membranes. Deeper insight into the factors

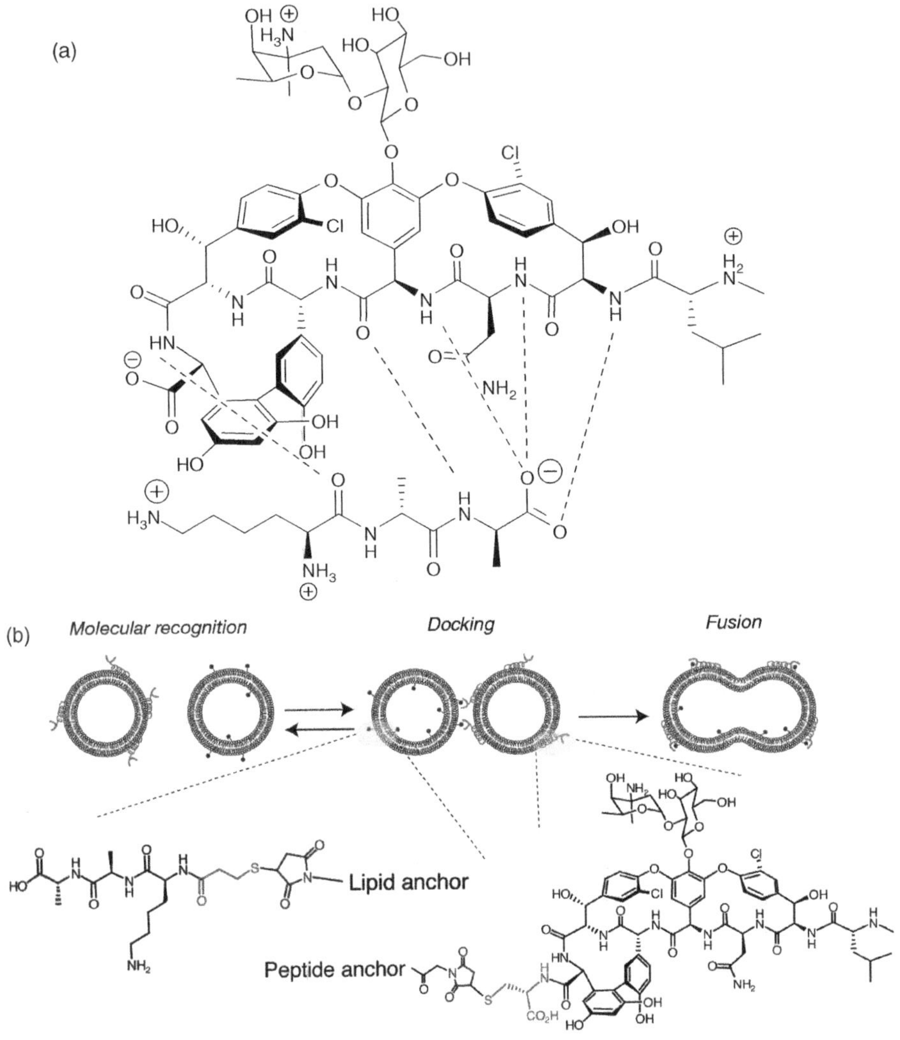

Figure 8.10 (a) Vancomycin (top) binds to Lys-D-Ala-D-Ala (bottom) via five hydrogen bonds. (b) Model of molecular recognition guided fusion. (Reproduced with permission. Copyright American Chemical Society: Ref. 48.)

that promote or inhibit membrane fusion may be of relevance to the development of sophisticated biomimetic drug and gene delivery systems.

8.4 STIMULI-RESPONSIVE SHAPE CONTROL OF VESICLES

One of the stunning features of biological membranes relates to their remarkable mechanical properties: phospholipid bilayer membranes are difficult to compress or stretch in the lateral direction, yet they are extremely flexible and can deform into perfectly planar as well as highly curved regions. Cells usually have a very narrow size distribution: all copies of a particular type of cell are nearly identical in size. Moreover, cells have a particular shape that is specific to their functions and they can adapt their shape in response to external stimuli. Remarkable examples include erythrocytes and dendritic cells. The shape of a cell is directed by the cytoskeleton, which could be described as a protein-based network that confers mechanical stability to the flexible cell membrane. A variety of proteins contain so-called BAR domains, which dimerize and bind to membranes and induce curvature.[50] In this section we discuss a number of biomimetic vesicles that can change their size or shape in response to an external stimulus such as pH change, irradiation with light of a specific wavelength, chemical reactions, or temperature change.

A key challenge in the development of biomimetic vesicles is the preparation of vesicles with a narrow size distribution. Although the average size of vesicles can be tailored with reasonable accuracy and reproducibility, conventional methods to prepare SUVs, LUVs, or GUVs typically yield a rather broad size distribution. In particular, it is difficult to prepare GUVs of a well-defined size and a narrow size distribution. Only very recently a promising method for the production of uniform, monodisperse GUVs was presented (Figure 8.11).[51] In short, monodisperse GUVs were generated in a microfluidic T junction under continuous flow. A lipid film was reconstituted in the junction by sequentially infusing water, oil, and water into the device. The cross flow at the T junction continuously thins, shears, and squeezes the membrane, and this membrane repeatedly releases multiple vesicles encapsulating uniform water droplets. This method has a number of important advantages: (1) it produces monodisperse GUVs; (2) the encapsulation is efficient and versatile enough that a variety of contents can be used; and (3) microfluidics reduces the volume of reagents and enables high-throughput vesicle production.

Many vesicles change size, shape, or permeability in response to pH change. An early example was reported by Chécot, Lecommandoux, Gnanou, and Klok, who prepared a pH-sensitive vesicle system from polybutadiene-*b*-poly(L-glutamic acid) (PB_{40}-*b*-PGA_{100}) in aqueous media.[52] The vesicles showed a fully reversible size variation from 100 to 160 nm in hydrodynamic radius monitored by DLS. This size change was attributed to the pH sensitivity of the secondary structure of the polypeptide segments. A more recent study was presented by Eisenberg and co-workers,[53] who described a pH-induced "breathing" feature of vesicles prepared from a triblock copolymer, namely, poly(ethylene oxide)$_{45}$-*block*-polystyrene$_{130}$-*block*-poly(2-diethylaminoethyl methacrylate)$_{120}$ (PEO_{45}-*b*-PS_{130}-*b*-$PDEA_{120}$). This block copolymer assembles in water at pH 10.4 to form vesicles with an average radius of 250 nm and a wall thickness of approximately 25 nm, measured by DLS and cryo-TEM, respectively. In detail, the vesicle wall was shown to consist of a sandwich of two, approximately 4 nm thick PS layers and one approximately 17 nm thick PDEA layer in the middle. With decreasing pH, the vesicle size as well as the thickness of all three layers increased. At pH 3.40 the wall thickness enlarged to a maximum of approximately 80 nm with the two PS layers cracked on the wall surfaces and one swollen PDEA layer

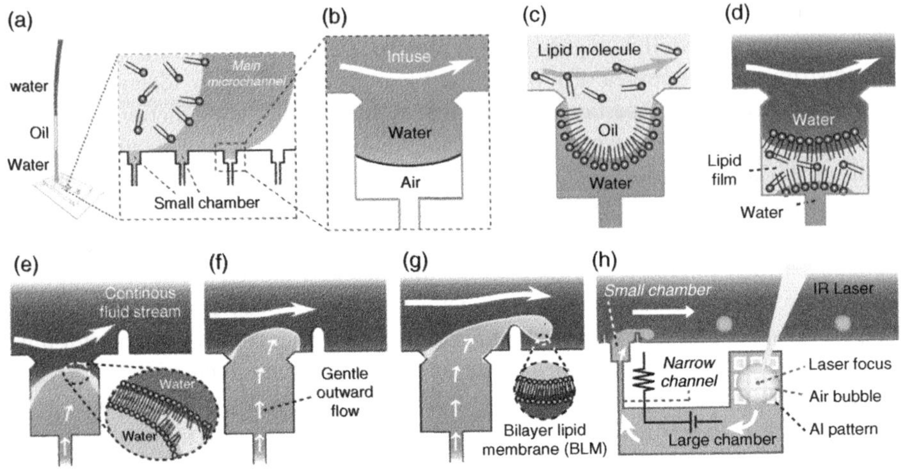

Figure 8.11 Microfluidic process for the generation of monodisperse unilamellar vesicles. (a) Water, oil with dissolved lipids, and water are sequentially infused into a microfluidic device, in which a main channel has many chambers in its walls. (b) Water fills the device while pushing air out through the PDMS wall. (c) Oil flushes away the water in the channel, but confines the remaining water in the chambers. (d) Water again flushes away the oil, and the residue forms an oil film in which amphiphilic lipid molecules form monolayers at the interface of water and oil. (e)–(g) Schematic depiction of the flow-driven unilamellar vesicle formation. (e) A cross flow at the microfluidic T junction thins the lipid film and drives the contact of monolayers to form a bilayer. (f) The gentle outward flow further bends out the bilayer. (g) Shear forces from the continuous fluid stream lead to the fission of the leading edge of the bilayer, that is, the generation of a unilamellar vesicle. (h) The system integrated with an optically generated microbubble. (Reproduced with permission. Copyright Wiley-VCH: Ref. 51.)

in the middle, extruding some chains through the cracks in the PS layers into the solution. Here, the vesicle size was amplified by 190%. This pH-dependent increase in thickness is due to the PDEA segment that swells by protonation and hydration. The PS layers constrain the swelling up to a pH slightly below 6. Then progressive swelling of the PDEA layer leads to cracks in the PS layers, and the size of the vesicles shows a sharp increase. In Figure 8.12 wedge-shaped sections of the cryo-TEM images at a pH range from 10.40 to 3.40 are shown. Intact walls are observed at pH 10.40 to 6.22, while apparent discontinuities were found at pH 5.65 and 3.40. These pH-induced changes are highly reversible and can repeatedly be cycled. Furthermore, the authors show that the vesicle wall exhibits an increasing permeability to protons from high to low pH.

In addition to size, shape transformations can also be induced by pH change (Figure 8.13). For instance, Kros and co-workers reported the reversible conversion of cyclodextrin vesicles into nanotubes.[54] This was achieved by decorating cyclodextrin vesicles with a $(LeuGlu)_4$ octapeptide functionalized with an adamantane anchor for host–guest interactions. The peptide self-assembled on the surface of the cyclodextrin vesicles and depending on the pH formed either a β-sheet or

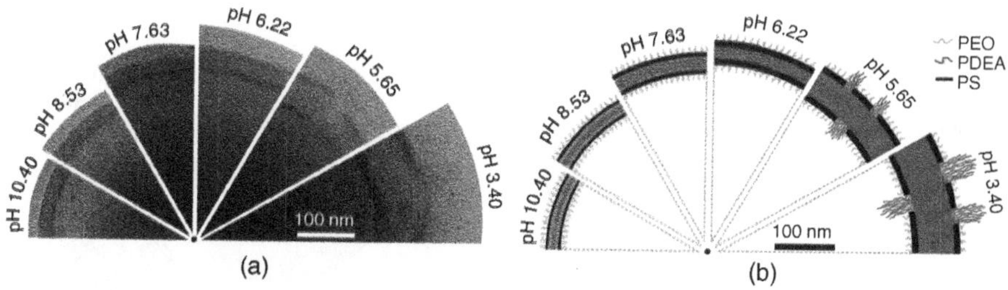

Figure 8.12 (a) Cryo-TEM images of vesicle wall structures at pH 10.4, 8.53, 7.63, 6.22, 5.65, and 3.40, respectively. (b) Schematic illustrations of the vesicle structures at corresponding pH values. (Reproduced with permission. Copyright American Chemical Society: Ref. 53.)

a random coil. This well-designed system combines three orthogonal interactions: (1) hydrophobic interactions in the cyclodextrin bilayers, (2) inclusion complex formation of β-cyclodextrin and adamantane, and (3) hydrogen bonding in β-sheet domains. At pH 7.4 the octapeptide does not form β-sheet domains in the presence of cyclodextrin vesicles, merely binding to the vesicle surface. By lowering the pH to 5.0, however, a transition from random coil to β-sheet is observed by circular dichroism (CD) spectroscopy. Furthermore, CD measurements revealed that the optimum molar ratio for β-sheet formation was 2:1 between vesicle monomer and oligopeptide. This implies that about 50% of the cyclodextrin moieties are available for the peptide with adamantane anchor, suggesting that the cyclodextrin vesicles preserve their bilayer structure and that the peptide is unable to permeate into the vesicles. The formation of ß-sheets caused the rearrangement of the cyclodextrin vesicles into thin fiber-like aggregates with a thickness of approximately 8 nm and a length of several hundred nanometers monitored by cryo-TEM. It was further shown that the system is reversible; by changing the pH back to 7.4 only spherical vesicles were observed again. Moreover, the pH-triggered release of a fluorescent cargo as a result of shape transformation has been shown by a quenching experiment. It should also be emphasized that the pH range of this process matches the decrease in pH that occurs upon endosomal uptake by cells. Hence, these experiments suggest that the peptide decorated cyclodextrin vesicles may be a useful vehicle for intracellular delivery of drugs or antigens that are encapsulated inside the vesicle or bound on the surface of the vesicle.

Vesicles can also change shape in response to an optical trigger (Figure 8.14). Azobenzenes are among the most studied photoresponsive molecules. The assembly of amphiphilic azobenzene derivatives in vesicles results in photoresponsive vesicles. A Japanese team demonstrated that an amphiphilic azobenzene can switch the shape of GUVs, as determined in real time by microscopic observation.[55] Photoisomerization induces a change in membrane fluctuation behavior or a morphological transition between ellipsoid and bud vesicle shapes, depending on the initial vesicle shape. The mechanism of this reversible photoswitching in the vesicle morphology is interpreted in terms of a change in the effective cross-sectional area of the

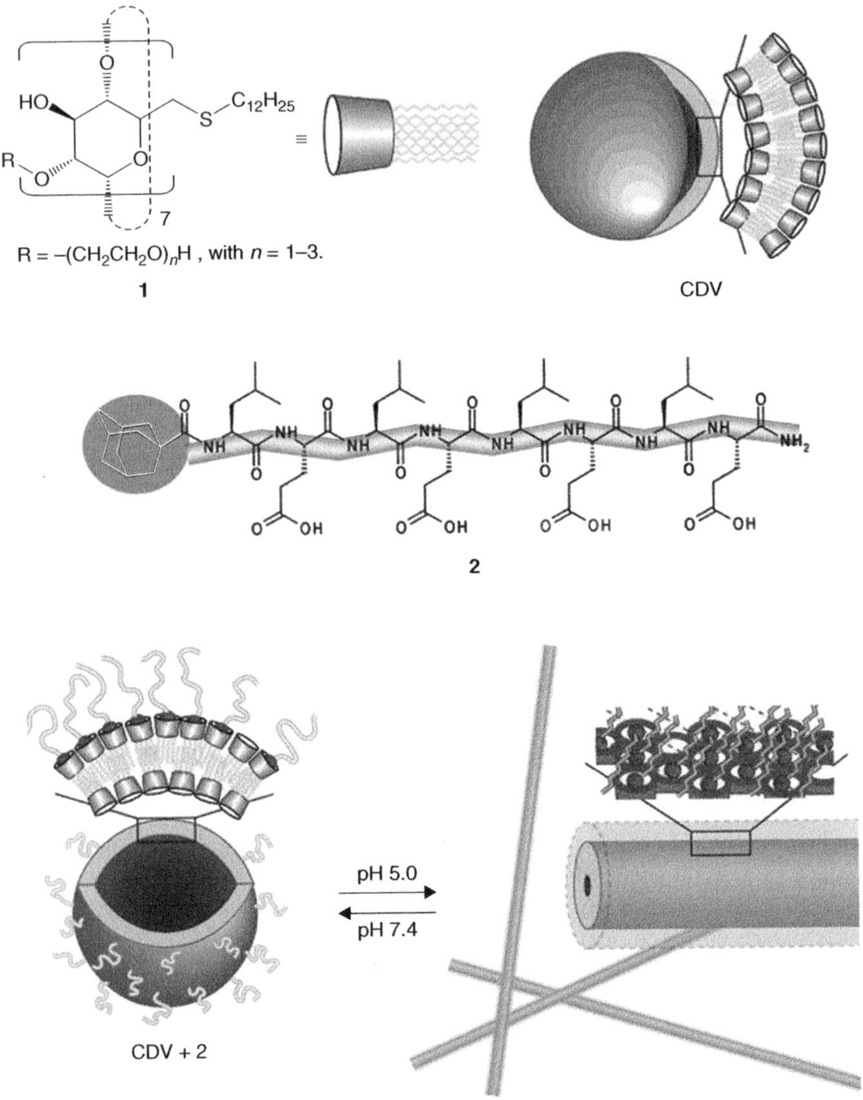

Figure 8.13 Molecular structures of amphiphilic β-cyclodextrin derivative, which self-assembles into vesicles and adamantane modified octapeptide, which binds to the vesicles by host-guest interaction. This peptide adapts a random coil conformation at pH 7.4 while bound to the vesicles. Upon acidification to pH 5.0, the peptide forms a β-sheet which induces a morphological change from a vesicle to a nanotube, with concomitant release of contents. (Reproduced with permission. Copyright American Chemical Society: Ref. 54.)

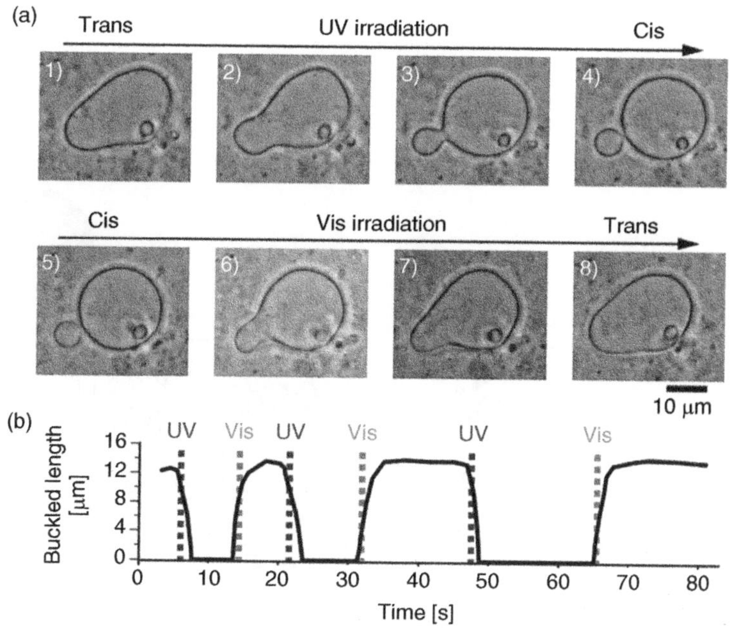

Figure 8.14 (a) Photoinduced reversible ellipsoid–bud transition in a cell-sized vesicle. (1–4) Transformation from ellipsoid to bud induced by UV light. (5–8) Reverse process from bud to ellipsoid induced by irradiation with green light. (b) Repetitive photoswitching of the morphology. The time development of the neck length on the buckled part in the ellipsoid–bud transformation is shown. (Reproduced with permission. Copyright American Chemical Society: Ref. 55.)

azobenzene upon photoisomerization. In a later study, the same group demonstrated the reversible light-induced exo- and endo-budding of GUVs.[56]

The application of synthetic vesicles for encapsulation and delivery may be limited due to the instability of their membranes. A variety of methods have been developed for stabilizing different kinds of membranes, mainly by free radical polymerization[6] as well as other chemical reactions. Deming and Holowka made use of the biomimetic crosslinking ability of dihydroxyphenylalanine (DOPA) amino acid residues (Figure 8.15).[57] DOPA moieties occur naturally in mussel adhesive proteins, which are well known for their crosslinking behavior upon oxidation.[58] After oxidation, the DOPA groups form DOPA–DOPA–quinone crosslinks. Here, the authors presented a system consisting of diblock copolypeptides that self-assemble into spherical vesicles, whose size can be controlled by extrusion. These peptide building blocks possess DOPA amino acid residues, which can be covalently crosslinked in water using $NaIO_4$ as oxidizing agent. In doing so, vesicles with improved membrane stability against freeze-drying, organic solvent, osmotic stress, and complex media were obtained.

A remarkable example of temperature-responsive vesicles was reported by Lee and co-workers.[59] They synthesized different kinds of so-called dumbbell-shaped

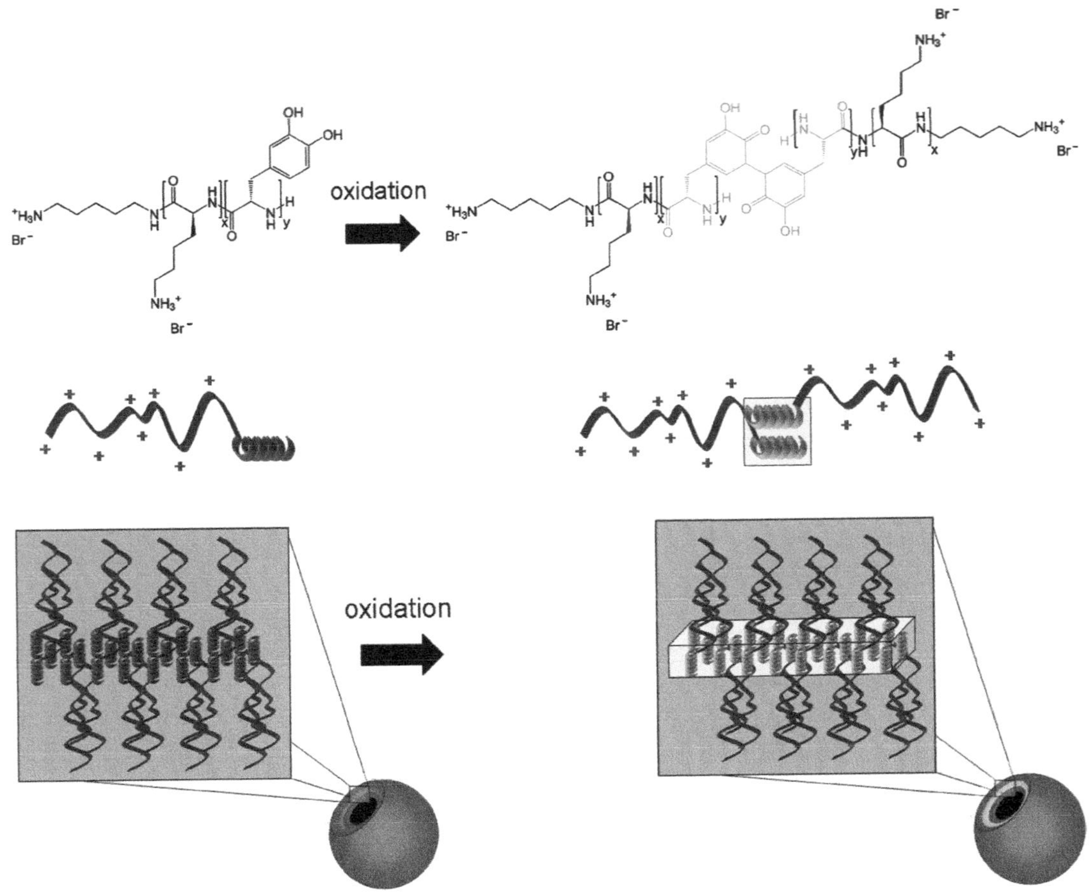

Figure 8.15 Structures and schematics of $K_{60}DOPA_{20}$ diblock copolypeptides and vesicles before and after oxidative crosslinking. (Reproduced with permission. Copyright Wiley-VCH: Ref. 57.)

rigid rod amphiphiles. These molecules consist of an aromatic rod segment that is grafted by hydrophilic polyether dendrons at one end and hydrophobic branches at the other end. It was found that one of these molecules was able to form spherical aggregates of micrometers in diameter with lateral nanopores in the shell in 0.01 wt % aqueous solution. The vesicle shell thickness of about 16 nm indicates that the rods are arranged in bilayer packing in which the hydrophobic alkyl chains are intercalated between the rod segments. The pores in the shell have a narrow size distribution with a typical diameter of 25 nm. Fascinatingly, the lateral pores closed completely upon heating to 65 °C without any changes in spherical shape. After 12 h at room temperature the shells started to form small openings and further annealing time led to an increased number of openings. Complete recovery was observed after 7 days resting at room temperature. The explanation for this temperature-induced gating is due to the oligo(ethylene oxide) dendritic residues, which have a lower critical solution temperature (LCST) behavior in aqueous media. Above the LCST, the ethylene oxide moieties are dehydrated and collapse into molecular globules, causing a decrease in the effective hydrophilic volume and as a result the pores close. Fluorescence microscopy and cryo-TEM were used to monitor the gating process, in addition ^{1}H NMR confirmed the dehydration process (Figure 8.16). The authors, moreover, showed that encapsulated fluorescence labeled DNA could be delivered into HeLa cells with release of DNA. This indicates that the vesicles can encapsulate a relatively large DNA molecule (6700 Da) and deliver it into the inside of the cell, which is similar to viral capsids that self-assemble into a protein coat for transporting viral genome controlled by a gating mechanism.[60]

In summary, although numerous examples of stimulus-responsive biomimetic vesicles have been reported, none of them approaches the sophistication and adaptiveness of living cells. On the other hand, if a synthetic vesicle exclusively responds to a single bioorthogonal signal (such as irradiation or local heating), it could possibly find application as a drug or gene delivery device. Furthermore, it is interesting to note that although there are several examples of vesicles with an "external" cytoskeleton that confers stability and/or directs shape, there are no reports on vesicles or polymersomes with a biomimetic "internal" cytoskeleton.

8.5 TRANSMEMBRANE SIGNALING AND CHEMICAL NANOREACTORS

The binding of a cell surface ligand or soluble small molecule, such as hormones or neurotransmitters, to specific receptors on the surface of another cell initiates the transduction of a signal from the outside to the inside of the cell and eventually causes chemical reactions inside the compartments of the cell.[61] In order to mimic these processes, artificial signaling systems have been designed that mimic the activity of G-protein coupled receptors (GPCRs) and tyrosine kinase receptors.[62, 63] Moreover, vesicles and polymersomes have been used as self-assembled nanoreactors in which a chemical reaction can take place.[64, 65]

In 2002 Hunter and co-workers provided the first example of artificial signal transduction across a bilayer membrane.[66] They designed an elegant system, which

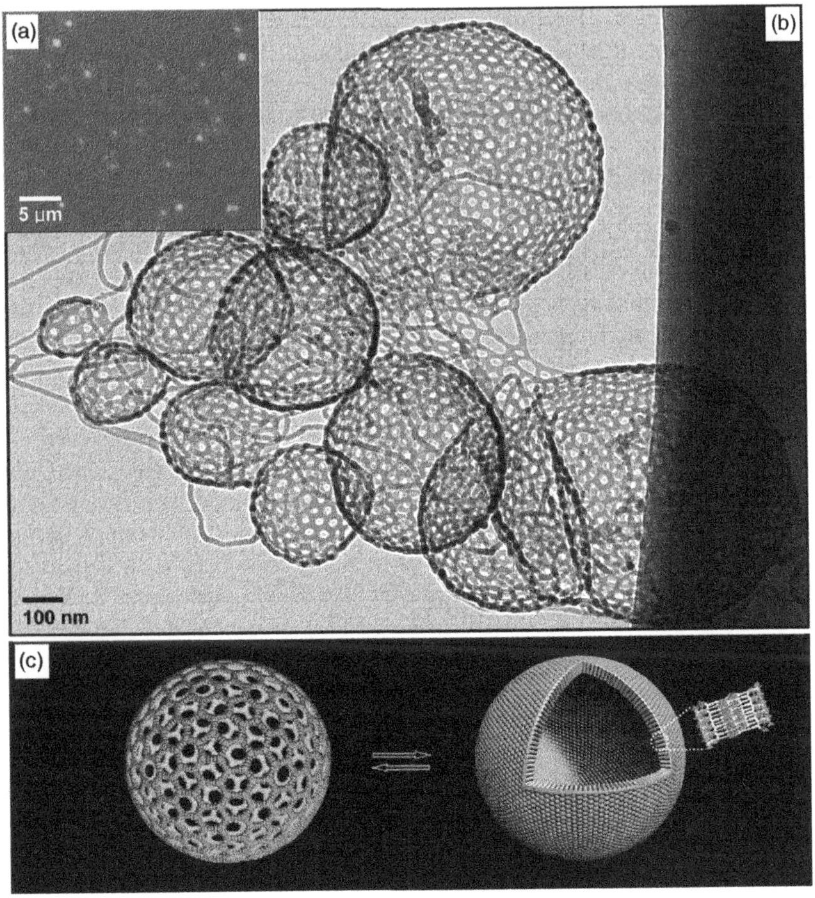

Figure 8.16 (a) Fluorescence micrograph and (b) cryo-TEM images of hollow spheres with a lateral nanoporous shell formed by self-assembly of rigid rode amphiphiles in aqueous solution (0.01 wt %). (c) Schematic representation of a reversible open/closed gating motion in the lateral nanopores of the capsules (green, polyether dendrons; yellow, aromatic segments; blue, hydrophobic branches). (Reproduced with permission. Copyright Wiley-VCH: Ref. 59.) (See insert for color representation of this figure.)

is able to conduct a molecular signal across a lipid membrane without physical transport of the messenger, and hereby successfully mimic biological systems, such as tyrosine kinase receptors. Two symmetric membrane-spanning molecules based on tail-to-tail cholesterol dimers were synthesized and incorporated into 200 nm EPC vesicles (Figure 8.17). Cholesterol was used because it inserts well into lipid membranes. The rigid alkyne-linker assured strong coupling between the external and internal ends. The head groups of the first symmetric molecule were functionalized with the amino acid cysteine as it is charged and has a thiol function. The second molecule was additionally functionalized with a 2,2′-dipyridyl disulfide group.

After incorporation of the molecules in the vesicle bilayer, first the external disulfide bonds were cleaved by the reducing agent tris(3-sulfonatophenyl)phosphane, which cannot cross the lipid membrane. Hereby the symmetric molecule is converted into its unsymmetrical analog with all the remaining disulfide units located on the inside of the vesicle. This activates the system for the sensing event based on the oxidation of two thiol groups on the outside of the membrane to a disulfide group. Hereby the signal-passing molecules are brought into close proximity and dimerize. Inside the vesicle the signaling event is accomplished as the reaction of a thiol moiety with an activated disulfide group releases a colored species. Actually, by addition of the external oxidant potassium ferricyanide, that is also incapable of crossing the membrane, an increase in the UV/Vis absorption at 341 nm was caused, which corresponds to the release of pyridine-2-thiol on the inside of the vesicle and proved that the system works. Thus, the authors concluded that molecular communication is possible across the membrane. Based on disulfide chemistry an unsymmetrical signal transduction system for sensor and signaling events without direct contact between the species involved was successfully designed.

More recently, the group of Schrader presented an alternative concept for unidirectional signal transduction triggered by the external addition of primary messenger molecules.[67] In this case, the transduction process can be monitored by a FRET effect inside the vesicle bilayer. The signal is induced on the outside of the membrane due to the recognition of the compact diammonium cation of diethylenetriamine (DET) by bisphosphonate dianions. Two transmembrane molecules were synthesized based on a similar tail-to-tail cholesterol dimer structure as used by Hunter and co-workers.[66] One end was functionalized with bisphosphonate dianions and the other end with tryptophan or a dansyl moiety, respectively (Figure 8.18). Both molecules were incorporated into 200 nm liposomes. By external addition of the primary messenger DET a strong FRET effect between the donor–acceptor pair could be observed, which indicated that a complex between the two transmembrane molecules was formed. Furthermore, the FRET effect was found to be proportional to the DET excess, whereby the signal strength reached its maximum at 40 molar equivalents with respect to the embedded transmembrane units. Additionally, it was shown that the read-out could be enhanced and transferred to the visible range by the addition of free eosin to the system. Thus, a stable multi-FRET system was obtained and the signaling was detectable by naked eye.

In 2009 Würthner and colleagues demonstrated that water-soluble perylene bisimide vesicles can be used as pH indicators.[68] The vesicles were prepared by the coassembly of wedge-shaped and dumbbell-shaped perylene bisimides and loaded with the water-soluble bispyrene-based donor 1,7-bis(1-methyl-pyrenyl)-1,4,7-triazaheptane. As the perylene bisimide membrane acts as an energy acceptor, the resulting system shows a FRET effect. The donor-loaded vesicles were in situ photopolymerized and were stable under acidic and basic conditions. TEM measurements of the loaded vesicles revealed an average diameter of 32–43 nm and a wall thickness of 6–8 nm, which is in accordance with a bilayer structure. Dynamic light scattering showed only a slight size change on pH variation. Intriguingly, these vesicles are sensitive to pH by displaying a pH-dependent fluorescence

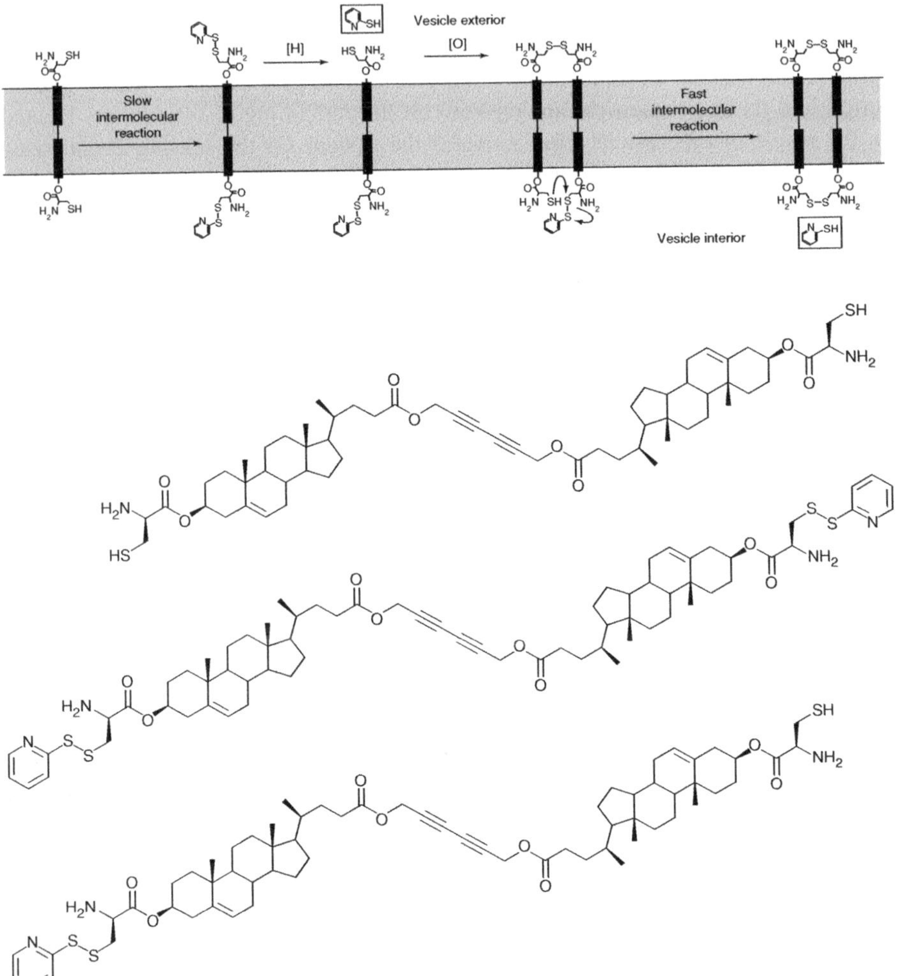

Figure 8.17 Sensing and signaling reactions and structure of the membrane-spanning molecules in a synthetic transmembrane signaling system. (Reproduced with permission. Copyright Wiley-VCH: Ref. 66.)

color change covering the whole visible light range, with white light emission at pH 9.0 (Figure 8.19). This can be explained by the pH-dependent structure of the bispyrene-based donor. Under acidic conditions the nitrogen atoms of the molecule were protonated, resulting in an electrostatic repulsion, and the pyrene units were present in their unstacked conformation. The situation was changed under basic conditions, where the molecule is uncharged. Here the pyrene units stacked due to $\pi-\pi$ interactions in aqueous solution. By exciting the vesicular sensor system (at 363 nm) at low pH the monomer fluorescence (370 –420 nm) was observed, while with increasing pH the monomer fluorescence decreases and the excimer emission

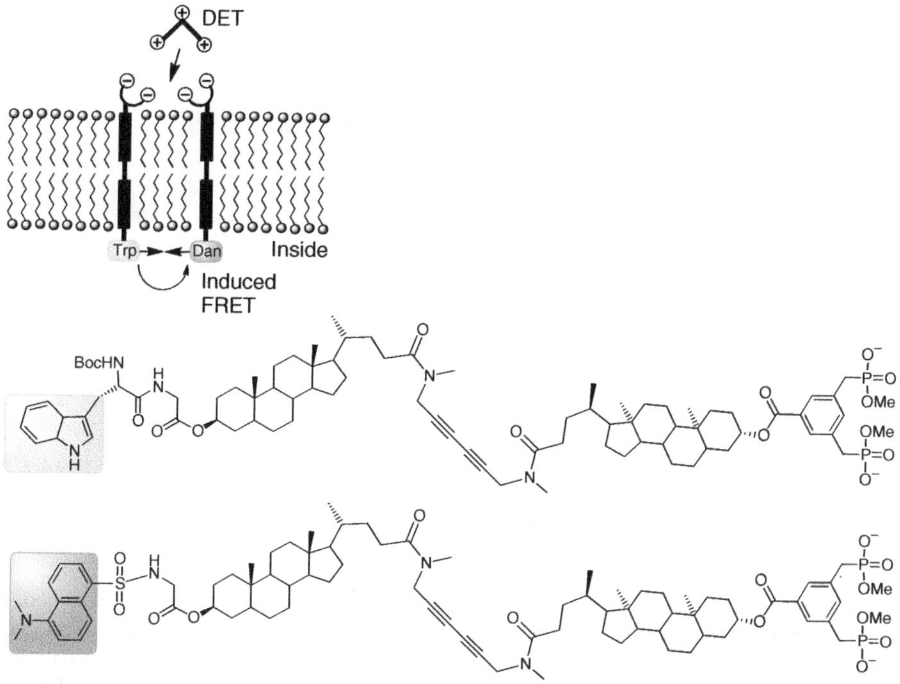

Figure 8.18 Artificial signal transduction with two transmembrane building blocks with recognition sites for a di- or tricationic primary messenger and effector sites for FRET signaling and their synthetic access from known steroidal cores. (Reproduced with permission from *Angew. Chem. Int. Ed.* **2009**, *48*, 8001. Copyright © 2009 Wiley-VCH.)

(460 –540 nm) of the stacked molecule came up. Furthermore, the excitation energy of the encapsulated donor is transferred to the bilayer perylene bisimide acceptor, at which the efficiency increased with pH and increasing acceptor emission was observed.

Not only signaling processes through the bilayer membrane of vesicles but also chemical reactions inside the interior of vesicles are of interest concerning biomimetic features of synthetic vesicles. The group of van Hest presented an outstanding example of stimuli-responsive polymersomes with gating pores (Figure 8.20).[69] They synthesized a stimuli-responsive block copolymer, poly (ethylene glycol)-*block*-poly(styrene boronic acid) (PEG-*b*-PSBA), and mixed it with the inert matrix-forming block polymer poly(ethylene glycol)-*block*-polystyrene (PEG-*b*-PS). TEM revealed the formation of polymersomes; furthermore, the average diameter was determined by DLS in the range of 200–500 nm, depending on the mixing ratio of PEG-*b*-PSBA to PEG-*b*-PS. It was shown that the two copolymers could coexist as a mixture in the membrane of the polymersomes without complete phase separation, whereby the optimal mixing ratio of the two block copolymers was determined to be $W_{PSBA} = 10\%$. Here the resulting polymersomes possessed

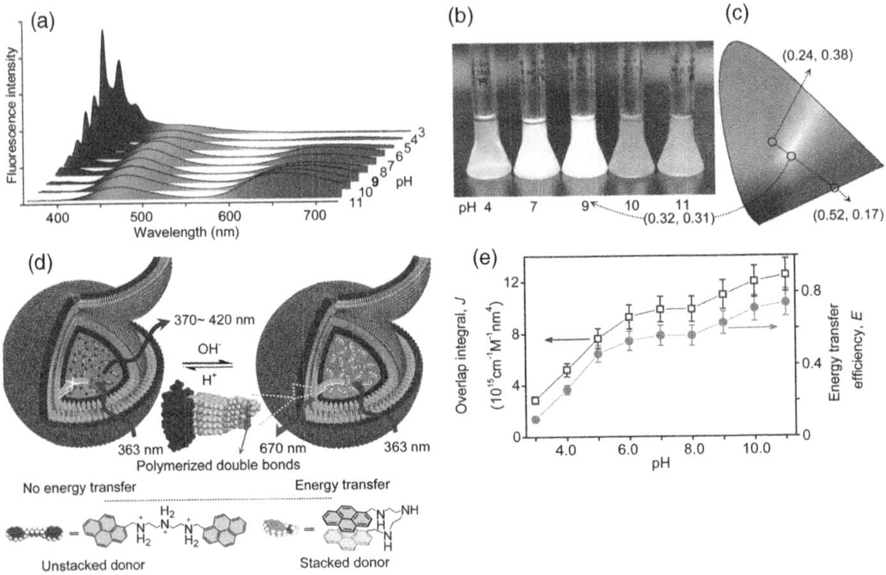

Figure 8.19 (a) Fluorescence spectra of donor-loaded polymerized vesicles in aqueous solution at pH 3.0–11.0. (b) Photograph of donor-loaded polymerized vesicles in aqueous solution at different pH under an ultraviolet lamp (366 nm). (c) CIE 1931 chromaticity diagram. The three points indicated by circles signify the fluorescence color coordinates for the donor excimers (0.24, 0.38), perylene membranes (0.52, 0.17), and white fluorescence coordinate (0.32, 0.31) for the donor-loaded polymerized vesicles at pH 9.0. (d) Schematic illustration of the donor-loaded polymerized vesicles with pH-tunable energy transfer. On average, 4.0×10^2 donor molecules are loaded into one perylene bisimide vesicle. The inner and outer layers of the vesicle consist of 5.2×10^3 and 8.4×10^3 perylene acceptor molecules, respectively. Their hydrophilic chains (blue) are exposed to water, with the hydrophobic part (orange) packed together and stabilized by polymerized double bonds (red). (e) pH-dependent energy-transfer efficiency (E, orange line) and overlap integral (J, blue line) of donor-loaded polymerized vesicles at pH 3.0–11.0. (Reproduced with permission. Copyright Nature Publishing Group: Ref. 68.) (See insert for color representation of this figure.)

evenly distributed phase-separated domains of the minor stimuli-responsive component in the inert block-copolymer matrix. In the presence of D-glucose or D-fructose the boronic acid moieties act as stimuli-responsive centers and become hydrophilic instead of hydrophobic. The PEG-*b*-PSBA chains dissemble into molecularly dissolved block copolymers and as a result the polymersomes become permeable. The formation of pores in the polymersomes after addition of sugar or base was demonstrated by the encapsulation of a small fluorescent dye. Furthermore, the enzyme *Candida antarctica* lipase B (CALB, 32 kDa, 5 μm) was encapsulated into polymersomes with different amounts of incorporated PEG-*b*-PSBA and the enzymatic activity toward a hydrolysis reaction inside the cavity of the polymersomes was observed. As expected, polymersomes with larger amounts of stimuli-responsive

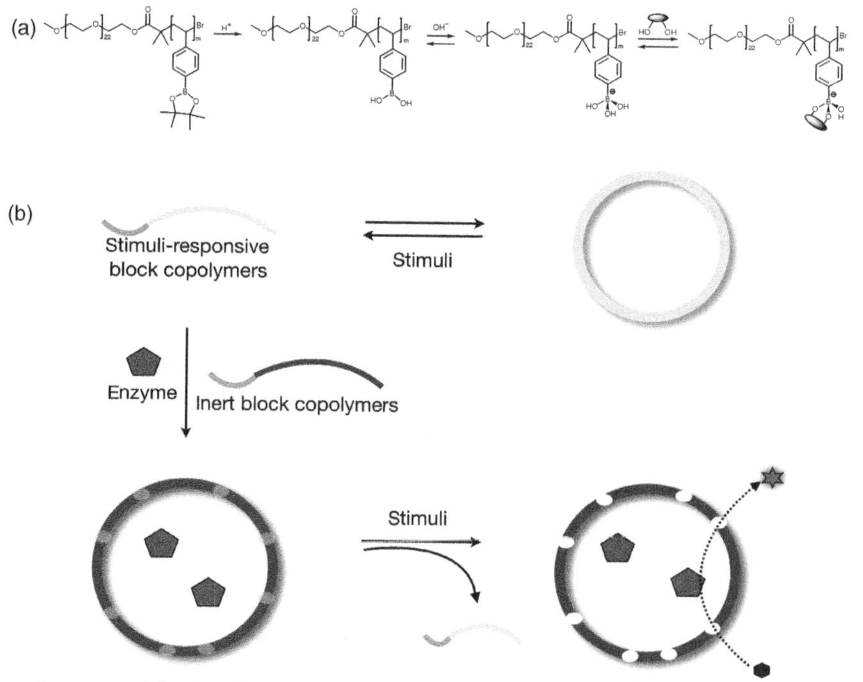

Figure 8.20 (a) Molecular structure of poly(ethylene glycol)-*block*-poly(styrene boronic acid) and its equilibrium with carbohydrates in water at pH >7. In the absence of carbohydrate, the block copolymer is amphiphilic and forms polymersomes. In the presence of carbohydrate, the block copolymer is hydrophilic and does not form polymersomes. (b) Schematic representation of the formation of permeable nanoreactors using the carbohydrate response of the block copolymers. The encapsulated enzyme catalyzes the hydrolysis of esters. (Adapted with permission. Copyright Wiley-VCH: Ref. 69.)

block copolymer possess a higher activity, which is due to the increased permeability of polymersomes after these domains are dissolved.

The formation of inorganic nanoparticles (NPs) in cells and microorganisms is known and it is suggested that enzymes or peptides may take part in the formation process.[70] Lipowsky and co-workers used giant vesicles as biomimetic compartments for the formation of CdS nanoparticles.[71] In order to identify the formation mechanisms of nanomaterials in confined compartments, they prepared vesicles from egg phosphatidylcholine (EPC) and encapsulated either $CdCl_2$ or Na_2S in millimolar concentrations. In order to distinguish the vesicles containing $CdCl_2$ or Na_2S, they were labeled with two different fluorescent lipids, Na_2S vesicles showed red fluorescence and $CdCl_2$ vesicles showed green fluorescence. Vesicle fusion was induced by an electric field and mixing of the two compartments caused the coprecipitation reaction $Na_2S + CdCl_2 \leftrightarrow CdS + 2NaCl$. The fluorescence of the resulting CdS particles in the interior of the fused vesicles could be monitored with confocal fluorescence microscopy (Figure 8.21).

As the volume of the GUV compartments is in the range of picoliters, small particles were obtained and the formation of irregular CdS sediments was suppressed. Hence, TEM measurements revealed the formation of CdS NPs with diameters ranging between 4 and 8 nm. Moreover, selected area electron diffraction (SAED) confirmed the single crystalline nature of the formed CdS NPs. The authors showed furthermore that another mixing method, slow content exchange, leads to the formation of bigger and polycrystalline CdS NPs with diameters around 50 nm. It was concluded that peptides or enzymes may not be essential for the possible mechanism of cell-based NP formation. The electrofusion approach furthermore may be used to form any kind of nanoparticle.

In summary, a number of interesting biomimetic transmembrane signaling systems have been proposed. However, in these systems a chemical trigger induces an optical signal rather than an (orthogonal) chemical response characteristic for biological signal transduction. Furthermore, it is evident that vesicles can be used as nanoreactors for the compartmentalization of chemical reactions. The challenge remains to design a system in which an external chemical trigger induces an internal cascade reaction in the confinement of a membrane vesicle.

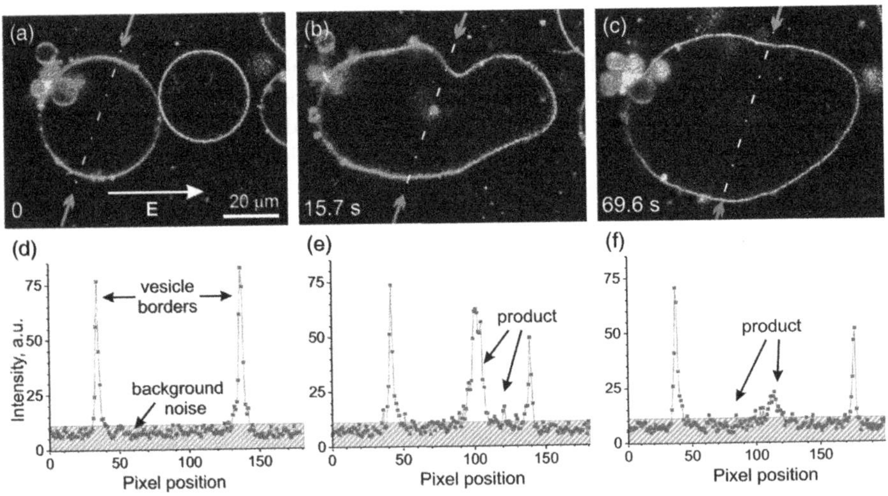

Figure 8.21 (a)–(c) Confocal scans of vesicles loaded with 0.3 m**M** Na_2S (red) and 0.3 m**M** $CdCl_2$ (green) undergoing fusion. (d)–(f) Intensity line profiles along the dash-dotted lines indicated by red arrows in (a)–(c), respectively. The direction of the field is indicated in (a). Before fusion [(a) and (d)], the vesicle interior shows only background noise similar to the external solution as indicated by the shaded zone in (d). After fusion [(b), (c), (e), and (f)], fluorescence from the product is detected in the interior of the fused vesicle. The time after applying the pulse is indicated on the micrographs. (Reproduced with permission. Copyright Wiley-VCH: Ref. 71.) (See insert for color representation of this figure.)

8.6 TOWARD HIGHER COMPLEXITY: VESICLES WITH SUBCOMPARTMENTS

Biological cells consist of multiple specialized organelles that carry out compartmentalized metabolic reactions. Artificial cells containing subcompartments with a specific, spatially separated cellular function are one step further toward a cell mimic. Therefore, "liposomes within liposomes" or "multivesicular vesicles" constitute a major milestone in the advancement of biomimetic vesicles.

A pioneering report on liposomes within liposomes from the Zasadzinski group[72] explores the biotin–streptavidin recognition motif discussed in Section 8.3.3. It was shown that clusters of vesicles that are held together by biotin–streptavidin interaction can be extruded to provide vesicle clusters of a rather well-defined size (ca. 1 μm). In addition, the vesicle clusters can subsequently be wrapped in a bilayer membrane that is literally unrolled from a Ca^{2+} stabilized phosphatidylserine cochleate cylinder. The cylinders unroll when Ca^{2+} is sequestered by EDTA, and the outer bilayer sticks to the vesicle cluster because it is functionalized with biotin. The authors provided impressive freeze-fracture TEM images of their "vesosomes" (Figure 8.22).

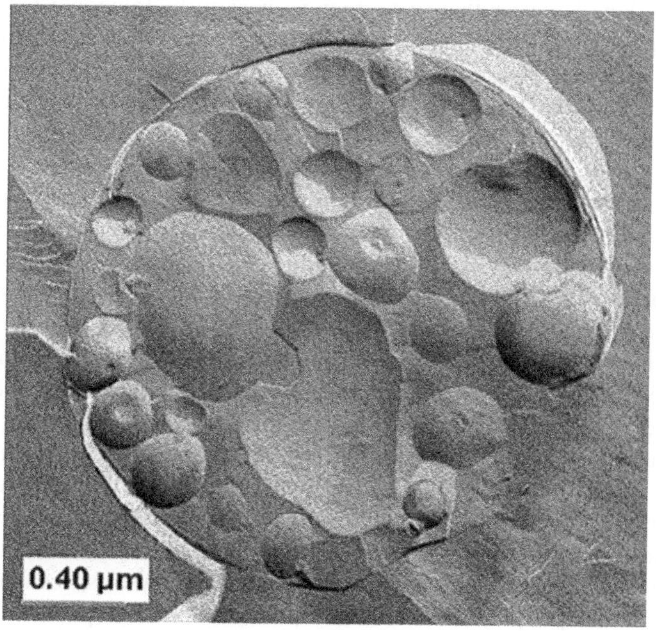

Figure 8.22 Freeze-fracture TEM of a "vesosome" (liposomes within a liposome) formed after mixing steptavidin-coated lipid cylinders and biotin-lipid coated liposome clusters. There is only one outer bilayer. The diameter of the internal liposomes is approximately 100 nm.

Chui et al.[73] presented an interesting example of a functional multivesicle assembly in 2008. The formation of such a supramolecular aggregate was made possible by a two-step double emulsion technique with a copolymer composed of acrylic acid (AAc) and distearin acrylate (DSA) units. Laser scanning confocal microscopy (LSCM) and TEM revealed the formation of unilamellar vesicles with a wall thickness of 25 nm. It was found that the size of the resulting vesicles is controllable with diameters ranging from 1 to 15 μm, depending on the parameters during formation and the amount of DSA in the copolymer. Due to this, the authors were able to prepare small vesicles in the first stage and suspend them in the water phase of the second double emulsion process designed to produce large vesicles (Figure 8.23). Hereby, multivesicle architectures were obtained, visualized by LSCM and TEM (Figure 8.24). Furthermore, the membrane of the inner and outer vesicles features pH-dependent transmembrane channels that are permeable for hydrophilic molecules, resulting from the molecular structure of the unilamellar vesicles. The membrane of the vesicles consists of a DSA bilayer, estimated to be at most 5 nm thick, and AAc regions located about 10 nm above and below the bilayer. The outermost AAc surface is charged and prevents the vesicles from aggregation. The inner layer of AAc is mostly not ionized, which stabilizes the vesicles by extensive hydrogen bonding and hydrophobic association. Furthermore, there are un-ionized AAc-rich regions parallel to the aligned lipid chains of DSA, which in the end are responsible for the formation of pH-responsive channels. The deprotonation of AAc-rich regions is easier than AAc regions near to hydrophobic DSA domains and an increase in pH value mainly ionizes the parallel AAc-rich regions. Consequently, the hydrogen bonds and hydrophobic associations of un-ionized AAc units are broken as soon as a critical density of ionized AAc units is present and the channels become permeable. As a proof of concept, the pH-dependent intake of the fluorescent protein calcein was demonstrated. At pH 5.0 the transport of the dye across the membrane was prohibited, while with an increase to pH 8.0 the molecule freely diffused into the vesicle and remained inside after decreasing the pH to 5.0 again. These synthetic vesicles with the ability to encapsulate different chemicals in subcompartments and release by external stimuli are a significant step toward mimicking eukaryotic cells.

Merkel and colleagues developed a biomimetic system modeling the behavior of Gram-negative bacteria under hyperosmotic stress.[74] The cell wall of Gram-negative bacteria consists of the inner cytoplasmic membrane, the outer membrane, and a crystalline murein wall. Under osmotic pressure, for example, during dehydration, the cell wall can activate a mechanical process in order to protect the cell. Here, the cell membrane is able to perform surface wrinkling, as well as the separation of the inner cytoplasmic membrane from the rigid outer membrane–murein wall complex, leading to the formation of the plasmolysis space. Furthermore, small endocytotic periplasmatic vesicles split from the cytoplasmic membrane, hereby reducing its surface.[75] In order to model such a membrane architecture Merkel and co-workers prepared double-shell vesicles (DSVs) by a two-step electroswelling method. In the first step GUVs of a slightly negative phospholipid mixture containing a red fluorescent dye were prepared, and in a second step a

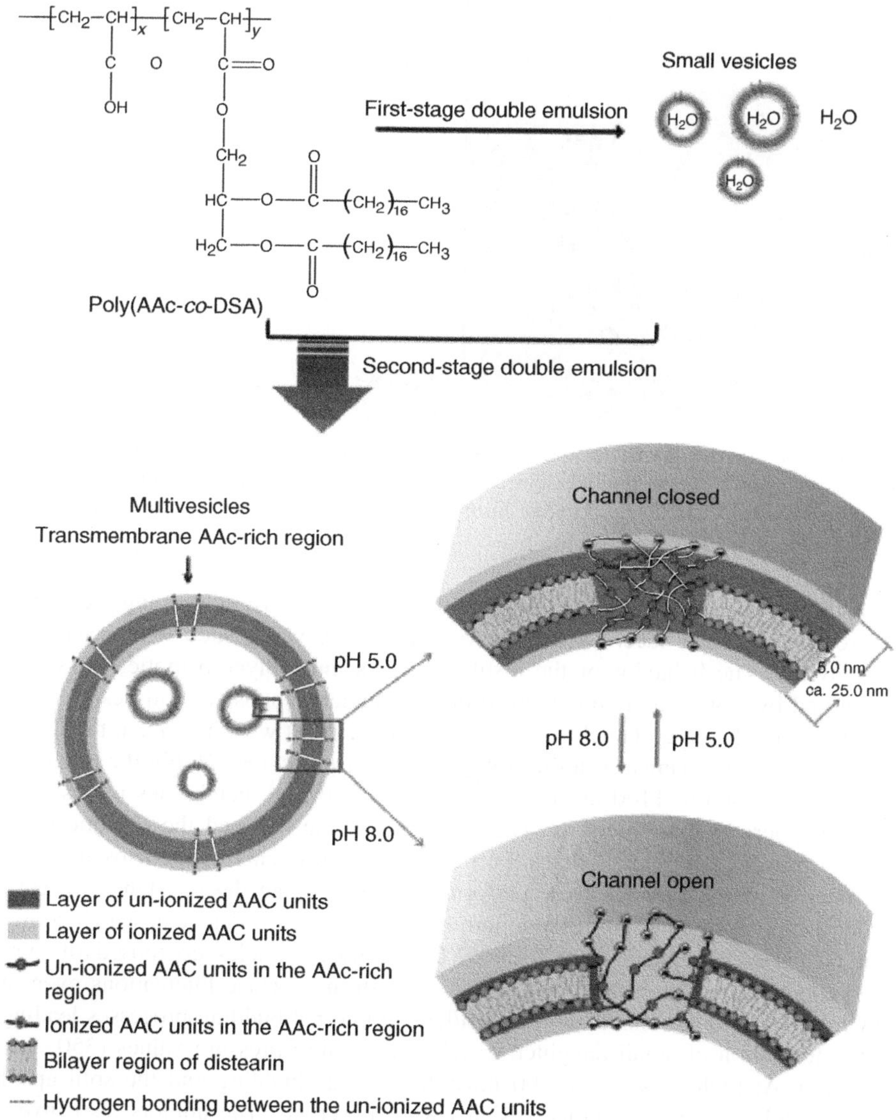

Figure 8.23 Illustration of multivesicle assemblies equipped with pH-responsive transmembrane channels from two-stage double emulsion of poly(AAc-*co*-DSA). The AAc-rich regions and the bilayer islets within the vesicle membrane are not drawn to scale. (Reproduced with permission. Copyright Wiley-VCH: Ref. 73.)

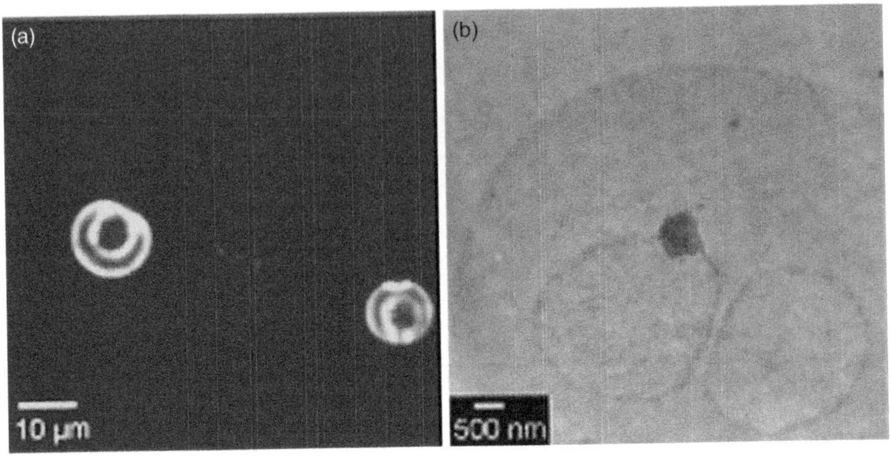

Figure 8.24 (a) Multivesicle assemblies (LSCM) and (b) a sectioned specimen (ca. 80 nm thickness) of the multivesicle assembly (TEM). (Reproduced with permission. Copyright Wiley-VCH: Ref. 73.) (See insert for color representation of this figure.)

strongly negatively charged lipid mixture marked with a dye fluorescing green was layered on the outer face. Lastly, the vesicle surface was coated with crystalline streptavidin. The behavior of the resulting DSVs under hyperosmotic stress was monitored by LSCM and compared to pure unilamellar phospholipid vesicles and streptavidin tethered unilamellar vesicles (Figure 8.25). It was shown that the DSVs exhibit a high mechanical stability toward osmotic pressures. Within the analyzed pressure range up to 1100 mosm/L only wrinkling of the membranes is observed but the spherical shape persists. The inner membrane released the osmotic pressure by forming daughter vesicles. Furthermore, experiments with enclosed dye in the interior of the vesicles were performed to investigate the resistance to leaks under osmotic pressure. No leakage out of the vesicles was detected, which shows that no permanent pores were formed. By contrast, vesicles composed of pure or streptavidin tethered phospholipids showed strong surface fluctuations even at low hyperosmotic pressures (2–3 mosm/L) and internal budding processes leading to the formation of small daughter vesicles at medium pressure values (350–700 mosm/L). At high pressures (1100 mosm/L) outside budding and the split-up of small vesicles from the mother vesicles were observed. The authors also investigated the behavior of DSV without a streptavidin layer under osmotic pressure. Here, the two layers underwent external budding, resulting in the formation of hundreds of tentacles. Overall, it was shown that the streptavidin-coated DSVs are very robust toward high osmotic pressures, supporting the hypothesis that the mechanical protection of Gram-negative cell walls is crucial for the osmotic regulatory system.

Caruso and co-workers went one step further toward the development of a functional artificial cell by designing "smart capsosomes."[76] Capsosomes are

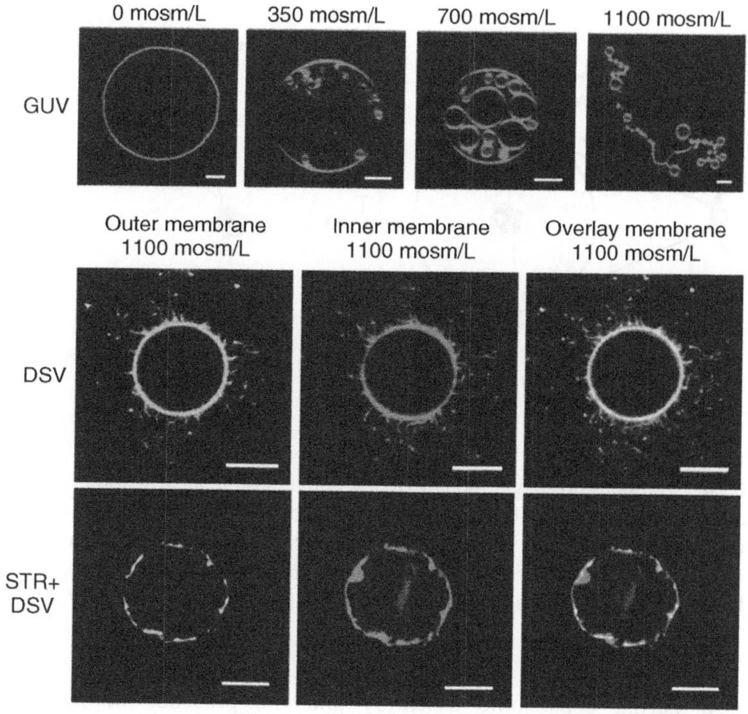

Figure 8.25 Upper row: Effect of hyperosmotic pressure on GUVs labeled with red dye. All images represent different vesicles. Middle row: Double-shell giant vesicle at 1100 mosm/L hyperosmotic pressure. Double-shell vesicles (DSVs) without surface proteins underwent an outside budding process under hyperosmotic conditions. The two membranes stuck together in newly formed nanotubes (yellow signal), or they formed separately new buds and tubes (green and red signals). Lower row: Vesicle coated with a crystalline streptavidin layer on the outer membrane surface (STR+DSV) showed a slight asymmetrical shape deformation without membrane budding (green signal), while the inner membrane released the osmotic pressure, separately forming daughter vesicles (red signal). Scale bars = 5 μm. (Reproduced with permission. Copyright American Chemical Society: Ref. 74.) (See insert for color representation of this figure.)

hierarchical liposome–polymer coassemblies containing thousands of liposomes. The capsules are obtained by a template synthesis. For this purpose a polymer precursor is deposited on a spherical silica template, followed by the sequential layering of liposomes and polymer separation layers. Lastly, a polymer capping layer is introduced and the template particle is removed. The size of the resulting particles is in the range of a few micrometers, depending on the number of liposome layers. The diameter of the incorporated liposomes is about 50 nm. Just like the organelles of a biological cell, the liposomes inside the resulting capsules can be used as functional subcompartments with the ability to conduct an enzymatic reaction in their interiors. In order to do so, enzyme-loaded liposomes

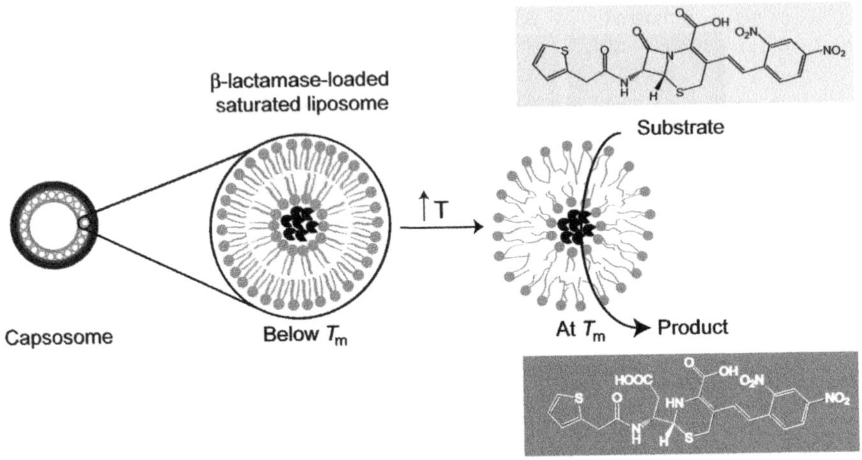

Figure 8.26 Temperature-triggered enzymatic conversion in a capsosome. An increase in temperature to the phase transition temperature (T_m) of the liposomes inside the capsosome results in a disordered liquid phase of the lipid membrane, allowing nitrocefin to cross the membrane to be hydrolyzed while retaining the β-lactamase inside the compartments. (Reproduced with permission. Copyright American Chemical Society: Ref. 76.)

were used during capsosome assembly and the presence of multiple intact enzyme-loaded liposomal subcompartments was confirmed by a Triton X triggered enzymatic colorimetric assay causing lysis of the liposomes. Furthermore, the authors showed that an increase of temperature to the phase transition temperature ($T_m = 41$ °C) of 1,2-dipalmitoyl-*sn*-glycero-3-phosphocholine (DPPC) liposomes caused an enhanced permeability of the lipid membrane, due to the formation of a disordered liquid phase. This was used as a trigger to initiate an enzymatic reaction without dissolving the liposomal subcompartments (Figure 8.26). Capsosomes with β-lactamase-loaded DPPC liposomes were prepared. Upon the increase of temperature to 41 °C in the presence of yellow nitrocefin, its red hydrolyzation product was observed by UV absorption. The enzymatic conversion is possible as the lipid membrane is permeable for the substrate and product. Moreover, the triggered reaction could be repeated over several cycles, without loss of functional activity of the enzymes, while the capsosomes preserved their structural integrity (Figure 8.27). Overall, capsosomes show a great potential toward the creation of artificial cells with small compartments featuring different specific properties.

In summary, a number of reports have described artificial cells with subcompartments. Either the outer membrane or the encapsulated compartments or both can be stimulus responsive. Key challenges are high-yield procedures to encapsulate small vesicles inside large ones as well as the selective encapsulation of several different types of vesicles inside a single large vesicle or polymer capsule. Ultimately, it should be possible to introduce the transmembrane signaling and nanoreactor concepts described in Section 8.5 into a multicompartment vesicle.

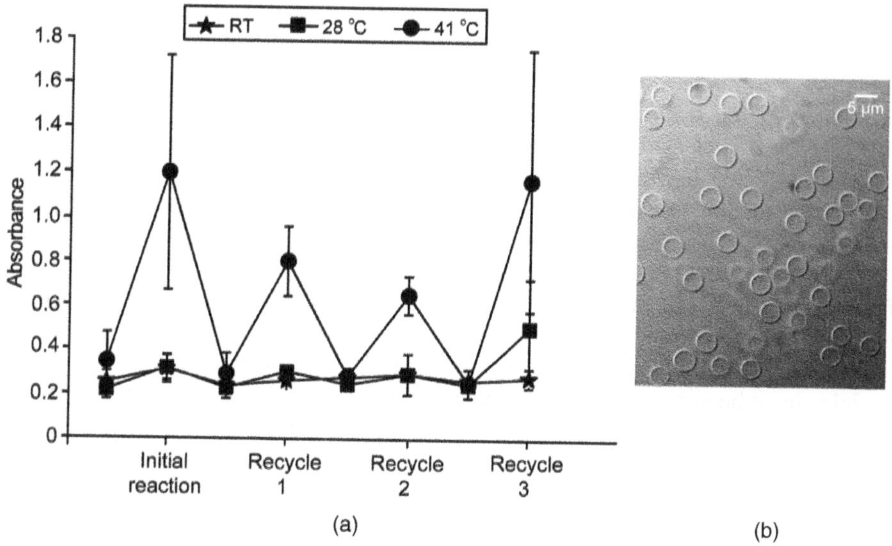

Figure 8.27 Temperature-triggered enzymatic reaction in capsosomes. (a) Absorbance readings of an enzymatic assay using capsosomes with β-lactamase-loaded DPPC liposomes ($C_{L(DPPC)\text{-}\beta}$) incubated at room temperature (23°C) (★), 28°C (■), or 41°C (•). The enzymatic conversion was only observed when the capsosomes were incubated at the phase transition temperature (T_m) of the liposomal subunits. The retention of the functional enzymes inside the liposomal subcompartments was confirmed by repetitively performing the temperature-induced assay. (b) Differential interference integrity (DIC) image of capsosomes after being reused four times. (Reproduced with permission. Copyright American Chemical Society: Ref. 76.)

8.7 CONCLUSION

Ever since their discovery in 1977, research on synthetic vesicles has been inspired by the remarkable properties of liposomes and biological membranes, and the relevance of vesicles as biomimetic model systems for biological membranes and their interactions is evident. Although even today the large majority of synthetic vesicles are composed of phospholipid-like amphiphilic molecules, in particular, block-copolymer vesicles have had a major impact in the development of robust vesicles. In the last decade, many reports have also demonstrated that vesicles can be prepared from molecules that are not amphiphilic. It has been shown that vesicles must not be based on hydrophobic interaction, but instead a wide range of other noncovalent interactions can give rise to the assembly of vesicles in aqueous solution. It is expected that many more "unconventional" vesicles will be described in the near future.

Specific molecular recognition between complementary recognition units can induce selective adhesion and fusion of vesicles, much like CAMs mediate the adhesion and fusion of biological membranes. Clearly, the multivalent noncovalent interactions of clusters of CAMs can overcome the surface repulsion that normally keeps membranes away from each other. However, to the best of our knowledge, there is no example yet of a specific biomimetic fusion of polymersomes, which are generally much more stable than vesicles composed of small amphiphiles. In this respect, it is clear that biological membranes are just thick and tight enough to pose a significant barrier, yet thin and flexible enough to be prone to deformation and fusion. Many researchers have chosen a reductionist approach toward mimicking protein–carbohydrate-induced adhesion as well as protein-induced fusion of membranes. Deeper insight into the factors that promote or inhibit membrane fusion may be of relevance to the development of sophisticated biomimetic drug and gene delivery systems.

Furthermore, although numerous examples of stimulus-responsive biomimetic vesicles have been reported, none of them approaches the sophistication and adaptiveness of living cells. Nevertheless, if a synthetic vesicle exclusively responds to a bioorthogonal signal (such as irradiation or local heating), it could possibly find application as a drug or gene delivery device. Vesicles have also been functionalized with an "external" cytoskeleton that confers stability and/or directs shape, but there are no reports on vesicles or polymersomes with a biomimetic "internal" cytoskeleton.

In biomimetic transmembrane signaling systems a chemical trigger induces an optical signal rather than an (orthogonal) chemical response characteristic for biological signal transduction. Vesicles can be used as nanoreactors for the compartmentalization of chemical reactions. The challenge remains to design a system in which an external chemical trigger induces a cascade reaction within the confines of a membrane vesicle.

Finally, a number of reports have described artificial cells with subcompartments. Either the outer membrane or the encapsulated compartments or both can be stimulus responsive. Key challenges are high-yield procedures to encapsulate small vesicles inside large ones as well as the selective encapsulation of several different types of liposomes inside a single large vesicle or polymer capsule.

Ultimately, the holy grail of biomimetic vesicles is the integration of molecular recognition, transmembrane signaling, and nanoreactor concepts into a well-defined multicompartment vesicle that would approach the sophistication of a living cell. Although it may be far-fetched to synthesize an artificial cell, it is certainly a realistic goal to design and assemble a smart, biomimetic nanocontainer that could function as a magic bullet in drug delivery. On the one hand, research on increasingly sophisticated biomimetic vesicles will give rise to innovative soft materials. On the other hand, a clever design as well as a detailed investigation of biomimetic vesicles will deepen our understanding of molecular recognition, signal transduction, and metabolic pathways in biological cells.

REFERENCES

1. Bangham, A. D.; Horne, R. W. *J. Mol. Biol*. **1964**, *8*, 660.
2. Kunitake, T.; Okahata, Y. *J. Am. Chem. Soc*. **1977**, *99*, 3860.

3. Israelachvili, J. N.; Mitchell, D. J.; Ninham, B. W. *J. Chem. Soc. Faraday Trans. 2*, **1976**, 1525.
4. Angelova, M. I.; Menger, F. M. *Acc. Chem. Res*. **1998**, *31,* 789.
5. Kunitake, T. *Angew. Chem. Int. Ed*. **1992**, *31,* 709.
6. Ringsdorf, H.; Schlarb, B.; Venzmer, J. *Angew. Chem. Int. Ed*. **1988**, *27,* 113.
7. Engberts, J. B. F. N.; Hoekstra, D. *Biochim. Biophys. Acta* **1995**, *1241,* 323.
8. Benvegnu, T.; Brard, M.; Plusquellec, D. *Curr. Opin. Colloid Interface Sci*. **2004**, *8,* 469.
9. Chong, P. L. G. *Chem. Phys. Lipids* **2010**, *163,* 253.
10. Voskuhl, J.; Ravoo, B. J. *Chem. Soc. Rev*. **2009**, *38,* 495.
11. Monserrat, K.; Graetzel, M.; Tundo, P. *J. Am. Chem. Soc*. **1980**, *102,* 5527.
12. Ghosh, P.; Khan, T.; Bharadwaj, P. K. *Chem. Commun*. **1996**, 189.
13. Tanaka, Y.; Miyachi, M.; Kobuke, Y. *Angew. Chem. Int. Ed*. **1999**, *38,* 504.
14. Ravoo, B. J.; Darcy, R. *Angew. Chem. Int. Ed*. **2000**, *39,* 4324.
15. Lee, H. K.; Park, K. M.; Jeon, Y. J.; Kim, D.; Oh, D. H.; Kim, H. S.; Park, C. K.; Kim, K. *J. Am. Chem. Soc*. **2005**, *127,* 5006.
16. Falvey, P.; Lim, C. W.; Darcy, R.; Revermann, T.; Karst, U.; Giesbers, M.; Marcelis, A. T. M.; Lazar, A.; Coleman, A. W.; Reinhoudt, D. N.; Ravoo, B. J. *Chem Eur. J*. **2005**, *11,* 1171.
17. Zhang, L.; Eisenberg, A. *Science* **1995**, *268,* 1728.
18. Dirks, A. J.; Nolte, R. J. M.; Cornelissen, J. J. L. M. *Adv. Mater*. **2008**, *20,* 3953.
19. Bertin, A.; Hermes, F.; Schlaad, H. *Adv. Polym. Sci*. **2010**, *224,* 167.
20. Discher, D. E.; Eisenberg, A. *Science* **2002**, *297,* 967.
21. LoPresti, C.; Lomas, H.; Massignani, M.; Smart, T.; Battaglia, G. *J. Mater. Chem*. **2009**, *19,* 3576.
22. Reynhout, I. C.; Cornelissen, J. J. L. M.; Nolte, R. J. M. *Acc. Chem. Res*. **2009**, *42,* 681.
23. Jeon, Y. J.; Bharadwaj, P. K.; Choi, S. W.; Lee, J. W.; Kim, K. *Angew. Chem. Int. Ed*. **2002**, *42,* 4474.
24. Ghosh, S.; Reches, M.; Gazit, E.; Verma, S. *Angew. Chem. Int. Ed*. **2007**, *46,* 2002.
25. Yoshikawa, I.; Sawayama, J.; Araki, K. *Angew. Chem. Int. Ed*. **2008**, *47,* 1038.
26. Anruka, Y.; Kishimura, A.; Oba, M.; Yamasaki, Y.; Kataoka, K. *J. Am. Chem. Soc*. **2010**, *132,* 1631.
27. Blumenthal, R.; Clague, M. J.; Durell, S. R.; Epand, R. M. *Chem. Rev*. **2003**, *103,* 53.
28. Jahn, R.; Lang, T.; Südhof, T. C. *Cell* **2003**, *112,* 519.
29. Ungar, D.; Hughson, F. M. *Annu. Rev. Cell Dev. Biol*. **2003**, *19,* 493.
30. Wickner, W. *Annu. Rev. Cell Dev. Biol*. **2010**, *26,* 115.
31. Marsden, H. R.; Elbers, N. A.; Bomans, P. H. H.; Sommerdijk, N. A. J. M.; Kros, A. *Angew. Chem. Int. Ed*. **2009**, *48,* 2330.
32. Marsden, H. R.; Tomatsu, I.; Kros, A. *Chem. Soc. Rev*. **2011**, *14,* 1572.

33. Chirovolu, S.; Walker, S.; Israelachvili, J.; Schmitt, F. J.; Leckband D.; Zasadzinski, J. A. *Science* **1994**, *264,* 1753.

34. Constable, E. C.; Meier, W.; Nardin C.; Mundwiler, S. *Chem. Commun*. **1999**, 1483.

35. Voskuhl, J.; Stuart, M. C. A.; Ravoo, B. J. *Chem. Eur. J*. **2010**, *16,* 2790.

36. Richard, A.; Marchi-Artzner, V.; Lalloz, M.-N.; Brienne, M. J.; Artzner, F.; Gulik-Krzywicki, T.; Guedeau-Boudeville, M.-A.; Lehn, J.-M. *Proc. Natl. Acad. Sci. U.S.A.* **2004**, *101,* 15279.

37. Webb, S. J.; Trembleau, L.; Mart R. J., Wang, X. *Org. Biomol. Chem*. **2005**, *3,* 3615.

38. Mart, R. J.; Liem, K. P.; Wang X.; Webb, S. J. *J. Am. Chem. Soc*. **2006**, *128,* 14462.

39. Causeret, M.; Taulet, N.; Comunale, F.; Favard, C.; Gauthier-Rouviere, C. *Mol. Biol. Cell* **2005**, *16,* 2168.

40. Cohen, M.; Joester, D.; Geiger, B.; Addadi, L. *ChemBioChem* **2004**, *5,* 1393.

41. Marchi-Artzner, V.; Gulik-Krzywicki, T.; Guedeau-Boudeville, M.-A.; Gosse, C.; Sanderson, J. M.; Dedieu J.-C.; Lehn, J.-M. *ChemPhysChem* **2001**, *2,* 367.

42. Hoekstra, D.; Düzgünes, N. *Methods Enzymol*. **1993**, *220,* 15.

43. Kashiwada, A.; Tsuboi, M.; Matsuda, K. *Chem. Commun*. **2009**, 695.

44. James, T. D.; Sandanayake, K. R. A. S.; Shinkai, S. *Angew. Chem. Int. Ed*. **1996**, *35,* 1910.

45. Stengel, G.; Zahn R.; Höök, F. *J. Am. Chem. Soc*. **2007**, *129,* 9584.

46. Stengel, G.; Simonsson, L.; Campbell, R. A.; Höök, F. *J. Phys. Chem. B* **2008**, *112,* 8264.

47. Gong, Y.; Luo, Y.; Bong, D. *J. Am. Chem. Soc*. **2006**, *128,* 14430.

48. Gong, Y.; Ma, M.; Luo, Y.; Bong, D. *J. Am. Chem. Soc*. **2008**, *130,* 6196.

49. Kahne, D.; Leimkuhler, C.; Lu, W.; Walsh, C. *Chem. Rev*. **2005**, *105,* 425.

50. Gallop, J. L.; McMahon, H. T. *Biochem. Symp. Soc*. **2005**, *72,* 223.

51. Ota, S.; Yoshizawa, S.; Takeuchi, S. *Angew. Chem. Int. Ed*. **2009**, *48,* 6533.

52. Chécot, F.; Lecommandoux, S.; Gnanou Y.; Klok, H. A. *Angew. Chem. Int. Ed*. **2002**, *41,* 1339.

53. Yu, S. Y.; Azzam, T.; Rouiller, I.; Eisenberg, A. *J. Am. Chem. Soc*. **2009**, *131,* 10557.

54. Versluis, F.; Tomatsu, I.; Kehr, S.; Fregonese, C.; Tepper, A. W. J. W.; Stuart, M. C. A.; Ravoo, B. J.; Koning R. I.; Kros, A. *J. Am. Chem. Soc*. **2009**, *131,* 13186.

55. Hamada, T.; Sato, Y. T.; Yoshikawa, K. *Langmuir* **2005**, *21,* 7626.

56. Ishii, K.; Hamada, T.; Hatakeyama, M.; Sugimoto, R.; Nagasaki, T.; Takagi, M. *ChemBioChem* **2009**, *10,* 251.

57. Holowka, E. P.; Deming, T. J. *Macromol. Biosci*. **2010**, *10,* 496.

58. Haemers, S; van der Leeden, M. C.; Koper, G. J. M.; Frens G. *Langmuir* **2002**, *18,* 4903.

59. Kim, J. K.; Lee, E.; Lim Y.; Lee, M. *Angew. Chem. Int. Ed*. **2008**, *47,* 4662.

60. Speir, J. A.; Munshi, S.; Wang, G.; Baker, T. S.; Johnson, J. E. *Structure* **1995**, *3,* 63.

61. Singer, S. J. *Science* **1992**, *255,* 1671.

62. Strader, C. D.; Fong, T. M.; Tota, M. R.; Underwood, D.; Dixon, R. A. F. *Annu. Rev. Biochem*. **1994**, *63,* 101.

63. Heldin, C.-H; Ericsson, J. *Science* **2001**, *294,* 2111.

64. Kim, K. T.; Meeuwissen, S. A.; Nolte, R. J. M.; van Hest, J. C. M. *Nanoscale* **2010**, *2,* 844.

65. Vriezema, D. M.; Aragones, M. C.; Elemans, J. A. A. W.; Cornelissen, J. J. L. M.; Rowan A. E.; Nolte, R. J. M. *Chem. Rev*. **2005**, *105,* 1445.

66. Barton, P.; Hunter, C. A.; Potter, T. J.; Webb S. J.; Williams, N. H. *Angew. Chem. Int. Ed*. **2002**, *41,* 3878.

67. Bernitzki K.; Schrader, T. *Angew. Chem. Int. Ed*. **2009**, *48,* 8001.

68. Zhang, X; Rehm, S; Safont-Sempere, M. M.; Würthner, F. *Nature Chem*. **2009**, *1,* 623.

69. Kim, K. T.; Cornelissen, J. J. L. M.; Nolte R. J. M.; van Hest, J. C. M. *Adv. Mater*. **2009**, *21,* 2787.

70. Mukherjee, P.; Ahmad, A.; Mandal, D.; Senapati, S.; Sainkar, S. R.; Khan, M. I.; Ramani, R.; Parischa, R.; Ajayakumar, P. V.; Alam, M.; Sastry, M.; Kumar, R. *Angew. Chem. Int. Ed*. **2001**, *40,* 3585.

71. Yang, P.; Lipowsky R.; Dimova, R. *Small* **2009**, *5,* 2033.

72. Walker, S. A.; Kennedy, M. T.: Zasadzinski, J. A. *Nature* **1997**, *387,* 61.

73. Chiu, H. C.; Lin, Y. W.; Huang, Y. F.; Chuang C. K.; Chern, C. S. *Angew. Chem. Int. Ed*. **2008**, *47,* 1875.

74. Csiszár, A.; Hoffmann B.; Merkel, R. *Langmuir* **2009**, *25,* 5753.

75. Schwarz, H.; Koch, A. L. *Microbiology* **1995**, *141,* 3161.

76. Chandrawati, R.; Hosta-Rigau, L.; Vanderstraaten, D.; Lokuliyana, S. A.; Städler, B.; Albericio, F.; Caruso, F. *ACS Nano* **2010**, *4,* 1351.

CHAPTER 9

Bioinspired Surfaces I: Gecko-Foot Mimetic Adhesion

LIANGTI QU and YAN LI

Center of Advanced Science and Engineering for Carbon (Case4Carbon), School of Chemistry, Beijing Institute of Technology, Beijing 100081, Peoples Republic of China

LIMING DAI

Department of Macromolecular Science and Engineering, Case School of Engineering, Case Western Reserve University, Cleveland, Ohio, USA

9.1 THE HIERARCHICAL STRUCTURE OF GECKO FEET

Flexible adhesive tapes are used widely in our daily lives for diverse purposes. However, they are rarely used for hanging heavy objects owing to the time- and rate-dependent viscoelasticity. Besides, the viscoelastic tapes do not work under vacuum environments such as in space. In order to overcome these drawbacks, Nature has created smart alternatives, including a gecko foot for adhesion without using any sticky viscoelastic liquid. As early as the 4th century B.C., it was observed that a gecko can run up and down a tree freely, even with its head facing down toward the ground.[1] Recent studies have discovered that the gecko's extraordinary climbing ability comes from a remarkable design of Nature for setal arrays with nanoscale β-keratin elastic hairs of a high aspect ratio at their feet and toes.[2] Each of the setal in the setal array consist of hundreds of spatulae at its tip. The hierarchical structure of the gecko toe pads is imaged in Figure 9.1 at macro-, meso-, micro-, and nanoscales.[3] Each of the gecko's four feet has five toes, and each toe has about 20 rows of sticky setal arrays with each setal array consisting of thousands of setal stalks, which amounts to approximately 200,000 setae per toe, and each seta terminating with 100–1000 spatulae. Typically, gecko setae are approximately 110 μm long and approximately 5 μm wide, inclining onto the gecko foot skin at an angle of about 45° with almost straight stalks at the base and curved branches of fine hairy fingers (i.e., spatula) at their top ends.

Bioinspiration and Biomimicry in Chemistry: Reverse-Engineering Nature, First Edition.
Edited by Gerhard F. Swiegers.
© 2012 John Wiley & Sons, Inc. Published 2012 by John Wiley & Sons, Inc.

The stalks bend and deform during climbing, which is believed to facilitate the switching between attachment and detachment.

9.2 ORIGIN OF ADHESION IN GECKO SETAE

The mechanism of adhesion in the dry adhesive pads of gecko has been a matter of continued debate for over a century. Many different mechanisms have been proposed from time to time, involving concepts of suction cups, capillary adhesion, friction, electrostatic attraction, microinterlocking, and van der Waals forces (vdW).[4–11] Among them, only the capillary adhesion and vdW force were acceptable by 1969. Subsequently, Kellar Autumn and co-workers performed the first direct force measurements of a single seta[4] by using a two-dimensional (2D) microelectromechanical system with force sensors and a wire as a force gauge. Their measurements revealed that a seta was ten times more "sticky" than even the predicted maximum adhesion for a whole animal, supporting an adhesive mechanism based on the vdW forces. Two years later, the same group obtained the experimental evidence for the vdW-induced dry adhension of gecko setae.[8] vdW interactions are ubiquitous in Nature. Although one of the weakest known forces between atoms and molecules, vdW forces can provide significant cohesive forces within solids as well as adhesive forces between solids when acting collectively. The discovery that the vdW interactions are responsible for the gecko foot adhesion has led to the development in both academic research and practical applications of gecko inspired dry adhesives.

For fundamental study, Johnson et al.[12] developed the Johnson–Kendall–Roberts (JKR) model by considering interactions between adhesive elastic spheres, in which the size of the contact area was determined via a balance between elastic and surface energies similar to Griffith's criterion[13] for crack growth in an elastic solid. Studies so far have further extended these theories to viscoelastic materials[14, 15] for coupled normal and shear loads[16] and biological attachments.[17–22] Some multiscale modeling/simulation of the deformation and adhesion characteristics[23, 24] of gecko seta can be found in Refs. 25–32. On the experimental front, those numous setae ensure a full contact between the toes of a gecko and the contact surface, thus resulting in a strong adhesion force. The gecko's foot has been demonstrated to show many significant adhesion characteristics due to its unique hierarchical structure. Some of the adhesion characteristics are listed below.[33]

1. *Anisotropic Adhesion*[34–36]. Gecko's toes can actively switch its toe direction for easy attachment and detachment.[37–39]
2. *Low Normal Detaching Force*[39]. The detaching force required for lifting the gecko–foot–toe from the contacted surface nearly equals zero.[40]
3. *High Compatibility*. Gecko exhibits high adhesion capability to wet or dry and molecularly smooth or very rough surfaces.[4, 41]
4. *Self-cleaning and Anticollapse Properties*[42]. Gecko's setae are made of elastic protein with a modulus of 2–4 GPa; β-keratin[43] can prevent

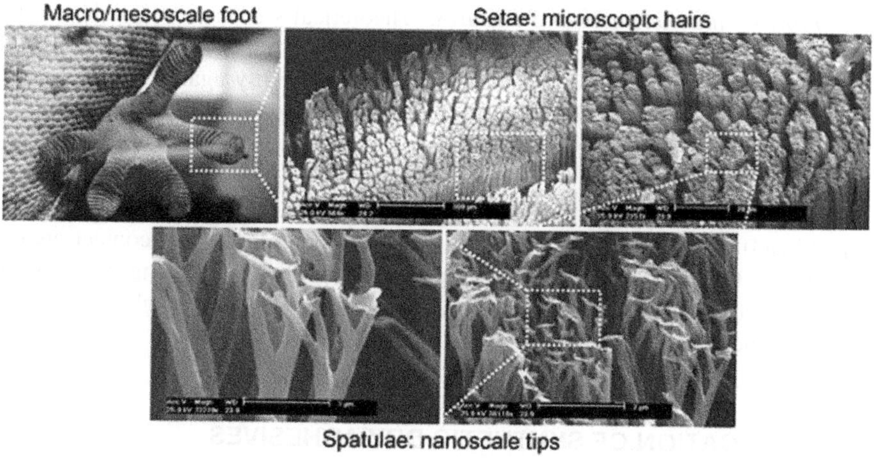

Figure 9.1 Multiscale combined hierarchical gecko foot hairs. As shown, millions of fine microscopic foot hairs (setae) on the attachment pads split into hundreds of nanoscale ends (spatulae). (Adapted with permission. Copyright Elsevier: Ref. 3.)

the impure contaminations from adsorbing. The complex hierarchical structures further enable the gecko foot to effectively remove contaminated particles.

9.3 STRUCTURAL REQUIREMENTS FOR SYNTHETIC DRY ADHESIVES

Just like the hierachically structured gecko feet described above, gecko-foot-mimetic synthetic dry adhesives need to meet the following structural requirements:[44, 45]

1. *Small Fibril Radius.* According to the contact splitting theory,[18, 46] the adhesion force is inversely proportional to the fibril radius, as supported by several recent studies that demonstrated a significant increase in adhesion force by patterning surfaces with reduced fibril radius.[46–48]
2. *High Aspect Ratio.* Based on the mechanism of crack propagation in rubbery materials, a high aspect ratio (small fibril radius and high fibril length) will increase the number of fibrils to contact the surface and decrease modulus[43, 46] for effective elastic energy dissipation.[49–51]
3. *Slanted Structures.* Shear and normal contact experiments revealed higher pull-off forces for slanted structures than for vertical fibrils.[52–55] A directional angle of the nanostructured fibers is a crucial factor for anisotropic, reversible dry adhesive (i.e., strong attachment and easy detachment)[35, 52–56] because an angled structure significantly lowers the effective modulus of the surface.[43, 56]

4. *Fibrils with Hierarchical Structures.* Theoretical studies have predicted that the hierarchical structures of a gecko's foot is essential to provide high mechanical stability, to provide enhanced adaptability against various rough surfaces,[17, 57] and to increase the energy penalty for crack propagation by longer fibers.[58]

A three-dimensional (3D) shape of the fibril tips (e.g., fibrils with mushroom-shaped head) can enhance the adhesion strength by increasing the contact area as compared to that of a spherical or a simple flat head.[50, 59–61] Among different 3D shape geometries, mushroom-shaped heads have so far given the highest adhesion-strength values of all polymeric fibers.[35, 50–52, 61, 62]

9.4 FABRICATION OF SYNTHETIC DRY ADHESIVES

Two main approaches based on micro/nanoscale casting and gas phase growth have been devised for fabicating synthetic dry adhesives from polymers and carbon nanotubes (CNTs), respectively.

For polymer-based dry adhesives, micro/nanoscale fibrils are produced mainly by lithography and microfabrication methods. In this case, the first step is to microfabricate masters containing arrays of holes with different dimensions,[46, 48, 49, 56, 62–65] such as anodized alumina or polycarbonate membranes, which are typically fabricated by photolithography,[46, 48, 63, 64, 66–68] etching,[67, 69, 70] or micromachining.[35, 53, 54, 60, 65, 67, 71–82] The fibrillar arrays can then be obtained by casting polymers on the microfabricated masters and can subsequently be modified by inking to produce different tip geometries[35, 52, 62, 66, 83] and/or coating with different surface chemistries.[54, 81, 84] For the CNT-based dry adhesives, however, chemical vapor deposition techniques are usually used to directly grow arrays of vertically aligned CNTs on appropriate substrates.[85–89]

9.4.1 Polymer-Based Dry Adhesives

9.4.1.1 Structures with a High Aspect Ratio The first synthetic nanohairs resembling those of a gecko pad were reported by Autumn et al. in 2002.[8] Figure 9.2 shows a schematic representation of the fabrication process. To start with, a flat wax surface was punched by an atomic force microscope (AFM) probe with a conical tip of apex radius of 10–20 nm and height of 15 μm. The punched surface was then filled with polymer, PDMS.[8] After curing, the molded polymer surfaces were detached from the wax by peeling. It was found that the perpendicular adhesive force for the PDMS spatulae with a tip radius of 230–440 nm was 181 ± 9 nN and 294 ± 21 nN for similarly structured polyester spatulae with a tip radius of ~350 nm.[8] For these synthetic hairs, 47–63% of the adhesion forces can be explained by vdW interactions in addition to polar interactions and surface roughness effects.[8, 90] Calculation based on JKR adhesion theory[12] further

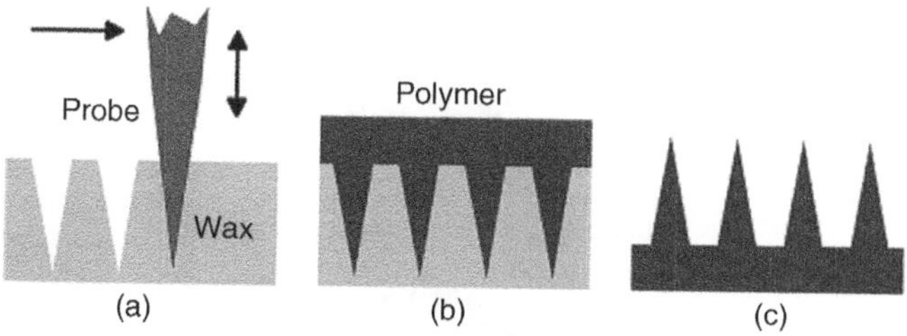

Figure 9.2 The synthetic gecko inspired nanohairs prepared by indenting a wax with an AFM tip and then molding polymer into that wax mold. (Adapted with permission. Copyright Brill: Ref. 49.)

indicated that adhesive force per unit area of the synthetic adhesives could be enhanced simply by reducing the size of spatulae and increasing their surface density.[8] Sitti and Fearing[90] have experimentally prepared adhesives by casting a polymer inside the pores of a nanoporous membrane and concluded that the basic design features of a hairy adhesive should include high aspect ratio of the fibers with micrometer to nanometer scale diameter, maximum possible hair density, maximum stiffness, low surface energy, and high tensile strength.

However, the PDMS material limited the mechanical stability of the high-aspect-ratio microfibers thus produced.[46, 91, 92] To achieve high-aspect-ratio structures for enhanced adhension, Geim et al.[93] presented a prototype of gecko tape by e-beam lithography and oxygen-plasma dry etching. These authors showed relatively high normal adhesion of 3 N/cm^2 for a 1 cm^2 patch of micropillars made of polyimide nanohairs with a diameter of 200–400 nm, height of 150–2000 nm, and periodicity of 400–4500 nm (Figure 9.3).[93] They also suggested that the flexibility of the back layer played a crucial role in obtaining adhesion similar to that of geckos; this would be in a good agreement with recent theoretical and experimental

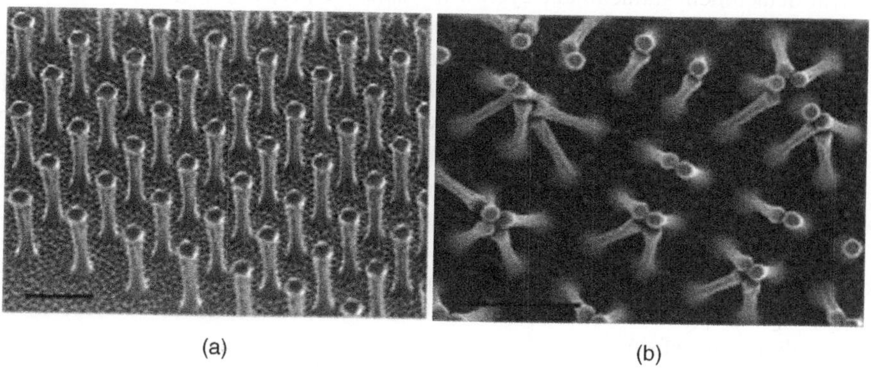

Figure 9.3 Gecko-inspired adhesive tape. (a) An array of micropillars made by micropatterning polyimide hairs. (b) Bunching of micropatterning polyimide hairs due to self-adhesion causes significant reduction in adhesive strength. Scale bars: 2 μm. (Adapted with permission. Copyright Nature Publishing Group: Ref. 93.)

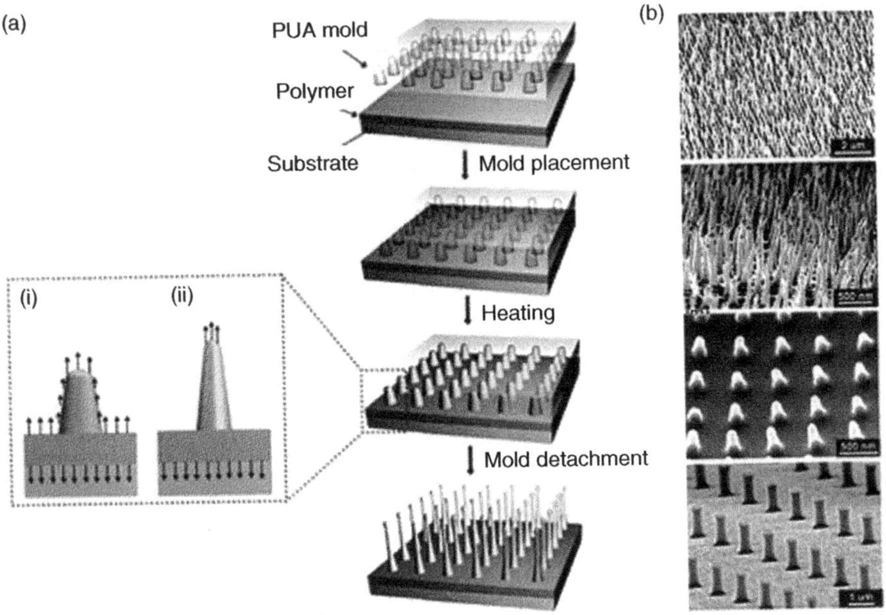

Figure 9.4 (a) A schematic illustration of the experimental procedure. (b) A large-area SEM image of the stretched PMMA nanohairs on PET film. (Adapted with permission. Copyright American Chemical Society: Ref. 96.)

studies.[60, 61] Thin back layers could equalize loads during pull-off and maximize adhesion. In contrast, thick back layers could deform during pulling, which leads to stress concentrations at the edge of the substrate.[94] The slow and expensive process of e-beam lithography is the main shortcoming of this method, although it allows for a high aspect ratio. Durability of the tape still remains an issue. Upon several detachment–attachment cycles, its adhesive property degraded because of hairs breaking and lateral bunching (Figure 9.3b).[95]

Following Geim's work,[93] alternative approaches, involving different materials, molds, and geometries of synthetic dry adhesives, have been developed to mimic gecko foot hairs with better properties. Examples include e-beam lithography, photolithography, and electrochemical etching. In particular, Jeong et al.[96] proposed a capillarity-driven rigiflex lithography aided by modulated interfacial tensions for polymer nanostructures on a solid substrate to obtain high-aspect-ratio polymer nanostructures (Figure 9.4). In this case, capillary forces induced deformation of the polymer melt into the void spaces of the mold, and the filled nanostructure was elongated upon removal of the mold due to tailored adhesive force at the mold/polymer interface.

In a similar but independent study, Greiner et al.[46] fabricated elastomeric model pillars with controlled geometry and material properties by soft-molding PDMS onto patterned SU-8 films on Si wafers (Figure 9.5). The elastomeric pillars thus prepared have dimensions ranging from 2.5 to 25 μm in radius and 2.5 to 80 μm in

Figure 9.5 High-aspect-ratio pillar arrays were made by soft-molding PDMS on patterned SU-8 films supported by Si wafers. (Adapted with permission. Copyright American Chemical Society: Ref. 46.)

height with the aspect ratio between 0.5 and 4. Their adhesion properties were systematically tested and compared with flat controls. It was found that micropatterned surfaces with an aspect ratio above 0.5 were more compliant than flat surfaces. The adhesion increased significantly with decreasing pillar radius and increasing aspect ratio of the patterned features. A preload-dependent adhesion was also demonstrated to be crucial for obtaining adhesives with tunable adhesion forces.

9.4.1.2 Directional Structure Apart from the aforementioned high aspect ratio and small radius, a directional angle of the nanostructure is another crucial factor for anisotropic (strong attachment and easy detachment) and reversible dry adhesions. Furthermore, an angled structure could significantly lower the effective modulus of a hair surface[43] and prevent structural bucking under a preload to render an apparently rigid material into a soft adhesive without self-matting problems.[54]

Polymer thin films provide a fertile ground for material processing, and hence the fabrication of angled polymer nanohairs with a high aspect ratio. Using optical lithography and polymer micromolding techniques, Aksak et al.[56] fabricated polyurethane microfibers with different hardness, angles, and aspect ratios. As shown in Figure 9.6, the resultant fibers have a diameter of 8 μm and a tilting angle of 25° with a good flexibility and a normal adhesion of 2.5–3.2 N/cm^2. Adhesion testing on a glass surface revealed that the fibers with a higher aspect ratio showed a stronger adhesion strength. However, it is a big challenge to successfully demold high-aspect-ratio SU-8 fiber arrays without fracturing by this method.

Yoon and co-workers developed an oblique metal deposition technique to obtain bent Janus nanopillars by postthermal annealing or e-beam irradiation.[97] In this technique, polymer nanopillars were first molded onto an etched silicon or silicon dioxide substrate with high-aspect-ratio nanoholes. Subsequently, metal layers of a few nanometers thickness (e.g., platinum) were obliquely deposited on the nanopillars using an inclined sample holder. The oblique deposition gave rise to "Janus-faced nanopillars," with one side covered by a metal layer while the other side was faced with the polymer. To achieve the directional structure, a simple

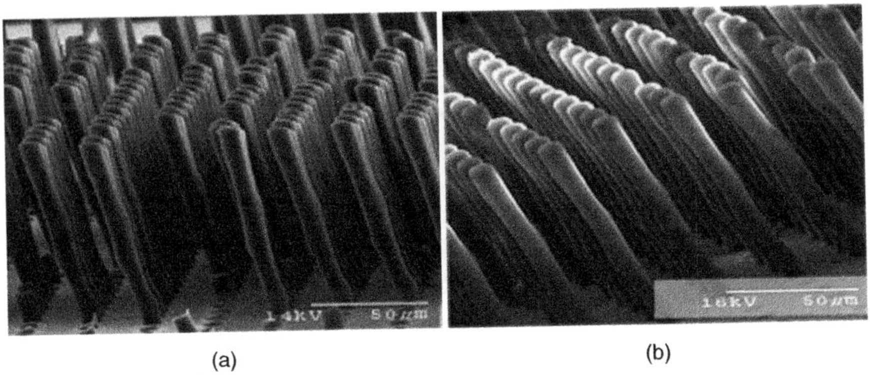

Figure 9.6 (a) SEM micrograph of angled SU-8 fiber array master template with an 8 μm diameter and a high aspect ratio. (b) Polyurethane molded microfibers with the same geometry. (Adapted with permission. Copyright American Chemical Society: Ref. 56.)

heating process was performed to cause the nanopillars to directionally bend against the metal side with a smaller thermal expansion coefficient (Figures 9.7a and 9.7b).[98] Similar to gecko-like adhesives, the resultant directional Janus nanopillars showed a greater adhesion hysteresis: the strong shear attachment when pulled from the bent direction (31 N/cm^2) contrasted with the easy detachment from the opposite direction (4.1 N/cm^2). Instead of the thermal annealing, these authors have also switched the bending direction of the nanopillars to the polymer side by employing e-beam irradiation (Figures 9.7c and 9.7d) as the metal layer prevented the polymer decomposition/crosslinking by e-beam irradiation.

Kim et al.[54] have used the e-beam exposure method to fabricate directional slanted hairs in nanoscale. Figure 9.8a shows a schematic illustration of the experimental procedure to fabricate stooped polymeric nanohairs with a high aspect ratio. As can be seen, vertical polyurethane acrylate (PUA) nanohairs were prepared by replicating the silicon master. After the Pt coating and exposure to e-beam at an angle of 30° for a period of time with an acceleration voltage, the vertical PUA nanohairs were transformed into stooped hairs (Figures 9.8a and 9.8b). The experimental data showed that the bending angle of the fabricated nanohairs was controlled by exposure time and acceleration voltages. The stooped nanohairs have a unidirectional feature of frictional adhesion with a remarkably high adhesion force (11 N/cm^2) in the forward direction (pulled in the stoop direction of the hairs) and a weak adhesion force (2.2 N/cm^2) in the opposite direction (pulled against the stoop direction of the hairs) (Figure 9.8c). In addition, the authors demonstrated that PUA nanohairs made by this method showed very good stability for over more than 100 attachment–detachment cycles.

Jeong et al.[99] systematically studied the effect of the leaning angle of gecko-inspired slanted polymer nanohairs on dry adhesion. They fabricated gecko-inspired slanted polymer nanohairs with various leaning angles (Figure 9.9) and analyzed their structural characteristics with particular emphasis on the performance of shear adhesion. It was demonstrated that the shear adhesion and adhesion hysteresis can

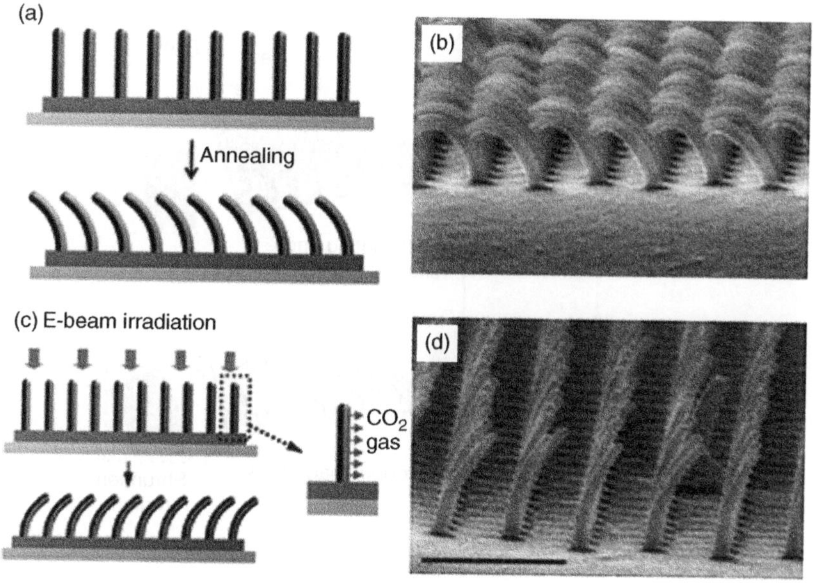

Figure 9.7 (a) A schematic illustration of the nanopillar bending against the metal side by annealing in convection oven. (b) A tilted SEM image showing bent nanopillars with 6 nm thick Pt layers coated on one side of the pillars, followed by thermal annealing at 120 °C for 30 min. (c) A schematic illustration showing the bending of nanopillars against the polymer side by e-beam treatment. A Pt layer coated on one side of the pillars prevents the polymer crosslinking/decomposition into CO_2 upon e-beam irradiation, resulting in the directional bending against the polymer side. (d) A tilted SEM image showing nanopillar bending after 6 s of e-beam irradiation. (Adapted with permission. Copyright Elsevier: Ref. 97.)

be greatly enhanced by increasing the leaning angle of nanohairs for both soft and hard materials due to an increased contact area and reduced structural stiffness.

On the other hand, Lee et al.[55] fabricated arrays of angled polystyrene (PS) microfibers by rolling a polyimide (25 μm) supported vertically aligned microfiber array against a clean microscope glass slide at 50°C. Figure 9.10 shows a schematic illustration of the fabrication process and SEM images for the resultant angled microfiber arrays. As can be seen, the average tilt angle of the unloaded microfibers from the surface normal is 45°, and the average center-to-center microfiber distance is 1.5 μm. Sliding of a clean glass surface against and along the microfiber direction without applying an external normal force produced an apparent shear stress of 0.1 and 4.5 N/cm^2, respectively, which demonstrated pure adhesion in one direction and high sliding friction in the opposite direction, similar to gecko setal arrays.

9.4.1.3 *Three-Dimensional Structure with Special Mushroom-Headed Tips*

Three-dimensional structures with specially shaped tips are particularly interesting for the adhesion performance of a fibrillar surface as spherical, conical, filament-like, band-like, sucker-like, spatula-like, flat, and toroidal tip

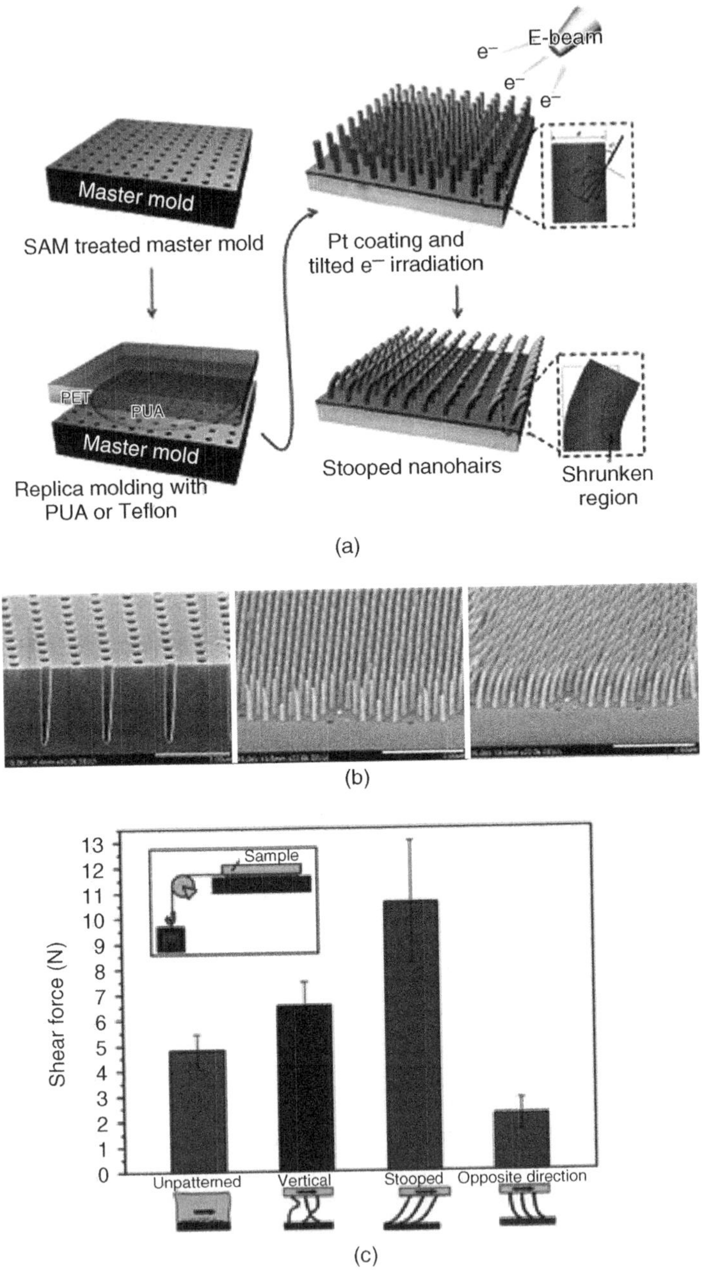

Figure 9.8 (a) Schematic illustrations of the fabrication of stooped nanohairs and physical mechanism for structural transformation. (b) SEM images of the master mold, replicated soft PUA pillars, and stooped hairs after the Pt coating and e-beam irradiation, respectively. (c) Shear forces for various cases with a dry adhesive patch of 1.0 cm × 1.0 cm. (Adapted with permission. Copyright Wiley-VCH: Ref. 54.)

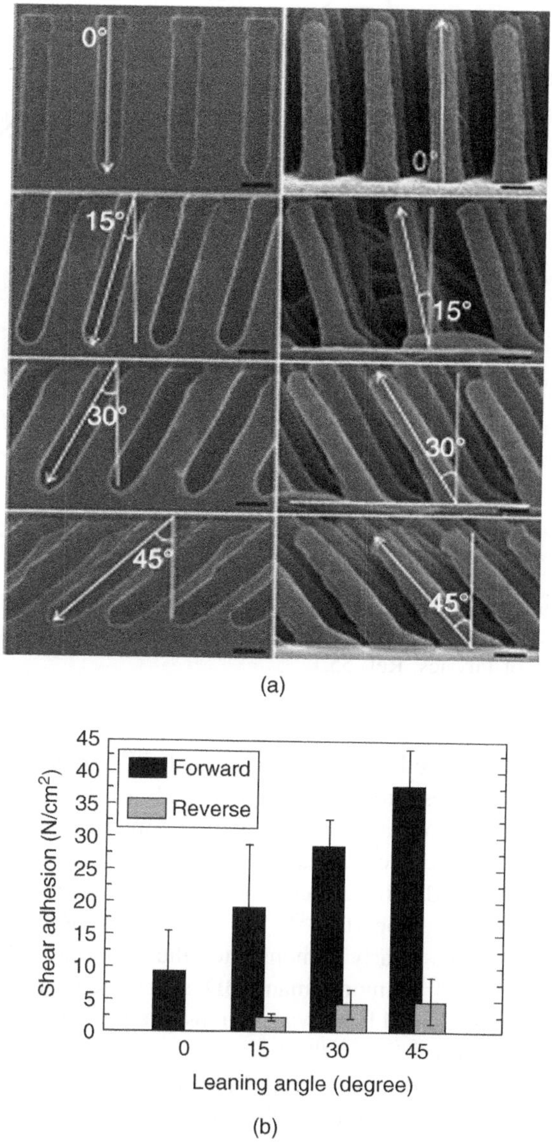

Figure 9.9 (a) SEM images of the Si master substrates (left) and formed PUA nanohairs (right) with different leaning angles, Scale bar = 400 nm. (b) Measured macroscopic shear adhesion force with an adhesive patch having PUA nanohairs with different leaning angles against a smooth glass surface upon removing preload of 0.3 N/cm. (Adapted with permission. Copyright American Institute of Physics: Ref. 99.)

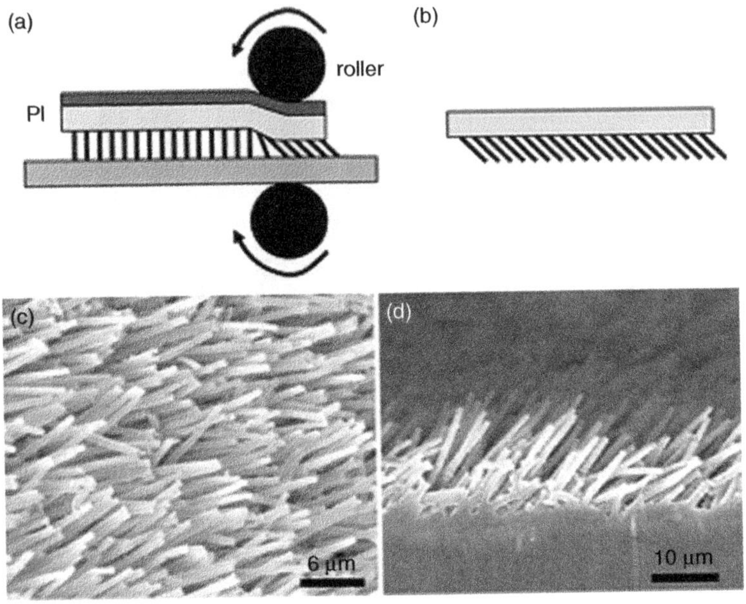

Figure 9.10 Schematic illustration of the fabrication process of angled microfiber arrays and SEM images of the produced fibrillar adhesive. (Adapted with permission. Copyright American Institute of Physics: Ref. 55.)

shapes have been observed in different animals.[100] Figure 9.11 shows SEM images of these structures on the attachment pads of a beetle, fly, spider, and gecko. Inspired by biological systems, various micro-to-nanoscale structures, including cylindrical or conical pillars (columns or posts) with flat, spherical, toroidal, or concave ends, have been designed and fabricated (Figure 9.12) for enhancing adhesion properties.[46, 50, 59–62, 101, 102] Recent theoretical and experimental studies have clearly demonstrated the important role of the tip shape to enhance adhesion.[59–62] Among many 3D shape geometries that have been studied, the mushroom-shaped head was demonstrated to yield the most significant enhancement of adhesion.[35, 50–52, 61, 62]

Thus, mushroom-shaped hairs have been fabricated by using lithographic methods from polyimide/polydimethylsiloxane (PDMS) in large scale[67, 75] or via micromolding followed by spatula tip formation via dipping,[79] as exemplified by Figure 9.13.

Although the mushroom-shaped fibers were identified several years back as the best option for dry adhesives with large normal adhesion strength, it has been a relatively recent experimental development to examine the correlation between the adhesion behavior and structure characteristics of the capped fibrils. Earlier fabrication processes kept the fiber height and size constant while varying the cap size. More recent technologies allowed for fabrication of the cap first with the underlying fiber geometry being defined using lithography by regulating the

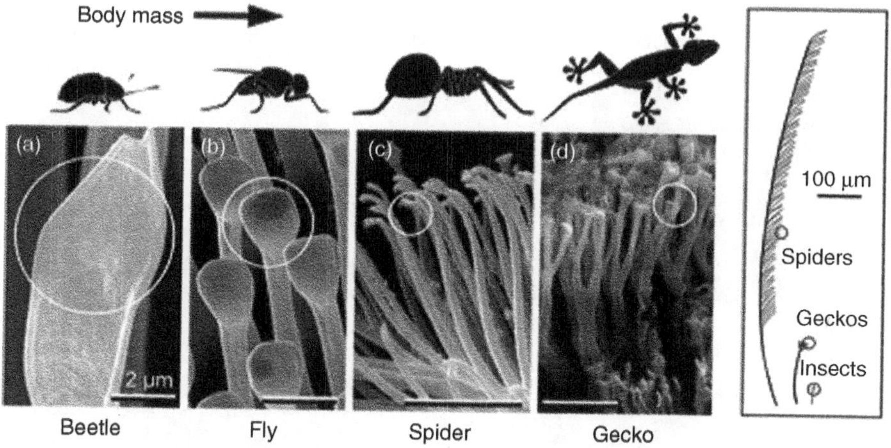

Figure 9.11 Terminal elements (circled areas) in animals with hairy design of attachment pads. Note that heavier animals exhibit finer adhesion structures. (Adapted with permission. Copyright National Academy of Sciences, U.S.A: Ref. 18.)

exposure dose and development time.[103] Figure 9.14 shows polyurethane replicas of original acrylic features defined with deep UV lithography.

In order to mimic gecko directional behavior, attempts have also been made to fabricate angled fibers with oriented mushroom tips for enhancing the anisotropic characteristics. In this context, Yao et al.[104] observed directional adhesion and shear interface strength in angled submillimeter diameter PDMS stalks with a terminal film. Kim et al.[105] have demonstrated submillimeter diameter angled polymer stalk arrays with angled ends for a climbing robot that exhibits desirable anisotropic shear forces. Michael et al.[35] fabricated tilted fibers and tips that combine the high interfacial strength of mushroom tipped micrometer-scale fibers with the directionality of fiber structures via lithography and a dipping technique from polyurethane (Figure 9.15). This design contains two independent tilted components, fiber and spatula. Fiber angles lay between 0° and 33° from horizontal. Spatula angles fell anywhere between 0° (horizontal) and 90°(vertical). It has been demonstrated that the combination of the tilted fibers and mushroom-shaped tips can allow for a significant load of 1 kg/cm^2 in shear direction. Furthermore, these adhesives exhibit directional characteristics of gripping when loaded in one direction, but self-releasing behavior when loaded in the opposite shear direction.

9.4.1.4 Multilevel Complex Hierarchical Structures So far, we have focused our discussion on adhesion with smooth substrates. Most practical surfaces are not perfectly smooth and have some degree of roughness. Adaptation to uneven and rough surfaces is a major feature of biological fibrillar adhesives. The gecko has evolved the ability to adhere to surfaces with varying roughnesses. The advantage of these fibrillar adhesives over flat unstructured adhesives for roughness adaptation is that each of the fibers deforms independently, allowing them to

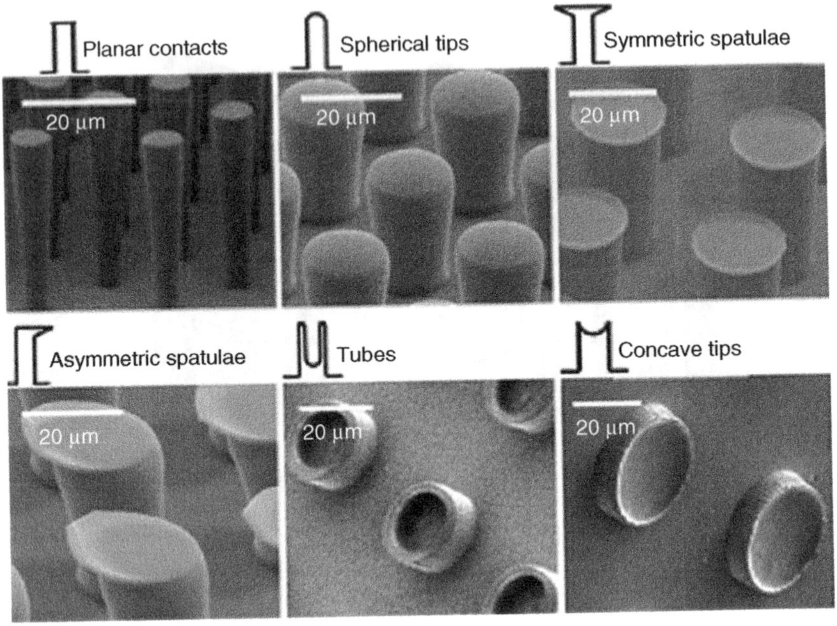

Figure 9.12 SEM images of typical tip geometries. (Adapted with permission. Copyright Wiley-VCH: Ref. 50.)

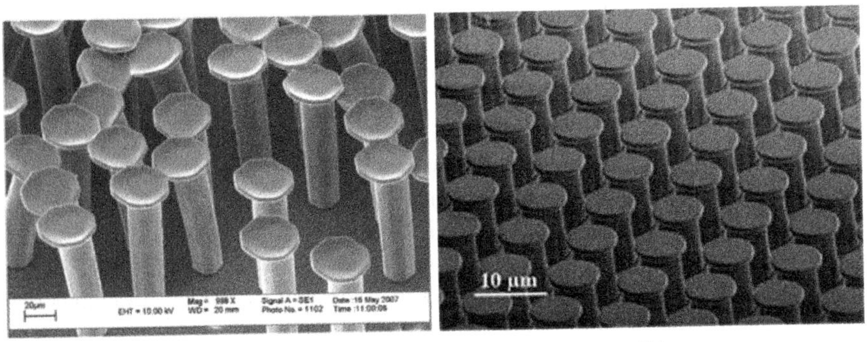

(a) (b)

Figure 9.13 (a) and (b) Examples of the mushroom-headed synthetic gecko spatula tips. (Adapted with permission. Copyright Elsevier: Ref. 67; Copyright 2009 Institute of Physics: Ref. 75, respectively.)

access deeper recessions to make an intimate contact with the surface. Therefore, the height of synthetic nanohair should be long enough to ensure adaptation to rough surfaces with varying amplitude and topography. However, the maximum height of polymer nanohairs is limited by the possible lateral collapse of hairy structures due to relatively low elastic modulus of polymers. Theoretical studies

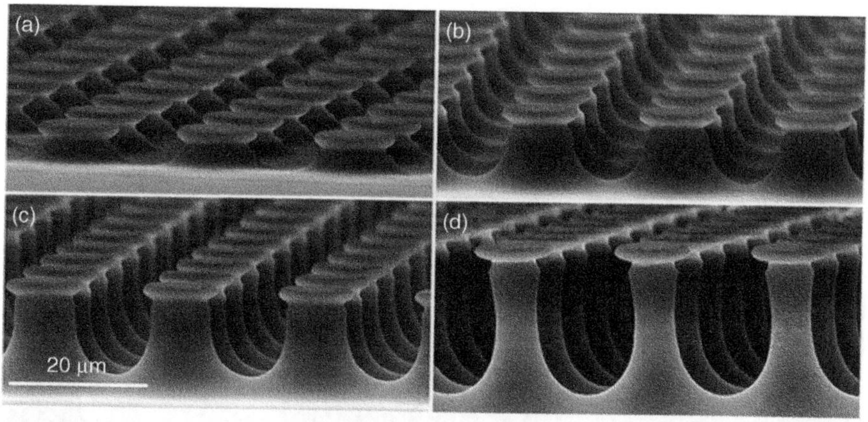

Figure 9.14 SEM images of ST-1060 polyurethane fibers cast from a negative silicone rubber mold produced by a commercial acrylic master mold: (a)–(c) are defined on a single mold with the same development time but with different exposure dose, (d) is formed with the same exposure dose as (c) but with an extra 50% development time. (Adapted with permission. Copyright Institute of Physics: Ref. 103.)

have revealed that the hierarchy of structures can give rise to increased adhesion strength against a rough surface either by reducing structural stiffness or by enhancing structural height without instability.[17, 57] Spring-based models predicted that appropriate multilevel hierarchical structures should exhibit a higher adhesive force and energy than a one-level structure for a given applied load due to improved adaptation and attachment ability.[106] Therefore, researchers have attempted to fabricate micro/nanoscale combined hierarchical fibers to more closely mimic the structure of the hairs on the gecko's foot, and hence illuminate its multifunctional adhesives properties.[82, 106]

In this regard, del Campo and co-workers[66, 78] fabricated well-defined arrays of hierarchical microfibrils by multistep photolithography combined with soft lithography using SU-8 (Figure 9.16). Double molding with PDMS rendered fibrils with the base pillars having a diameter of 50 μm and a height of 200 μm, and the top pillars having a diameter of 5 μm and heights ranging from 2.5 to 10 μm. However, the adhesion was not satisfactory since the pillars formed on the base structures did not have desirable features for the adhesion enhancement in this particular case.[78]

Hierarchical polymeric microfibrils can also be fabricated by other methods.[107–109] As shown in Figure 9.17, PMMA hierarchical arrays have been prepared by using branched porous alumina templates, which allowed the formation of nanofibril arrays with nanofibrils of 60 nm in diameter and length-to-diameter aspect ratios as high as 100:1 on their tips.

More interestingly, Jeong et al.[60] reported a method for fabricating angled, hierarchically patterned high-aspect-ratio polymer nanohairs to generate high-performance directionally sensitive dry adhesives. As can be seen in Figure 9.18, the slanted polymeric nanostructures were first molded from an etched polySi substrate containing slanted nanoholes. An angled etching technique was then

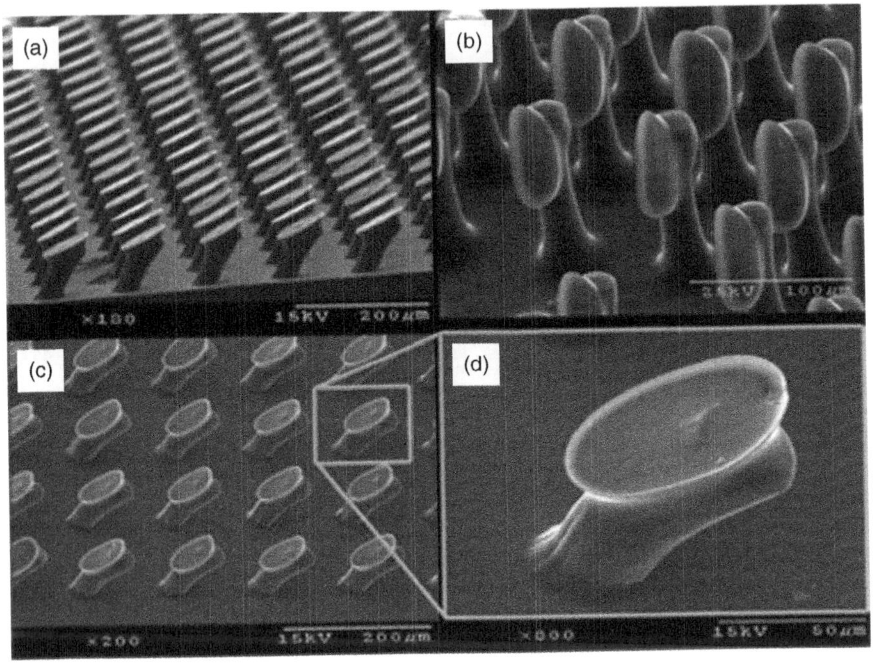

Figure 9.15 SEM images of arrays of 35 μm diameter angled polyurethane microfibers with angled mushroom tips. Tip orientation can be controlled to form tips with varying angles: (a) 34°; (b) 90°; (c, d) 23°. Details of the tip can be seen in (d). (Adapted with permission. Copyright Wiley-VCH: Ref. 35.)

developed to fabricate slanted nanoholes with flat tips by inserting an etch-stop layer of silicon dioxide. This unique etching method was equipped with a Faraday cage system to control the ion incident angles in the conventional plasma etching system. Thus, the polymeric nanohairs were fabricated with tailored leaning angles, sizes, tip shapes, and hierarchical structures. The replicated, slanted nanohairs with a controlled leaning angle and bulged flat top showed excellent directional adhesion, exhibiting strong shear attachment (~26 N/cm^2 in maximum) in the angled direction and easy detachment (~2.2 N/cm^2) in the opposite direction. In addition to single scale nanohairs, monolithic, micro–nanoscale combined hierarchical hairs were also fabricated in this study by using a two-step UV-assisted molding technique (Figure 9.18a). These hierarchical nanoscale patterns maintained their adhesive force even on a rough surface (roughness <20 μm) whereas simple nanohairs lost their adhesion strength under the same condition.

By using a two-step molding method, Murphy et al.[110] have also fabricated multilevel complex hierarchical structures from polyurethane. First, the tips of an array of micromolded vertical-aligned or angled polymer microfibers were coated with a layer of liquid polymer by contact against a thin reservoir layer on a donor substrate (Figures 9.19a and 9.19b). After coating, the wet fiber tips are placed on an etched silicon wafer (Figure 9.19c). This master template wafer has

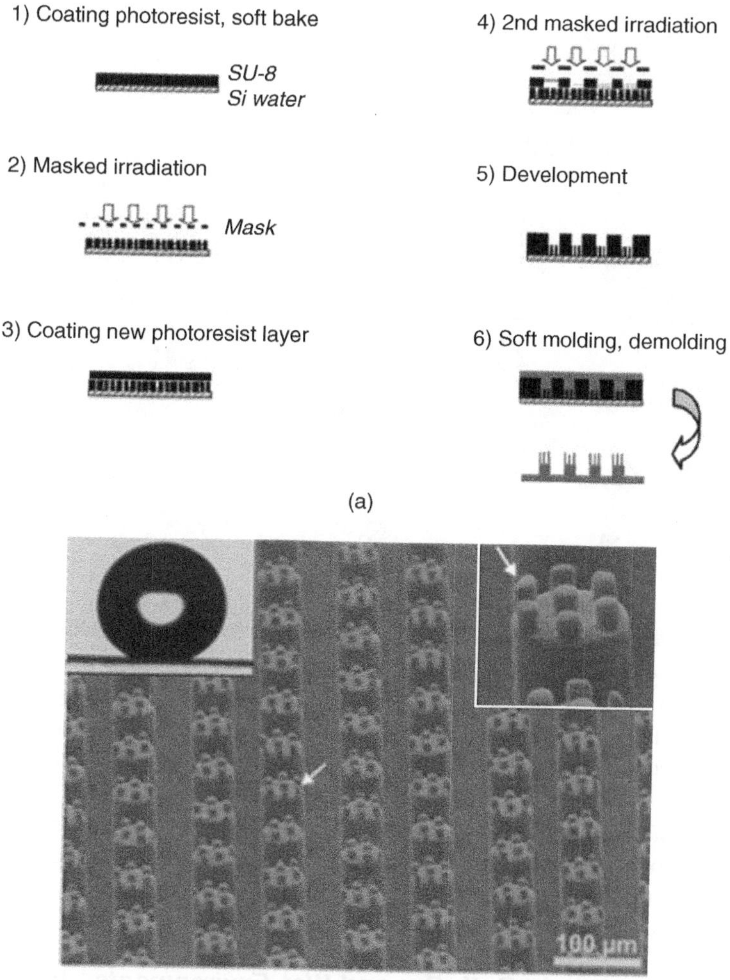

(a)

(b)

Figure 9.16 (a) Schematic of the processing steps for the fabrication of hierarchical PDMS pillars through two-step photolithography and soft molding. (b) SEM image showing an array of hierarchical pillars fabricated by soft molding Sylgard 184 on SU-8 photolithographic templates. The right insert shows a close view of the hierarchical structure, the left a water droplet resting on the structure (contact angle 160°). (Adapted with permission. Copyright Wiley-VCH: Ref. 78.)

micron-scale-diameter cylindrical holes with widened tips formed by deep reactive ion etching utilizing the notching effect. Capillary forces draw the liquid polymer into the cavities, forming branches on the fiber ends. The sample is then left to cure at room temperature before the mold is removed using dry etching. When etching is complete, the final hierarchical structures remain (Figure 9.19d).[110] Figure 9.20 shows the scanning electron micrographs of polyurethane hierarchical

Figure 9.17 SEM micrographs of (a) hierarchical microfibril array featuring ~10 μm wide and ~70 μm long microfibrils, each branching into a ~60 nm wide and ~0.5 μm long nanofibril array; (b, c) magnified top view and (d) oblique view of the nanofibril array. (Adapted with permission. Copyright Institute of Physics: Ref. 109.)

fibers with hierarchical mushroom tips. These hierarchical structures were found to exhibit both increased adhesion and interface toughness, suggesting that a hierarchical structure could adhere with higher strength to uneven surfaces with the roughness amplitude on the same scale as the length of the base fibers.

9.4.1.5 Adhesives in Both Dry and Wet Environments Recently, there has been much interest in developing synthetic adhesives acting in both wet[111, 112] and dry conditions.[46, 49, 59, 64, 113] Although certain synthetic adhesives mimicking the nanoscale surface features of setae have been demonstrated to maintain adhesive performance over many cycles,[4, 34, 93, 114] gecko-foot-mimetic adhesion largely diminished upon full immersion in water.[10, 115] Lee et al.[84] have recently reported a hybrid biologically inspired adhesive consisting of an array of nanofabricated polymer pillars coated with a thin layer of a synthetic polymer that mimics the wet adhesive proteins found in mussel holdfasts (Figure 9.21). It was found that the wet adhesion of the nanostructured polymer pillar arrays increased nearly 15-fold when coated with the mussel-mimetic polymer. Furthermore, the system maintained its adhesive performance for over a thousand contact cycles in both dry and wet environments. This hybrid adhesive, which combines the salient design elements of both gecko and mussel adhesives, should be useful for reversible attachment and deattachment to a variety of surfaces in diverse environments.

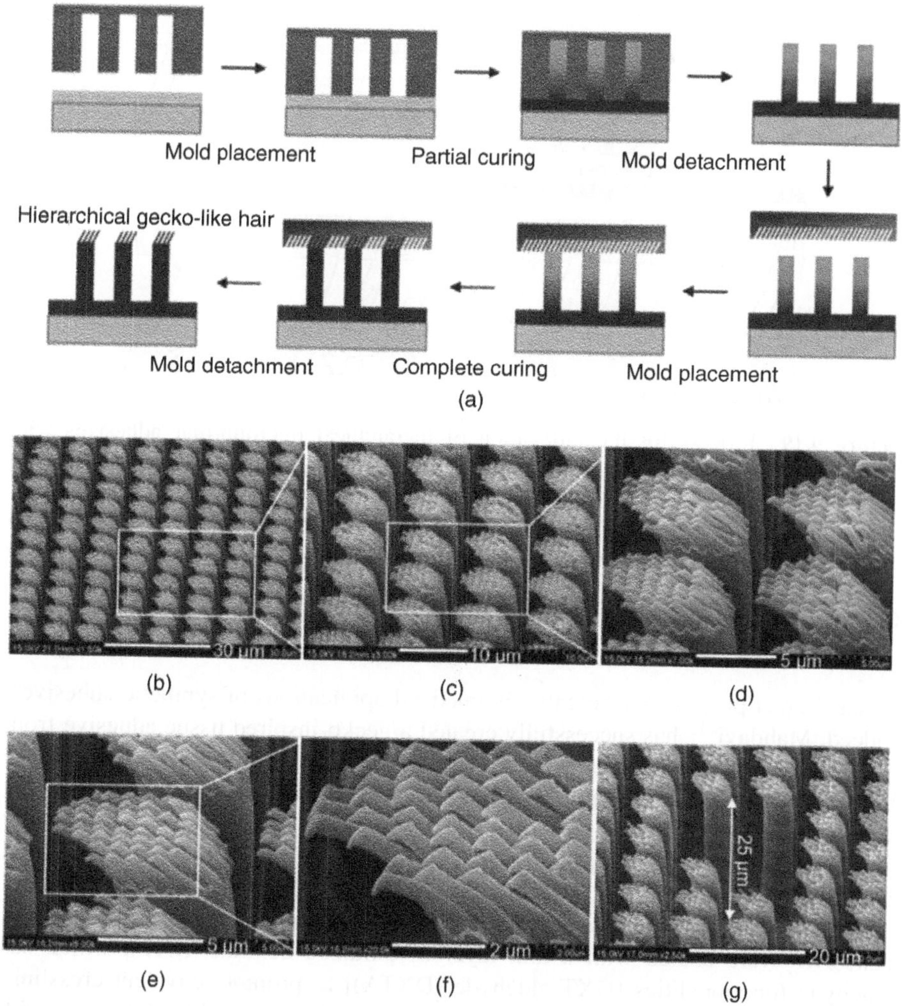

Figure 9.18 (a) A schematic illustration for the fabrication of dual-scale hierarchical gecko-like hairs by two-step UV-assisted capillary force lithography. (b) A titled SEM image of two-level hierarchical PUA hairs formed over a large area. (c–g) Magnified, titled images of (b), showing well-defined high aspect ratio and angled nanohairs with bulged flat top formed on 5 μm pillars (5 μm spacing, 25 μm height). (Adapted with permission. Copyright National Academy of Sciences, U.S.A: Ref. 60.)

A biodegradable and biocompatible wet gecko-inspired tissue adhesive has also been fabricated by using poly(glycerol-*co*-sebacate acrylate) (PGSA) to modify the surface of a nanotopography mimicking gecko feet.[112]

The translation of existing gecko-inspired adhesives for medical applications is a complex task as multiple parameters (e.g., biocompatibility, biodegradation, strong adhesive tissue bonding, as well as compliance and conformability to tissue surfaces) must be optimized. The demonstrated both dry and wet adhesions

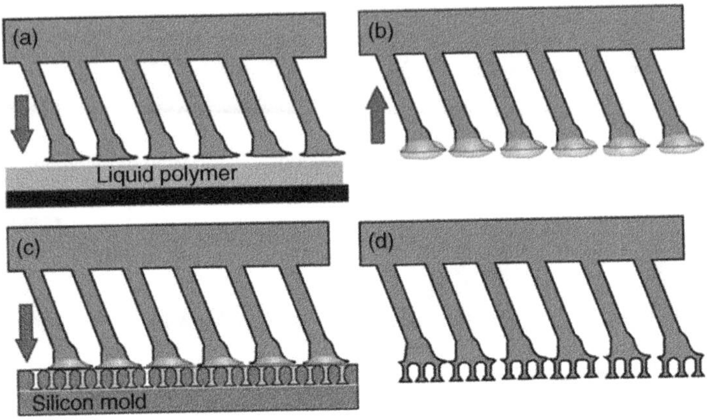

Figure 9.19 Process for the fabrication of hierarchical microfibrillar adhesives with a controlled fiber tip shape. (a) Base fibers with mushroom tips are dipped into a donor liquid polyurethane layer. (b) Some of the liquid polymer is retained by the tips. (c) The fiber array is placed onto an etched silicon mold, where the liquid from the tips is drawn into the negative features. (d) After the polyurethane has cured, the silicon mold is etched away with a dry etching process. (Adapted with permission. Copyright American Chemical Society: Ref. 110.)

indicate great promise for potential biomedical applications of synthetic adhesives. Indeed, Mahdavi[112] has successfully created a gecko-inspired tissue adhesive from a biocompatible and biodegradable elastomer combined with a thin tissue-reactive biocompatible surface coating (Figure 9.22). Tissue adhesion was optimized by varying dimensions of the nanoscale pillars, including the ratio of tip diameter to pitch and the ratio of tip diameter to base diameter. Aldehyde-functionalized polysaccharides have been used effectively in animal models to bond hydrogel materials to tissue proteins with minimal host inflammation. Therefore, PGSA adhesives were coated with a thin layer of oxidized dextran (DXT), which has aldehyde functionalities [DXT aldehyde (DXTA)] to promote covalent crosslinking with tissue. Coating these nanomolded pillars of biodegradable elastomers with a thin layer of oxidized dextran was found to significantly increase the interfacial adhesion strength on porcine intestine tissue in vitro and in the rat abdominal subfascial in vivo environment. This gecko-inspired medical adhesive could have potentials for sealing wounds and for replacement or augmentation of sutures or staples.

For potential biomedical applications, repeated adhesion and/or self-cleaning capabilities are also important. Using principles inspired by the study of naturally occurring sticky systems such as the micro- and nanoscale fibers on the toes of geckos and the adhesive proteins secreted by marine animals such as mussels, Glass et al.[81, 116] exploited a multistep fabrication process consisting of optical lithography, micromolding, polymer synthesis, dipping, stamping, and photopolymerization to produce uniform arrays of polyurethane elastomeric microfibers with mushroom-shaped tips coated with a thin layer of lightly crosslinked p(DMA-*co*-MEA),

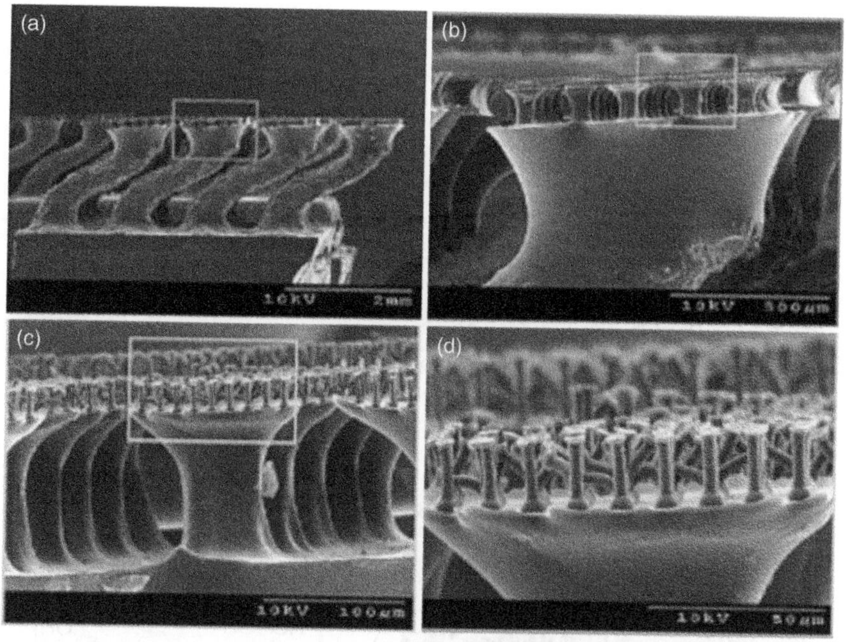

Figure 9.20 SEM micrographs of three-level hierarchical polyurethane fibers: (a) 400 μm diameter curved base fibers; (b) base fiber tip with midlevel 50 μm diameter fibers; (c) midlevel fibers in detail; (d) terminal third level fibers at the tip of the midlevel fibers are 3 μm in diameter and 20 μm in height and have 5 μm diameter flat mushroom tips. (Adapted with permission. Copyright American Chemical Society: Ref. 110.)

a DOPA-containing mussel-inspired polymer [poly-(dopamine methacrylate-*co*-2-methoxyethyl acrylate)], for adhesion repeatedly in fully submerged wet environments. The process diagram for creating the DOPA-containing polymer-coated, patterned fiber arrays is illustrated in Figures 9.23a–9.23g, while Figures 9.23h and 9.23i show optical and SEM images for the resulting fibrillar adhesive.

Furthermore, Ko et al.[117] fabricated the interesting hybrid Ge/parylene nanowire connectors, which enables wet and dry adhesion properties. The superhydrophobic surface of the nanowire connectors enabled the wet, self-cleaning of contaminant particles from the surface, similar to the lotus effect. Lu et al.[118] reported interesting, strong adhesives of polythipphene (PTH) films on various smooth surfaces such as glass, mica, and GaAs after drying them from their wet states under ambient conditions. The normal and shear dry adhesion forces of the films on glass were measured to be as high as 80 ± 8 and 174 ± 10 N/cm^2, respectively. These extraordinary strong adhesion forces are attributed to the high strength and stiffness of PTH and the high contact fraction (79%) of PTH nanotips on the smooth surfaces induced by the wet to dry process.

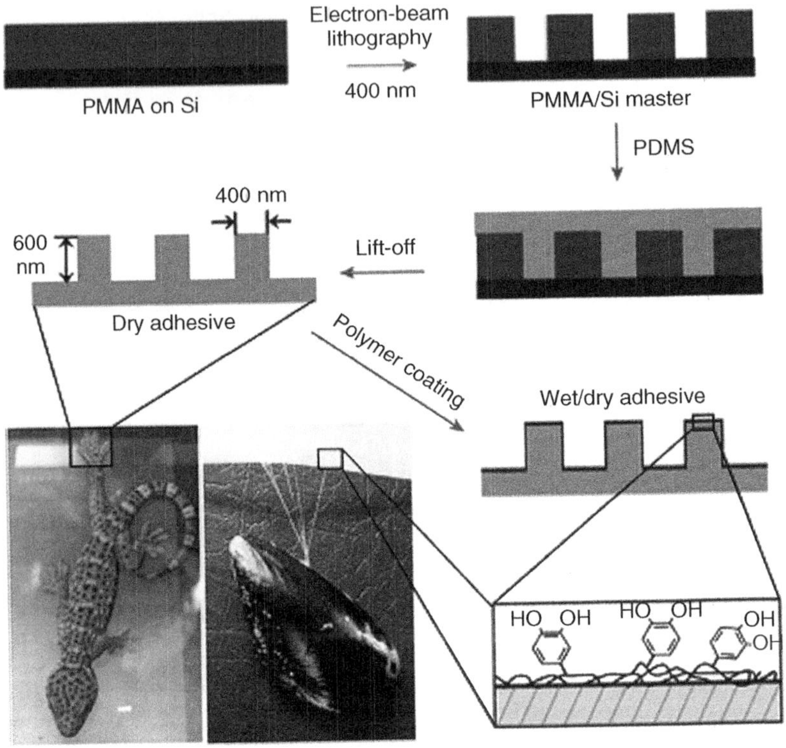

Figure 9.21 Rational design and fabrication of wet/dry hybrid nanoadhesive. Electron-beam lithography was used to create an array of holes in a PMMA thin film supported on Si (PMMA/Si master). PDMS casting onto the master is followed by curing, and lift-off resulted in gecko-foot-mimetic nanopillar arrays. Finally, a mussel-adhesive-protein-mimetic polymer is coated onto the fabricated nanopillars. The topmost organic layer contains catechols, a key component of wet adhesive proteins found in mussel holdfasts. (Adapted with permission. Copyright Nature Publishing Group: Ref. 84.)

9.4.1.6 Adhesives with Self-Cleaning Properties One of the unique characteristics of geckos that differentiate them from other climbing animals is the self-cleaning property of their spatulae. Dirt particles detach from a gecko's spatulae as the gecko walks along it.[42] Interestingly, gecko-inspired self-cleaning has also been demonstrated for synthetic structures.[117, 119] For instance, Lee and Fearing[119] fabricated a high-aspect-ratio polypropylene fibrillar adhesive, which showed the self-cleaning capability. In contrast to a conventional pressure-sensitive adhesive (PSA), the synthetic fibrillar adhesive recovered, after contamination, about 33% of the shear adhesion of clean samples after multiple contacts with a clean, dry surface (Figure 9.24). However, particles larger than 2.5 μm could not be self-cleaned due to the multiple contacts with too many fibers.

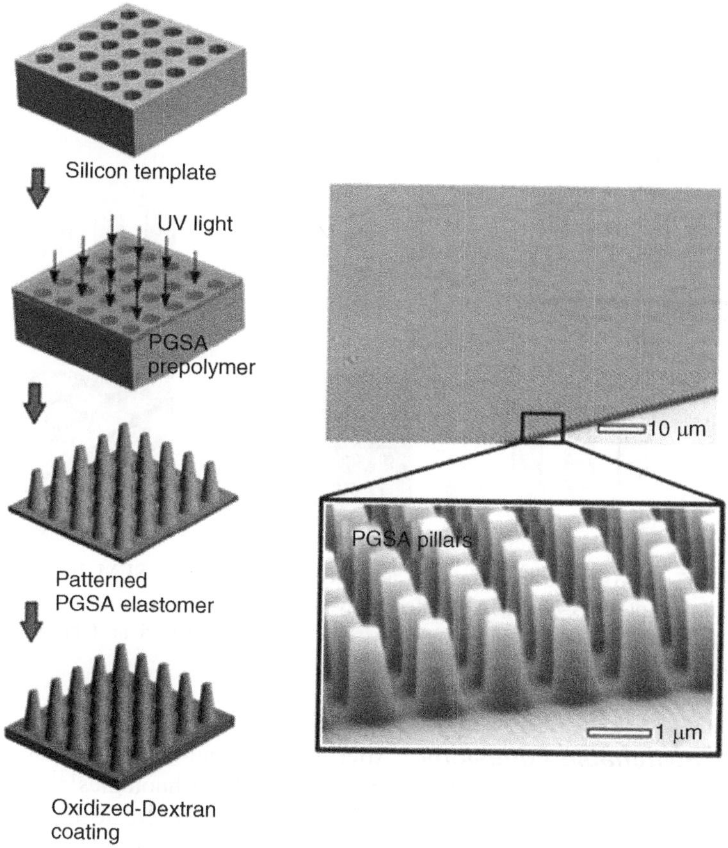

Figure 9.22 Development of biodegradable synthetic gecko patterns. Nanomolding of the PGSA prepolymer is accomplished by photocuring the prepolymer under UV light followed by removal of the pattern and subsequent spin coating of DXTA on the surface of the pillars. SEM demonstrated excellent pattern transfer and fidelity. (Adapted with permission. Copyright National Academy of Sciences, U.S.A: Ref. 112.)

Just like the dry self-cleaning discussed above, Kim et al.[73] reported that hydrophilic polyurethane mushroom-shaped microfiber arrays possess wet self-cleaning ability (Figure 9.25). In comparison with a flat surface made of the same polyurethane, the fiber array exhibited almost 100% wet self-cleaning without any degradation of adhesive strength. They attribute this cleaning ability to the mushroom-shaped tip ending geometry of the fiber array, which causes the fiber array to be apparently hydrophobic even though the fiber material is hydrophilic. These results suggest that tip ending shape is one of the significant design parameters for developing contamination-resistant polymer fibrillar adhesives.

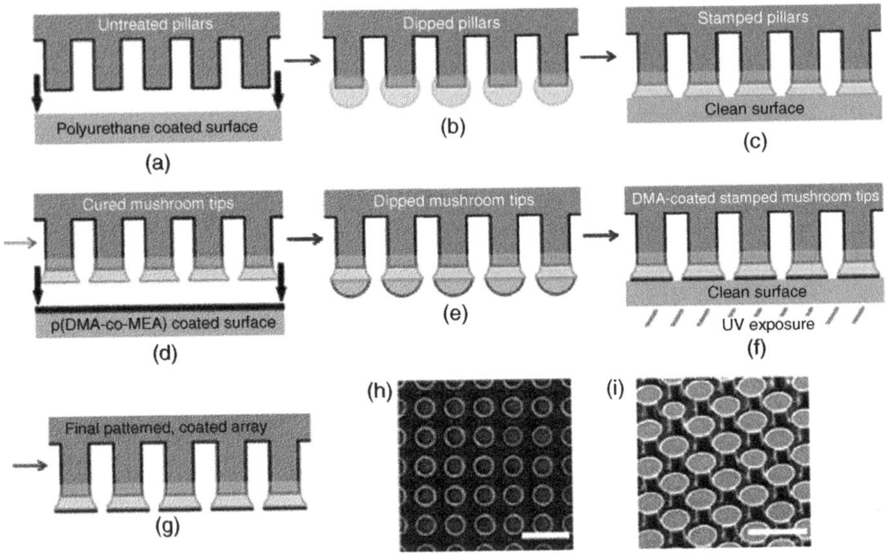

Figure 9.23 (a)–(g) Process diagram and (h) optical and (i) SEM images of fabricated patterned, p(DMA-*co*-MEA)-coated arrays of fibers with mushroom tips. Scale bar: 200 μm in (h) and (i). (Adapted with permission. Copyright American Chemical Society: Ref. 116.)

9.4.1.7 Switchable Adhesions After a catch-bond-based nanoadhesive sensitive to shear stress,[120] the combination of patterning technologies[121] and responsive polymer materials[122, 123] has allowed researchers to create microstructured surfaces with switchable adhesion. Application of an external field (e.g., temperature or magnetic fields) caused changes in the topographical design, influencing the final adhesion performance. Reddy et al.[101] have used shape memory thermoplastic elastomers to obtain the microstructured adhesive surface by soft molding the material at its highest transition temperature (T_{perm}). Arrays of vertical micropillars with diameter between 0.5 and 50 μm and heights between 10 and 100 μm were fabricated (Figure 9.26a). Mechanical deformation of this topography at the lower transition temperature (T_{trans}) followed by cooling to room temperature in the deformed position yielded a temporary nonadhesive surface consisting of pillars in a tilted position (Figure 9.26b). By reheating above T_{trans}, the patterned surface switched from the temporary nonadhesive state to a permanent adhesive surface with at least a 200-fold increase in pull-off force (Figure 9.26c). Xie and coworkers[124, 125] reported a self-peeling reversible dry adhesive system with a unique built-in adhesion reversal mechanism. It consists of a smooth (nonstructured) dry adhesive layer and a shape memory polymer layer, with the latter introducing a heat-triggered "self-peeling" adhesion reversal mechanism.

A hierarchical micro/nanostructure has also been developed that uses hybrid MEMS and nanofabrication techniques to produce a reversible adhesive system through the application of a magnetic field.[126] The reversible adhesive system

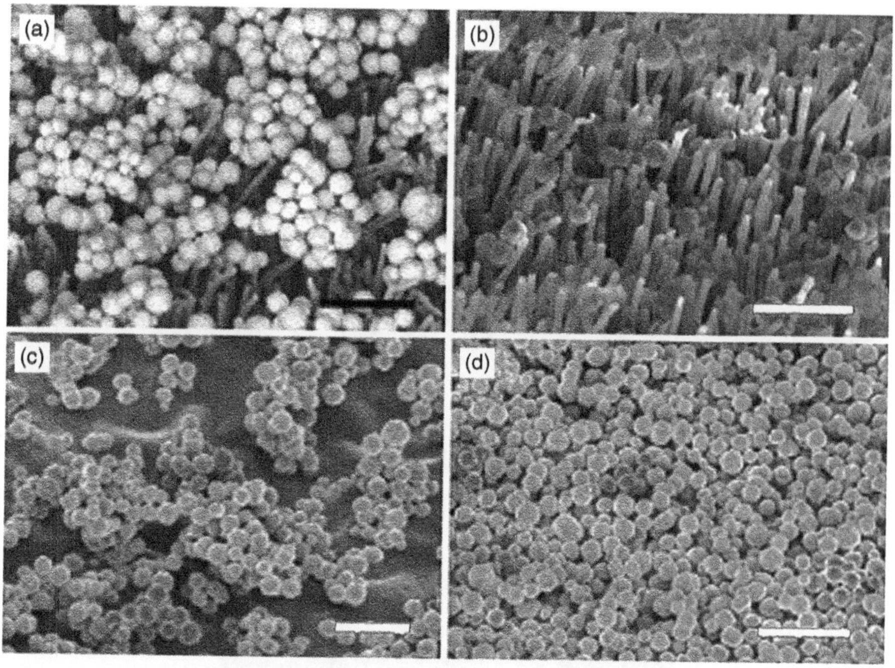

Figure 9.24 SEM images of the polypropylene fibrillar adhesive and conventional pressure-sensitive adhesives (PSAs). (a) Fibrillar adhesive contaminated by gold microspheres. (b) Fibrillar adhesive after 30 contacts (simulated steps) on clean glass substrate. (c) Conventional PSA contaminated by microspheres. (d) Conventional PSA after contact on a clean glass substrate. All scale bars correspond to 10 μm. Microspheres on fibrillar adhesive are removed by simulated steps, but microspheres on PSA cover more area after the steps. (Adapted with permission. Copyright American Chemical Society: Ref. 119.)

was comprised of paddle-shaped cantilevers (10 μm × 130 μm) fabricated in electroplated nickel, which were coated with vertically aligned polymeric nanorods created by using an active ion etching process. This adhesive system was very similar to the real gecko-foot system (Figure 9.27). The nickel cantilevers, when placed in a magnetic field, reorient themselves so that the terminal pad of the structure, responsible for adhesion, rotates to face away from an adhering surface, decreasing the adhesion force by reducing the available adhesive area. The adhesion force in the "off" state (with magnetic field) was approximately 40 times lower than in the "on" state (without magnetic field).

The adhesion can also be reversibly regulated by applying mechanical strain.[127, 128] Recently, Jeong et al.[127] reported a simple yet robust method of fabricating a large-area (3 × 3 cm^2), stretchable dry adhesive with micropillars (diameter, 20 μm; height, 20 μm; and pillar density, 6.25 × $10^4/cm^2$) in the form of a wrinkled PDMS sheet (thickness, 1 mm) as shown in Figures 9.28a and 9.28b. Regularly ordered surface wrinkles were generated by a strain mismatch due to oxygen plasma treatment on a prestrained (or extended) PDMS sheet and

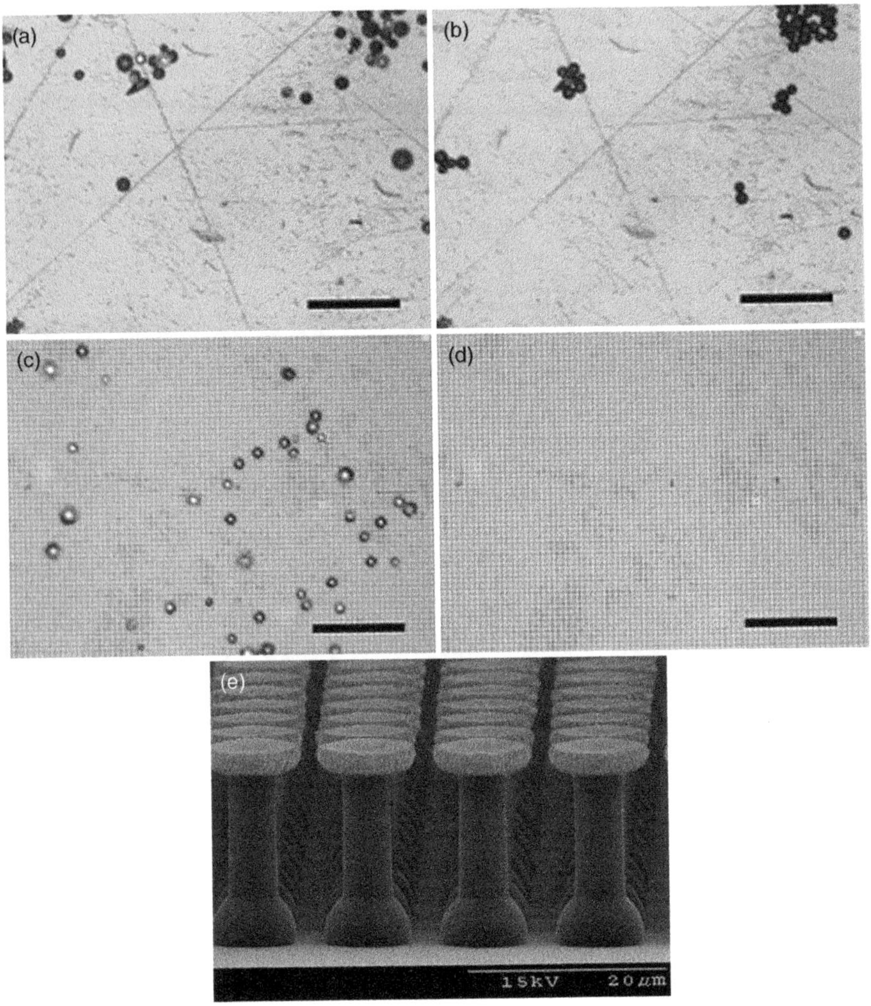

Figure 9.25 Optical microscopic images of surfaces contaminated with silica particles [(a) and (c)] and surfaces washed with DI water [(b) and (d)] for a flat surface [(a) and (b)] and a fiber array (c,d), respectively. (Scale bars: 200 μm.) (e) SEM image of a polyurethane mushroom-shaped microfiber array. (Adapted with permission. Copyright American Chemical Society: Ref. 73.)

subsequent strain release (Figure 9.28c). Relatively strong normal (10.8 N/cm^2) and shear adhesion (14.7 N/cm^2) forces could be obtained in the "on" state for a fully extended (strained) PDMS sheet (prestrain of 3%), whereas the forces in the "off" state could rapidly be reduced to nearly zero once the prestrain was released (prestrain of 0.5%) (Figure 9.29). These authors finally demonstrated that the wrinkled PDMS sheet has good durability over more than 100 cycles of attachment and detachment.[127]

Figure 9.26 SEM images of Tacoflex pillars with $r = 10$ μm, $h = 100$ μm, and 20 μm interpillar spacing. (a) Original pattern. (b) Top and side views of tilted pillars after deformation at 70 °C and fixation in the deformation state by cooling below T_{trans}. (c) Top and side views of recoverd pillars after reheating at 70 °C. (Adapted with permission. Copyright Wiley-VCH: Ref. 101.)

Nadermann et al.[129] demonstrated that a structure with a fibrillar surface terminated by a continuous film can be switched between two metastable states. The first state, in which the film is stretched between fibrils, has strongly enhanced adhesion compared to an unstructured flat control. In the second state, the film collapses onto the substrate between fibrils and is held up away from the substrate at the fibrils, resulting in a surface with a periodic array of bumps having much reduced adhesion. The interface could be switched mechanically between these two

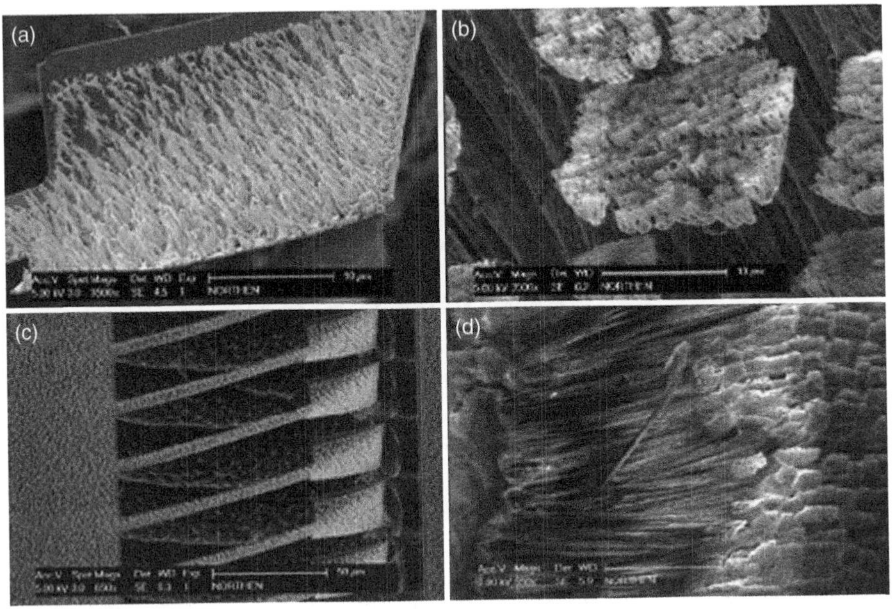

Figure 9.27 Electron micrographs of synthetic structures (left) and the analogous gecko structures (right). (a) Paddle surface coated with evenly spaced uncondensed aligned vertical polymer nanorods (left), and (b) the branched terminus of a seta into spatulae (right), same magnification and scale bar 10 μm. (c) Freestanding nickel cantilevers and paddles coated with nanorods (left) and an array of setae (d) (right), same magnification and scale bar 50 μm. (Adapted with permission. Copyright Wiley-VCH: Ref. 126.)

states repeatedly, thus providing a means for active control of surface mechanical properties.

9.4.2 Carbon-Nanotube-Based Dry Adhesives

In addition to polymers as fabrication materials for synthetic, gecko-foot-mimetic adhesives, vertically aligned carbon nanotubes (VA-CNTs, both single-walled and multiwalled) have recently been used for mimicking gecko adhesives. Because of their extraordinary high aspect ratio, mechanical, electrical, and thermal properties, high tensile and flexural strength, and high elastic modulus, VA-CNTs showed great potential for dry adhesive applications.[130–133] Several groups have actively investigated dry adhesives based on VA-CNTs generated by chemical vapor deposition (CVD). In particular, Yurdumakan et al.[85] investigated the adhesion behaviors of CVD-generated vertically aligned multiwalled carbon nanotube (VA-MWNT) arrays on quartz or silicon substrates being transferred into a PMMA matrix. The adhesive properties of the MWNT arrays was measured by the tip of a scanning probe microscope (SPM), which was pushed into the nanotube array and then retracted. Adhesive strength was then determined from the maximum negative force in the force–displacement curve recorded during retraction.[85] The

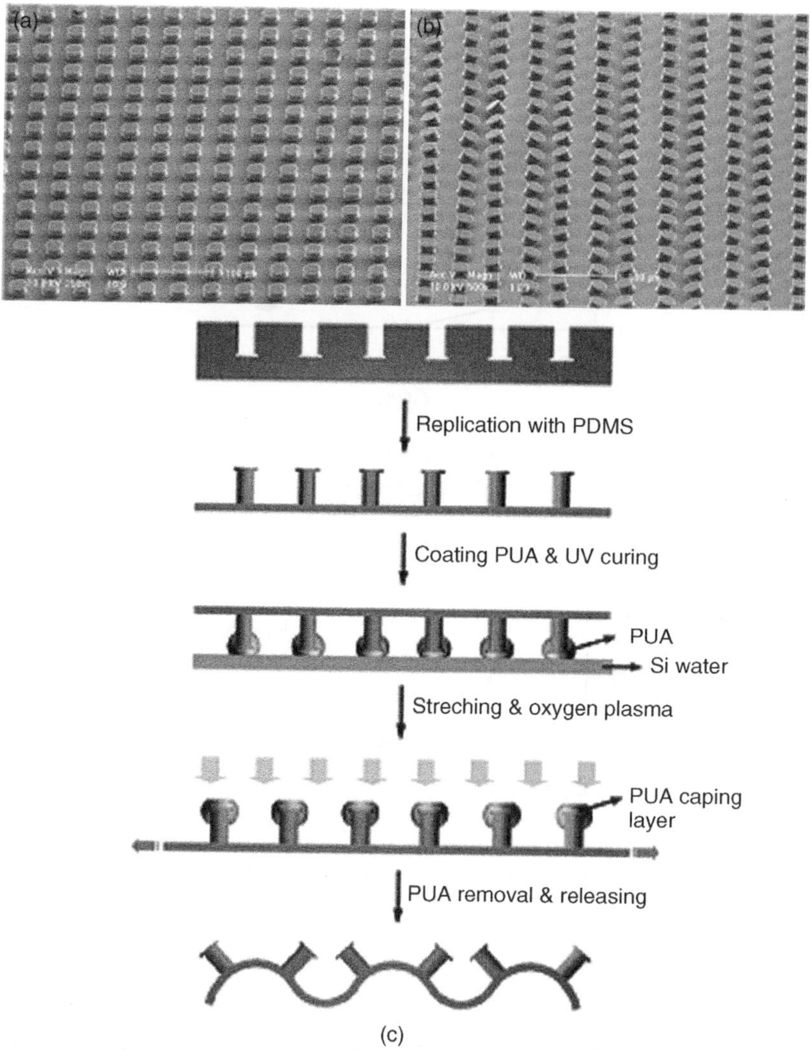

Figure 9.28 Titled SEM images of the PDMS adhesive patch with micropillars (a) before and (b) after the removal of the prestrain. (c) Schematic illustration for fabricating a stretchable dry adhesive with micropillars. (Adapted with permission. Copyright American Chemical Society: Ref. 127.)

conservatively estimated force/area of this array ($1.6 \pm 0.5 \times 10^{-2}$ nN/nm^2) was 200 times higher than that of a gecko's seta (10^{-4} nN/nm^2).[4] The side contact of fibers with substrates over a large contact area could provide a stronger adhesion force than that of a tip contact.[134, 135] Although impressive, this adhesion strength cannot be observed for a macroscopic contact. In microscopic measurement, the probe penetrates into the nanotube brush, resulting in side contacts between the

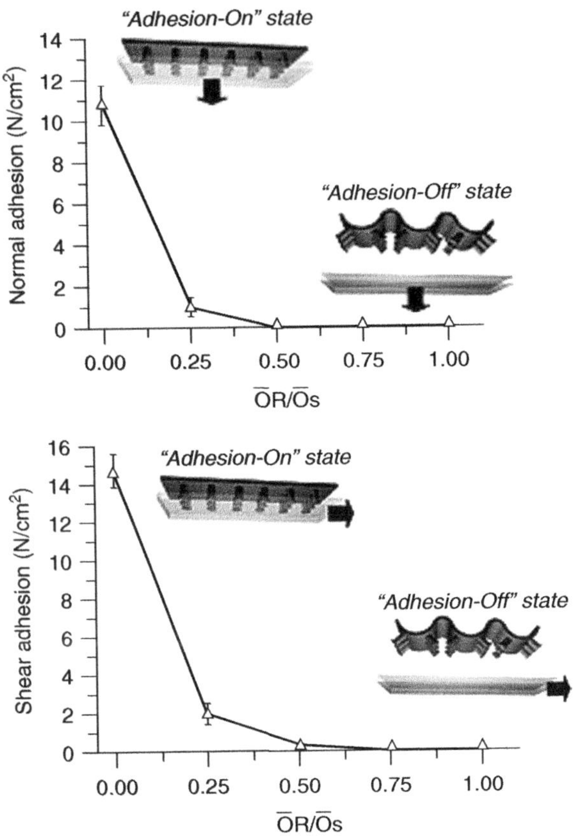

Figure 9.29 Measurement of macroscopic (a) normal and (b) shear adhesion forces with a PDMS adhesive patch having micropillars at different strain levels against the smooth glass surface under a preload of 0.1 N/cm^2. (Adapted with permission. Copyright American Chemical Society: Ref. 127.)

probe tip and nanotubes. The multiple side contacts could drastically enhance the observed adhesion. However, a macroscopic object, instead of penetrating the fiber brush, will largely lie on the tip of the fibers resulting in predominantly point contact. As a result, observed adhesion will be relatively low. For example, Zhao et al.[87] tested the macroscopic adhesion of VA-MWNT arrays with different surface features. They found the maximum adhesive strengths were less than 12 N/cm^2 in the normal direction and about 8 N/cm^2 in the shear direction with glass surface.

Compared with VA-MWNTs, the smaller nanotube diameter, higher packing density, and more perfect, electron rich, $\pi - \pi$ conjugated carbon structure characteristic of their single-walled counterparts could also make VA-SWNT arrays useful for gecko-foot-mimetic dry adhesives. Indeed, Qu and Dai[133] have demonstrated that the VA-SWNT arrays produced by using a combined method of PECVD and fast heating process exhibited a much higher macroscopic adhesive

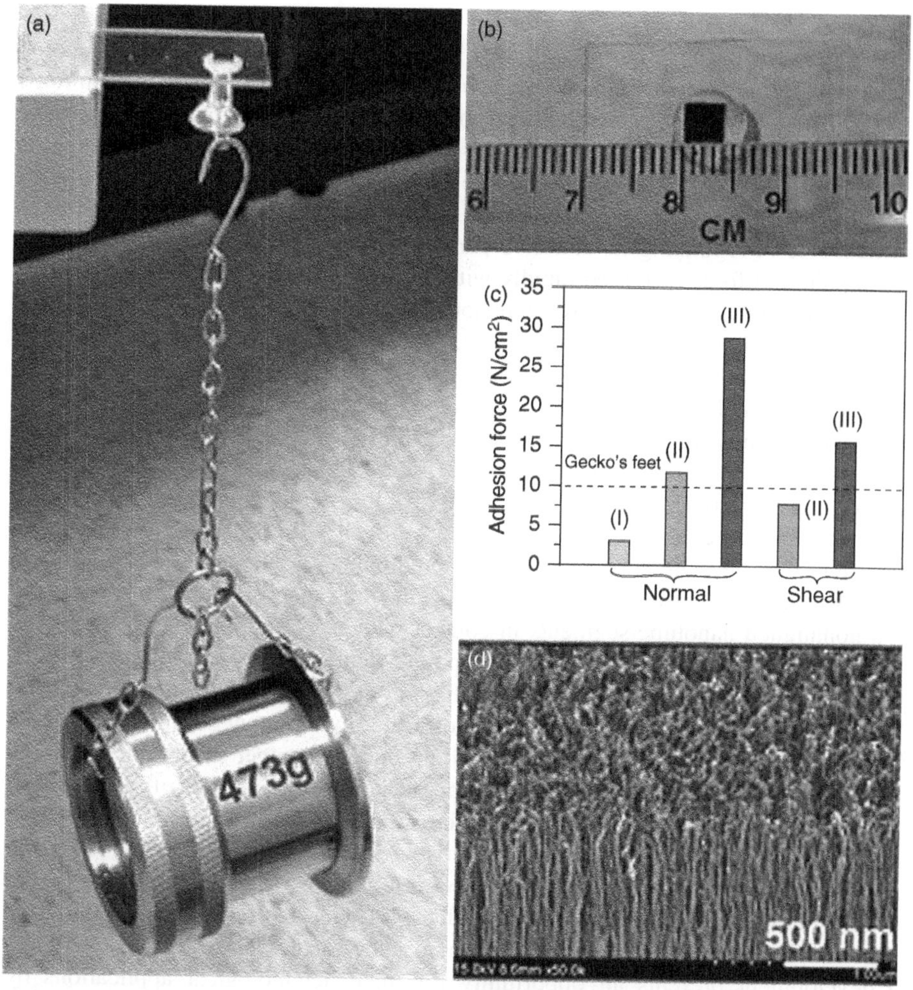

Figure 9.30 (a) A photo showing a stainless steel adapter hanging on a VA-SWNT dry adhesive film (4 mm × 4 mm). (b) VA-SWNT dry adhesive film on a horizontally placed glass surface. (c) A comparison of the maximum achievable adhesion forces for (i) microfabricated polymer hairs,[93] (ii) VA-MWNTs,[87] and (iii) the as-grown aligned VA-SWNTs. The dashed line represents the adhesion force for gecko feet. (d) A side-top view SEM image of the VA-SWNT film under a high magnification. (Adapted with permission. Copyright Wiley-VCH: Ref. 133.)

force of 29 N/cm^2 than that of the VA-MWNTs and natural gecko feet (10 N/cm^2) (Figure 9.30). Furthermore, these VA-SWNT dry adhesives showed excellent thermal resistance due to the unique thermal and electric properties intrinsically associated with SWNTs.

Theoretical studies have indicated that an optimum adhesion could be achieved by the combination of a size reduction and shape optimization with hierarchical structures.[17, 19] Qu et al.[88] have further used the hierarchical structures of VA-MWNT arrays with a straightly aligned body and curly tangled end segments at the top for adhesive applications. These authors reported a high shear adhesion force (~100 N/cm^2) for strong shear binding-on but a much lower normal adhesion force (~10 N/cm^2) for easy normal lifting-off (Figure 9.31).[88] Since the vdW force is mainly responsible for the adhesive force between the nanotube film and the glass slide, the structure at and near the top surface of the VA-MWNT film plays a critical role in regulating its adhesive performance. As shown in Figure 9.31, the shear adhesion force increased rapidly with increasing tube length due to the shear-induced alignment of the nonaligned nanotube top layer to dramatically increase the contact area (Figure 9.31d and Figure 9.32). In contrast, the normal adhesion force is almost insensitive to the nanotube length as a result of point contact (Figure 9.32). During the initial contact, the top nonaligned nanotube segments (Figure 9.32a) adopted randomly distributed "line" contact with the glass substrate (Figure 9.32b). Upon shear adhesion force measurement (Figure 9.32c), the applied shear force caused the nonaligned nanotube segments to align along the shear direction on the glass substrate (Figure 9.32c) and the vertically aligned nanotube trunks to tilt along the shear direction, leading to a predominantly aligned line contact with the glass surface. During the normal adhesion force measurements, however, the top nonaligned nanotube segments in contact with the glass substrate were peeled from the substrate through a "point-by-point" detaching process (Figure 9.32d), requiring a much lower force than that for pulling off the entire nanotube array. These failure modes have also been demonstrated by computer simulations.[88, 136] The line contact detachment is expected to produce a stronger shear adhesion force than the normal adhesion force governed by the point-by-point peel-off detachment. An alternative sticking and detaching of the VA-MWNT on various substrates with different flexibilities and surface characteristics, including glass plates, PTFE film, rough sandpaper, and PET sheet, has also been demonstrated. These findings pave the way to construct aligned CNT dry adhesives with a strong shear adhesion for firm attachment and relatively weak normal adhesion for easy detachment. Such a development presents an opportunity for many technological applications by mimicking the walking of a living gecko.

Ge et al.[86] have developed a synthetic gecko tape by transferring micropatterned CNT arrays onto flexible polymer tape based on the hierarchical structure found on the foot of a gecko lizard (Figures 9.33a–9.33d). The gecko tape can support a shear stress (36 N/cm^2) nearly four times higher than the gecko foot and sticks to a variety of surfaces, including Teflon (Figure 9.33e). Both the micrometer-size setae (replicated by nanotube bundles) and nanometer-size spatulae (individual nanotubes) are necessary to achieve macroscopic shear adhesion and to translate the weak vdW interactions into high shear forces. These authors demonstrated a macroscopic flexible patch that can be used repeatedly with peeling and adhesive properties better than the natural gecko foot.

Subsequently, the same authors further showed that these CNT-based synthetic tapes exhibited self-cleaning capability.[137] In addition to water-cleaning (Figure 9.34), these synthetic tapes could also be cleaned by a contact mechanism

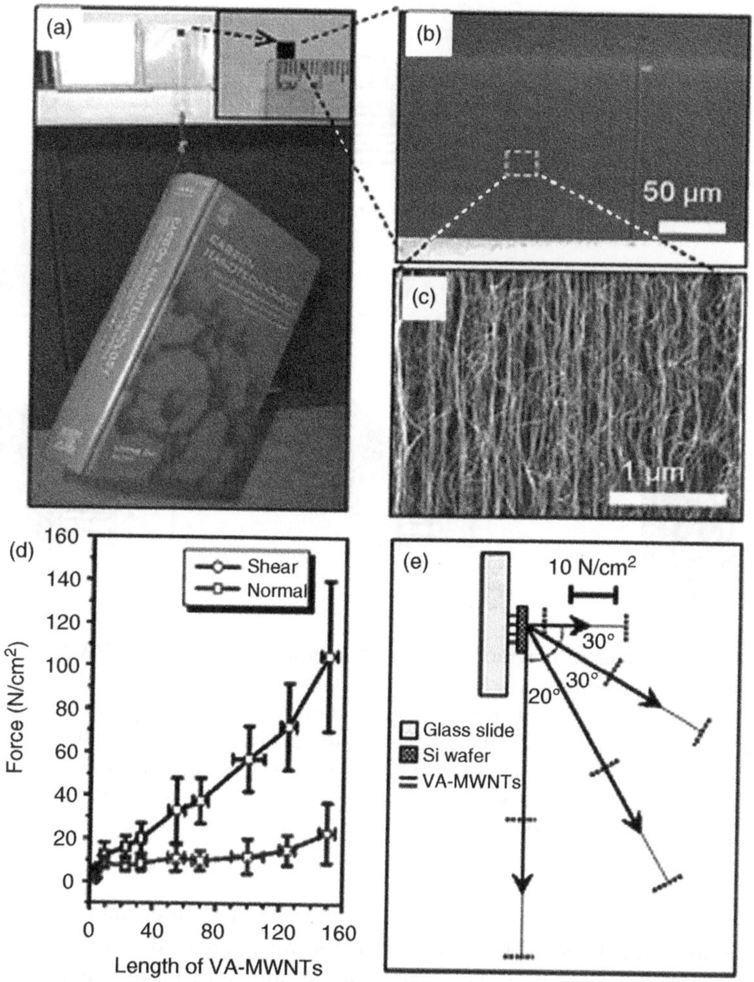

Figure 9.31 (a) A book with a mass of 1480 g is held in shear by a 4 mm × 4 mm carbon nanotube sample. Panels (b) and (c) show SEM images of the VA-MWNT array film under different magnification. (d) The average shear and normal adhesion forces measured from more than 20 samples of the same class, with deviations of length and force indicated by the error bars. (e) The adhesion strength of VA-MWNTs with length 100 ± 10 μm at different pull-away directions. The red arrows represent the average forces measured for more than 20 samples, whereas the two perpendicular blue dot lines define possible deviations of the force measured for different samples of the same class. (Adapted with permission. Copyright AAAS: Ref. 88.)

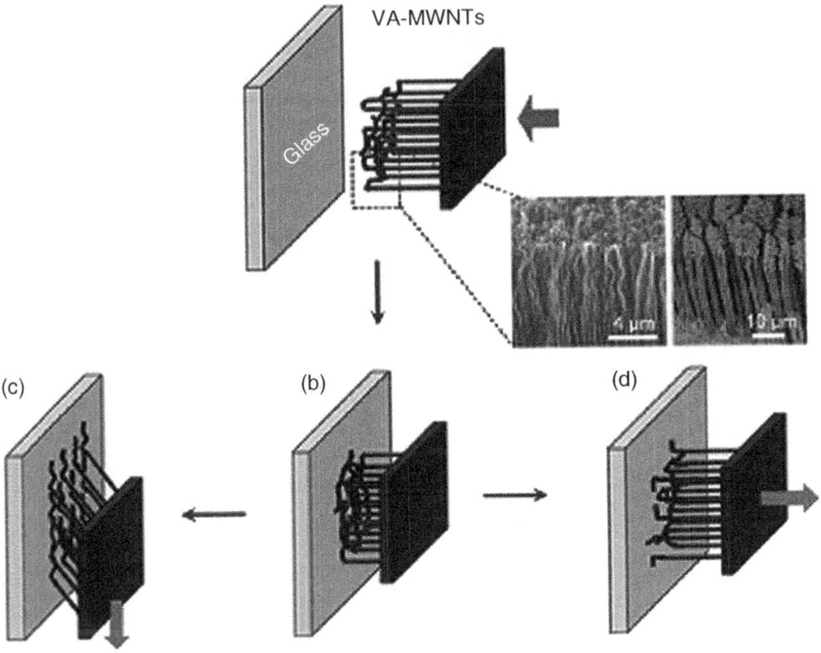

Figure 9.32 Schematic diagrams for the morphological change of VA-MWNT arrays during adhesion measurements. Panels (a) and (b) show the top nonaligned nanotube segments adopted randomly distributed "line" contact with the glass substrate during the initial contact. Panel (c) shows shear adhesion force stretching the nonaligned nanotubes on the substrate to form the line contact. Panel (d) shows normal adhesion force leading to the nonaligned nanotubes point-by-point peel-off from the substrate. (Adapted with permission. Copyright AAAS: Ref. 88.)

similar to that exhibited by the gecko (Figure 9.34). After mechanical cleaning, the shear strength recovers to 90% (and 60% for water-cleaned samples) of the values measured before soiling. The self-cleaning ability of these synthetic tapes for retaining their shear resistance makes them an excellent choice for gecko-inspired adhesives for a wide range of potential applications.[138]

9.5 OUTLOOK

We have summarized recent progress in the development of artificial polymer-based and CNT-based dry adhesives. Through innovative molecular and structural designs, they show excellent adhesion strength (even higher than that of a gecko), smart directional adhesion as well as rough surface adaptability attractive for various potential applications, ranging from robotics to biomedical devices to aerospace systems.[60, 105, 112, 139, 140] Polymer-based methods offer a

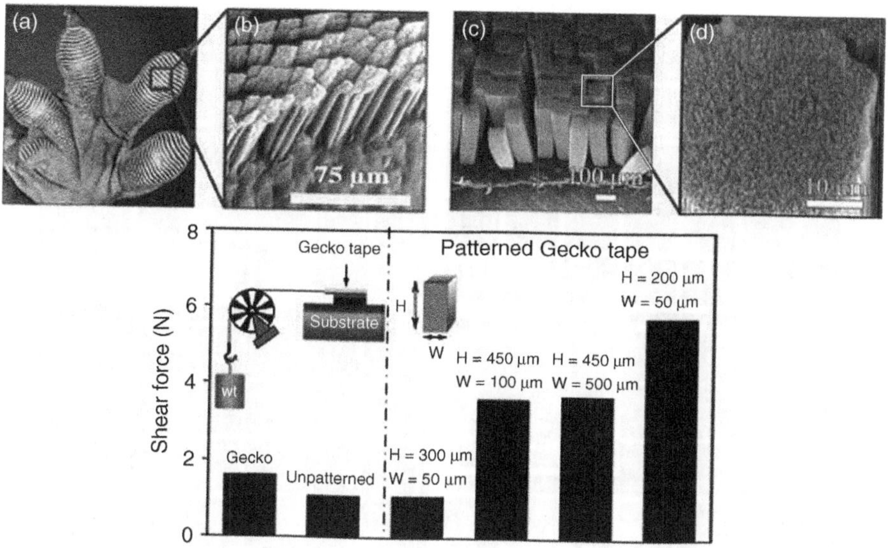

Figure 9.33 Microfabricated aligned MWNT setae and spatulae. (a) Optical picture of gecko foot showing that the setae are arranged in many lobes along the foot. (b) SEM image of natural gecko setae terminating into thousands of smaller spatulae. (c) and (d) Side views and higher-magnification SEM image of the 100 μm setae. (e) Weight supported by synthetic gecko tapes is compared with the force supported by a live gecko for a 0.16 cm^2 area. The shear force supported by unpatterned and patterned gecko tapes on mica substrate. Inset shows the geometry used in the shear measurements. (Adapted with permission. Copyright National Academy of Sciences, U.S.A: Ref. 86.)

cost-effective and scalable approach to fabricating gecko-mimicking nanohairs with tailored geometry (angle, radius, height, shape of tip, and hierarchy) and tunable material properties (modulus, surface energy, etc.).[3] However, the poor low/high-temperature tolerance and low mechanical strength of polymer materials limited the resolution and aspect ratio of polymer nanostructures, leading to a low adhesion strength. On the other hand, CNT-based dry adhesives usually have high level adhesion strengths due to the superior structural features, such as high aspect ratio, extremely small radius (10 nm), and high modulus (1000 GPa), characteristic of CNT materials.[86, 88] Consequently, CNT is considered the most promising material for mimicking gecko feet. However, CNT-based dry adhesives have been much less discussed in the literature than the polymer-based dry adhesives, presumably due to the complicated process for CNT growth and patterning.[85–88, 137] Besides, the high-temperature requirement for growth of carbon nanotubes precludes their direct growth on polymers or most other flexible surfaces. However, the CNT arrays can be transferred to a variety of flexible substrates. This, together with recent advances in the field of nanofabrication techniques, could offer smart artificial dry adhesives for a large variety of applications.

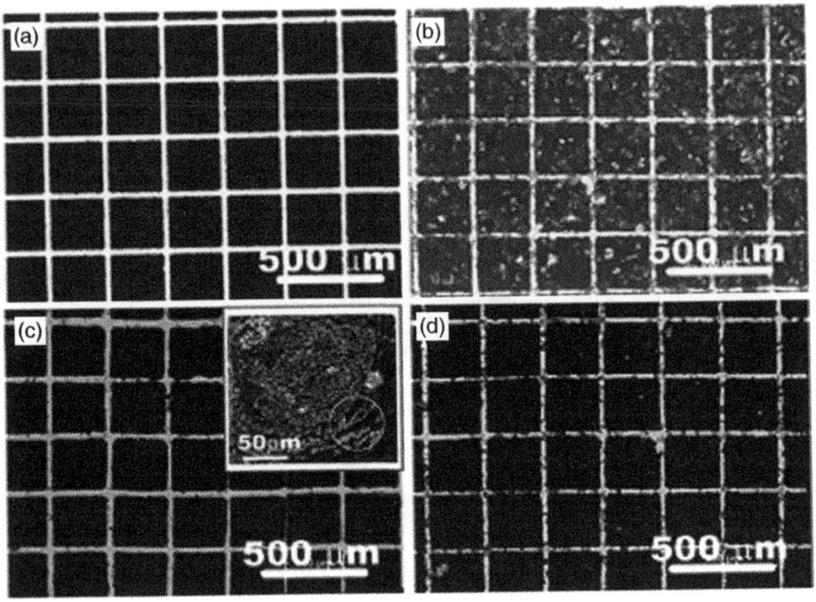

Figure 9.34 Optical images showing dusted samples after different treatments. (a) Image of pristine 250 μm sample. (b) Image of dusted 250 μm sample. (c) Image of dusted sample cleaned with water. The inset shows a higher magnification image of the CNT pillar and the microcracks that are developed during drying are seen inside the region marked by a white circle. (d) Image of dusted sample cleaned by applying vibrations. (Adapted with permission. Copyright American Chemical Society: Ref. 137.)

ACKNOWLEDGMENTS

The authors acknowledge partial support from NSF (CMMI-1047655), AFOSR (FA-9550-12-1-0069 under the Polymer Chemistry Task in the Directorate of Chemistry and Life Sciences; Dr. Charles Lee–program Manager), AFOSR-MURI under the Low Density Materials Program (Dr. Joycelyn Harrison–program manager), NSFC-NSF MWN (NSF-DMR 1106160) SRF for ROCS, SEM (20100732002), NSFC (21004006), and the program for the new century excellent talents in University (NCET-10-0047).

REFERENCES

1. http://classics.mit.edu/Aristotle/history_anim.html.
2. Autumn, K.; Hsieh, T.; Zesch, W.; Chan, W. P.; Fearing, R.; Full, R. J. *Am. Zoologist* **1999**, *39,* 105A.
3. Hoon, E. J.; Kahp, Y. S. *Nano Today* **2009**, *4,* 335.
4. Autumn, K.; Liang, Y. A.; Hsieh, S. T.; Zesch, W.; Chan, W. P.; Kenny, T. W.; Fearing, R.; Full, R. J. *Nature* **2000**, *405,* 681.

5. Maderson, P. F. A. *Nature* **1964**, *203,* 780.
6. Irschick, D. J.; Austin, C. C.; Petren, K.; Fisher, R. N.; Losos, J. B.; Ellers, O. *Biol. J. Linn. Soc.* **1996**, *59,* 21.
7. Stork, N. E. *J. Exp. Biol*. **1980**, *88,* 91.
8. Autumn, K.; Sitti, M.; Liang, Y. A.; Peattie, A. M.; Hansen, W. R.; Sponberg, S.; Kenny, T. W.; Fearing, R.; Israelachvili, J. N.; Full, R. J. *Proc. Natl. Acad. Sci. U.S.A.* **2002**, *99,* 12252.
9. Federle, W.; Baumgartner, W.; Holldobler, B. *J. Exp. Biol*. **2004**, *207,* 67.
10. Huber, G.; Mantz, H.; Spolenak, R. *Proc. Natl. Acad. Sci. U.S.A.* **2005**, *102,* 16293.
11. Tian, Y.; Pesika, N.; Zeng, H.; Rosenberg, K.; Zhao, B.; McGuiggan, P.; Autumn, K.; Israelachvili, J. *Proc. Natl. Acad. Sci. U.S.A.* **2006**, *103,* 19320.
12. Johnson, K. L.; Kendall, K.; Roberts, A. D. *Proc. R. Soc. London A* **1971**, *324,* 301.
13. Griffith, A. A. *Philos. Trans. R. Soc. London A* **1921**, *221,* 163.
14. Hui, C. Y.; Baney, J. M.; Kramer, E. J. *Langmuir* **1998**, *14,* 6570.
15. Haiat, G.; Huy, M. C. P.; Barthel, E. *J. Mech. Phys. Solids* **2003**, *51,* 69.
16. Kim, K. S.; McMeeking, R. M.; Johnson, K. L. *J. Mech. Phys. Solids* **1998**, *46,* 243.
17. Yao, H.; Gao, H. *J. Mech. Phys. Solids* **2006**, *54,* 1120.
18. Arzt, E.; Gorb, S.; Spolenak, R. *Proc. Natl. Acad. Sci. U.S.A.* **2003**, *100,* 10603.
19. Gao, H.; Yao, H. *Proc. Natl. Acad. Sci. U.S.A.* **2004**, *101,* 7851.
20. Huber, G.; Gorb, S.; Spolenak, R.; Arzt, E. *Biol. Lett*. **2005**, *1,* 2.
21. Persson, B. N. J. *J. Chem. Phys*. **2003**, *118,* 7614.
22. Spolenak, R.; Gorb, S.; Gao, H.; Arzt, E. *Proc. R. Soc. A* **2005**, *461,* 305.
23. Kamperman, M.; Kroner, E.; del Campo, A.; McMeeking, R. M.; Arzt, E. *Adv. Eng. Mater*. **2010**, *12,* 335.
24. Kroner, E.; Maboudian, R.; Arzt, E. *Adv. Eng. Mater*. **2010**, *12,* 398.
25. Sauer, R. A. *Comp. Meth. Biomech. Biomed. Eng*. **2009**, *12,* 627.
26. Sylves, K.; Maute, K.; Dunn, M. L. *Struct. Multidisc. Optim*. **2009**, *38,* 455.
27. Zhao, B. X.; Pesika, N.; Zeng, H. B.; Wei, Z.; Chen, Y. F.; Autumn, K.; Turner, K.; Israelachvili, J. *J. Phys. Chem. B* **2009**, *113,* 3615.
28. Yamaguchi, T.; Gravish, N.; Autumn, K.; Creton, C. *J. Phys. Chem. B* **2009**, *113,* 3622.
29. Long, R.; Hui, C. Y. *Proc. R. Soc. A* **2009**, *465,* 961.
30. Chen, B.; Wu, P. D.; Gao, H. *Proc. R. Soc. A* **2008**, *464,* 1639.
31. Yamaguchi, T.; Gravish, N.; Autumn, K,; Creton, C. *J. Phys. Chem. B* **2009**, *113,* 3622.
32. Zhao, B. X.; Pesika, N.; Rosenberg, K.; Tian, Y.; Zeng, H. B.; McGuiggan, P.; Autumn, K.; Israelachvili, J. *Langmuir* **2008**, *24,* 1517.
33. Zhang, H.; Guo, D. J.; Dai, Z. D. *Chinese Sci. Bull*. **2010**, *55,* 1843.
34. Tian, Y.; Pesika, N.; Zeng, H. B. *Proc. Natl. Acad. Sci. U.S.A.* **2006**, *103,* 19320.
35. Michael, P. M.; Burak, A.; Sitti, M. *Small* **2009**, *5,* 170.

36. Lee, J.; Majidi, C.; Schubert, B.; Ronald, S. F. *J. R. Soc. Interface* **2008**, *5,* 835.
37. Autumn, K.; Dittmore, A.; Santos, D.; *J. Exp. Biol*. **2006**, *209,* 3569.
38. Schubert, B.; Lee, J.; Majidi, C.; *J. R. Soc. Interface* **2008**, *5,* 845.
39. Autumn, K.; Peattie, A. N. *Integr. Comp. Biol*. **2002**, *42,* 1081.
40. Gao, H. J.; Wang, X.; Yao, H. M.; Gorb, S.; Arzt, E. *Mech. Mater*. **2005**, *37,* 275.
41. Autumn, K.; Gravish, N. *Philos. Trans. R. Soc. A: Math. Phys. Eng. Sci*. **2008**, *366,* 1575.
42. Hansen, W.; Autumn, K. *Proc. Natl. Acad. Sci. U.S.A.* **2005**, *102,* 385.
43. Autumn, K.; Majidi, C.; Groff, R. E.; Dittmore, A.; Fearing, R. *J. Exp. Biol*. **2006**, *209,* 3558.
44. Chui, B. W.; Kenny, T. W.; Mamin, H. J.; Terris, B. D.; Rugar, D. *Appl. Phys. Lett*. **1998**, *72,* 1388.
45. Autumn, K. *MRS Bull*. **2007**, *32,* 473.
46. Greiner, C.; del Campo, A.; Arzt, E. *Langmuir* **2007**, *23,* 3495.
47. Chung, J. Y.; Chaudhury, M. K. *J. R. Soc. Interface* **2005**, *2,* 55.
48. Crosby, A. J.; Hageman, M.; Duncan, A. *Langmuir* **2005**, *21,* 11738.
49. Metin, S.; Ronalds, F. *J. Adhes. Sci. Technol*. **2003**, *17,* 1055.
50. del Campo, A.; Greiner, C.; Alvarez, I.; Arzt, E. *Adv. Mater*. **2007**, *19,* 1973.
51. del Campo, A.; Alvarez, I.; Filipe, S.; Wilhelm, M. *Adv. Funct. Mater*. **2007**, *17,* 3590.
52. Murphy, M. P.; Aksak, B.; Sitti, M. *J. Adhes. Sci. Technol*. **2007**, *21,* 1281.
53. Santos, D.; Spenko, M.; Parness, A.; Kim, S.; Cutkosky, M. *J. Adhesion Sci. Technol*. **2007**, *21,* 1317.
54. Kim, T.; Jeong, H. E.; Suh, K. Y.; Lee, H. H. *Adv. Mater*. **2009**, *21,* 2276.
55. Lee, J.; Fearing, R. S.; Komvopoulos, K. *Appl. Phys. Lett*. **2008**, *93,* 191910.
56. Aksak, B.; Murphy, M. P.; Sitti, M. *Langmuir* **2007**, *23,* 3322.
57. Bhushan, B.; Peressadko, A. G.; Kim, T. W. *J. Adhes. Sci. Technol*. **2006**, *20,* 1475.
58. Persson, B. N. J. *J. Adhes. Sci.Technol*. **2007**, *21,* 1145.
59. Kim, S.; Sitti, M. *Appl. Phys. Lett*. **2006**, *89,* 261911.
60. Jeong, H. E.; Lee, J. K.; Kim, H. N.; Moon, S. H.; Suh, K. Y. *Proc. Natl. Acad. Sci*. **2009**, *106,* 5639.
61. Kim, S.; Aksak, B.; Sitti, M. *Appl. Phys. Lett*. **2007**, *91,* 221913.
62. del Campo, A.; Greiner, C.; Arzt, E. *Langmuir* **2007**, *23,* 10235.
63. Lamblet, M.; Verneuil, E.; Vilmin, T.; Buguin, A.; Solberzan, P.; Leger, L. *Langmuir* **2007**, *23,* 6966.
64. Glassmaker, N. J.; Jagota, A.; Hui, C. Y.; Kim, J. *J. R. Soc. Interface* **2004**, *1,* 23.
65. Peressadko, A.; Gorb, S. N. *J. Adhes*. **2004**, *80,* 247.
66. del Campo, A.; Greiner, C. *J. Micromech. Microeng*. **2007**, *17,* R81.
67. Davies, J.; Haq, S.; Hawke, T.; Sargent, J. P. *Int. J. Adhes. Adhes*. **2009**, *29,* 380.
68. Boesel, L. F.; Greiner, C.; Arzt, E.; del Campo, A. *Adv. Mater*. **2010**, *22,* 2125.
69. Northen, M. T.; Turner, K. L. *Nanotechnology* **2005**, *16,* 1159.

70. Kim, D. S.; Lee, S. H.; Ahn, C. H.; Lee, J. Y.; Kwon, T. H. *Lab Chip* **2006**, *6*, 794.
71. Kim, S.; Sitti, M.; Jang, J. H.; Thomas, E. L. *IEEE Nano* **2008**, 780.
72. Kim, T.; Pang, C.; Suh, K. Y. *Langmuir* **2009**, *25*, 8879.
73. Kim, S.; Cheung, E.; Sitti, M. *Langmuir* **2009**, *25*, 7196.
74. Kim, S.; Sitti, M.; Xie, T.; Xiao, X. C. *Soft Matter* **2009**, *5*, 3689.
75. Sameoto, D.; Menon, C. *J. Micromech. Microeng*. **2009**, *19*, 115002.
76. Parness, A.; Soto, D.; Esparza, N.; Gravish, N.; Wilkinson, M.; Autumn, K.; Cutkosky, M. *J. R. Soc. Interface* **2009**, *6*, 1223.
77. Sameoto, D.; Menon, C. *J. Micromech. Microeng*. **2009**, *19*, 115026.
78. Greiner, C.; Azrt, E.; del Campo, A. *Adv. Mater*. **2009**, *21*, 479.
79. Aksak, B.; Murphy, M. P.; Sitti, M. *IEEE Int. Conf. Robotics Automation* **2008**, 3058–3063.
80. Cheung, E.; Sitti, M. *Langmuir* **2009**, *25*, 6613.
81. Glass, P.; Chung, H.; Washburn, N.; Sitti, M. *Langmuir* **2009**, *25*, 6607.
82. Jeong, H. E.; Lee, S. H.; Kim, P.; Suh, K. Y. *Colloids Surf. A* **2008**, *313/314*, 359.
83. Sitti, M.; Cusick, B.; Aksak, B.; Nese, A.; Lee, H. I.; Dong, H.; Kowalewski, T.; Matyjaszewski, K. *ACS Appl. Mater. Interfaces* **2009**, *1*, 2277.
84. Lee, H.; Lee, B. P.; Messersmith, P. B. *Nature* **2007**, *448*, 338.
85. Yurdumakan, B.; Raravikar, N. R.; Ajayan, P. M.; Dhinojwala, A. *Chem. Commun*. **2005**, 3799.
86. Ge, L.; Sethi, S.; Ci, L.; Ajayan, P. M.; Dhinojwala, A. *Proc. Natl. Acad. Sci. U.S.A.* **2007**, *104*, 10792.
87. Zhao, Y.; Tong, T.; Delzeit, L.; Kashani, A.; Meyyappan, M.; Majumdar, A. *J. Vac. Sci. Technol. B* **2006**, *24*, 331.
88. Qu, L. T.; Dai, L. M.; Stone, M.; Xia, Z. H.; Wang, Z. L. *Science* **2008**, *322*, 238.
89. Maeno, Y.; Nakayama, Y. *Appl. Phys. Lett*. **2009**, 94, 012103.
90. Sitti, M.; Fearing, R. S. *Proceedings of the IEEE Nanotechnology Conference. Washington* **2002**, 137.
91. Roca-Cusachs, P.; Rico, F.; Martinez, E.; Toset, J.; Farre, R.; Navajas, D. *Langmuir* **2005**, *21*, 5542.
92. Hui, C. Y.; Jagota, A.; Lin, Y. Y.; Kramer, E. J. *Langmuir* **2002**, *18*, 1394.
93. Geim, A. K.; Dubonos, S. V.; Grigorieva, I. V.; Novoselov, K. S.; Zhukov, A. A.; Shapoval, A. A. *Nat. Mater*. **2003**, *2*, 461.
94. Long, R.; Hui, C. Y.; Kim, S.; Sitti, M. *J. Appl. Phys*. **2008**, *104*, 044301.
95. Jagota, A.; Bennison, S. J. *Integr. Comp. Biol*. **2002**, *42*, 1140.
96. Jeong, H. E.; Lee, S. H.; Kim, P.; Suh, K. Y. *Nano Lett*. **2006**, *6*, 1508.
97. Jeong, H. E.; Kim, T.; Kang, T. J.; Yoon, H.; Char, K.; Suh, K. Y.; Tahk, D. *Nano Today* **2009**, *4*, 385.
98. Kwak, M. K.; Jeong, H. E.; Kim, T.; Yoon, H.; Suh, K. Y. *Soft Matter* **2010**, *6*, 1849.
99. Jeong, H. E.; Lee, J. K.; Kwak, M. K.; Moon, S. H.; Suh, K. Y. *Appl. Phys. Lett*. **2010**, *96*, 043704.

100. Spolenak, R.; Gorb, S.; Gao, H. J.; Arzt, E. *Proc. R. Soc. A* **2005**, *461,* 305.
101. Reddy, S.; Arzt, E.; del Campo, A. *Adv. Mater*. **2007**, *19,* 3833.
102. Schubert, B.; Majidi, C.; Groff, R. E.; Baek, S.; Bush, B.; Maboudian, R.; Fearing, R. S. *J. Adhes. Sci. Technol*. **2007**, *21,* 1297.
103. Sameoto, D.; Menon, C. *Smart Mater. Struct*. **2010**, *19,* 103001.
104. Yao, H., Rocca, G. D.; Guduru, P.; Gao, H. *J. R. Soc. Interface* **2008**, *5,* 723.
105. Kim, S.; Spenko, M.; Trujillo, S.; Heyneman, B.; Santos, D.; Cutkosky, M. R. *IEEE Trans. Rob*. **2008**, *24,* 65.
106. Kim, T. W.; Bhushan, B. *Ultramicroscopy* **2007**, *107,* 902.
107. Jeong, H. E.; Lee, S. H.; Kim, J. K.; Suh, K. Y. *Langmuir* **2006**, *22,* 1640.
108. Jeong, H. E.; Lee, S. H.; Kim, P.; Suh, K. Y. *Colloid Surf. A* **2008**, *359,* 313.
109. Kustandi, T. S.; Samper, V. D.; Ng, W. S.; Chong, A. S.; Gao, H. *J. Micromech. Microeng*. **2007**, *17,* N75.
110. Murphy, M. P.; Kim, S.; Sitti, M. *ACS Appl. Mater. Interfaces* **2009**, *1,* 849.
111. Lee, H.; Dellatore, S. M.; Miller, W. M.; Messersmith, P. B. *Science* **2007**, *318,* 426.
112. Mahdavi, A. *Proc. Natl. Acad. Sci. U.S.A*. **2008**, *105,* 2307.
113. Gorb, S.; Varenberg, M.; Peressadko, A.; Tuma, J. *J. R. Soc. Interface* **2007**, *4,* 271.
114. Northen, M. T.; Turner, K. L. *Sens. Actuat. A* **2006**, *130/131,* 583.
115. Sun, W., Neuzil, P., Kustandi, T. S., Oh, S.; Samper, V. D. *Biophys. J*. **2005**, *89,* L14.
116. Glass, P.; Chung, H.; Washburn, N. R.; Sitti, M. *Langmuir* **2010**, *26,* 17357.
117. Ko, H.; Zhang, Z. X.; Chueh, Y. L.; Ho, J. C.; Lee, J.; Fearing, R. S.; Javey, A. *Adv. Funct. Mater*. **2009**, *19,* 3098.
118. Lu, G. W.; Hong, W. J.; Tong, L.; Bai, H.; Wei, Y.; Shi, G. Q. *ACS Nano* **2008**, *2,* 2342.
119. Lee, J.; Fearing, R. *Langmuir* **2008**, *24,* 10587.
120. Forero, M.; Thomas, W. E.; Bland, C.; Nilsson, L. M.; Sokurenko, E. V.; Vogel, V. *Nano Lett*. **2004**, *4,* 1593.
121. Gates, B. D.; Xu, Q. B.; Love, J. C.; Wolfe, D. B.; Whitesides, G. M. *Annu. Rev. Mater. Res*. **2004**, *34,* 339.
122. Alexander, C.; Shakesheff, K. M. *Adv. Mater*. **2006**, *18,* 3321.
123. Lendlein, A.; Kelch, S. *Angew. Chem. Int. Ed*. **2002**, *41,* 2034.
124. Xie, T.; Xiao, X. C. *Chem. Mater*. **2008**, *20,* 2866.
125. Wang, R. M.; Xie, T. *Langmuir* **2010**, *26,* 2999.
126. Northen, M. T.; Greiner, C.; Arzt, E.; Turner, K. L. *Adv. Mater*. **2008**, *20,* 3905.
127. Jeong, H. E.; Kwak, M. K.; Suh, K. Y. *Langmuir* **2010**, *26,* 2223.
128. Lin, P. C.; Vajpayee, S.; Jagota, A.; Hui, C. Y.; Yang, S. *Soft Matter* **2008**, *4,* 1830.
129. Nadermann, N.; Ning, J.; Jagota, A.; Hui, C. Y. *Langmuir* **2010**, *26,* 15464.
130. Dresselhaus, M. S.; Dresselhaus, G.; Jorio, A. *Annu. Rev. Mater. Res*. **2004**, *34,* 247.

131. Kim, P.; Shi, L.; Majumdar, A.; McEuen, P. L. *Phys. Rev. Lett*. **2001**, *87,* 215502.

132. Dai, L. M. *Carbon Nanotechnology, Recent Developments in Chemistry, Physics, Materials Science and Device Applications*, Elsevier, Amsterdam, 2006.

133. Qu, L. T.; Dai, L. M. *Adv. Mater*. **2007**, *19,* 3844.

134. Lee, J.; Schubert, B.; Majidi, C.; Fearing, R. S. *J. R. Soc. Interface* **2008**, *5,* 835.

135. Majidi, C. S.; Groff, R. E.; Fearing, R. S. *J. Appl. Phys*. **2005**, *98,* 103521.

136. Hu, S. H.; Jiang, H. D.; Xia, Z. H.; Gao, X. S. *ACS Appl. Mater. Interfaces* **2010**, *2,* 2570.

137. Sethi, S.; Ge, L.; Ci, L. J.; Ajayan, P. M.; Dhinojwala, A. *Nano Lett*. **2008**, *8,* 822.

138. Ge, L.; Ci, L.; Goyal, A.; Shi, R.; Mahadevan, L.; Ajayan, P. M.; Dhinojwala, A. *Nano Lett*. **2010**, *10,* 4509.

139. Ko, H.; Zhang, Z. X.; Ho, J. C.; Takei, K.; Kapadia, R.; Chueh, Y. L.; Cao, W. Z.; Cruden, B. A.; Javey, A. *Small* **2010**, *6,* 22.

140. Robert, B. *Assembly Automation* **2008**, *28/4,* 282.

CHAPTER 10

Bioinspired Surfaces II: Bioinspired Photonic Materials

CUN ZHU and ZHONG-ZE GU

State Key Laboratory of Bioelectronics, Southeast University, Nanjing, Peoples Republic of China 210096

10.1 STRUCTURAL COLOR IN NATURE: FROM PHENOMENA TO ORIGIN

Nature provides inexhaustible wealth to humankind. The controlled propagation of light in Nature constitutes an astonishing diversity in coloration that leads to a colorful world. The coloration in Nature, especially that exhibited by active organisms, presents amazing characteristics and has drawn great interest for a long period of time. Many living creatures make use of coloration to adapt to the surrounding environment. The colorations are usually used as signals to warn their enemies through mimicry or to mislead their natural enemies through camouflage, or to transmit information (Figure 10.1).

During the long eons of evolution, animals have created many mechanisms to develop their body color. There are, in general, two types of color: (1) pigmental color and (2) structural color. Pigmental color mainly relies on the selective absorption of light by chemical chromophores to achieve an exhibition of color. Structural color is usually generated by geometric structures, which rely on control of the transportation of visible light using reflection units located on their skin or surface. Compared with pigmental color, structural color is much more efficient in energy consumption and also in the use of light. Generally, coloration in Nature mostly derives from manipulating the flow of light. But the most vivid and brightest colors in Nature come from "structural color"; that is, it arises from surfaces with periodic structures in micro- and nanoscales.

Photonics associated with periodic structures that contribute to the bright iridescent color produced by fish, birds, and insects has attracted great scientific interest for over a century. Considerable research has been conducted on the study of the mechanism and the imitation of these structures.[1–6] Generally, most of the

Bioinspiration and Biomimicry in Chemistry: Reverse-Engineering Nature, First Edition.
Edited by Gerhard F. Swiegers.
© 2012 John Wiley & Sons, Inc. Published 2012 by John Wiley & Sons, Inc.

Figure 10.1 Various colorations provided by living creatures in Nature. (a) Blue *Morpho* butterfly. (b) Peacock. (c) Longhorn beetles *Tmesisternus isabellae*. (Reproduced with permission from Ref. 81. Copyright © 2009, the Optical Society of America.) (d) Myxomycetes *Diachea leucopoda*. (Reproduced with permission. Copyright the Optical Society of America: Ref. 3.) (See insert for color representation of this figure.)

structural color in Nature is believed to originate from thin film or multilayer interference, the diffraction or scattering effect of light, or from combinations thereof.[6] One typical example is the *Morpho* butterfly, a well-known iridescent insect living in Central and South America. Comprehensive investigations have been devoted to a structural analysis of *Morpho* butterfly wings and the theoretical explanation of their iridescent color, with the help of advanced characterization technique.[7–11] When observed with a microscope, both sides of the transparent epithelial membrane of their wings consist of a large number of overlapping scales having two different scale types, known as *ground scale* and *cover scale* (Figure 10.2). The ground scale is mainly formed by flat rectangular lamellae of size ~50–100 mm in width and 150–200 mm in length. These are arranged in order and attached to the transparent epithelial membrane. These scales usually form network structures involving rows aligned parallel to each other and crosslinked by the cover

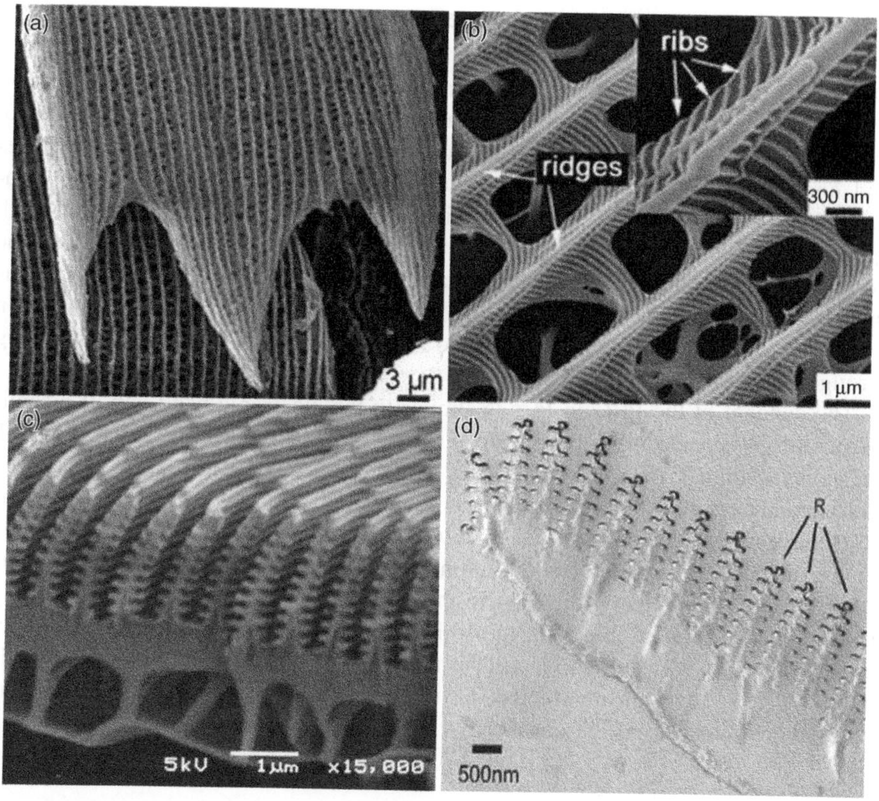

Figure 10.2 (a)–(c) SEM images and (d) TEM image of the wings of *Morpho* butterflies. They are composed of two scales with different size, known as the ground scale and the cover scale. Both these two scales have many equidistant ridges and are crosslinked by ribs. (Panels (a) and (b) reproduced with permission. Copyright Wiley-VCH: Ref. 8. Panel (c) reproduced with permission. Copyright The Royal Society: Ref. 10. Panel (d) reproduced with permission. Copyright Nature Publishing Group: Ref. 7.)

scale on their surface. The cover scale is composed of periodical plates with a submicron dimension, called cross ribs. The spacing between the main ridges and cross ribs lies in the range of 0.5 to ~5.0 mm, depending on the species of the butterfly. Both the ground scale and cover scale have many equidistant ridges, but the ground scale plays a key role for the origination of structural color in most *Morpho* species. The brilliant blue color of the *Morpho* butterfly mainly originates from these periodical, hierarchical structures on their wings, which contribute to the diffraction and interference effect of light between air and cuticle, rather than pigments or dyes, which selectively absorb the visible light. Nanostructures of this type, that Nature has created, offer plenty of inspiration for the design of advanced biomimetic materials and optical devices.

10.2 BIOINSPIRED PHOTONIC MATERIALS

During the last decades, much effort has been directed toward the construction of devices that can confine light with particular wavelengths and that can control the direction of light propagation, based on the study of surface nanostructures and control mechanisms of the *Morpho* butterfly and other similar living creatures. Photonic crystals (PCs) are one kind of well-known artificial material with spatially ordered lattices that exhibit brilliant structural colors.[12, 13] Due to the periodic arrangement of the dielectric materials, a remarkable property, known as the photonic band gap (PBG), appears. This property leads to light with certain wavelengths or frequencies located in the PBG being prohibited from propagating through the PCs (Figure 10.3).

The concept of the PBG in PCs is similar to the electronic band gap of a semiconductor. When electrons propagate through semiconductors, they interact with the atoms arranged in an ordered lattice, which results in the formation of allowed and forbidden energy states. A similar situation exists in the PCs. When photons propagate through the PCs, they interact with the ordered periodic structures formed by the dielectric materials. Thus, photons with certain frequencies that fall in the PBG become prohibited by the PCs. In this way, the flow of light can conveniently be controlled. When visible light transmits through a PC, light with certain wavelengths matched with the PBG will be reflected by the periodic structure, while others will pass through unaffected. Distinct structural colors associated with the reflected frequencies will be observed.

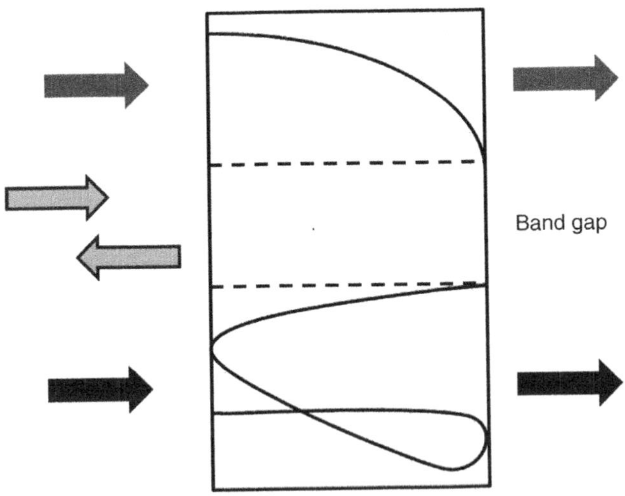

Figure 10.3 Schematic illustration of the photonic band gap in photonic crystals. When the visible light transmits through the PCs, light with specific frequency that falls in the PBG will be reflected by the periodical structures, while others pass through it without being affected.

Another interesting phenomenon in PCs is the property of viewing angle dependence, which means that different structural colors will be observed when viewing at different orientations. This phenomenon is caused by the diversity of lattice constants formed by the PC at different viewing angles. PCs can mainly be divided into three types according to the arrangement of the dielectric periodic units: one-dimensional (1D) PCs, two-dimensional (2D) PCs, and three-dimensional (3D) PCs.

10.2.1 The Fabrication of Photonic Materials

Colloidal crystals composed of monodisperse silica nanospheres, polymer nanospheres, or composite nanospheres, with the diameter ranging from several hundred nanometers to several micrometers, are one kind of promising materials for the fabrication of photonic materials. A variety of methods for fabricating photonic materials have been reported so far,[14–25] including *top–down* micromachining,[16, 17] *bottom–up* self-assembly,[18–20] holographic lithography,[21–23] laser-guided stereolithography,[24] and electrophoretic deposition.[25] As only the 3D PCs can meet the requirements for controlling the flow of light in all directions, the fabrication of 3D PCs has been extensively investigated, especially those with the PBG in the visible region.

Sedimentation is considered one of the most convenient and versatile ways to generate 3D crystalline lattice.[26–29] This technique incorporates several complex processes including Brownian motion, gravitational settling, and crystallization. The control of several parameters is necessary for, and beneficial in, the fabrication process. These parameters include the settling velocity and the concentration, size, and density of the colloidal crystals. Despite the simplicity of this method, it is hard to control the morphology of the resulting structures because it involves the formation of cracks and lattice mismatch in the epitaxial growth process between colloidal spheres. The number of assembled layers is also not easily controllable. The sedimentation process should, moreover, be slow enough for the crystallization process to form a well-ordered 3D lattice. It may take several days, even several months, to obtain a suitable result, which means it is time consuming.

To overcome these limitations, a much quicker "lifting" method has been developed for the construction of 3D crystalline lattice in recent years.[30] This approach was derived from the "vertical deposition" technique reported by Colvin and coworkers[31] In this process, the substrate is first immersed in a suspension of colloidal crystals and then lifted at a constant speed. During the lifting process, the self-assembly of colloidal crystals takes place at the air–liquid interface. The film produced by this method has a close-packed face-centered-cubic (fcc) lattice with the {111} facet parallel to the substrate. Due to the constant lifting speed, the thickness of film obtained is relatively uniform; this can be conveniently modulated from single layer to multiple layers by precisely controlling the lifting speed and the concentration of the colloidal solution. A schematic illustration of the lifting method is shown in Figure 10.4. As can be seen, colloidal crystal films with well-controlled and highly ordered 3D periodical structures can be obtained by

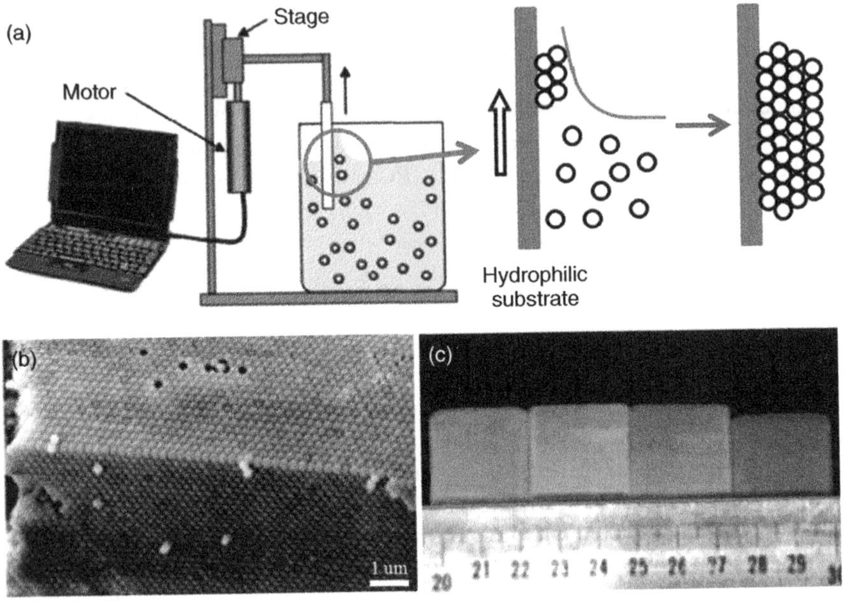

Figure 10.4 (a) Schematic illustration of the lifting method. During the lifting process, the self-assembly of colloidal crystals takes place at the air–liquid interface, due to the capillary force and the evaporation of solvent. The lifting speed can be precisely controlled by the computer. (b) SEM images of the obtained highly ordered PC films. (c) PC films with various distinct brilliant structural color fabricated by using colloidal crystals with different diameters. (Reproduced with permission. Copyright the American Chemical Society: Ref. 31.) (See insert for color representation of this figure.)

this technique. Furthermore, the band gap and structural color exhibited may be controlled by using colloidal crystals with different diameters.

10.2.2 The Design and Application of Photonic Materials

As described above, photonic crystals have the ability to control the propagation of light, whose signal can be captured and monitored through the spectrum. Taking advantage of their optical properties, significant achievements have been made in their practical application over the past several years.[32–42] This section presents an overview of the development of photonic materials. We focus on the design and application of photonic materials in the fields of biomimetic materials, sensors, optical devices, and display devices.

10.2.2.1 Waveguide Applications In order to rigorously control the propagation of light with specific frequencies, photonic materials have become the focus of much scientific research. However, although monodisperse colloidal crystals naturally tend to form ordered 3D crystalline lattices during the self-assembly process, defects like vacancies, cracks, or boundaries are difficult to avoid. In fact, they always accompany the self-assembly process. Thanks to these defects, various different functionalities may be endowed upon PCs. For example, a point defect in a photonic material that acts as a cavity has the ability to trap photons; a line defect can control and direct the propagation route of photons. These defects disrupt the periodicity of the crystalline lattice and create specific optical states within the band gap. Therefore, light coupling to these states can be localized within the defect regions and propagate under control. When light transmits through photonic materials, the propagation of those having frequency within the PBG can be guided by the defects since the ordered structures around the defects block the escape of this light. Light of other frequencies remains undisturbed and travels through the material normally. A schematic illustration is shown in Figure 10.5.

Designing Defects in Photonic Materials. Generally, defects can be divided into two types: *intrinsic defects* and *extrinsic defects*. Intrinsic defects always occur during the self-assembly process of PCs. Extrinsic defects are artificially introduced during or after the formation of PCs. By introducing defects into PCs, functional photonic devices such as optical waveguides, switches, and microlaser devices have been developed.[43–45]

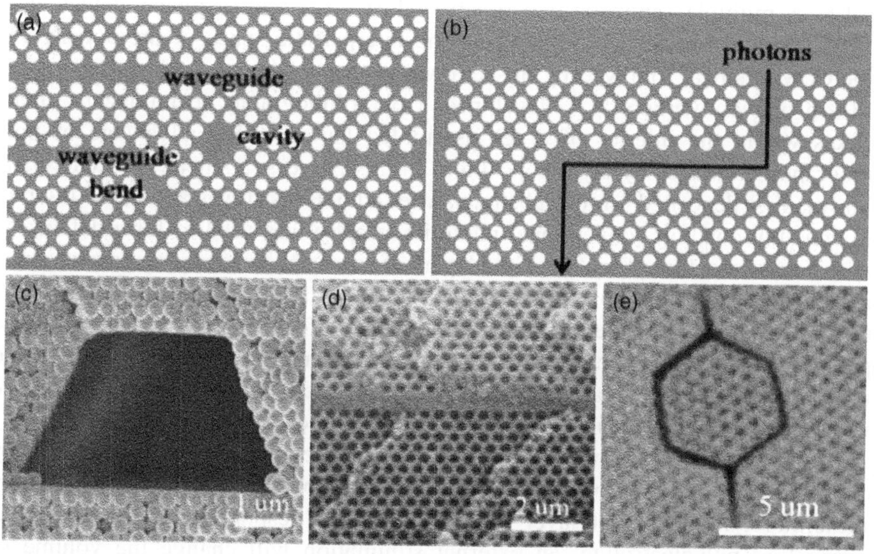

Figure 10.5 (a) Schematic illustration of the different kinds of defect that are induced in PCs for the fabrication of waveguide materials and (b) the propagation of photons with the frequency located in the PBG in the photonic materials with defects. (c) SEM image of trapezoid-shaped defects embedded in the silica colloidal crystals (Reproduced with permission. Copyright Wiley-VCH: Ref. 50.) (d) Planar defects embedded in colloidal crystals. (Reproduced with permission. Copyright Wiley-VCH: Ref. 52.) (e) Two direction bend waveguide fabricated through lithography in the PCs. (Reproduced with permission. Copyright Wiley-VCH: Ref. 45.)

A large number of methods have been developed for the introduction of defects into PCs in recent years, including lithography, electrochemical etching, direct writing, colloidal assembling, and layer-by-layer deposition.[46] Braun and co-workers[47, 48] reported a laser-induced photopolymerization method to form microscale line defects in a silica-based crystalline lattice. They filled the interparticle space among the self-assembled silica spheres with a photosensitive monomer. A laser was then used to scan the specific region, causing the photosensitive monomer that had been infiltrated to start becoming polymerized. After the removal of unexposed monomer, a microscale line defect was introduced into the silica lattice.

Layer-by-layer deposition is another commonly used strategy to introduce defects. Generally, a layer of colloidal crystal film is first deposited on the substrate, followed by the deposition of a defect layer. Sequential growth of the colloidal crystal layer results in a planar defect within the self-assembled crystalline lattice. Materials such as polymers, polyelectrolyte, colloidal crystals with different size or refractive index, are usually used to form the defect layer. Selective doping layers of different dielectric materials also introduce disturbances to the periodic lattice.

López's group utilized this method with the help of a chemical vapor deposition (CVD) technique to create a homogeneous silica defect layer inside a polystyrene opal.[49] A multiple sandwich-like inverse silica opal, with a planar defect layer embedded, was obtained by infiltrating the interparticle space of the polystyrene spheres with silica, followed by the removal of the polystyrene spheres.

Ozin and co-workers[51, 52] developed a photolithography technique to create defects within the interior of a self-assembled lattice taking advantage of both lithography and layer-by-layer deposition. In this process, a silica colloidal crystal film was first deposited on the substrate. After a photoresist was applied to the film through spin-coating, conventional photolithography was used to form defects on the surface of the film. Then, multiple layers of silica spheres were again assembled on the photoresist film to generate patterned defects embedded inside the silica lattice. The removal of the silica spheres created defects embedded in the inverse opal structures, the size and morphology of which can conveniently be modulated during the fabrication. Zhao and co-workers also used similar approach to induce the formation of a defect in the colloidal lattice based on photolithography.[50]

During the fabrication process, the thickness of the defect layer can be precisely controlled at its introduction, to thereby control the optical properties of the system. Furthermore, if the defect layer is composed of a responsive material such as hydrogel or polyelectrolyte, an external stimulation will change the volume or thickness, thereby creating a tunable optical property of the defect layer.

Applications of Photonic Materials with Defects. The fabrication of photonic crystal fibers is one of the practical applications that arises out of the introduction of defects into PCs.[54, 55] It is achieved by introducing single defects into photonic crystals that are used for guiding light through a 3D periodical lattice. Two main kinds of photonic crystal fibers are involved: (1) total internal reflection photonic crystal fibers and (2) photonic band gap fibers. The former has a core surrounded by an array of holes. The core of the latter is usually hollow, inside of which losses can greatly be reduced thanks to confinement of the light, thereby keeping scattering and absorption to a rather low level.

Another application is the fabrication of laser devices, which can be obtained when the defect layer is composed of a fluorescent laser dye. Optical investigations by Ozin and co-workers showed that emission at the PBG is strongly prohibited and narrow luminescence peaks appear exactly at the wavelengths of the defect transmission states.[53] Thus, these waveguide photonic materials have specific properties and promising application in optic communications, laser generation, nonlinear devices, highly sensitive sensors, and high-power transmission.[56, 57]

10.2.2.2 Surface Wettability Control: Superhydrophobicity and Superhydrophilicity Besides the brilliant structural color displayed by the wings of *Morpho* butterfly, another interesting phenomenon, known as *surperhydrophobicity*, has also drawn much attention. As described earlier, the wings of the *Morpho* butterfly have scales of two different sizes: *ground scales*, which are responsible for the origin of the structural color, and *cover scales*, which act as a waterproof layer to prevent water from penetrating their wings. This waterproofing property and phenomenon not only exist on the wings of butterflies but also in other living creatures; for example, the legs of water striding insects and the leaves of lotus and rice.[58]

Theoretical Investigation of Surface Wettability. Considerable efforts have been directed toward the study of wettability on solid surfaces, in both theoretical investigations and their practical applications. As early as 1805, Young proposed that a kind of specific energy exists between interfaces, called *surface energy*. Thus, the status of a liquid on a solid surface depends on the balance of the surface tension among the solid, liquid, and gas interface. Under ideal conditions, the contract angle can be calculated from Young's equation:[60]

$$\cos\theta = \frac{\gamma_{SV} - \gamma_{SL}}{\gamma_{LV}} \tag{10.1}$$

where θ is the contact angle, and γ_{SV}, γ_{SL}, and γ_{LV} are the surface tension between solid–vapor, solid–liquid, and liquid–vapor interface, respectively. When θ is larger than 90°, the surface could be considered as hydrophobic. The surface is hydrophilic when θ is smaller than 90°, as shown in Figure 10.6. It should be noticed that Young's equation is based on the assumption that the solid surface is flat and only applies for the situation where $\gamma_{SL} - \gamma_{SV} < \gamma_{LV}$.

Following Young's equation, Wenzel proposed an equation for the calculation of the contact angle on a rough surface,[61] as follows:

$$\cos\theta' = r\cos\theta \tag{10.2}$$

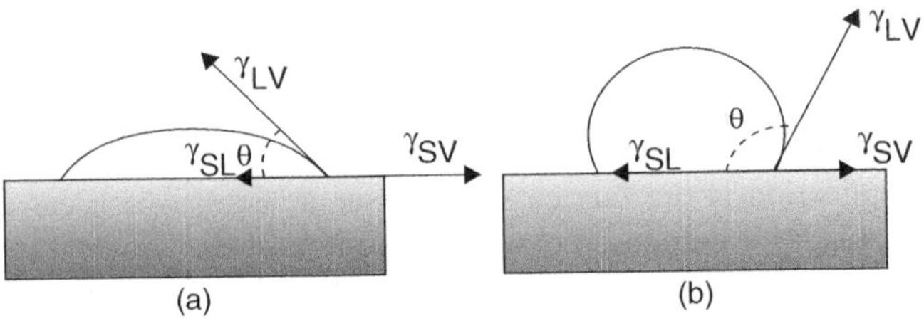

Figure 10.6 The status of liquid on flat solid surface with (a) hydrophilic and (b) hydrophobic property.

where θ' is the contact angle on the rough surface, θ is the contact angle on a flat solid surface, and r is the roughness factor, which is larger than 1. From this equation one can conclude that when $\theta > 90^\circ$, the hydrophobic properties are enhanced when the roughness of the hydrophobic surface is increased. When $\theta < 90^\circ$, the hydrophilic properties are enhanced when the roughness of the hydrophilic surface is increased. This theory works well in situations where the roughness can be infiltrated by the liquid droplets. In the case where the droplets cannot penetrate into the roughness and bridge the surface protuberances, the droplets can be considered to be located on a composite surface composed of solid and air, with the contact angle calculated from the Cassie–Baxter equation:[62]

$$\cos\theta' = f\cos\theta - (1 - f) \tag{10.3}$$

where f refers to the area fraction of the solid–liquid interface, while $(1 - f)$ refers to that of solid–air interface. The Wenzel and Cassie–Baxter models are shown in Figure 10.7.

Superhydrophobic Surfaces. Generally, the wettability of droplets on a solid surface is mainly determined by the average free energy per unit area beneath the droplet, the roughness coefficient, and the structure of surface protuberances. Based on our understanding of the generation of hydrophobic properties, various kinds of artificial superhydrophobic surfaces displaying both structural color

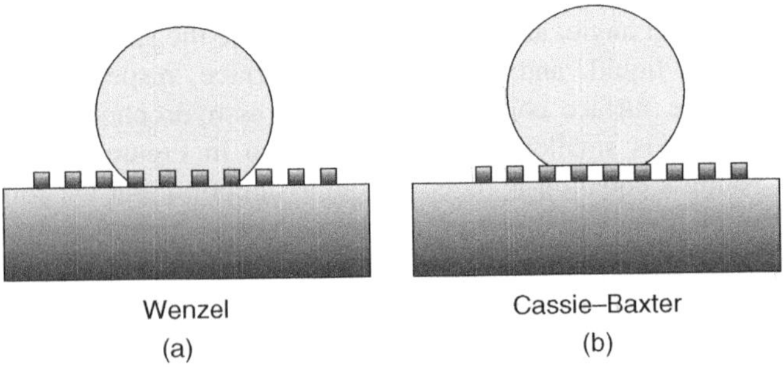

Figure 10.7 The Wenzel and Cassie–Baxter models.

and hydrophobic properties can be achieved by fabricating dual-scale roughness structures on the substrate and taking advantage of these hierarchical structures to increase the contact angle (as inspired by the wings of *Morpho* butterfly).

Sato and co-workers reported a dipping method to fabricate uniform inverse opal films with rough surfaces.[63] The suspension used for the fabrication of the inverse opal film contained spheres of two different sizes: polystyrene spheres with a diameter of several hundred nanometers and silica nanoparticles with a diameter of only 6 nm. First, the opal film was created using the lifting method. This film was then calcined at 450 °C to remove the polystyrene spheres and solidify the silica nanoparticles. A rough surface with inverse opal structures was thereby obtained with the small silica nanoparticles filling in the voids among the polystyrene spheres. The superhydrophobic inverse opal film was finally obtained by modifying the surface of the film with fluoroalkylsilane. The contact angle of this superhydrophobic inverse opal film was 155°. It clearly showed that rough surfaces contribute greatly to the hydrophobic properties.

As is well known, ordered opal films usually cannot be obtained using deposition methods involving mixtures of nanospheres with different size. But in this research, they found that well-ordered opal films could be generated when the diameter ratio of the small particles was less than 0.15. This finding is significant for the self-assembly of colloidal crystals and the fabrication of hierarchical structures or rough surfaces.

Xu and co-workers presented a facile route to fabricate superhydrophobic surfaces via solidification of emulsion droplets that contain the colloidal nanospheres in silicone oil.[64] With the help of shearing force created by stirring, silica nanospheres packed into beads that were generated from the emulsion droplets after the evaporation of solvent. After calcining at 450 °C, steel sieves were used to narrow the size distribution and the beads were then treated with fluoroalkylsilane to enhance the hydrophobicity. The contact angle obtained from the rough surface formed by these silica beads was 162°, indicating that the surface was superhydrophobic, as shown in Figure 10.8.

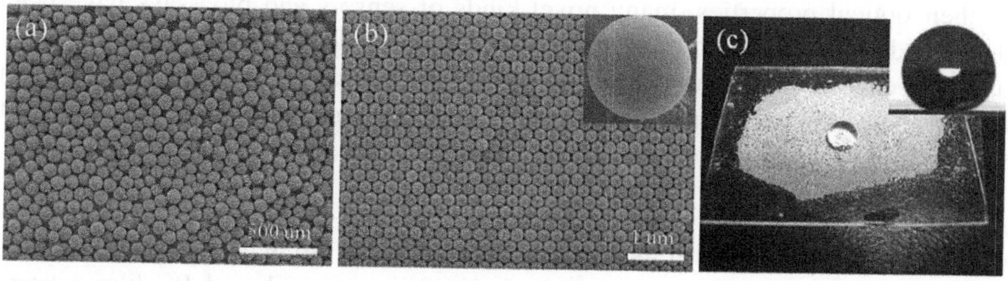

Figure 10.8 (a) and (b) SEM images of the rough surface made of colloidal crystal beads with dual-scale hierarchical structures. (a) Enlarged view of the rough surface; (b) the surface of a colloidal crystal bead. The insert is the SEM image of a single colloidal crystal bead. (c) The status of water droplet on the rough surface. The contact angle of the water droplet measured on this surface is around 162°, demonstrating that the surface shows ideal superhydrophobic property. (Reproduced with permission. Copyright the American Chemical Society: Ref. 64.)

One important characteristic of a surperhydrophobic surface is its *self-cleaning* property. On the surface of surperhydrophobic structures, the water contact angle is greater than 150°. Water will therefore form spherical droplets due to the surface tension, thereby effectively minimizing the contact area of the solid–liquid interface. As dust can easily attach to the surface of the water droplets, they will be carried away from the surperhydrophobic surface, leading to self-cleaning. This effect arises from the presence of hierarchical structures of different scales located on the surface, and is known as the *lotus effect*.[59]

Superhydrophilic Surfaces. Superhydrophilicity is another important surface characteristic, which leads to a totally different arrangement of water on the solid surface. When water flows on a superhydrophilic surface, it can spread out on the surface, resulting in a very low contact angle. Since titanium dioxides (TiO_2) exhibit photocatalytic and photoinduced superhydrophilic properties, they have been used to prepare superhydrophilic surfaces with structural color.

Sato and co-workers reported the fabrication of TiO_2 inverse opal films with superhydrophilic properties, by the use of a vertical lifting method.[65] After formation of a polystyrene opal film, a hydrophilic treatment was applied to the surface of the polystyrene using a 7% solution of polyethylene imine in ethanol. TiO_2 nanoparticles with a diameter of 15 nm were infiltrated into the voids of the polystyrene spheres, followed by calcining at 500 °C to remove the polystyrene spheres and form TiO_2 inverse opal films. Irradiation of the resulting rough surface with UV light resulted in a contact angle of nearly 0°. When water was dropped on the TiO_2 inverse opal film, it infiltrated into the film due to capillary forces and no water droplets formed on the surface. The stability of this superhydrophilic TiO_2 inverse opal films is excellent. Even after storage in the dark for a period of several months, the contact angle can still be kept below 1° or 2°.

10.2.2.3 Sensors and Bioassays As described previously, one of the important properties of the PCs is the PBG generated by well-ordered periodical structures, which has the ability to reflect light with certain wavelengths. By making use of their optical properties, many novel kinds of sensors and bioassays have been developed from such photonic materials.[66–72] In these systems, PCs are usually used to generate or transmit signals from the biological recognition event, which brings about the possibility of fabricating simple, highly sensitive, and low-cost sensors and bioassays.

Photonic Sensors. PC film forms the basis of one of the new type of sensors that have been developed in recent years. Such sensors are mainly divided into two types according to the detection method: (1) fluorescence-based PC film sensors and (2) label-free PC film sensors.

For the former, which is widely used in scientific research, molecules are labeled with dyes or fluorescent tags, so that the status of the targeted molecules can be recognized by the fluorescence signals. In this situation, the PC films improve the sensitivity of the signal through the enhancement of excitation light whose wavelength is matched with the PBG.

For the latter, the diffraction peaks are usually used as an indicator. The change of the Bragg diffraction peak exhibited by the PC film could easily be distinguished and detected by optical measurement, due to the physical or chemical changes of the film (e.g., changes in the refractive index or the periodicity of the 3D lattice caused by reactions of the components). Thus, there is no need to track the status of the molecules with tags, giving rise to the term "label-free" method.

Li's group demonstrated one kind of sensitive sensor for the detection of anions based on the combination of both ionic liquids and PCs, whose signal can be directly recognized by the naked eye[73] (Figure 10.9). The silica colloidal crystal arrays were used as templates to obtain inverse opal films. The ionic liquid monomer, methyl methacrylate monomer, the crosslinker ethylene glycol dimethacrylate, and the initiator azobisisobutryonitrile (AIBN) were first dissolved in a solvent mixture of methanol and chloroform and then infiltrated into the voids of the silica spheres. After photopolymerization under UV light and the removal of the silica template, a photonic ionic liquid film with inverse opal structures was obtained. The hydrophilicity and hydrophobicity of ionic liquids can easily be adjusted through counteranion exchange, resulting in a change in the solubility of ionic liquids in solvents such as water and organics. Utilizing this property of the ionic liquids, the resulting film can respond to different kinds of anion solutions to exhibit various colors, arising from the change of the solubility and the refraction index.

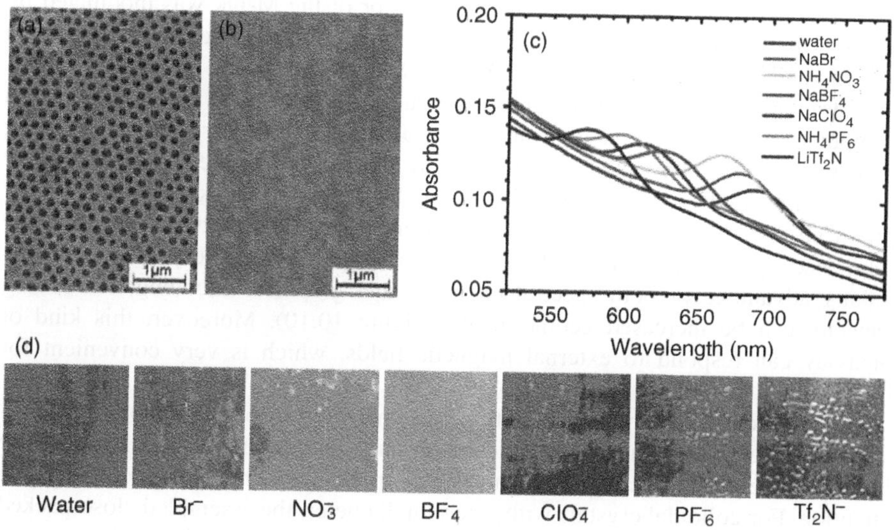

Figure 10.9 SEM images of the anion-sensitive ionic liquid based inverse opal film with (a) opened pore structures and (b) closed pore structures. (c) Stop band shift of the film in response to various kinds of anion aqueous solutions. (d) Color presented by the film when soaking into different kinds of anion aqueous solutions. Different colors will be exhibited in response to the anions due to the change of the solubility and the refraction index, which can be directly recognized by the naked eye. (Reproduced with permission. Copyright Wiley-VCH: Ref. 73.) (See insert for color representation of this figure.)

Asher and co-workers recently developed responsive sensors based on the photonic crystal film.[74–77] These sensors were generated by embedding monodisperse nanospheres with non-close-packed ordered crystalline structures in a matrix of responsive hydrogel. The hydrogels have a reversible swelling process in response to external stimulation such as chemicals, pH, or temperature, which lead to a change of interparticle distance and lattice constant of the ordered structures, thereby changing the optical properties of the PC film. Following this approach, they have successfully designed a glucose sensor using glucose-sensitive hydrogel PC film for the detection of the glucose concentration in tear fluid.[77] The color exhibited by the film changes along with a shift in the reflection peak as the glucose concentration varies. Further work has been conducted on the sensitivity and response speed of glucose at low concentration, so that it is now possible to apply this responsive photonic material to clinical diagnosis, such as monitoring the approximate glucose levels of inpatients who suffer from diabetes mellitus or hyperglycemia.

Photonic Bioassays. Photonic crystals also provide a significant contribution in the field of biomolecular detection and bioassays. Kwon's group utilized magnetic colloidal crystals to create free-floating, structurally colored particles for multiplex high-throughput bioassays.[79] Since the magnetic particles can respond to an external magnetic field, they first place the M-ink, a mixture composed of superparamagnetic Fe_3O_4/SiO_2 colloids, a liquid solvent, and a photocurable resin, into a channel made by polydimethylsiloxane (PDMS). Then the color of the M-ink was modulated by varying the intensity of an external magnetic field. When the desired color for a specific barcode was achieved, the UV light was used to freeze the ordered chain-like structures of the superparamagnetic particles in the M-ink with the help of a computer-controlled spatial light modulator as mask. The color tuning and fixing process could be completed in a very short time, approximately 0.1 second per bit, meaning that color-barcoded microparticles could be quickly and effectively produced. Thus, various barcodes with different color and pattern can easily be fabricated by simply changing the induced mask and the magnetic field intensity. Taking advantage of both spectral encoding and graphical encoding, the coding capacity can be increased geometrically (Figure 10.10). Moreover, this kind of bioassay can respond to external magnetic fields, which is very convenient for isolation and screening.

With the development of fabrication and characterization technology, the possibility of detecting multiple analytes in a single sample, as a way to meet the increasing demands in clinical diagnosis and applications, has drawn considerable attention. For colloidal crystals with a certain diameter, the assembled close-packed periodical lattice has a predetermined band gap, and the optical signal expressed by the characteristic reflection peak is definite. If one specific spectrum from the colloidal crystals with a certain diameter corresponds to one code, it is possible to realize multiple codes by using PCs with different reflection peaks. Moreover, this optical signal is much more stable than fluorescent dyes or organic tags, as it originates from the PBG of PCs, which derive from a physical phenomenon rather than a chemical change. These properties make photonic crystals suitable for encoding.

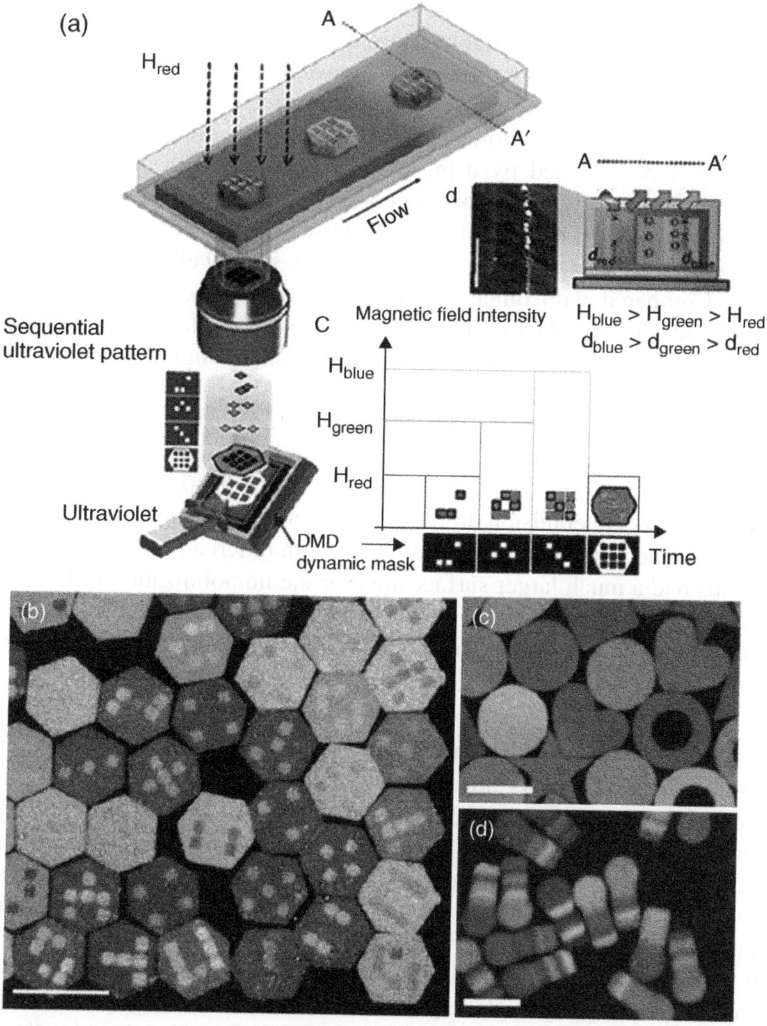

Figure 10.10 (a) Schematic illustration of high-throughput bioassays generated from M-ink with the help of an external magnetic field and a computer-controlled spatial light modulator as mask. Taking advantage of both spectral encoding and graphical encoding, various barcodes can be obtained (b–d). (b) Hexagon-type 2D color-barcoded microparticles; (c) microparticles with various shapes and colors; (d) bar-type 1D color-barcoded microparticles. (Reproduced with permission. Copyright Nature Publishing Group: Ref. 79.) (See insert for color representation of this figure.)

Gu and co-workers reported a facile method of fabricating monodisperse colloidal crystal beads with controllable size and optical properties for multiplex encoding bioassays, using microfluidic technology.[67, 68] These beads were obtained by self-assembly and solidification of monodisperse colloidal spheres in the emulsion droplets generated from the microfluidic device. By taking advantage of microfluidic technology, the generated emulsion droplets containing colloidal crystals are uniform and size-controllable. Thus, a controlled and continuous fabrication of monodisperse colloidal crystal beads can conveniently be achieved. Furthermore, treatment of these beads at high temperatures can effectively enhance their mechanical strength. A series of colloidal crystal beads with different diffraction peak positions could be obtained by using colloidal crystals with different diameters. These can be used as biomolecular carriers for multiplex encoding.

As the beads were composed of monodisperse self-assembled colloidal crystals, the surface and bulk of the beads were comprised of well-ordered 3D periodic structures. Compared with other encoding approaches, these photonic beads are more stable, sensitive, and easier to use in code design. Utilizing the optical signals of the photonic beads, label-free encoding and real-time, high-throughput detection could be realized.

In their later research, alternative kinds of colloidal crystal beads with inverse opal structures were developed for label-free multiplex bioassays (Figure 10.11).[78] These were fabricated by the self-assembly of silica nanoparticles together with polystyrene spheres in emulsion droplets dispersed in silicone oil. After the removal of the polystyrene spheres, inverse opal beads with ordered porosity were obtained. These beads had a much larger surface area for the immobilization of biomolecular materials. Label-free bioassays were realized based on the shift in the reflection peaks resulting from the change of refractive index caused by the specific analyte binding.

10.2.2.4 Tunable Optical Devices Although various kinds of brilliant color exist in Nature, these colors are not always static. There are a large number of examples of tunable colors in Nature. Many animals such as fish, insects, and amphibians can change their body color in response to external stimulation, in order to adapt themselves to the surrounding environment.[80, 81] For example, *chameleons* can adjust the absorption efficiency of different types of chromatophores in their skins through migration and redistribution of dye among these chromatophores to present various colors. The *damselfish* can modulate the distance between adjacent reflecting units by slightly stretching or shrinking its skin to exhibit different colors according to the external conditions. The adult leaf beetle *charidotella egregia* induces color changes by controlling the filling of pigments via injection or draining of a fluid between the bottom of its elytron and the upper porous multilayer. This modulates the interference and absorption of light (Figure 10.12). Generally, these phenomena are cause by slightly stretching or shrinking the skin of animals that have special arrays of light reflecting cells on the surface of their skin, or by injection or removal of a fluid from their skin. These tiny alterations at the surface are sufficient to change the wavelength of light that is reflected and thereby change the coloration that is exhibited.

In recent years, the demand for materials with tunable optical properties in response to external stimulation has increased, due to their promising practical applications in color displays, sensors, and so on. Photonic crystals—one kind of functional material with periodical structures that can exhibit brilliant structural color—have widely been used in the design of devices with tunable optical properties. Responsive photonic crystals can be realized when the changes in photonic properties are reversible. In accordance with Bragg's law, two major strategies have been developed to achieve control of the stop band and the exhibited color: through either (1) control of the average refractive index of the dielectric material or (2) control of the periodicity of the crystalline lattice. In order to modulate the average refractive index of a photonic crystal, materials such as liquid crystals or photochromic dyes are usually infiltrated into the interparticle spaces. For controlling the periodicity of the ordered structures, stimulus-responsive materials are usually used in the fabrication of photonic crystals. They may be used either directly for the construction of building blocks of the photonic crystal or as a matrix for the

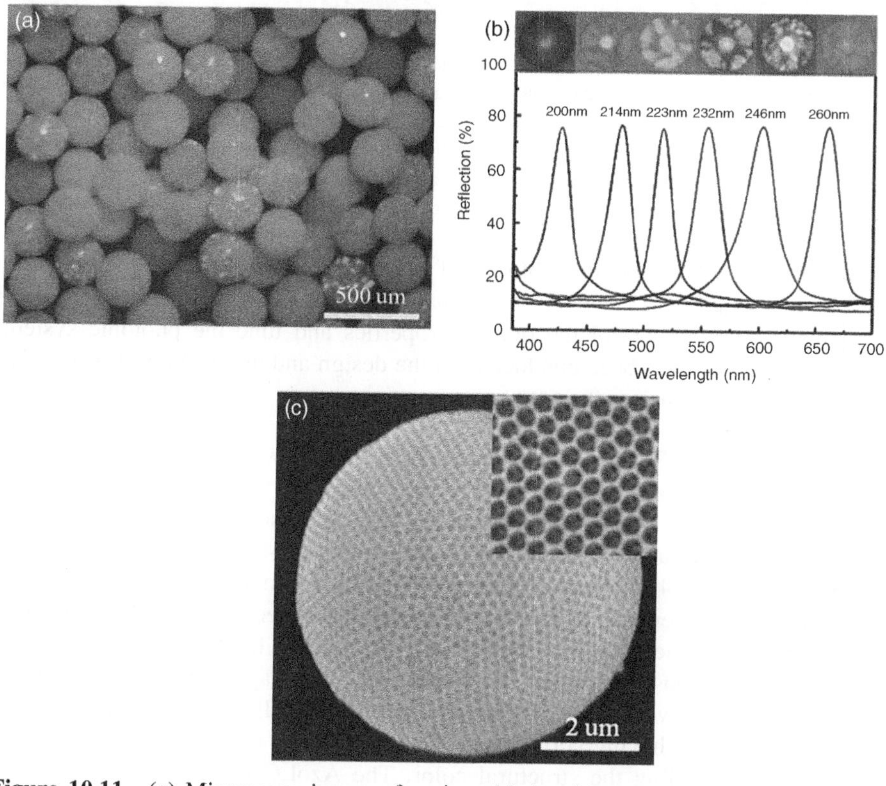

Figure 10.11 (a) Microscope image of various kinds of inverse opal beads fabricated through microfluidic technology by using polystyrene spheres with different size as sacrificial templates. (b) Reflection spectra of these beads. These optical signals are quite stable and have close relationships with the size of colloidal crystals used for the fabrication. (c) SEM images of the inverse opal beads. The insert shows that they have a porous surface. Further investigation found that both the surface and bulk of the beads are composed of well-ordered 3D periodical structures, providing a rather high specific surface area for the potential application in the bioassays. (Reproduced with permission. Copyright Wiley-VCH: Ref. 78.) (See insert for color representation of this figure.)

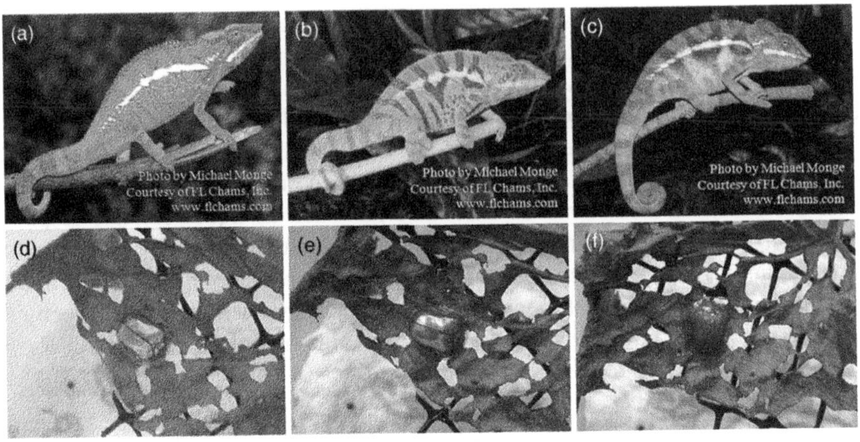

Figure 10.12 Living creatures with tunable colors in Nature. They can change their body color in response to the external stimulation to adapt themselves to the surrounding environment. (a)–(c) Chameleon. (Reproduced with permission from Michael Monge, Copyright by FL Chams, Inc.) (d)–(f) Beetles *charidotella egregia*. (Reproduced with permission. Copyright the American Physical Society: Ref. 80.) (See insert for color representation of this figure.)

immobilization of the building blocks of photonic crystals. External stimulation, such as temperature, pH, chemicals, mechanical force, or electric or magnetic field are usually used to control the optical properties and tune the photonic system. Considerable work has been conducted on the design and application of responsive photonic materials in recent years.

Light-Responsive Photonic Materials. Sato and co-workers reported a photo-tunable photonic film whose reflection peak could be controlled by the phase transition of liquid crystals between nematic phase and isotropic phase.[82, 83] It was fabricated by infiltrating a mixture of liquid crystalline 4-pentyl-4′-cyanobiphenyl (5CB) and 4-butyl-4′-methoxyazobenzene (AzoLC) into the void of the inverse opal films. In the nematic phase, the liquid crystal molecules are aligned parallel to the surface of the spherical void structures causing light to be scattered due to the heterogeneous anisotropy, leading to the films appearing opaque. In the isotropic phase, however, the anisotropy disappeared, allowing the film to selectively diffract light having certain wavelengths. Hence, the liquid crystals play a key role in controlling the structural color. The AzoLC in the film can undergo a trans- to cis-photoisomerization when irradiated by light, leading to the phase transition of 5CB from the nematic phase to the isotropic phase. Therefore, control of the film color can be realized with the help of light.

Gu and co-workers also developed similar film by taking advantage of UV-sensitive liquid crystal 4-butyl-4′-methoxyazobenzene (BMAB), mixed with the 5CB to form a composite polystyrene opal film (Figure 10.13).[84] By utilizing polystyrene spheres with different diameter, various colors and patterns could be exhibited, with light used as the switch.

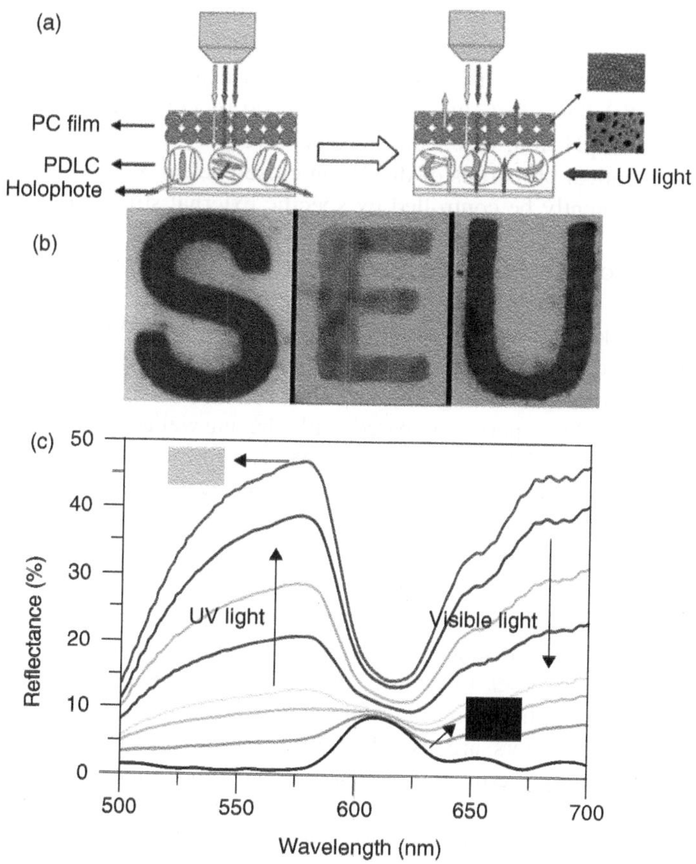

Figure 10.13 (a) Schematic illustration of a reversible photonic device composed of PC film and liquid crystals, which can conveniently be modulated by UV and visible light; (b) characteristics presented by the photonic devices; (c) reflection spectra change of the film under the irradiation of UV light. These changes can be reversed to the original state under the irradiation of visible light. The insets show the colors before and after UV irradiation. (Reproduced with permission. Copyright the American Institute of Physics: Ref. 84.) (See insert for color representation of this figure.)

Temperature-Responsive Photonic Materials. Asher and co-workers demonstrated a polymer-based responsive photonic crystal by embedding non-close-packed colloidal crystals in the matrix of a thermosensitive hydrogel of poly(*N*-isopropylacrylamide) (PNIPAM).[85] As PNIPAM is a thermosensitive polymer that has a reversible volume phase transition between a hydrated state and a dehydrated state around its lower critical solution temperature, an increase in temperature caused the PNIPAM matrix to shrink with an accompanying decrease in the total volume. This leads to a decrease of the interparticle distance in the 3D ordered structures, causing a blue shift of the reflection peak. On the other hand, when the temperature is decreased, the swelling of PNIPAM matrix results in a red shift of the reflection peak. This reversible swelling and shrinking process can effectively switch the diffraction of light, which covers almost the entire visible region. Similar effects have been developed by constructing these matrices with temperature, pH, ion, or chemical responsive polymers[86–88] (Figure 10.14). Therefore, the optical properties of these non-close-packed 3D ordered lattices can easily and conveniently be controlled by specific external stimulation.

Chemical-Responsive Photonic Materials. Xia's group developed a patterned photonic system using poly(dimethylsiloxane)/polystyrene composite colloidal crystal film by taking advantage of a swelling process.[89, 90] When organic solvent was applied to the surface of the film, the selected area of poly(dimethylsiloxane) (PDMS) matrix swelled, leading to a red shift of the reflection peak. A different color occurred in those parts compared with the unswelled part (Figure 10.15). After the evaporation of the solvent, the PDMS matrix turned back to the original state and the color patterns disappeared. Using this swelling method, it is possible to control the photonic band gap by using different solvents or by controlling the swelling rate of the matrix.

Mechanical-Force-Responsive Photonic Materials. An external mechanical force is also a convenient and effective way to control the optical properties of photonic systems. In order to enable mechanical deformation, elastic polymers are typically used for the design and fabrication of such photonic materials. The monomer elastic polymers are usually infiltrated into the void of the photonic structures, followed by thermopolymerization or photopolymerization.

Fudouzi and Sawada fabricated an elastic silicone film with reversible optical properties through embedding polystyrene spheres in a PDMS matrix.[91] When the film was stretched, the thickness was decreased in the direction perpendicular to the stretching direction, leading to a reduction of the lattice distance and a blue shift in the reflection peak (Figure 10.16). The tuning range of the reflection peak is determined by the stretching ratio and the volume fraction of polymer content. Thus, the optical property could easily be modulated within a certain range, by an external stretching force.

Electrical-Field-Responsive Photonic Materials. Ozin's group demonstrated the fabrication of electrically controlled photonic films.[92–94] A voltage-dependent photonic film was obtained by embedding the opal structures formed by silica colloidal crystals in the matrix of polyferrocenylsilane gel. When a potential induced by the electrical field was applied to the composite film, the polymer gel started to swell or shrink in response to the induced potential, resulting in a shift of the reflection peak. The swell or shrink rate was determined by the voltage, while the status of the film depended on the induced potential, either oxidative or reductive. In later research, they removed silica spheres in the composite film to form an inverse opal film, which effectively increased the tuning range and response rate. It also allowed for a beneficial decrease in the driving voltage. Thus, reversible

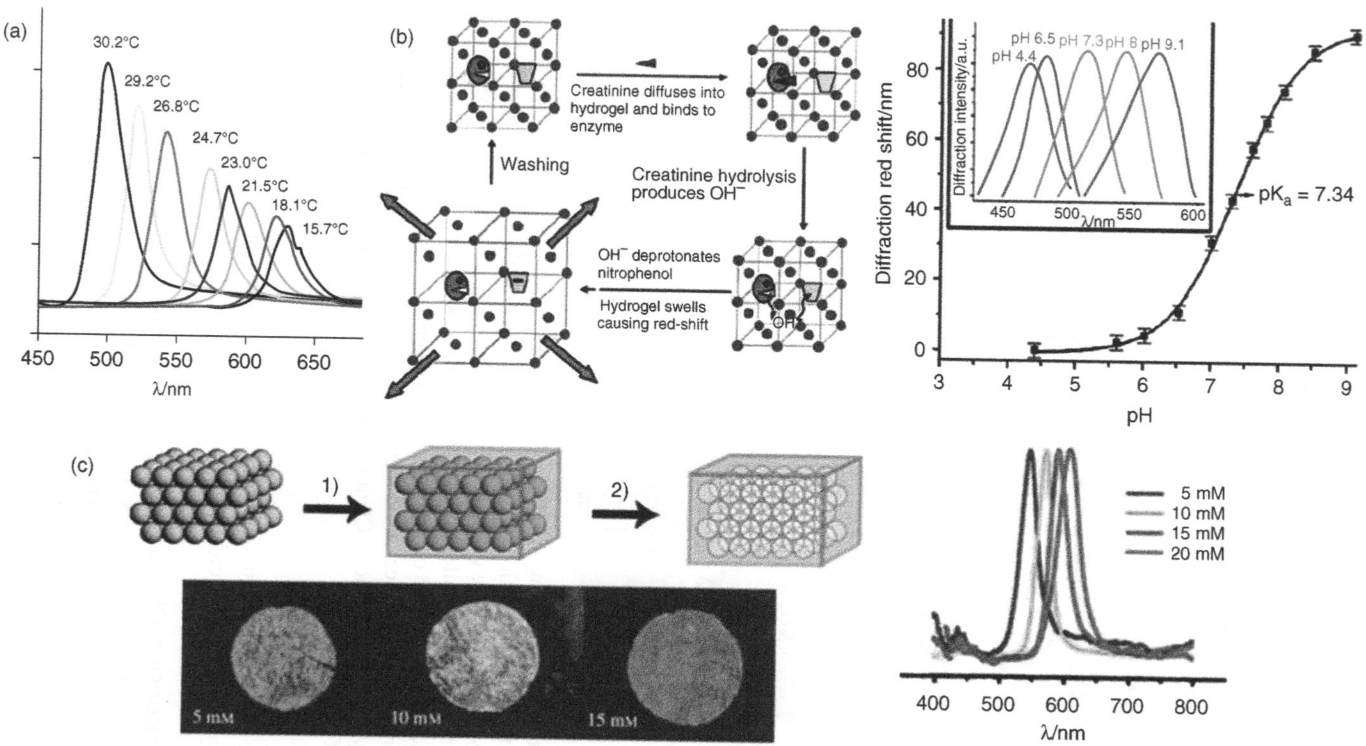

Figure 10.14 Responsive photonic materials-based PCs and their responsiveness to specific external stimulation such as temperature, pH, and chemicals. (a) Temperature dependence of the reflection spectra of the porous NIPA gel made using close-packed silica colloidal crystals as template. (Reproduced with permission. Copyright the American Chemical Society: Ref. 86.) (b) Schematic illustration of pH-responsive polymerized crystalline colloidal array (left) and the pH dependence of reflection spectra of this intelligent sensing array (right). (Reproduced with permission. Copyright the American Chemical Society: Ref. 88.) (c) Photographs (left) and reflection spectra (right) of the periodically ordered interconnecting porous poly(NIPA-*co*-AAPBA) gel in response to different concentrations of glucose. (Reproduced with permission. Copyright Wiley-VCH: Ref. 87.) (See insert for color representation of this figure.)

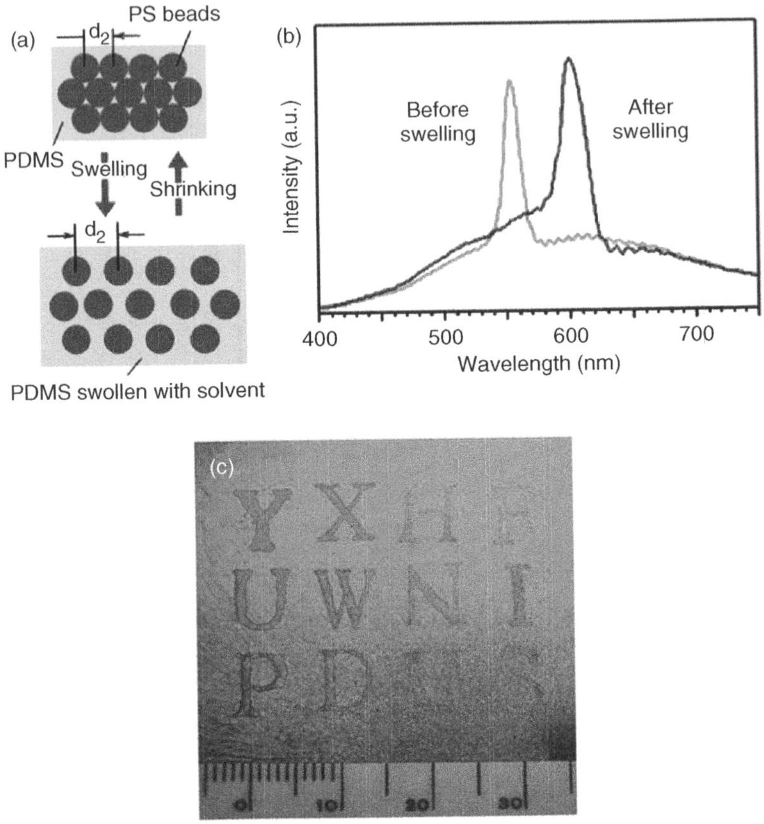

Figure 10.15 Patterned photonic films fabricated by taking advantage of the swelling process of polymer matrix. (a) Schematic illustration of the swelling and shrinking process of tunable colloidal crystals. The lattice constant will be increased by swelling the PDMS matrix with an appropriate solvent, while it will shrink back to the original state after the evaporation of solvent. (b) The change of reflection spectra before and after the swelling process. (c) Letters printed on the surface of the matrix using a rubber stamp. (Reproduced with permission. Copyright the American Chemical Society: Ref. 89.) (See insert for a color representation of this figure.)

control of structural color in almost the whole visible region can be realized in response to an induced electronic field (Figure 10.17).

Magnetic-Field-Responsive Photonic Materials. Adding magnetic properties to the photonic crystals is also a strategy for the intelligent control of the stop band. Magnetic colloidal crystals have distinguished themselves in the design of tunable photonic materials in recent years. Without an external magnetic field, magnetic colloidal crystals suspended in solutions display Brownian motion arising from interparticle electrostatic repulsive forces from the charged groups on the surface of the nanospheres. The magnetic spheres are kept in disordered status

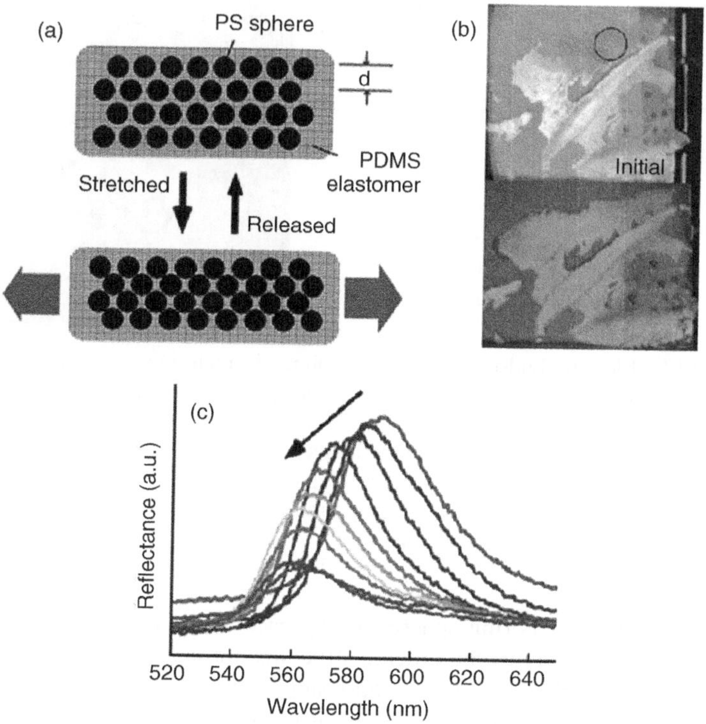

Figure 10.16 External mechanical-force-responsive photonic films fabricated by embedding colloidal crystals in PDMS matrix. (a) Schematic illustration of the reversible elastic deformation of the composite colloidal crystal film. The lattice constant can easily be modulated by the stretch rate of the film. (b) Digital photographs of the composite colloidal crystal film before (top) and after (down) stretch. (c) The change of reflection spectra during the stretching process. (Reproduced with permission. Copyright the American Chemical Society: Ref. 91.) (See insert for color representation of this figure.)

at this time. When an external magnetic field is applied, these spheres respond to the magnetic field by forming non-close-packed chain-like ordered structures that parallel to the direction of the external magnetic field. Chains of this type, formed in solution by the monodisperse magnetic colloidal crystals with a certain diameter, have the ability to diffract light and exhibit distinct structural color. The interparticle distance is a result of the balance between the magnetic attractive force and the interparticle electrostatic repulsive force, as was reported by Asher and co-workers.[95] In the direction perpendicular to that of the magnetic field, the magnetic spheres are affected by interparticle electrostatic repulsive forces and magnetic repulsive forces between magnetic dipoles, which keep them away from each other to prevent immediate aggregation. In the direction parallel to that of the magnetic field, they are mainly affected by interparticle electrostatic repulsive forces and magnetic attractive forces. When the intensity of the magnetic field is strong enough, a balance between the attractive and repulsive forces starts to form,

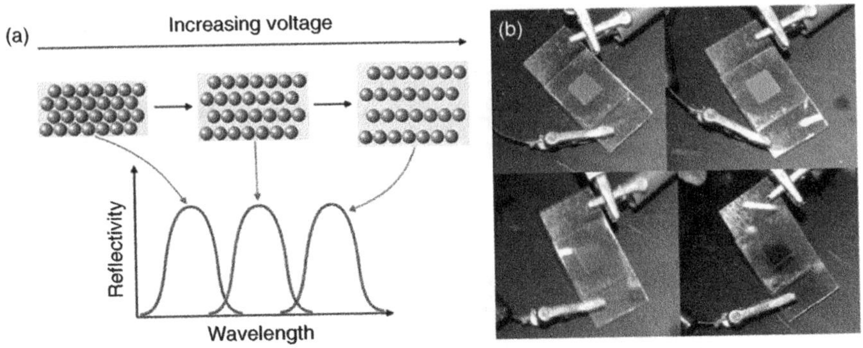

Figure 10.17 Electric-field-induced tunable photonic materials. (a) Schematic illustration of the operation of voltage-tunable full-color opal film. As the film was obtained by embedding the silica colloidal crystals into the matrix of polyferrocenylsilane gel, modulation of the voltage causes a change in the lattice constant. (b) Photographs of the tunable opal film in response to the voltage change. This kind of film has a tuning range covering the whole visible region. (Reproduced with permission. Copyright Elsevier: Ref. 37.) (See insert for color representation of this figure.)

which causes the neighboring spheres to approach each other. Thus, it is possible to control the periodicity of the ordered structures through control of the magnetic attractive force, which has a close relationships with the intensity of the magnetic field. In this way one can realize control of the interparticle distance and thereby also the structural color that is displayed.

Philipse and co-workers reported the fabrication of monodisperse magnetic polymethyl methacrylate spheres that exhibit field-induced colloidal crystallization.[96, 97] These magnetic crystals presented iridescent structural colors due to the inhomogeneous field gradient offered by the magnet, which was used to induce the arrangement of the magnetic crystals.

Yin's group developed a high-temperature hydrolysis process for the synthesis of superparamagnetic colloidal crystal clusters with a tuning range that covers the whole visible region when induced by an external magnetic field.[98, 99] As these magnetic clusters mainly consist of magnetite capped with a layer of polyacrylic acid (PAA), they have a rather high magnetic content and saturated magnetization. Therefore, they have a quicker response than the magnetite/polymer composite does in an external magnetic field. With the help of magnetic stimulation, they can self-assemble into ordered lattices to diffract visible light (Figure 10.18).

In a following study, they modified the magnetic clusters with a thin layer of silica.[100, 101] These Fe_3O_4@SiO_2 particles disperse well in different solvents and present ideal tunable optical properties, which has expanded their prospect for practical application. Based on this mechanism, many optical devices have been developed, such as microcarriers, bioassays, sensors, and color displays.[101–104]

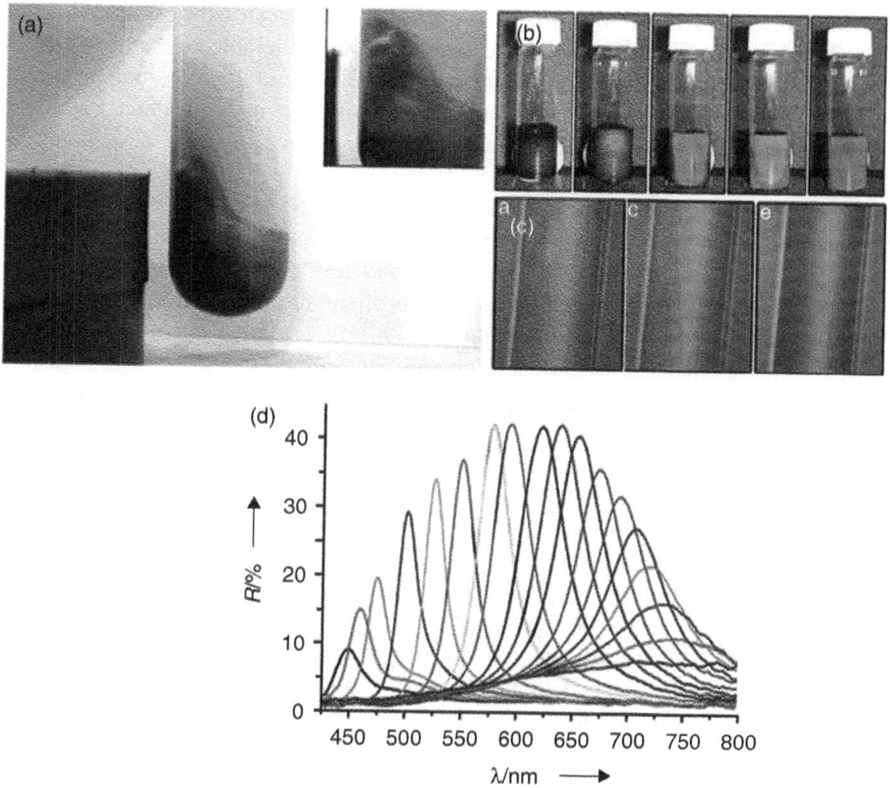

Figure 10.18 Magnetically tunable photonic crystals. With the help of an external magnetic field, they have a tuning range covering the whole visible region. (a) Magnetic colloidal crystals exhibiting various Bragg colors due to the inhomogeneous field gradient, which compresses or relaxes the crystal lattice. (Reproduced with permission. Copyright the American Chemical Society: Ref. 96.) (b) Photographs of magnetic colloidal crystals in response to an external magnetic field. The magnetic crystals will exhibit various colors while gradually altering the magnet–sample distance. (c) Optical microscope images of magnetic colloidal crystal solution enclosed in a glass capillary under an increasing magnetic field. (d) External magnetic field intensity dependence of the reflection spectra of the magnetic colloidal crystals. (Panels (b) and (d) reproduced with permission. Copyright 2007 Wiley-VCH: Ref. 98. Panel (c) reproduced with permission. Copyright the American Chemical Society: Ref. 99.) (See insert for color representation of this figure.)

10.3 CONCLUSION AND OUTLOOK

The enlightenments endowed by Nature are endless. We human beings benefit substantially from mimicry of Nature, which brings revolutionary change to our daily life. Coloration—one of the most amazing gifts bestowed by Nature—has provided plenty of inspiration to us, especially in respect of the structural color that originates from ordered, periodic structures. Considerable research has been conducted on the origin and mechanisms of the structural color that is exhibited by living creatures. The fabrication of biomimetic materials that use these mechanisms in one way or another now forms the source of many new, practical applications.

Photonic crystals are well-known artificial materials that have the ability to form spatial periodic crystalline lattices that reflect light of a certain wavelength. They offer an ideal material for the design and fabrication of novel, functional, biomimetic devices that can exhibit distinct, brilliant structural color. They present their vitality in various fields, including not only biology and chemistry, but also in fields such as material science and energy. By taking advantage of the specific properties of photonic crystals, such as their spatially ordered periodic structures as well as their photonic band gap and stable optical property, many devices have been constructed. These have promising prospects in practical applications such as waveguides, surface wettability control, bioassays, sensors, optics, and displays. It should be emphasized that the examples described in this chapter comprise a mere representative subset of what researchers have achieved to date. A growing number of novel devices and applications have been developed based on the manipulation of photonic crystals in recent years.

Although we have already enjoyed the convenience that photonic materials bring to us, great efforts still need to be devoted to their improvement and practical applications. For example, in order to apply photonic materials in responsive and tunable devices, the switching speed should be fast, with the response time short. This means that real-time, responsive and dynamic tunable optical properties must be achieved. In addition, precise control of the tuning process, which has a close relationship to the optical properties, should also be taken into consideration for the design of photonic devices.

Most of the photonic devices that have been developed to date remain in the laboratory stage, due to limitation such as high cost of fabrication, complex and tedious operation steps, or harsh conditions of use. In order to widely apply photonic materials into practical application, fabricated devices should be industrialized and commercialized. Thus, it is necessary to reduce the cost and simplify the fabrication and the use of these devices by unskilled operators. Moreover, new techniques for the preparation of photonic materials in higher quality and large scale need to be developed. It is not easy to obtain photonic systems with well-ordered crystalline lattices in large scale at present.

The field therefore offers both a challenge but also an opportunity for us. In our opinion, with the development of advanced technology, biomimetic photonic materials will come to enjoy a much brighter future, both in their theoretical investigations and their practical application. Such developments will effectively promote interdisciplinary cooperation and bring more convenience to our daily lives.

REFERENCES

1. Parker, A. R. *J. Opt. A*, **2000**, *2,* R15.
2. Vigneron, J. P.; Colomer, J. F.; Rassart, M.; Ingram, A. L.; Lousse, V. *Phys. Rev. E* **2006**, *73,* 021914.
3. Inchaussandague, M.; Skigin, D.; Carmaran, C.; Rosenfeldt, S. *Opt. Express* **2009**, *17,* 16183.
4. Brink, D. J.; van der Berg, N. G.; Botha, A. J. *Appl. Opt*. **2002**, *41,* 717.
5. Prum, R. O.; Cole, J. A.; Torres, R. H. *J. Exp. Biol*. **2004**, *207,* 3999.
6. Kinoshita, S.; Yoshioka, S. *ChemPhysChem*. **2005**, *6,* 1442.
7. Potyrailo, R. A.; Ghiradella, H.; Vertiatchikh, A.; Dovidenko, K.; Cournoyer, J. R.; Olson, E. *Nat. Photonics* **2007**, *1,* 123.
8. Weatherspoon, M. R.; Cai, Y.; Crne, M.; Srinivasarao, M.; Sandhage, K. H. *Angew. Chem. Int. Ed*. **2008**, *47,* 7921.
9. Ding, Y.; Xu, S.; Wang, Z. L. *J. Appl. Phys*. **2009**, *106,* 074702.
10. Yoshioka, S.; Kinoshita, S. *Proc. R. Soc. B* **2006**, *273,* 129.
11. Vukusic, P.; Sambles, J. R. *Nature* **2003**, *424,* 852.
12. Yablonovitch, E. *Phys. Rev. Lett*. **1987**, *58,* 2059.
13. John, S. *Phys. Rev. Lett*. **1987**, *58,* 2486.
14. Yin, Y.; Lu, Y.; Gates, B.; Xia, Y. *J. Am. Chem. Soc*. **2001**, *123,* 8718.
15. Foresi, J. S.; Villeneuve, P. R.; Ferrera, J.; Thoen, E. R.; Steinmeyer, G.; Fan, S.; Joannopoulos, J. D.; Kimerling, L. C.; Smith, H. I.; Ippen, E. P. *Nature* **1997**, *390,* 143.
16. Lin, S. Y.; Fleming, J. G.; Hetherington, D. L.; Smith, B. K.; Biswas, R.; Ho, K. M.; Sigalas, M. M.; Zubrzycki, W.; Kurtz, S. R.; Bur, J. *Nature* **1998**, *394,* 251.
17. Noda, S.; Tomoda, K.; Yamamoto, N.; Chutinan, A. *Science* **2000**, *289,* 604.
18. Ozin, G. A.; Yang, S. M. *Adv. Funct. Mater*. **2001**, *11,* 95.
19. Vlasov, Y. A.; Bo, X. Z.; Sturm, J. C.; Norris, D. J. *Nature* **2001**, *414,* 289.
20. Xia, Y.; Gates, B.; Yin, Y.; Lu, Y. *Adv. Mater*. **2000**, *12,* 693.
21. Campbell, M.; Sharp, D.; Harrison, M.; Denning, R.; Turberfield, A. *Nature* **2000**, *404,* 53.
22. Scrimgeour, J.; Sharp, D. N.; Blanford, C. F.; Roche, O. M.; Denning, R. G.; Turberfield, A. J. *Adv. Mater*. **2006**, *12,* 1557.
23. King, J. S.; Graugnard, E.; Roche, O. M.; Sharp, D. N.; Scrimgeour, J.; Denning, R. G.; Turberfield, A. J.; Summers, C. J. *Adv. Mater*. **2006**, *12,* 1561.
24. Wanke, M. C.; Lehmann, O.; Muller, K.; Wen, Q.; Stuke, M. *Science* **1997**, *275,* 1284.
25. Holgado, M.; Garcia-Santamaria, F.; Blanco, A.; Ibisate, M.; Cintas, A.; Miguez, H.; Serna, C. J.; Molpeceres, C.; Requena, J.; Mifsud, A.; Meseguer, F.; Lopez, C. *Langmuir* **1999**, *15,* 4701.

26. Davis, K. E.; Russel, W. B.; Glantschnig, W. J. *Science* **1989**, *245,* 507.
27. Pusey, P. N.; van Megen, W. *Nature* **1986**, *320,* 340.
28. Park, S. H.; Qin, D.; Xia, Y. X. *Adv. Mater.* **1998**, *10,* 1028.
29. Gates, B.; Qin, D.; Xia, Y. *Adv. Mater.* **1999**, *11,* 466.
30. Gu, Z. Z.; Fujishima, A.; Sato, O. *Chem. Mater.* **2002**, *14,* 760.
31. Jiang, P.; Bertone, J. F.; Hwang, K. S.; Colvin, V. L. *Chem. Mater.* **1999**, *11,* 2132. The conception of "vertical deposition" technique was first reported by Nagayama and co-workers, for the fabrication of a 2D array: Dimitrov, A. S.; Nagayama, K. *Langmuir* **1996**, *12,* 1303.
32. Sato, O.; Kubo, S.; Gu, Z. Z. *Acc. Chem. Res.* **2009**, *42,* 1.
33. Galisteo-López, J. F.; Ibisate, M.; Sapienza, R.; Froufe-Pérez, L. S.; Blanco, Á.; López, C. *Adv. Mater.* **2011**, *23,* 30.
34. Joannopoulos, J. D.; Villeneuve, P. R.; Fan, S. *Nature* **1997**, *386,* 143.
35. Xia, Y.; Gates, B.; Li, Z. Y. *Adv. Mater.* **2001**, *13,* 409.
36. López, C. *Adv. Mater.* **2003**, *15,* 1679.
37. Ozin, G. A.; Arsenault, A. C. *Mater. Today* **2008**, *11,* 44.
38. Alexeev, V. L.; Sharma, A. C.; Goponenko, A. V.; Das, S.; Lednev, I. K.; Wilcox, C. S.; Finegold, D. N.; Asher, S. A. *Anal. Chem.* **2003**, *75,* 2316.
39. Gates, B.; Xia, Y. *Adv. Mater.* **2001**, *13,* 1605.
40. Gu, Z. Z.; Uetsuka, H.; Takahashi, K.; Nakajima, R.; Onishi, H.; Fujishima, A.; Sato, O. *Angew. Chem. Int. Ed.* **2003**, *115,* 922.
41. Fudouzi, H.; Xia, Y. *Langmuir* **2003**, *19,* 96513.
42. Li, H.; Marlow, F. *Chem. Mater.* **2005**, *17,* 3809.
43. Thylén, L.; Qiu, M.; Anand, S. *ChemPhysChem* **2004**, *5,* 1268.
44. Birner, A.; Wehrspohn, R. B.; Gösele, U. M.; Busch, K. *Adv. Mater.* **2001**, *13,* 377.
45. Scrimgeour, J.; Sharp, D. N.; Blanford, C. F.; Roche, O. M.; Denning, R. G.; Turberfield, A. J. *Adv. Mater.* **2006**, *18,* 1557.
46. Braun, P. V.; Rinne, S. A.; García-Santamaría, F. *Adv. Mater.* **2006**, *18,* 2665.
47. Lee, W.; Pruzinsky, A.; Braun, P. V. *Adv. Mater.* **2002**, *14,* 271.
48. Pruzinsky, S. A.; Braun, P. V. *Adv. Funct. Mater.* **2005**, *15,* 1995.
49. Palacios-Lidón, E.; Galisteo-López, J. F.; Juárez, B. H.; López, C. *Adv. Mater.* **2004**, *16,* 341.
50. Yan, Q.; Zhou, Z.; Zhao, X. S.; Chua, S. J. *Adv. Mater.* **2005**, *17,* 1917.
51. Vekris, E.; Kitaev, V.; von Freymann, G.; Perovic, D. D.; Aitchison, J. S.; Ozin, G. A. *Adv. Mater.* **2005**, *17,* 1269.
52. Tétreault, N.; Mihi, A.; Míguez, H.; Rodríguez, I.; Ozin, G. A.; Meseguer, F.; Kitaev, V. *Adv. Mater.* **2004**, *16,* 346.
53. Fleischhaker, F.; Arsenault, A. C.; Schmidtke, J.; Zentel, R.; Ozin, G. A. *Chem. Mater.* **2006**, *18,* 5640.
54. St. J. Russell, P. *Science* **2003**, *299,* 385.
55. Knight, J. C. *Nature* **2003**, *424,* 847.

56. Reeves, W. H.; Knight, J. C.; St. J. Russell, P.; Roberts, P. J. *Opt. Express* **2002**, *10,* 609.

57. Ortigosa-Blanch, A.; Knight, J. C.; Wadsworth, W. J.; Arriaga, J.; Mangan, B. J.; Birks, T. A.; St. J. Russell, P. *Opt. Lett*. **2000**, *25,* 1325.

58. Neinhuis, C.; Barthlott, W. *Ann. Bot*. **1997**, *79,* 667.

59. Gao, X. F.; Jiang, L. *Nature* **2004**, *432,* 36.

60. Young, T. *Trans. R. Soc*. **1805**, *95,* 65.

61. Wenzel, R. N. *Ind. Eng. Chem*. **1936**, *28,* 988.

62. Cassie, A.; Baxter, S. *Trans. Faraday. Soc*. **1944**, *40,* 546.

63. Gu, Z. Z.; Uetsuka, H.; Takahashi, K.; Nakajima, R.; Onishi, H.; Fujishima, A.; Sato, O. *Angew. Chem., Int. Ed*. **2003**, *42,* 894.

64. Sun, C.; Gu, Z. Z.; Xu, H. *Langmuir* **2009**, *25,* 12439.

65. Gu, Z. Z.; Fujishima, A.; Sato, O. *Appl. Phys. Lett*. **2004**, *85,* 5067.

66. Zhao, Y. J.; Zhao, X. W.; Gu, Z. Z. *Adv. Funct. Mater*. **2010**, *20,* 2970.

67. Zhao, X. W.; Cao, Y.; Ito, F.; Chen, H. H.; Nagai, K.; Zhao, Y. H.; Gu, Z. Z. *Angew. Chem. Int. Ed*. **2006**, *45,* 6835.

68. Sun, C.; Zhao, X. W.; Zhao, Y. J.; Zhu, R.; Gu, Z. Z. *Small* **2008**, *4,* 592.

69. Zhao, Y. J.; Zhao, X. W.; Tang, B. C.; Xu, W. Y.; Li, J.; Hu, J.; Gu, Z. Z. *Adv. Funct. Mater*. **2010**, *20,* 976.

70. Li, J.; Zhao, X. W.; Zhao, Y. J.; Gu, Z. Z. *Chem. Commun*. **2009**, 2329.

71. Li, M. Z.; He, F.; Liao, Q.; Liu, J.; Xu, L.; Jiang, L.; Song, Y. L.; Wang, S.; Zhu, D. B. *Angew. Chem. Int. Ed*. **2008**, *47,* 7258.

72. Ganesh, N.; Block, I. D.; Cunningham, B. T. *Appl. Phys. Lett*. **2006**, *89,* 023901.

73. Hu, X.; Huang, J.; Zhang, W.; Li, M.; Tao, C.; Li, G. *Adv. Mater*. **2008**, *20,* 4074.

74. Alexeev, V. L.; Das, S.; Finegold, D. N.; Asher, S. A. *Clin. Chem*. **2004**, *50,* 2353.

75. Lee, K.; Asher, S. A. *J. Am. Chem. Soc*. **2000**, *122,* 9534.

76. Asher, S. A.; Sharma, A. C.; Goponenko, A. V.; Ward, M. M. *Anal. Chem*. **2003**, *75,* 1676.

77. Holtz, J. H.; Asher, S. A. *Nature* **1997**, *389,* 829.

78. Zhao, Y. J.; Zhao, X. W.; Hu, J.; Xu, M.; Zhao, W. J.; Sun, L. G.; Zhu, C.; Xu, H.; Gu, Z. Z. *Adv. Mater*. **2009**, *21,* 569.

79. Lee, H.; Kim, J.; Kim, H.; Kim, J.; Kwon, S. *Nat. Mater*. **2010**, *9,* 745.

80. Vigneron, J. P.; Pasteels, J. M.; Windsor, D. M.; Vértesy, Z.; Rassart, M.; Seldrum, T.; Dumont, J.; Deparis, O.; Lousse, V.; Biró, L. P.; Ertz, D.; Welch, V. *Phys. Rev. E* **2007**, *76,* 031907.

81. Liu, F.; Dong, B. Q.; Liu, X. H.; Zheng, Y. M.; Zi, J. *Opt. Express* **2009**, *17,* 16183.

82. Kubo, S.; Gu, Z. Z.; Takahashi, K.; Ohko, Y.; Sato, O.; Fujishima, A. *J. Am. Chem. Soc*. **2002**, *124,* 10950.

83. Kubo, S.; Gu, Z. Z.; Takahashi, K.; Fujishima, A.; Segawa, H.; Sato, O. *J. Am. Chem. Soc*. **2004**, *126,* 8314.

84. Han, G. Z.; Xie, Z. Y.; Zheng, D.; Sun, L. G.; Gu, Z. Z. *Appl. Phys. Lett*. **2007**, *91,* 141114.

85. Weissman, J. M.; Sunkara, H. B.; Tse, A. S.; Asher, S. A. *Science* **1996**, *274,* 959.
86. Takeoka, Y.; Watanabe, M. *Langmuir* **2003**, *19,* 9104.
87. Nakayama, D.; Takeoka, Y.; Watanabe, M.; Kataoka, K. *Angew. Chem*. **2003**, *115,* 4329.
88. Sharma, A. C.; Jana, T.; Kesavamoorthy, R.; Shi, L.; Virji, M. A.; Finegold, D. N.; Asher, S. A. *J. Am. Chem. Soc*. **2004**, *126,* 2971.
89. Fudouzi, H.; Xia, Y. *Langmuir* **2003**, *19,* 9653.
90. Fudouzi, H.; Xia, Y. *Adv. Mater*. **2003**, *15,* 892.
91. Fudouzi, H.; Sawada, T. *Langmuir* **2006**, *22,* 1365.
92. Arsenault, A. C.; Miguez, H.; Kitaev, V.; Ozin, G. A.; Manners, I. *Adv. Mater*. **2003**, *15,* 503.
93. Arsenault, A. C.; Puzzo, D. P.; Manners, I.; Ozin, G. A. *Nat. Photonics* **2007**, *1,* 468.
94. Puzzo, D. P.; Arsenault, A. C.; Manners, I.; Ozin, G. A. *Angew. Chem. Int. Ed*. **2009**, *48,* 943.
95. Xu, X.; Friedman, G.; Humfeld, K. D.; Majetich, S. A.; Asher, S. A. *Chem. Mater*. **2002**, *14,* 1249.
96. Sacanna, S.; Philipse, A. P. *Langmuir* **2006**, *22,* 10209.
97. Sacanna, S.; Rossi, L.; Kuipers, B. M. W.; Philipse, A. P. *Langmuir* **2006**, *22,* 1822.
98. Ge, J.; Hu, Y.; Yin, Y. *Angew. Chem. Int. Ed*. **2007**, *46,* 7428.
99. Ge, J.; Hu, Y.; Zhang, T.; Huynh, T.; Yin, Y. *Langmuir* **2008**, *24,* 3671.
100. Ge, J.; He, L.; Goebl, J. S.; Yin, Y.; *J. Am. Chem. Soc*. **2009**, *131*(*10*), 3484.
101. Ge, J.; Yin, Y. *Adv. Mater*. **2008**, *20,* 3485.
102. Zhu, C.; Chen, L.; Xu, H.; Gu, Z. Z. *Macromol. Rapid Commun*. **2009**, *30,* 1945.
103. Kim, H.; Ge, J.; Kim, J.; Choi, S.; Lee, H.; Lee, H.; Park, W.; Yin, Y.; Kwon, S. *Nat. Photonics* **2009**, *3,* 534.
104. Zhu, C.; Xu, W.; Chen, L.; Zhang, W.; Xu, H.; Gu, Z. Z. *Adv. Funct. Mater*. **2011**, *21,* 2043.

CHAPTER 11

Biomimetic Principles in Macromolecular Science

WOLFGANG H. BINDER, MARLEN SCHUNACK, FLORIAN HERBST, and BHANUPRATHAP PULAMAGATTA

Lehrstuhl Makromolekulare Chemie, Fakultät f. Naturwissenschaften II, Institut f. Chemie, Martin-Luther-University Halle-Wittenberg, Von-Danckelmann-Platz 4, D-06120 Halle, Germany

11.1 INTRODUCTION

> And God said, "Let the earth bring forth living creatures according to their kinds—livestock and creeping things and beasts of the earth according to their kinds." And it was so.
>
> —Genesis 1:24

As a theologist would preach, so a scientist would soberly state that, while less prosaic, evolution has never been able to separate living and nonliving matter. When structural features of a living cell are subjected to close inspection, polymeric matter of all kinds can be seen acting together on a surprisingly tiny scale; using biopolymers for ordering, self-assembly, recognition, and reproduction processes.

Thus, the linear biopolymers called *proteins* are the main structural elements of enzymes and structural tissue, using only 21 amino acids as repetitive building blocks (known as *monomers* in synthetic polymer science). It needs only the shuffling of these 21 monomeric units by God (or evolution) to build all of the essential helices, folded β-sheets, and superfolded, highly organized large protein complexes of dimensions 10–100 nm.

Being a significantly more primitive molecule, DNA as a complex, helically folded, self-assembling (supramolecular) polymer needs only four nucleobases attached to a repetitive backbone of D-ribofuranose and phosphate units to store all the genetic information specific to humans.

Bioinspiration and Biomimicry in Chemistry: Reverse-Engineering Nature, First Edition.
Edited by Gerhard F. Swiegers.
© 2012 John Wiley & Sons, Inc. Published 2012 by John Wiley & Sons, Inc.

By contrast, the synthetic polymer scientist has hundreds of monomeric molecules at hand, which, when polymerized into often poorly defined macromolecules of varying chain length (i.e., high polydispersity) and extremely low (if any) sequence specificity, assemble into only partially regular structures such as microphase separated or partially crystalline arrangements.

So, why write a chapter about biomimetic polymers, when human attempts to mimic Nature in the form of synthetically fabricated polymer molecules can only be described as ineffective to say the least?

As it turns out, the past decades have seen immense progress in the design, fabrication, and application of "bioinspired" polymers, often taking advantage of "biomimetic building principles" for polymer design.[1] Thus, for example, significant developments in the field of medical polymers have led to a deep understanding of the interaction between polymer surfaces and the immune system of mammals.[4] These developments have culminated in the control of stem-cell differentiation by management of polymeric elasticity,[2, 3] which in turn is highly important in tissue engineering.[4]

Another brilliant example involves the design of self-healing materials, where the idea of blood vessels and their subsequent repair after rupture[5] has been mimicked by polymer scientists using capsules filled with reactive polymers, that are able to heal by catalytic crosslinking reactions.[6]

A plethora of structural design using polymers[7] has been taken over from Nature in fields such as biomineralization[8], artificial spider silk[9], elastic polymeric nanofibers[10] and nanocomposites,[11] lightweight design of (cellular) materials, or self-cleaning surfaces by control of micro- and nanostructure.[12]

This chapter therefore hopes to convince the reader that some concepts in polymer science are brilliant at mimicking Nature in terms of structure, function, and concept. As these features of biomimetic polymers are too multitudinous to be discussed here, only the following aspects are covered in this chapter:

- Principles of polymer science for synthesizing polymers and comparing these methodologies to natural processes. Major differences in the structural concept and also the mode of chain growth are discussed.
- Structures in biomimetic polymers that resemble structures in Nature. The basic aim is *not* to focus on synthetic polymers bearing peptide and proteinaceous elements to guide the assembly into "bio-like" ordered structures, but to demonstrate the structural diversity of synthetic polymers forming regular structures *similar* to helices, β-strands, and other natural (supramolecular) assemblies.
- Motility and movement via synthetic polymers, featuring mechanical work exerted by linear polymers and polymer networks, including lubrication.
- Mimicking binding and subsequent catalysis via synthetic polymeric structures in analogy to the binding pockets of enzymes.
- Generating materials that show self-healing properties, one of the main features of living matter and one of the humanity's greatest desires.

All other topics are *not* covered in this chapter, including fiber-reinforced materials, hybrid polymers, ion channels, poly- and oligo(saccharides), as well as any kind of polymers with a directly linked biomolecule, containing DNA-like or peptide, amino acid, or related biostructures.

11.2 POLYMER SYNTHESIS VERSUS BIOPOLYMER SYNTHESIS

11.2.1 Features of Polymer Synthesis

The synthesis of chain-like polymers is defined as the repetitive addition of monomers onto a growing polymer chain, thereby leading to the formation of an elongated chain by a classical chain-growth mechanism. The principles of such chain-growth polymerization reactions are based on (1) the usually exothermic reaction of each monomer addition ($\Delta H < 0$) and (2) a loss in reaction entropy ($\Delta S < 0$), which is explained by the loss of at least one translational degree of freedom when incorporating the monomer into the polymer chain.

Figure 11.1 gives some biochemical polymerization reactions and their transition states or catalytic cycles, in relation to well-known synthetic counterparts. The figure shows either the main catalytic cycle or the main transition state of elongation within protein biosynthesis (Figure 11.1a), RNA/DNA polymerization (Figure 11.1b), and poly(saccharide) biosynthesis (Figure 11.1c). All biochemical polymerization reactions rely on the following major features during chain growth:

1. The use of energetically highly activated monomeric units [such as aminoacyl-t-RNA as activated amino acids in protein biosynthesis, nucleotide triphosphates as building blocks of DNA/RNA, and UDP sugars as activated glycosyl donors in oligo- and poly(saccharide) synthesis] to achieve a thermodynamically favored reaction ($\Delta G < 0$) at each chain elongation step.
2. The use of a templated polymerization via hydrogen-bonding interactions of nucleotides, to generate sequence specificity using an already existing polymer strand and thereby directing the synthesis of the newly formed strand [in peptide (Figure 11.1a) and DNA/RNA (Figure 11.1b) synthesis].
3. The use of biocatalysts, which are usually highly specific enzyme complexes, to direct the polymerization reaction, controlling regio- and stereochemistry of the newly formed bonds. Important for this endeavor is the avoidance of undesired side reactions, such as chain-transfer reactions or termination reactions.

Biochemical polymerization reactions typically avoid highly reactive intermediates by use of active phosphates. Most polymerization reactions in Nature can therefore be classified as polycondensations, where phosphate ions are released during the polymerization reaction in each step. By contrast, synthetic polymers are usually formed by highly reactive intermediates like anions/cations (Figure 11.1d), or radicals (Figure 11.1e), or via polycondensation (Figure 11.1f) using elevated

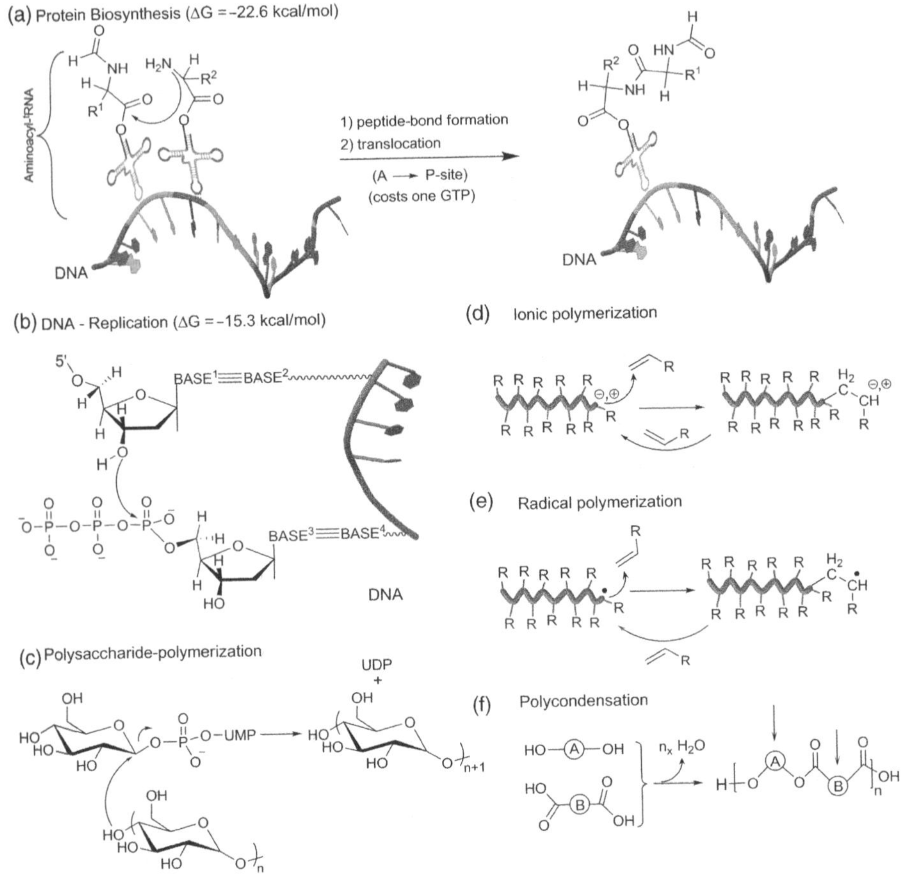

Figure 11.1 Transition state or catalytic cycle of biochemical (a–c) and synthetic (d–f) polymerization reactions.

reaction temperatures or catalytic polymerization reactions that preclude conditions close to physiological environments. Additionally, the difference in binding energy between a double bond and two single bonds generates the driving force for the polymerization. This is totally different from biochemical polymerization reactions.

11.2.2 "Living" Chain Growth

A special case of polymerization reactions is classified as "living" polymerization reactions,[13] where special reaction kinetics allow the presence of highly reactive intermediates and thus a linear (uniform) growth of the polymer chains, together with a lack of side reactions (such as chain-transfer or termination reactions).[14]

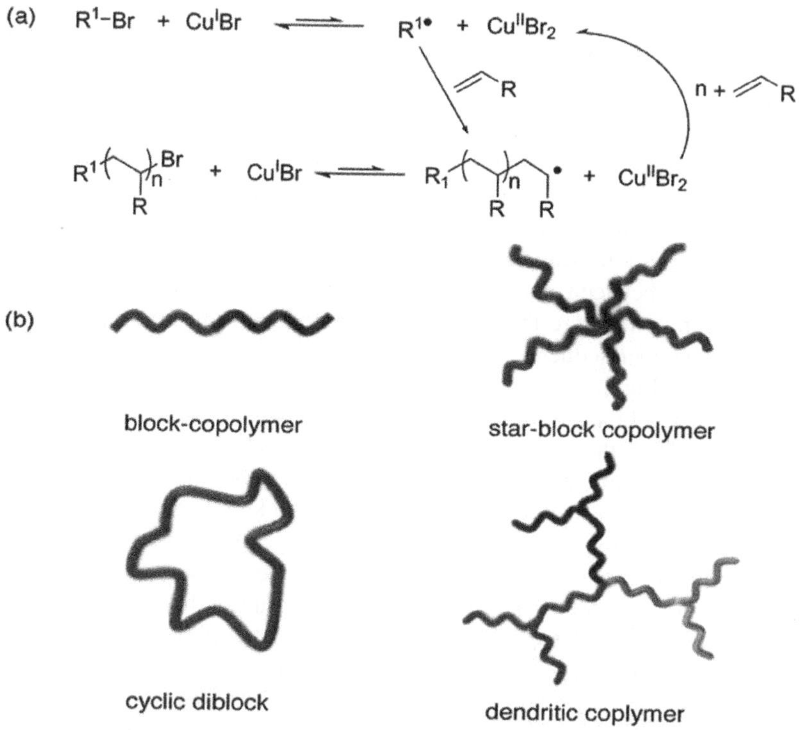

Figure 11.2 (a) Principle of a "living" radical polymerization reaction [atom transfer radical polymerization (ATRP)]. (b) Structural features of polymer architectures prepared via living polymerization methods.

Figure 11.2a shows a typical example of living (controlled) radical polymerization [atom transfer radical polymerization (ATRP)],[15] where an equilibrium between an active (free radical) species (R·) and a dormant (covalently bonded) species (R–Br) allows a significant reduction of the radical species by a factor of $\sim 10^6$. A Cu(I)-catalyzed electron-transfer process is responsible for conducting the polymerization reaction at moderate temperatures. Similar concepts to those of a living polymerization have been described for living radical polymerization [nitroxide mediated polymerization (NMP),[16] reversible addition fragmentation transfer (RAFT[17])]; ring opening metathesis polymerization (ROMP),[18] and living cationic polymerization.[19] An additional feature of these types of polymerization reactions is the possibility to restart the polymerization once new monomer is added, after the initial monomer has been fully consumed. Currently, the concept of living polymerization can be used to prepare, various polymer architectures (such as block, star, and graft polymers) with high structural fidelity as shown in Figure 11.2b.[20]

11.2.3 Aspects of Chain Length Distribution in Synthetic Polymers: Sequence Specificity and Templating

In contrast to biochemically synthesized polymers, synthetic polymers are not sequence specific and do show polydispersity [or chain length distribution, which is given by the ratio of the weight average (M_w) versus the number average (M_n) molecular weight]. In Nature, polymers usually use templates to control those two important features, thus being able to duplicate DNA/RNA or transform a sequence of DNA into a protein sequence on the ribosomes via the use of highly specific hydrogen bonds between nucleobases (see Figures 11.1a and 11.1b for details). The resulting polymers are then uniform (monomodal) with respect to their chain length and monomer sequence.

As shown in Figure 11.3a, template polymerization process requires a primer strand (T), which binds the monomeric units via noncovalent forces, thus positioning them along the primary strand for the actual polymerization reaction.[21] Experimentally, a template-driven polymerization can be proved by an increase of the polymerization rate with the concentration of the template (T). Sequence specificity as well as a singular chain length can be the consequence if the underlying scaffolding process displays sufficient specificity (i.e. an association energy high enough to overcome the thermal energy, leading to unscaffolded polymerization). Several review papers have addressed these issues, leading to some more controlled (radical) polymerization reactions with a reduced polydispersity (M_w/M_n) compared to the untemplated process.[22, 23]

Usually, the selectivity of the synthetic processes is not sufficient to generate sequence-specific polymers in reasonable fidelity, as the competitive nontemplated polymerization reactions are significant. However, some highly developed examples have been reported by use of redox polymerization of monomers (pyrrole, thiophene), taking place on a DNA template to arrange monomers close to a helical scaffold; subsequently inducing electropolymerization of the monomers close to the DNA scaffold.[23]

Datta and Schuster[24] have reported one of the first working examples of a synthetic polymerization reaction encoded by a duplex strand of DNA (see Figure 11.3b). A DNA scaffold consisting of modified cytosines is used to direct the oxidative polymerization of poly(aniline) (PANI) oligomers by use of a peroxidase-mediated (mild) polymerization reaction. Basically, a high level of sequence-specific polymer is obtained by this comparatively complex method.

Another strategy to prepare sequence-specific polymers via a synthetic polymerization process relies on the extremely rapid polymerization of stoichiometrically added, different N-substituted maleimide monomers, which show a rapid addition during the ATRP process (see Figure 11.3c).[25] Critical for (partial) success is a strong differential reactivity of the respective maleimide monomers, allowing a rapid monomer addition so that each growing polymer chain ideally reacts with only one monomeric unit. Despite the fact that only one or two of the respective monomers are added within the growing chain, a moderate level of sequence

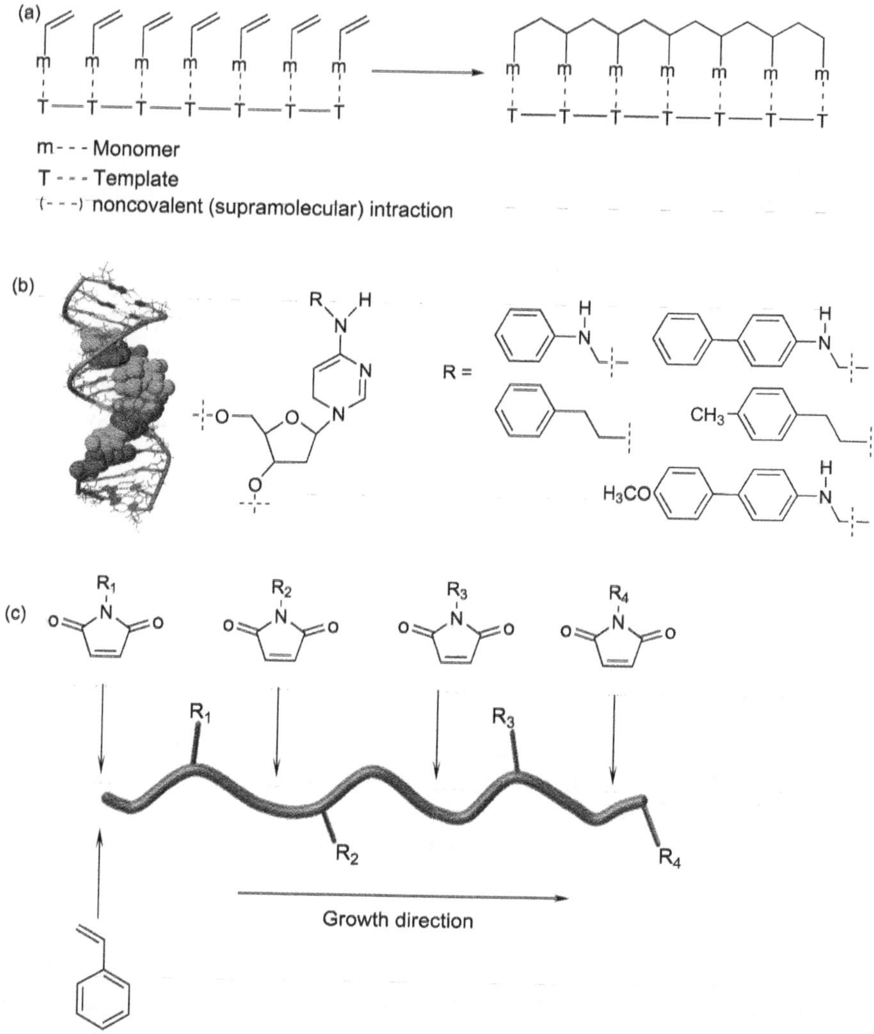

Figure 11.3 Templated polymerization reactions. (a) Principle of a templated polymerization reaction, relying on the scaffold (T), the noncovalent force (—) between the template, and the monomer (M). (b) DNA-mediated scaffold polymerization of poly(aniline) via redox-mediated polymerization. (c) Approach to (partially) sequence selective polymers by use of functionalized maleimides polymerized via ATRP.

specificity could be obtained. Therefore, the quest for a truly biomimetic polymerization, leading to the in vitro chemical synthesis of a monodisperse, sequence-specific polymer consisting of nonbiochemical monomers in a nonrepetitive process, has not been reached and remains a vision for the future.

11.3 BIOMIMETIC STRUCTURAL FEATURES IN SYNTHETIC POLYMERS

11.3.1 Helically Organized Polymers

Among β-sheets and β-turns the helix is one of Nature's main secondary structure elements. Its most prominent representative is DNA with its double helical structure, as first discovered by Watson and Crick in 1953.[26] As there are many interesting functions carried out by helical biomacromolecules, including molecular recognition, information storage, and catalysis, chemists are interested to adapt Nature's ideas for the synthesis of helically organized polymers. The helix is an inherently chiral structure and so right- and left-handed helices are nonidentical mirror images. Materials with excess of one of the two species may be optically active and thus also interesting for the separation of enantiomers and asymmetric catalysis.

Beside the necessity to find methods to introduce an excess of one helical sense, the second quest is to generate materials with stable compact solution structures like those present in DNA. In 1955, Natta and co-workers found that highly stereoregular isotactic polypropylene (PP) has a helical structure in the solid state, due to the presence of crystalline domains.[27] However, in solution PP is totally dynamic in character and consists only of short helical segments separated by frequently occurring helical reversals among disordered and random coil conformations; this results in an optically inactive solution.

Optically active and helix-forming polymers can be relatively easily prepared by using either chiral monomers, monomers with chiral side groups or chiral additives. Such systems will not be discussed in this chapter.

One approach to the synthesis of helical polymers is *asymmetric polymerization*, in which chirality is introduced into a polymer chain via the polymerization reaction. As a subclass of asymmetric polymerization, *helix-selective polymerization* involves generating helical polymers whose chirality is based on a helical conformation with an excess of one screw-sense. If the right- or left-handed helix is synthesized preferentially, the resulting polymer can be optically active. An example of a helix-selective polymerization is the synthesis of poly(triphenylmethyl methacrylate) [poly(TrMA), **3**] first reported by Okamoto et al.[28] (see Figure 11.4).

Poly(TrMA) was synthesized via living anionic polymerization of the monomer TrMA (**2**) using a (−)-sparteine-*n*-BuLi (**1**) complex as initiator. The resulting polymer was highly isotactic (>99%) and the helical structure persisted in the liquid state since the helical conformation is stabilized by steric repulsion of the bulky triphenylmethyl groups. Although a chiral initiator complex was necessary to realize a helical conformation with an excess helical sense in solution, helical poly(TrMA) was prepared from an achiral monomer. The polymer contained no chiral components; thus, chirality was only caused by the helicity. By replacing the bulky side groups with less steric demanding methyl groups (resulting in a highly isotactic PMMA) (**4**), the hindrance for rotation of the polymer chain is lowered drastically. As a result, the predominant helical conformation, and thus the optical

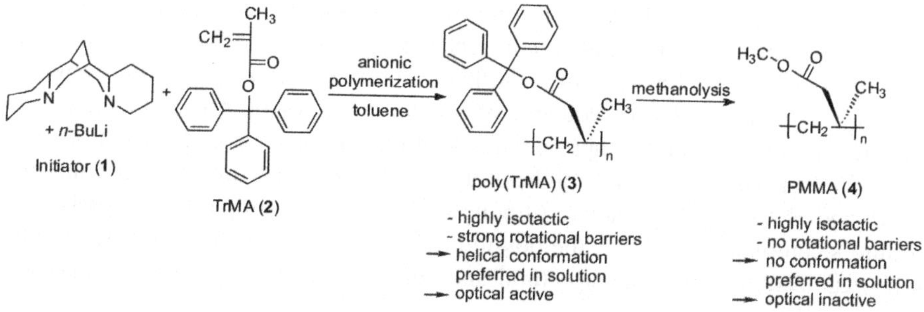

Figure 11.4 Synthesis of poly(triphenylmethyl methacrylate) [poly(TrMA)].

activity is lost. Further examples of constrained helices are polyisocyanates and polycarbodiimides.[29]

Another approach to helical synthetic materials are the so-called foldamers.[30] Although foldamers are oligomers rather than polymers, they fold into a conformationally ordered state (e.g. a helix) in solution. These structures are stabilized by a collection of noncovalent interactions between nonadjacent monomer units. Due to the dynamic, reversible character of the noncovalent interactions their structure in solution is dynamic, including folding and unfolding reactions into a preferred conformation (or a set of congruent conformations).

Lehn and co-workers reported on the self-generation of an extended helical structure from achiral, alternating pyridine–pyrimidine heterocycles (see Figure 11.5b).[31] The system is based on the preference of 2,2′-bipyridine to adopt a transoid (planar) conformation in solution, while the cisoid (nonplanar) form is about 5.7 kcal mol^{-1} less stable (Figure 11.5a). The helical conformation derives

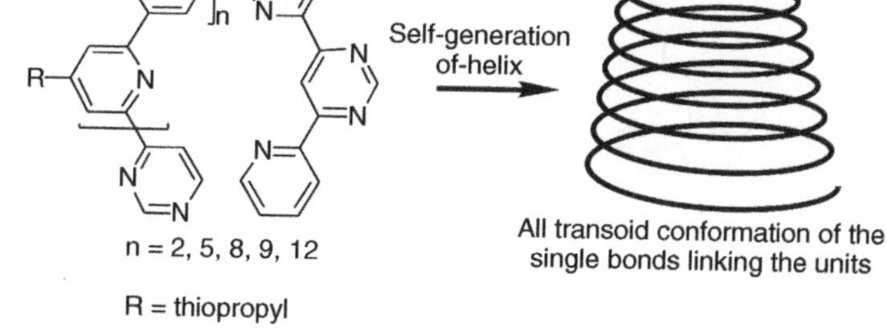

Figure 11.5 Pyridine–pyrimidine heterocycles.

from the steric repulsion between the =CH-groups and the electrostatic interaction stemming from nitrogen atoms on adjacent repeat units, as well as the all-transoid conformation of the linked units. Further important factors for helix-formation are the correct number and sequence of heterolytic aromatic rings and the proper linking of the rings at the appropriate positions. In this case the creation of a chiral, helical structure from polyheterolytic strands results in a racemic mixture of *M* and *P* forms of the helix. Although the system is simple in comparison to Nature's complexity, it shows the possibility of forming a helical superstructure by self-organization, where the structural and conformational information is encoded in the polyheterolytic strand.

Besides the above-mentioned single-stranded helices, double- as well as multistranded helices were developed. Lehn and co-workers used helical metal complexes containing organic strands intertwined around the metal centers (Figure 11.6).[32]

Metal coordination complexes of this type are also called "helicates." A string of 2,2′-bipyridine groups, separated by a spacer unit (flexible ether bond) form a double helical structure (Figure 11.6b) by the addition of copper(I) ions. The driving force is the (geometrically preferred) complexation of the tetrahedral Cu(I) ions by the 2,2′-bipyridine groups. By using thymidine-substituted ligands (Figure 11.6a),

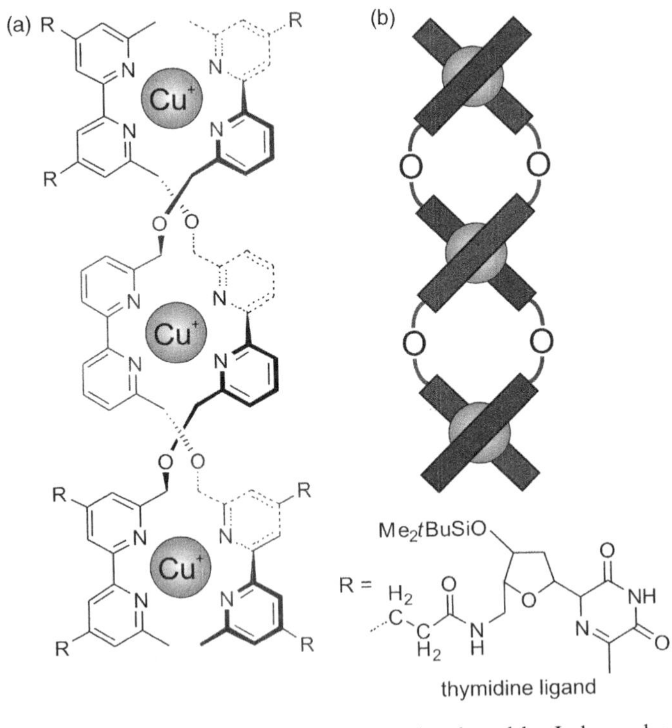

Figure 11.6 Helical metal complexes (helicates) developed by Lehn and co-workers.[32]

the resulting helical complex resembles DNA, although in contrast to the DNA, the thymidine residues are attached to the outer surface of the double helic.

Although many different approaches have been reported for the formation of artificial synthetic helices, there is still a lack of structural perfection and functionality in the mimicry of Nature's complexity.

11.3.2 β-Sheets

While various principles based on synthetic polymers have been applied to the formation of helical structures, there is a lack of systems for the formation of artificial β-sheets made from synthetic polymers. Most of the reported β-sheets originate from hybrid materials, consisting of a synthetic (e.g., polymer chain) and a biological moiety (e.g., oligo- or polypeptides).[33–35] Although even common polymerization reactions like ATRP[36] or the azide/alkyne "click" reaction[37] were applied to the synthesis of these materials, the monomers are peptide-containing and thus the principle is rather *peptidomimetic* than *biomimetic* from the viewpoint of a synthetic polymer chemist.

However, a few systems were reportedly made from synthetic polymers. Kendhale et al.[38] reported isotactic *N*-alkyl acrylamide oligomers (Figure 11.7a) and proved that their structures were β-sheets in the solid state (Figure 11.7b) as well as in solution. Although this material is not a polymer in a narrow sense since it is prepared by a nonpolymerizing synthetic strategy and only four units were connected, the polymeric analog would be an interesting candidate for artificial synthetic β-sheets if the lack of a suitable polymerization technique is solved.

Another polymer-related attempt was reported by Uemura et al.[39] using ligand 5 for the formation of a supramolecular polymer by complexation of silver(I) ions in the crystalline state (Figure 11.8). The nonsymmetric coordination environment around Ag^{I} atoms induces directionality into the chains (Figure 11.8b), while adjacent chains were aligned via complementary hydrogen bonds between the amide moieties of ligand 5 (Figure 11.8a). Although this arrangement is conceptually

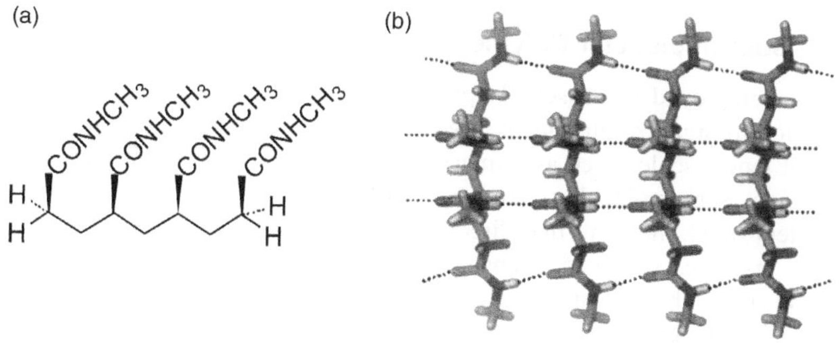

Figure 11.7 (a) Chemical structure of *N*-alkyl acrylamide; (b) single X-ray structure. (H bonds are highlighted with black dots). (Reproduced with permission. Copyright the Royal Society of Chemistry: Ref. 38.)

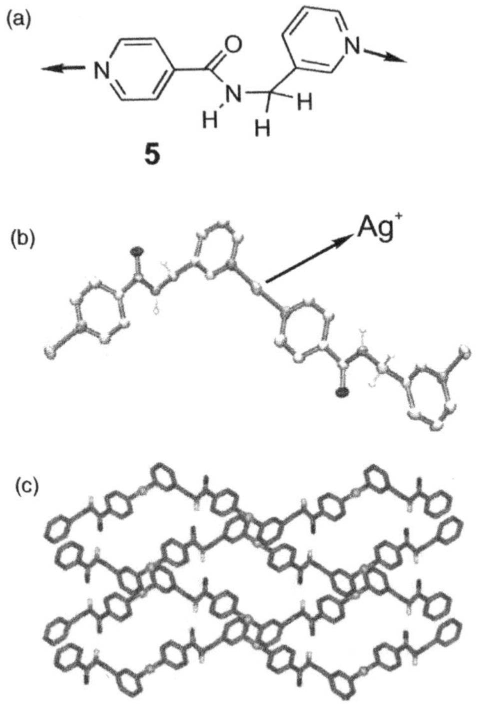

Figure 11.8 (a) Ligand with amide moiety; arrows indicate binding to Ag^+ ion; (b) formation of chains due to interactions with Ag^+ ions; (c) alignment of adjacent chains due to hydrogen bonds. (Reproduced with permission. Copyright Wiley-VCH Verlag GmbH & Co. KGaA: Ref. 39.)

similar to the β-sheet motif of proteins, it is limited in this case to the crystalline state.

11.3.3 Supramolecular Polymers

Nature has employed self-assembly processes quite efficiently from the beginning of evolution to organize biological molecules in the hierarchy of architectures over a range of length scales such as primary, secondary, and tertiary structures in DNA and proteins. Supramolecular chemistry can be described as the chemistry behind the covalent bond, meaning that directed secondary interactions were used to promote, for instance, a defined arrangement of small molecules (as in Section 11.6, self-healing). The broad variety of these interactions includes dipole–dipole interactions, $\pi-\pi$ stacking, ionic interactions, metal complexes, and hydrogen bonds. These noncovalent interactions are reversible, highly directed (key–lock systems) in many cases, and highly dynamic. The aforementioned characteristics and interactions are found in biological systems as well. Thus, the possibility to design defined supramolecular systems gives the chance to mimic properties and functionalities of biological systems.

The principle of supramolecular chemistry was introduced by J. M. Lehn[40] into the field of polymer chemistry. Lehn and co-workers[41] performed experiments using bifunctional, low molecular weight molecules bearing either the Hamilton receptor (**6**) or barbituric acid group (**7**) (see Figure 11.9a). The Hamilton receptor and barbituric acid undergo a strong and directed association (known as a "key–lock" system). By use of an inert solvent (such as tetrachloroethane), the bifuntional compounds **6** and **7** in an equimolar ratio form supramolecular gels, as confirmed by viscosity and NMR measurements. The magnitude of association can be derived

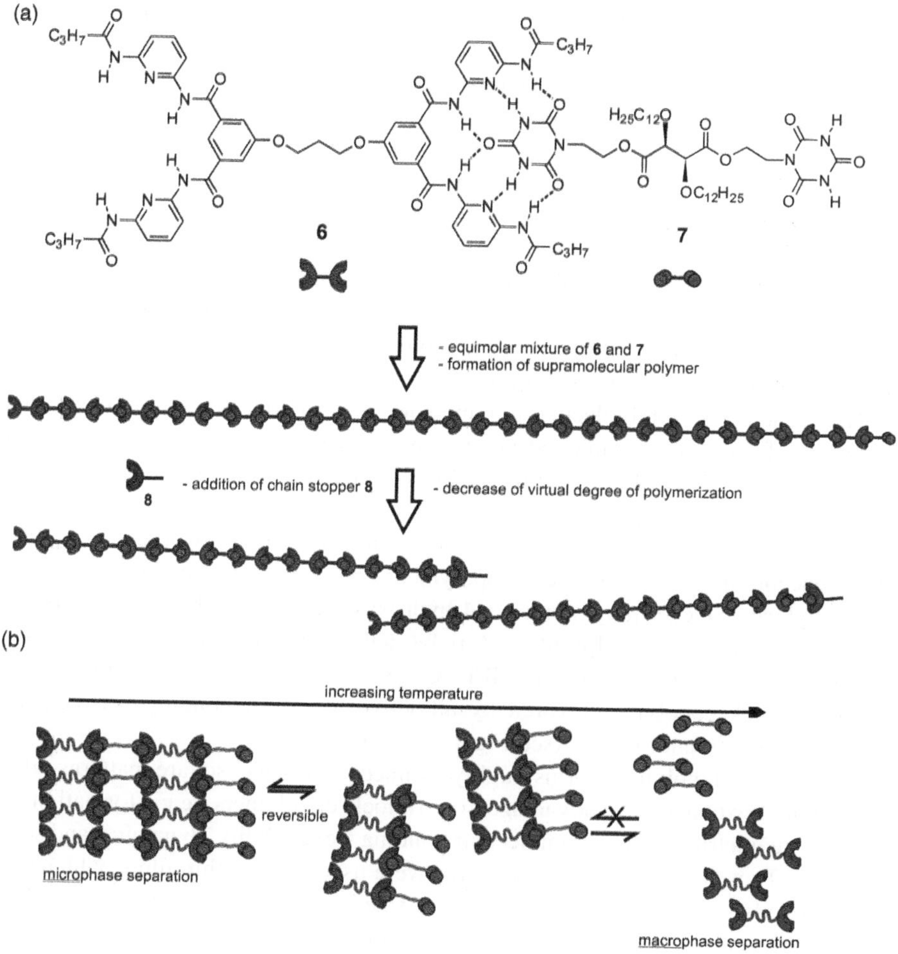

Figure 11.9 Formation of a linear supramolecular polymer and influence of chain stoppers. (a) Formation of a linear supramolecular and decrease of the DP by addition of chain stopper **8**. (b) Influence of temperature on a supramolecular BCP.

from the broadness of the N–H NMR resonances. Although the polymer is made of reversibly attached low molecular weight molecules, it can be characterized in terms of a virtual degree of polymerization (DP) and, thus, in terms of a virtual molecular weight. The DP strongly depends on the association rate constant of the two functional groups **6** and **7** and is therefore dependent on the temperature and polarity of the solvent.

The reversibility and dynamics of the supramolecular polymer were proved by addition of chain stopper molecule **8** in different concentrations. Since **8** is monofunctional, its addition causes an immediate and distinct decrease of the virtual DP (see Figure 11.9a), as evidenced by a prompt drop of the viscosity and confirmed by NMR measurements.

Since hydrogen bonds are one of the most prominent interactions for the formation of supramolecular polymers,[42] it was also applied to the synthesis of various building blocks capable of forming supramolecular block copolymers (BCPs)[43–46] (also see Section 11.3.4). The attachment of functional groups, capable of undergoing strong and directed hydrogen bonding,[43] opens the possibility of combining macrophase separating polymers to form microphase separated BCPs.[47] As shown in a general scheme (Figure 11.9b) the attractive interactions cause alignment of the two blocks into a defined BCP-like structure. By increasing the temperature and, thus, lowering the magnitude of aggregation, the size of the structure decreases. The initial size is recovered to the same extent, when the temperature is decreased. The reversibility is irrevocably lost when the temperature is increased to the point where hydrogen bonding is not possible anymore. Without this attractive interaction, irreversible macrophase separation takes place.

The capacity of supramolecular polymers to undergo directed assembly can also be used to form more complex structures in comparison to the above-mentioned simple, linear or sheet-like assembly. Self-assembled structures are well known in Nature, for example, molecules like lipids form both nanostructures (vesicles and micelles) and thin films (membranes). Hawker and co-workers[48] applied this concept by using diblock copolymers stabilized via hydrogen bonds. Poly(ethylene oxide)-*b*-poly(styrene-*r*-4-hydroxystyrene) (**9**) [PEO-*b*-P(S-*r*-4HS)] and poly(styrene-*r*-4-vinylpyridine)-*b*-poly(methyl methacrylate) (**10**) [P(S-*r*-4VP-*b*-PMMA)] were used as an A-B/B′-C system, which forms a supramolecular A-B-C triblock copolymer favored by attractive hydrogen bonding between the modified poly(styrene) blocks (see Figure 11.10).

The resulting polymer forms highly defined nanoscale square patterns, due to the combination of the advantages of triblock copolymers (e.g., formation of complex geometries) and supramolecular bonding (reversibility, dynamics, and self-ordering). These nanoscale square patterns were used to form thin films, which can be employed as a template for fabrication of size integrated circuits by simple photodegradation of the PMMA block. Such circuits were considerably smaller than those produced by common and expensive photolithography, revealing the possibility to drastically increase the amount of stored data. This example shows how Nature's concepts can be applied to modern technological challenges.

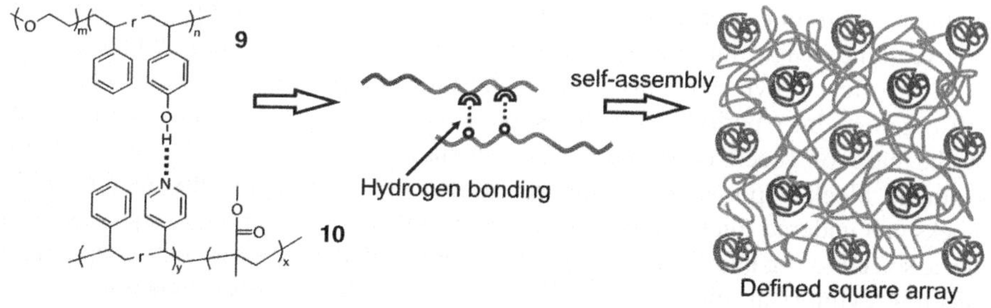

Figure 11.10 Formation of a defined square array due to attractive hydrogen bonding.

However, there is one main problem when applying synthetic supramolecular polymers for biological applications: water. Water is omnipresent in all biological systems and especially affects the designed hydrogen-bonding interaction since it can compete with the hydrogen-bonding moieties and reduce the extent of association drastically (e.g., see Figure 11.9 between molecules **6** and **7**).

Despite this, there are many applications of synthetic supramolecular polymers in biomedical systems using interactions other than hydrogen bonding.[49, 50] One common example is protonated poly(ethylenimine)[51] (PEI) (see Figure 11.11). The polymer cation belongs to a class called pH-sensitive polymers. It buffers the pH, in the case of PEI, over a wide range. Furthermore, PEI has one special attribute: it will spontaneously adhere to and condense DNA to form PEI/DNA complexes. Due to the positive charge along the PEI, it can balance out the negative charge of the DNA and condense it to PEI/DNA complexes of ~100 nm in size. These complexes can be transported into a cell, thus being a feasible candidate for gene delivery systems and gene therapy.

Besides the above-mentioned examples, there are multiple other systems where supramolecular interactions were applied, for instance, in delivery of therapeutics, bioseparation, tissue engineering, and cell culture.[52, 53]

11.3.4 Self-Assembly of Block Copolymers

The cell membrane and its compartments are made by self-organization of bilayer amphiphilic lipid molecules. This is also one of the most important examples of self-assembled molecular structures in Nature. It is the compartmentalization of cells that enables great control over enzyme reaction order and protection of a cell against its harmful contents, and also serves as scaffolds to precisely decorate with biomolecules that act as recognition elements. It has been a challenge to

$$H_3C-CH_2-NH\left[CH_2-CH_2-NH\right]_n CH_2-CH_2-\overset{+}{N}H_3$$

Figure 11.11 Poly(ethylenimine) (PEI) in its linear form (but usually used in its more complex, branched form).

synthetically mimic such biological structures in Nature and in this section we try to give some insight into synthetically fabricated structures using polymers.

Most often, polymer chains behave like tangled spaghetti but sometimes they self-organize into periodic structures on a nanoscale (5–100 nm), particularly in the case of block copolymers. Like natural amphiphilic molecules (lipids), synthetic block copolymers comprised of two or more chemically incompatible blocks can undergo microphase separation and self-organize in dilute aqueous or in block-selective organic solution to generate aggregated structures such as spherical micelles, cylindrical micelles, and membranes (vesicles or polymersomes).

Block copolymers are broadly classified into *linear block copolymers*, comprised of polymer chains arranged in sequential manner, *nonlinear block copolymers* like star or branched block copolymers, consisting of two or more linear BCPs or homopolymers bonded at a common branch point, and *cyclic block copolymers* as already described in Section 11.2. Figure 11.2b schematically demonstrates different block copolymer architectures. Key factors influencing such self-organization are molecular architecture, chemical composition (monomers), molar mass distribution, and supramolecular interactions.

With the advent of various controlled and living polymerization techniques, synthesis of polymers with a variety of molecular architecture and chemical compositions is made simple. As the shape of self-assembled amphiphilic aggregates in solution is governed by the relative size of the hydrophobic to hydrophilic component, this relative size in turn determines the curvature at the interface, which is further related to the packing parameter (P). The definition of the packing parameter is represented in Eq. 11.1 and Figure 11.12.

$$P = \frac{v}{al} \tag{11.1}$$

In this formula, v is the volume of the hydrophobic component, a is the area at hydrophilic and hydrophobic interface, and l is the length of the hydrophobic

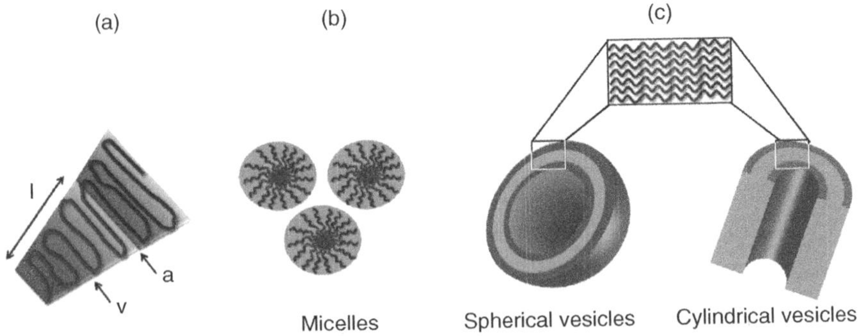

Figure 11.12 (a) Geometric scheme determining the packing parameter (P). Different morphological shapes like (b) spherical micelles and (c) bilayer spherical and cylindrical vesicles formed by amphiphilic block copolymers in solution.

part normal to the interface. So it is the packing parameter that finally determines the form of aggregated structures like micelles or vesicles. Spherical micelles, cylindrical micelles, and bilayer vesicles are preferred when the packing parameter scales to one-third ($\frac{1}{3}$), one-half ($\frac{1}{2}$), and unity (1), respectively.[54]

Biomimetic membranes are formed from vesicles, which have a spherical shell structure with a hollow core and shell comprised of a self-assembled bilayer of amphiphilic molecules such as lipids (liposome) or surfactants. A decade ago Discher and Eisenberg[55, 56] for first time prepared block copolymer vesicles of poly(styrene)-*b*-poly(acrylic acid) (PS-*b*-PAA) in dimethylformamide (DMF). Bilayer tubular vesicles made of poly(styrene)-*b*-poly(ethylene oxide) (PS-*b*-PEO) in DMF have also been reported. Various block copolymers comprised of polymers like poly(lactic acid) (PLA), poly(caprolactone) (PCL), poly(butadiene) (PB), poly(methylacrylate) (PMA), and others have been investigated. A series of experiments from Antonietti and Förster[57] decreased the hydrophilic to hydrophobic block ratio, thereby increasing the interfacial area in a series of poly(butadiene)-*b*-poly(ethylene oxide) (PB-*b*-PEO) block copolymers, which showed a structural transition from spherical to cylindrical micelles and finally resulted in vesicles (Figure 11.13).

An analogy in terms of structural similarity can be drawn between bilayer polymersomes and the natural cell membranes. Attempts to further extend this analogy have involved the inclusion of proteins, lipids, and other biological molecules aiming toward their usage as synthetic cell elements in cellular recognition, targeted delivery, and biosensors.[58] In contrast to lipid vesicles, polymersomes are formed by block copolymers. The membrane properties of polymersomes are related to the chemical composition, the block length ratio, and the molecular weight of the block copolymers. Membrane thickness correlates to the molecular weight (MW) of the hydrophobic block by $(M_h) : d \sim (M_h)^a$, where $a \sim 0.5$.[59] Besides the membrane thickness, the block copolymer molecular weight reduces the lateral mobility of polymer chains within the membrane, which accounts for the better resistance to

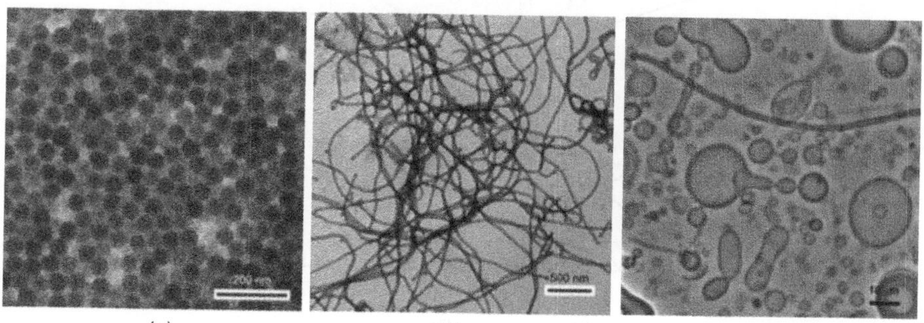

(a) (b) (c)

Figure 11.13 Transition electron micrographs of the structural transition from (a) spherical to (b) cylindrical micelles and (c) vesicles in a series of poly(butadiene)-*b*-poly(ethylene oxide) (PB-*b*-PEO) block copolymers.[57] (Reproduced with permission. Copyright Wiley-VCH Verlag GmbH & Co. KgaA: Ref. 57.)

dissolution and lower membrane permeability. Thus, block copolymer polymersome membranes possess enhanced rigidity and toughness and are more robust than conventional vesicles made of low molar mass lipid molecules.[60, 61]

This superior mechanical and thermodynamic stability of polymersomes compared to their lipid analogs increased their potential in a wide variety of applications like responsive membranes, encapsulation, and nanoreactors. The membrane-elastic behavior of PEO-*b*-PEE "giant" polymersomes has been studied by a micropipette aspiration method[62] and it was found that comparable membrane elasticity to the fluid state of lipid membranes can be observed. Apparent membrane viscosity investigations have revealed that membrane fluidity decreases with an increase in molecular weight (Figure 11.14).[63]

Another most important property of amphiphilic membranes is their selective permeability to hydrophobic and hydrophilic molecules and various investigations in this regard have been done to understand the transport phenomena of various species through the walls, which will facilitate polymersomes exploitation. Eisenberg and co-workers[64] loaded a drug, doxorubicin, into a vesicular system based on poly(styrene-*b*-acrylic acid) (PS-*b*-PAA) and found that due to the much higher glass transition temperature (T_g) of PS in comparison to lipids, the rate of drug diffusion through the polymer membrane was significantly slower.

Viruses have evolved to deliver the required cargo to the target cell at a suitable time, by disassembling when triggered by an appropriate body condition like pH or temperature. These viral capsids are often sensitive to the environmental variables and usually conventional liposome systems prove to be inherently leaky and have shorter half-life time during circulation. In contrast, more robust polymersomes when used as encapsulants for target delivery can circulate in vivo for extended time periods without leaking.

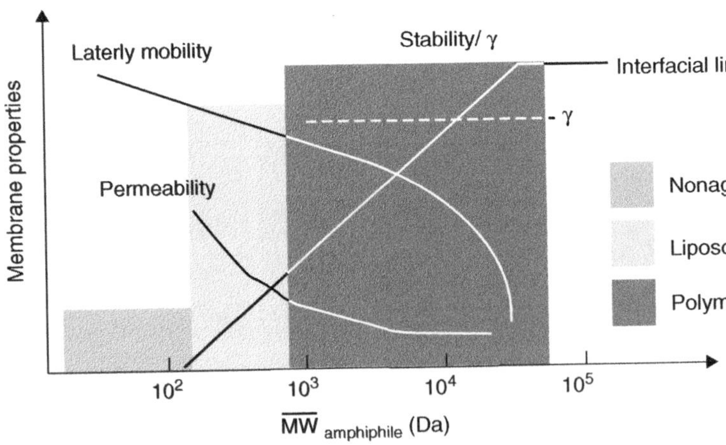

Figure 11.14 Schematic plot representing the physical properties of a membrane with respect to molecular weight of the amphiphile molecule. (Reproduced with permission. Copyright John Wiley & Sons: Ahmed, F.; Photos, P. J.; Discher, D. E. *Drug Dev. Res.* **2006**, *67*, 4.)

Polyester-based degradable polymersomes have been formed from PEG-*b*-PLA or PEG-*b*-PCL and studied thoroughly for the degradation and release of encapsulated therapeutics. The mechanism of releasing encapsulated therapeutics from polymersomes is based on hydrolytic degradation of hydrophobic polyester blocks such as polylactic acid (PLA) or polycaprolactone (PCL). Quantitative studies on doxorubicin and protein encapsulation during hydration shows efficiencies comparable to conventional liposomes and proves the potential of polymersomes as controlled release encapsulates.[65] As an example, Discher and co-workers[66] have demonstrated a smart approach to rapidly shrinking a growing tumor by means of pH-sensitive and degradable poly(ethyleneglycol)-*b*-poly(lactic acid) polymersomes loaded with anticancer drugs that disintegrated into membrane-lytic micelles within hours at 37 °C and low pH. The pH-triggered release within the endolysosomes was possible due to their acidic pH.

Several biological membrane processes such as protein integration, fusion, and DNA encapsulation can be mimicked successfully using synthetic polymersomes by interfacing with biological structures or ligands (Figure 11.15a). Meier and co-workers[67] demonstrated that inserted channel proteins can dock with viruses and facilitate transfer-loading of viral DNA into polymersomes (Figure 11.15b). Polymersomes made of PEO-*b*-PEE show similar functions like stealth liposomes,[68] which circulate in the bloodstream for ~20 hours and end up being engulfed by phagocytic cells of the liver and spleen.

Polymersomes acting as nanoreactors that mimic bioreactors have been prepared by Meier and Nardin[69] using poly(2-methyloxazoline)-poly(dimethylsiloxane)-poly(2-methyloxazoline), PMOXA-*b*-PDMS-*b*-PMOXA triblock copolymer with encapsulated enzyme. Incorporation of membrane proteins within polymersome walls to create channels allows the selective harvesting and release of specific molecules on demand.

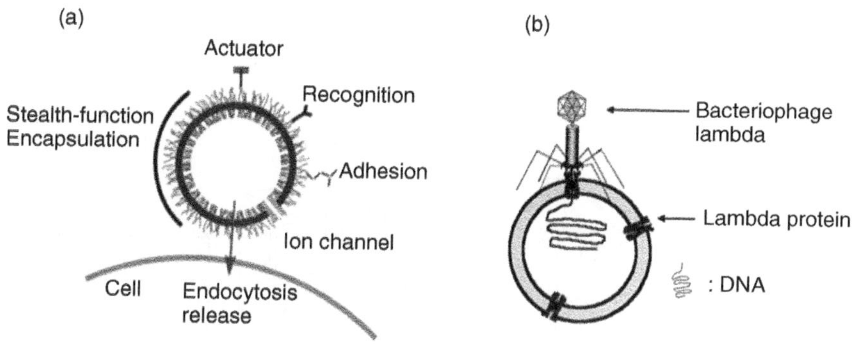

Figure 11.15 (a) Schematic of biological membrane function of polymersomes attached with biological structures or ligands. (Reproduced with permission. Copyright Wiley-VCH Verlag GmbH & Co. KgaA: Ref. 57.) (b) Representation of virus-assisted loading of DNA into polymersome containing channel proteins in its membrane. (Reproduced with permission. Copyright the National Academy of Sciences U.S.A.: Ref. 67.)

Microphase separation and self-assembly in block copolymers also occurs in the bulk state, generating a large variety of morphologies. Microphase segregation and phase behavior in block copolymer melts is governed by certain parameters, such as degree of polymerization (N) (i.e., the number of repeat units in a polymer chain) and segment–segment (Flory–Huggins) interaction parameter (χ), which influence the thermodynamics of the system.[70]

The product χN is the parameter that is of interest in the case of block copolymer phase behavior. There is a critical value for χN at ~10.5, as predicted by Leibler[71] using mean field approximation. For χN less than ~10.5, the block copolymer chains are miscible and disordered (Figure 11.16). For χN above ~10.5, they phase-separate into periodically ordered structures; this transition is widely termed an "order to disorder transition" (ODT). As the parameter χ is related to inverse temperature, the ODT in block copolymers can also be traversed by changing the temperature. Hence, the blocks in block copolymers are miscible and phase-separated above and below the critical temperature.

Figure 11.16 shows the different block copolymer morphologies formed in the case of simple, linear, AB-type diblock copolymers. Symmetric volume fractions of both blocks generate lamellar arrangement of the blocks and with asymmetric volume composition of blocks, hexagonally arranged cylindrical, spherical, or gyroid-like structures have been observed. The domain size depends on the polymer chain length and is normally in the range of 10–60 nm. Block copolymers with a rich variety of architectures (triblock, star block, etc.), chemical compositions, and functionalities have been synthesized and investigated for much more complex and various morphologies (see Section 11.2). As it is not feasible to discuss all block copolymers reported to date, some examples are named here: these include polystyrene-*b*-poly(butadiene) (PS-PB), poly(2-vinyl pyridine)-*b*-polystyrene (P2VP-PS), polystyrene-*b*-poly(methyl methacrylate) (PS-PMMA), polystyrene-*b*-poly(isoprene) (PS-PI), poly(butadiene)-*b*-poly(ethylene oxide), PI-PS-PDMS, PS-PB-PMMA, and poly(norbornene) block copolymers.[72]

Nanoporous membranes for the filtration of viruses have been prepared by T. P. Russel and co-workers using the cylindrical microdomain-forming PS-*b*-PMMA

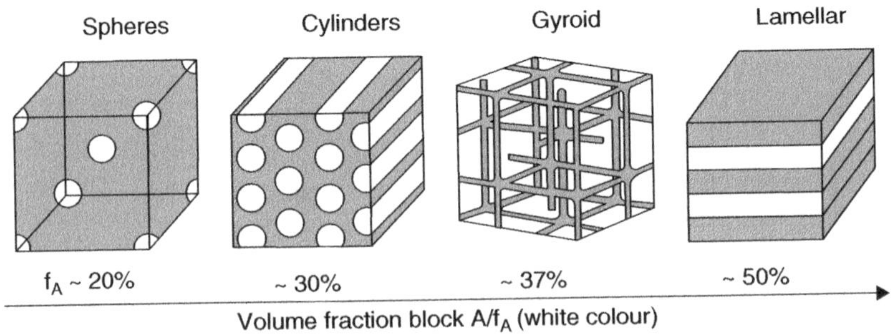

Figure 11.16 Schematic picture showing geometry of morphologies in diblock copolymers.

block copolymer by etching away the cylindrical PMMA domains in PS matrix.[73] Various other applications like masks for nanolithography, sensor arrays, and nanostructure fabrication using block copolymer as templates have also been reported.[74]

11.4 MOVEMENT IN POLYMERS

Commuting from one place to another is an important attribute and has become part and parcel of life. Every mode of commute has its own mechanism of movement and consumes a certain energy. This energy may be supplied inherently or via an external source. In Nature much of the cell activity and its functions are based on the directed transport of cargos such as macromolecules and membranes. It is essential for such cargo to be transported to the right place at the right time since any disruption in this respect can result in developmental defects and diseases. Hence, looking at what lies behind biological motions provides many interesting aspects and knowledge about biomolecular movement mechanisms.

An impressive example of biomolecular movement is the actin-based myosin and the microtubule-based kinesin proteins. Myosin performs biological activities like muscle contraction, cytokinesis, cell movement, vesicle transport, and certain signal transduction pathways.[75] Kinesin is involved in membrane transport, mitosis and meiosis, messenger RNA, and microtubule polymer dynamics.[76] The energy supply for these motions is obtained from the hydrolysis of ATP inducing the force-generating conformational change of the bound head group in myosin and the neck-linker in the case of kinesin. The detailed movement mechanism and functions of myosin and kinesin protein families are discussed elsewhere.[77, 78] Mimicking motions induced by force-generating conformational and volume changes in synthetic polymers, by synthetic chemistry and designed architecture, has great potential for powering nano- and microscale devices. The ability of polymers to move actively in response to an external stimulus such as heat or light has been observed in a few polymers and an attempt to create more of such smart or intelligent polymer materials with required functionalities is being fiercely pursued in recent years. In all of these, stimuli-responsive effects on the molecular level are converted into macroscopic movement.

11.4.1 Polymer Gels and Networks as Chemical Motors

Osada et al.[79] reported a polymer gel system with chemo-mechanical properties, which produces gel motility similar to muscles. Polyelectrolyte gel systems absorb water to swell and deswell upon application of electric potential by water exudation. This repetitive process of water intake and discharge induces motion of the polymer gel in the water by converting electrical energy into mechanical energy.[80, 81] The reverse mechanism—mechanical deformation-induced electrical potential—has been observed for polyelectrolyte gels made of crosslinked poly(acrylic acid); it can produce electrical potentials as large as a few millivolts.[82] Mechanical deformation of the polymer gel induces spontaneous ionization of carboxylic acids at a local level to produce an electrical potential.

This property of gels reminds us of touch-sensing biological soft tissue systems such as human fingers, which also convert mechanical energy into electrical signals transmitted via the nervous system for further processing. Mechanoelectrical properties of polyelectrolyte gels are similar to those of tactile perception in living organisms. In addition, they also match the softness, modulus, elasticity, and rheological characteristics of natural tissue.[79]

Osada et al.[83] prepared a polymer gel system capable of converting chemical energy into mechanical motions (chemomechanical property), which can serve as an actuator or an artificial muscle. The system is based on an electrokinetic molecular assembly reaction of surfactant molecules on the hydrogel made of crosslinked poly(2-acrylamido-2-methylpropane) sulfonic acid (PAMPS). The cationic surfactant molecules bind to the surface of an anionic polymer gel network, inducing local shrinkage by reducing the osmotic pressure difference between the gel interior and the solution outside. Surfactant binding is directed selectively to one side of the gel by applying an electric field to induce contraction and curvature of a polymer gel strip suspended in solution in a ratchet mechanism. Reversing the direction of the electric field causes contraction on the opposite side of the strip, thereby generating a worm-like motion in the gel strip at a velocity of up to 25 cm/min. Figure 11.17 shows the successive movement of gel-looper strip at respective time intervals.

A new class of chemical motor based on crosslinked amphiphilic copolymer gels prepared from monomers of *n*-steryl acrylate (SA) and acrylic acid (AA) was reported, again by Osada and group.[84] The crosslinked PSA-*co*-PAA gel swollen in an organic solvent like THF undergoes translational and rotational motions in water over an extended time. The driving force for the motion originates from the surface tension of the spreading organic solvent, which is pumped out of the gel due to high osmotic and hydrostatic pressure in the gel (see Figure 11.18).

Northen and Woodbury[85] recently reported light-induced movement of solvent-swollen porous polymer microstructures composed of different methacrylate copolymers functionalized with the photocleavable group 4-nitroveratryloxycarbonyl (NVOC). Upon laser excitation of NVOC-Gly polymer structures in organic solvent, the NVOC is photocleaved, resulting in shrinkage of the polymer and release of solvent and photocleavage products. Asymmetric illumination causes differential shrinkage of the microstructures, resulting in bulk movement. This suggests possible applications for light-driven micromechanical motion and optically patterned release of reagents.

Lower critical solution temperature (LCST) behavior of certain polymers is another significant property that has been exploited largely for biomedical applications. Polymers of this type undergo a thermally induced, reversible phase transition; they are soluble in a solvent (water) at low temperatures but become insoluble and phase separated as the temperature is raised above the critical temperature (LCST). Poly(*N*-isopropylacrylamide) (PNIPAAm) is the most popular temperature-responsive polymer since it exhibits a sharp phase transition in water (LCST) at around 32 °C.[86] The detailed mechanism and application of LCST polymers and polymer networks have been discussed in a review by Gil et al.[87]

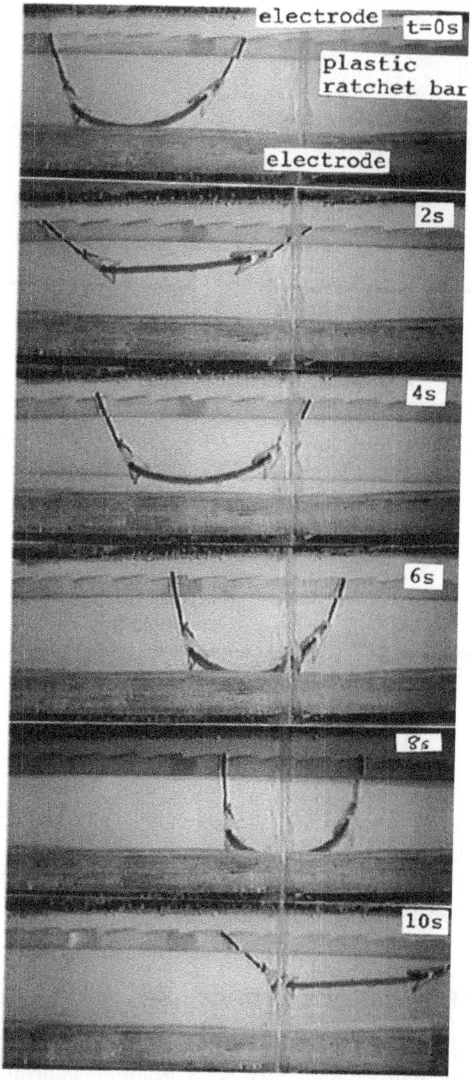

Figure 11.17 Series of pictures showing the polymer gel-looper strip device suspended from a plastic ratchet bar in a solution of *N*-(-*n*-dodecyl) pyridinium chloride containing sodium sulfate. It moves like a worm by repeatedly curling and straightening. (Reproduced with permission. Copyright Nature Publishing Company: Ref. 83.)

on stimuli-responsive polymers and their bioconjugates. Basically, the use of this polymer enables one to trigger mechanical work by reaching the LCST, thus inducing a contraction of the polymer. Similar to muscular systems, thermal energy can thus be transformed into mechanical work.

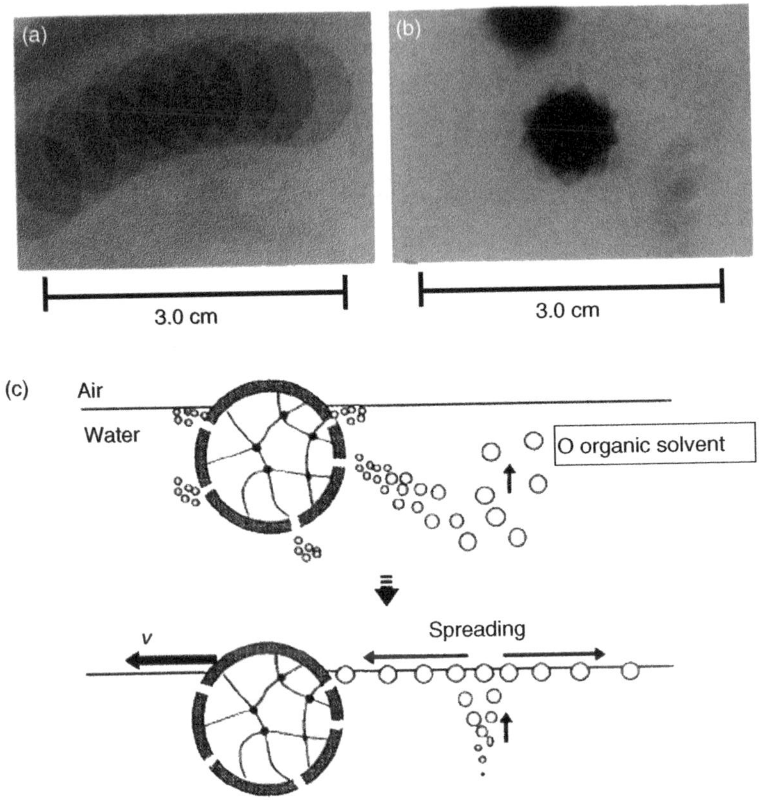

Figure 11.18 Pictures of (a) translational motion and (b) angular motion of disk-shaped and cubic-shaped copolymer gels. (c) Schematic illustration of gel movement induced by the release and surface spreading process of organic solvent in water. (Reproduced with permission. Copyright American Chemical Society: Ref. 84.)

11.4.2 Polymer Brushes and Lubrication

Two contacting surfaces in relative motion always experience a certain mechanical friction between them, which results in gradual wear-and-tear of the surface and sometimes even catastrophic failure. Thus, there arises a need for lubrication between such surfaces to make their relative motion smooth, with a low coefficient of friction. In Nature some biological surfaces such as synovial joints (hip, knee, shoulder, ankle, and finger joints) exhibit extremely low coefficients of friction in the range of 0.001–0.03 even though the relative motions are complex and the loads are surprisingly high.[88]

A simplified representation of a synovial joint is shown in Figure 11.19. The bones are covered by a thin layer of articular cartilage bathed in synovial fluid confined by synovial membrane. It is the synergistic effect of synovial fluid (hydrodynamic film lubrication) and cartilage tissue (boundary lubrication) that reduces friction during movement.[88, 89] Synovial fluid is a dialysate of blood plasma with added

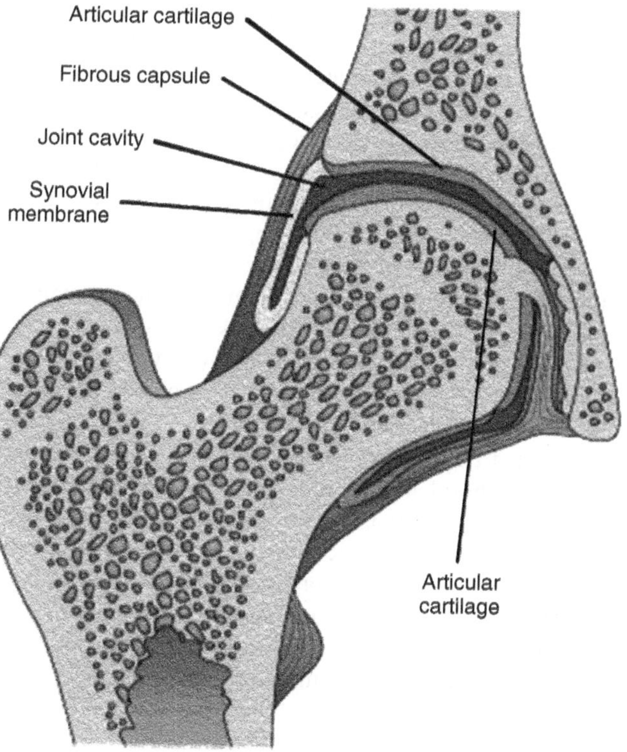

Figure 11.19 Synovial hip joint. (Reproduced with permission from Ref. 88.)

hyaluronic acid, complex proteins, glycoproteins with bottlebrush structure, and other additives providing the required viscoelastic properties[90] to form a lubricating film by immobilizing large amounts of water molecules. However, water alone cannot achieve such low friction values due its lower viscosity inhibiting the formation of boundary film at high pressures. The articular cartilage is a soft but incompressible natural composite composed of water (approximately 75%) enmeshed in a network of collagen fibers and high molecular weight proteoglycans[91] with a bottle-brush structure providing cushioning by taking the load.

Inspired by the lubricating ability of such bottlebrush structured molecules in synovial joints, several investigators have examined the reduction of the friction coefficient under wet conditions by grafting hydrophilic polymers and polyelectrolytes onto surfaces.[92, 93]

Takahara and co-workers[94, 95] investigated the tribological properties of high-density polymer brushes based on poly(methyl methacrylate) (PMMA). Hydrophilic polymers were synthesized via surface-initiated living polymerization of the monomer onto a silica surface. The wear resistance of the resulting brush surface under specific load was measured and the effect of different solvents on sliding friction of tethered brushes was studied.

Spencer and colleagues have undertaken an extensive study on surface modification and the tribological behavior of poly(L-lysin)-*g*-poly(ethylene glycol) (PLL-*g*-PEG)[96, 97] while seeking for a biomaterial coating for medical devices such as catheters, heart valves, stents, and biosensors to avoid protein absorption on the oxide surface of such devices. Their study aimed at grafting PEG onto a PLL backbone to achieve a comb-like structure. It later examined boundary lubrication between an oxide surface and poly(L-lysin)-*g*-poly(ethylene glycol) in aqueous media. They found that increasing both the length of PEG side chains and the grafting ratio improved the tribological performance by increasing the thickness of the boundary film and thus reducing friction. Polymer brushes containing 2-methacryloyloxyethyl phosphorylcholine or poly(MPC) is another poly electrolyte that has been widely investigated. Superhydrophilic poly(MPC) due to the existence of phosphorylcholine units in its structure is not only a blood-compatible polymer but also has protein resistant properties.[98]

Ho et al.[99] evaluated frictional properties of a copolymer consisting of 2-methacryloyloxyethyl phosphorylcholine and *n*-butyl methacrylate [poly(MPC-*co*-BMA)] as a potential coating material for catheter applications. Considering the high rate of consumption of biomedical devices, surface modification of devices that are in contact with soft body tissue or blood such as catheters is very important. Reducing the frictional force and average surface roughness by attaching hydrophilic polymers besides providing a smooth catheterization may also decrease the possibility of physical trauma and inflammation. In this respect, polyurethane was coated by poly(MPC-*co*-BMA), which resulted in enhanced boundary lubrication and decreased frictional forces.

In 2004, Kawaguchi and co-workers reported using poly(MPC) to modify contact areas in an artificial joint.[100] In joint replacement surgery, wear of the bearing surface in prosthetic joints is the main problem because in most cases the lifetime of the replacement is shorter than the patient's lifetime. Therefore, it is essential to consider all the parameters which influence the life expectancy of the replacement to avoid revision surgery. Mimicking the natural joints, they tried to find a formulation close to synovial fluid ingredients. Hence, MPC was grafted onto the surface of the polyethylene component of artificial hip joint (see Figure 11.20) to create an artificial articular cartilage and synovial fluid effect.[100]

Takahara, Kobayashi, and colleagues[101, 102] investigated macroscopic frictional behavior of 2-methacryloyloxyethyl phosphorylcholine in aqueous media from the viewpoint of practical engineering applications. Poly(MPC) was prepared via surface-initiated ATRP of monomer from silicon wafer. The results indicated an extremely low coefficient of friction and dependency of macroscopic frictional properties of polymer brushes on the humidity and solvent quality. Recently, they evaluated tribological properties of some other hydrophilic polymers immobilized on Si surfaces under wet conditions and compared them with poly(MPC) performances under the same conditions.

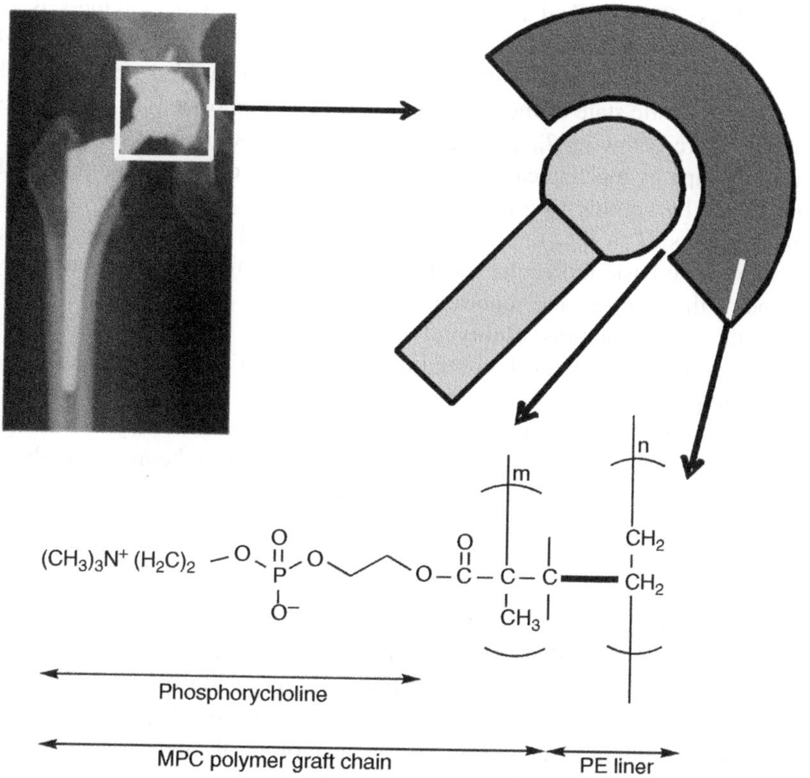

Figure 11.20 An X-ray of a replaced hip joint shown in the figure upper left. The schematic figure in the upper right depicts the relationship between the femoral head and the PE liner. MPC is bound to the PE liner by the covalent bond with a photoinduced graft polymerization technique. (Reproduced with permission. Copyright Nature Publishing Group: Ref. 100.)

11.4.3 Shape-Memory Polymers

The process of regaining an original shape from a temporary structural deformation in a material body by applying simple external stimuli such as heat, light, magnetic field, electric field, or a mechanical force is called the *shape memory effect*. Polymers that exhibit such shape memory effects are termed *shape memory polymers* (SMPs). They are considered to be smart or actively moving polymers. Temperature is the most commonly used external stimulus for shape recovery by polymers. The shape memory effect in a polymer is related to its structure, composition, architecture, morphology, and also the programmed processing.

Polymer materials are processed and molded under the application of mechanical force and temperature, above a critical transition temperature (T_{trans}), to obtain a required shape. This critical transition temperature (T_{trans}) in polymers is either the glass transition temperature (T_g) for amorphous polymers or/and the melting temperature (T_m) for semicrystalline polymers. Such a molded permanent shape formed in the elastic state of the material is retained (memorized) as the SMP cools to a temperature below its T_{trans}. The deformation can be induced on such molded permanent shape by mechanical force at $T > T_{trans}$. This deformed temporary shape can be frozen by cooling to a temperature below its T_{trans}. Again on applying the thermal stimulus ($T > T_{trans}$), the SMP regains its initial original shape. A detailed discussion of the basic molecular mechanism of shape memory function is beyond the scope of this chapter and reported elsewhere.[103]

Since the shape memory ability of the polymer is entropy driven, the polymer chains always prefer the less strained random coil configuration. SMPs usually are composed of multiphase structure containing a hard (fixed) phase and a soft (reversible) phase, with the hard phase composed of crystalline, glassy domains, chain entanglements, or crosslinked net points (chemical or physical). The reversible phase, on the other hand, is the major constituent of SMPs and is responsible for elasticity during deformation. It undergoes strain recovery upon applying the external stimulus.

This unique shape memory behavior of polymers has made them suitable in biomedical applications such as body implants or scaffolds for tissue growth, which are normally large and bulky. They can be placed into the body in a compressed temporary configuration by minimally invasive surgery and subsequently converted to their original shape by external stimulus in vivo. Biomedical applications demand biocompatibility (as implants), degradation in biological environments (for scaffolds, internal sutures), and specific mechanical properties.[104]

Biocompatible and degradable shape memory thermoplastic elastomers have been prepared by macrodiols of lactones and cyclic diesters with low molar mass diols by cocondensation using bifunctional coupling agents as shown in Figure 11.21a.[103] The resulting polymers are linear multiblock copolymers containing crystallizable poly(*p*-dioxane) as the hard, fixed segment and a switchable soft segment such as crystallizable poly(ε-caprolactone) ($T_m \sim 46$ and 64 °C) and amorphous poly[(L-lactide)-*co*-(glycolide)] ($T_g \sim 35 - 50$ °C), whose thermal transition temperatures (T_g or T_m) lie in the range of room or body temperature.

Shape memory polymers have been designed and tested for various biomedical applications. Lendlein and Langer[105] have investigated multiblock copolymers as smart sutures in rats, where the loosely sutured incision on belly tissue and abdominal tissue was actuated by raising the temperature above T_{trans} to close the wound tight enough (see Figure 11.22). Biodegradability and compatibility tests show positive results; the mechanical properties are comparable to the mechanical stress in the soft tissue.

Maitland and co-workers[106, 107] have developed an intravascular, laser-activated therapeutic device using shape memory polymer (SMP) which can be inserted into blood vessels in its temporary shape via minimally invasive surgery. Upon activating with a laser, the SMP device coils to its original shape and mechanically

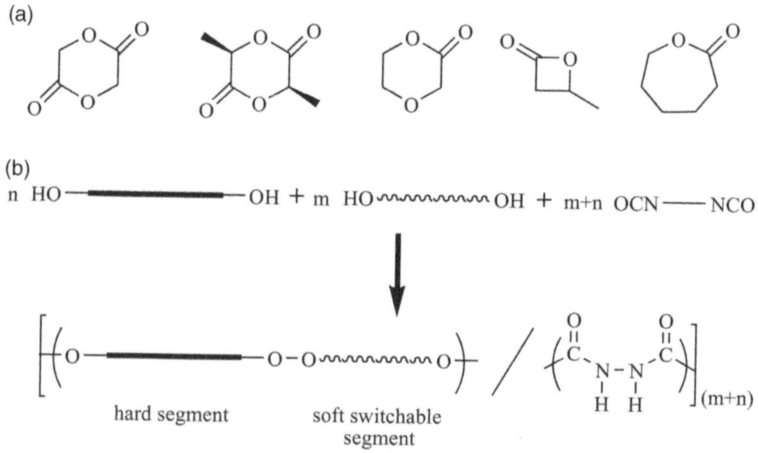

Figure 11.21 (a) Chemical structures of lactones and cyclic diesters that are used to prepare macrodiols. (b) Schematic synthetic pathway for shape memory multiblock copolymers by cocondensation reaction.

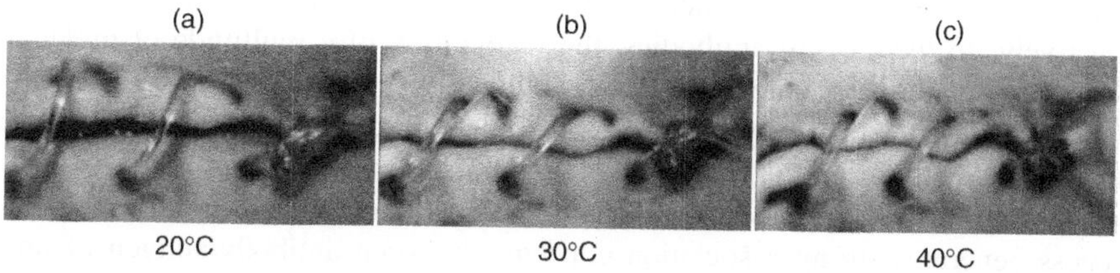

Figure 11.22 Series of photos showing the shrinkage of suture fiber made of degradable shape memory polymer for wound closure in rats while increasing the temperature from 20 to 40°C (left to right). (Reproduced with permission. Copyright The American Association for the Advancement of Science: Ref. 105.)

retrieves the blood clot (thrombus) and restores blood flow to the brain (see Figure 11.23).

Biodegradable intragastric implants[108, 109] made of shape memory polymers are used to tackle an obesity problem by inflating after only a small amount of food is consumed to curb the appetite of the patient. Polyurethane foam with a T_g switching transition has been proposed as a shape- and size-determining tool for the human ear canal, so that the hearing device can be prepared with an exact fit. Shape memory polymer composites such as polyurethane with tantalum fillers are exploited as stents (intracranial aneurysm coils) for prevention of strokes.[110]

An important field of application for shape memory polymers is in active medical devices and implants. The advent of new shape memory polymer materials that are biodegradable and can be actuated with stimuli other than heat is opening up new possibilities for various applications.

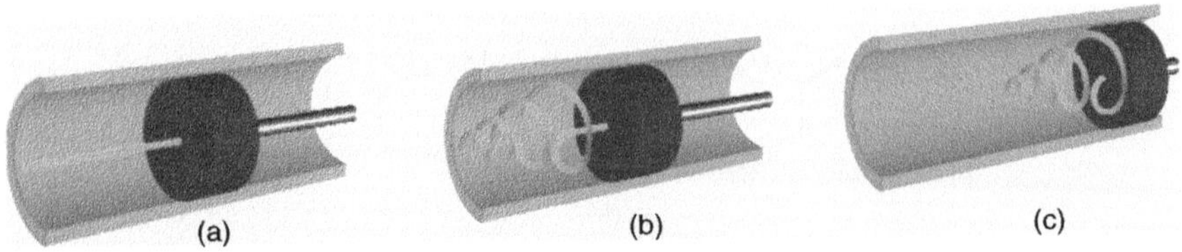

Figure 11.23 Schematic depicting a clot removal in a blood vessel using the laser-activated intravascular thrombectomy device made of shape-memory polymer coupled to an optical fiber. (a) The device is delivered through a catheter distal to the blood clot in its temporary straight rod form. (b) The device is later transformed into its permanent corkscrew form by laser heating. (c) The deployed device is retracted to capture the thrombus. (Reproduced with permission from the Optical Society: Small, W.; Wilson, T.; Benett, W.; Loge, J.; Maitland, D. *Opt. Express* **2005**, *13* (20), 8204–82139).

11.5 ANTIBODY-LIKE BINDING AND ENZYME-LIKE CATALYSIS IN POLYMERIC NETWORKS

The concept for selective binding and catalysis in Nature is most effectively realized by antibodies and enzymes, which can recognize small and large molecules highly selectively. In the case of antibodies, the recognition of a multitude of molecular features in Nature is achieved by an evolutionary selection process, where the binding of the molecules (antigens) to a (cellular) library is the central mechanism of finding and selecting the most suitable binding site within an antibody molecule (which in turn is then produced by the cell, leading to a monoclonal antibody). This process generates strong association constants between antibody/antigen of up to $K_{assn} \sim 10^{10}$ M^{-1}. The principle of binding, followed by selection to generate specific (and also catalytically active) binding sites, can be achieved by molecular imprinted polymers (MIPs) (see Figure 11.24).[111–114]

Thus, the template (which may be a low molecular weight molecule, a receptor molecule, a transition-state analog, or even a whole cell) is incubated with a monomer and a crosslinker that are subjected to a polymerization process. Often radical polymerization (using acrylates, styrenes) or ROMP processes (using cyclic monomers) are used to form a dense network around the template molecule, where—due to the formation of supramolecular contacts/associates by selective interaction between the monomers and the templates—the shape of the template is "imprinted" into the MIP. Subsequent release of the template molecule generates the free, now template-binding site within the MIP. As easy as it sounds, major difficulties have been reported for MIPs, both in terms of reproducibility and multiple binding constants. The use of a high amount of crosslinker (~70 mol %) is important, as the resulting network need a high density in order to (1) remain sufficiently rigid and (2) really reflect the shape of the templating molecule. Of course, the supramolecular interaction between the templating molecule and the monomers must also be sufficiently strong, taking (cleavable) covalent bonds into consideration.

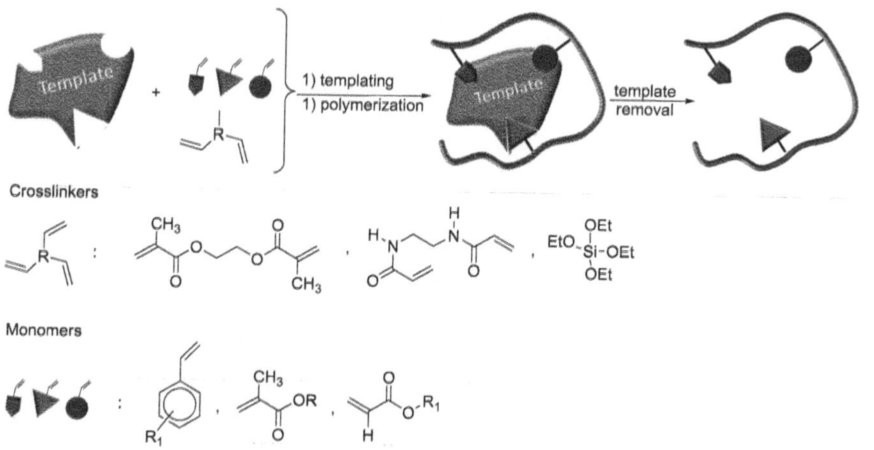

Figure 11.24 Synthesis of molecular imprinted polymers (MIPs) via monomer organization around the template and subsequent polymerization. After template removal, the generated hollow space in the MIP can act as selective recognition or catalytic site.

Typical molar ratios between the templating molecule to the monomer are 1:4, in order to achieve a sufficient imprinting process. If covalent approaches are used up to 90% of all vacated sites are newly accessible, whereas with noncovalently formed networks, only ~15% of the sites can be accessed by new substrates.[112]

In terms of the polymerization conditions during MIP formation,[115] a higher molecular weight of the crosslinked polymers and the use of a lower initiator concentration together with a higher degree of ordering within the crosslinked MIP–polymer (i.e., induced by magnetic fields) is advantageous for producing a better MIP.

Due to their ease of preparation and the possibility for reuse, MIPs have become a central topic in analytical science, most of all in separation chemistry and in enantiomeric separation.[116] Thus, a large variety of sensors[111, 117] for biomolecules such as antibiotics,[118] carbohydrates,[119] proteins,[120] and even whole cells[121, 122] have been developed.

More important than pure binding of substrates to MIPs is their catalytic activity, acting either as a transition-state analog or as a selective layer for another catalytic process.[111, 123, 124] Figure 11.25 shows a series of recent examples, demonstrating the rich possibilities that are accessible by MIP-based catalysis.

In principle, two different aspects have been investigated recently, which can be designated as "biomimetic": (1) the use of transition-state analogs to induce a specific catalytic reaction and (2) the use of an MIP for substrate selection within a catalytic process.

Using the first principle, an *aldolase*[125] (Figure 11.25a) as well as a *carboxypeptidase*[126] (Figure 11.25b) can be mimicked. In both cases a template molecule resembling the transition state of the planned catalytic reaction is used, leading to significant rate enhancements of ~20–80. However, as stated in both cases, the

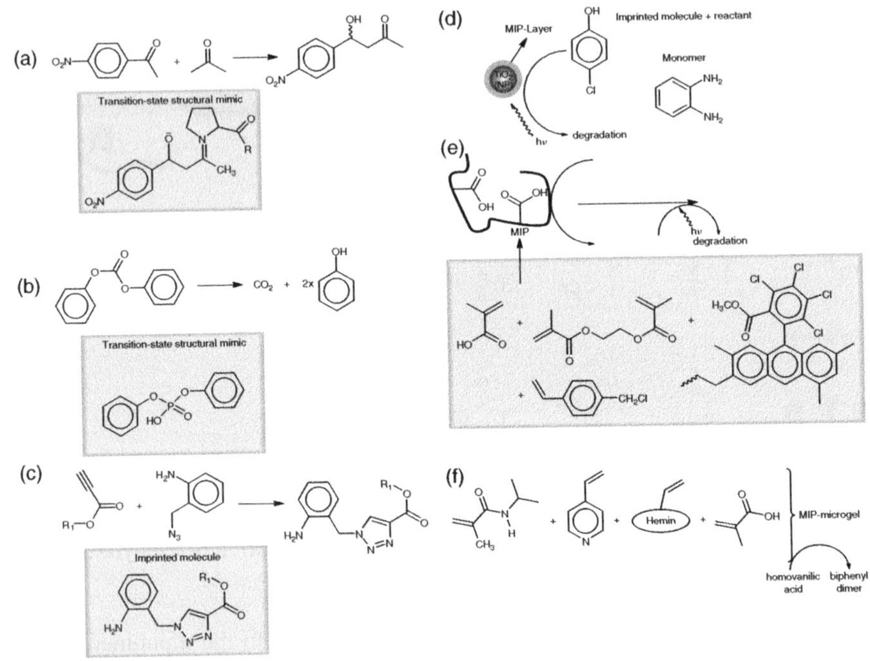

Figure 11.25 Catalysis by MIPs. (a) MIP active as *carboxypeptidase A*; (b) MIP mimicking a type I *aldolase*; (c) regioselectivity enhancement in dipolar cycloaddition reactions; (d) selective photochemical oxidation of 4-chlorophenol catalyzed by a combined MIP/TiO_2 nanoparticle; (e) photochemical oxidation of 4-chlorobenzene by rose-bengal and an MIP; (f) a pH-sensitive MIP-microgel active as *peroxidase* catalyst.

underlying rate constants are "polyclonal" binding sites, thus reflecting the formation of many different catalytic sites, some more efficient, some being significantly less efficient than the others. Surprisingly, in both cases[125, 126] Michaelis–Menten-type kinetics was observed, despite the presence of multiple catalytic sites.

A more chemical example relies on the imprinting of the end product of a 1,3-dipolar cycloaddition reaction (see Figure 11.25c).[127] The regioselectivity (formation of the 1,4-triazole) of this cycloaddition reaction can be enhanced significantly, if an MIP previously imprinted with a structurally similar 1,4-triazole is used as catalyst of this reaction.

Another, more sophisticated MIP is positioned as a layer on a catalytically active nanoparticle[128] or photochemically active dye[129] (see Figures 11.25d and 11.25e). Thus, the photooxidation of 4-chlorophenol can be catalyzed by nanoparticles made from TiO_2. In order to achieve the necessary selectivity toward 4-chlorophenol, an MIP layer imprinted with 4-chlorophenol was polymerized around the TiO_2 nanoparticle, resulting in a more selective photooxidation due to the selective substrate transport to the TiO_2 surface. A similar example[129] has been reported using an attached rose-bengal dye, which acts as a photocatalyst for the oxidative degradation of chlorophenols.

A further expansion of catalytically active MIPs can be achieved when microgels are used instead of the normally imprinted polymers. Thus (see Figure 11.25f), a thermosensitive microgel made from a poly(N-isopropylacrylamide) gels in copolymerization with acrylamide, 4-vinylpyridine, and styryl-modified hemin can be generated, which—based on the earlier discussed lower-critical-solution effects—displays a strongly pH-dependent catalytic activity due to the reversible protonation/deprotonation of the hemin moiety. The catalytic activity of horseradish peroxidase can be mimicked, again displaying Michaelis–Menten-type kinetics.

To sum up, MIPs are potentially useful catalytic materials, with a selective binding capacity and a moderate catalytic effect. Due to their simplicity, however, they may be useful biomimetic catalysts for future applications.

11.6 SELF-HEALING POLYMERS

As polymers have become an integral part of almost all areas of life, scientists and engineers have been studying the feasibility of autonomic healing (self-healing) of microcracks, generated by mechanical forces. Mimicking Nature plays a key role in the development of new, bioinspired self-healing materials.

The inspection of the healing process of bones (see Figure 11.26a) serves as a useful basis for the development of a self-healing polymer system and is one example of biomimicry in this respect.[130] This healing process in bones consists of multiple stages like deposition and assembly of material, as illustrated in Figure 11.26b. The network of blood vessels in the bone is ruptured by a fracture event; this leads to (I) internal bleeding and formation of the fibrin clot, followed by (II) the development of unorganized fiber mesh, which seems to be a preorganization of the bone but not hardened. Finally, (III) the fibrocartilage is calcified and (IV) converted into the fibrous bone, followed by the last step (V) of initiating autonomic healing of the bone, the transformation into the lamellar bone.

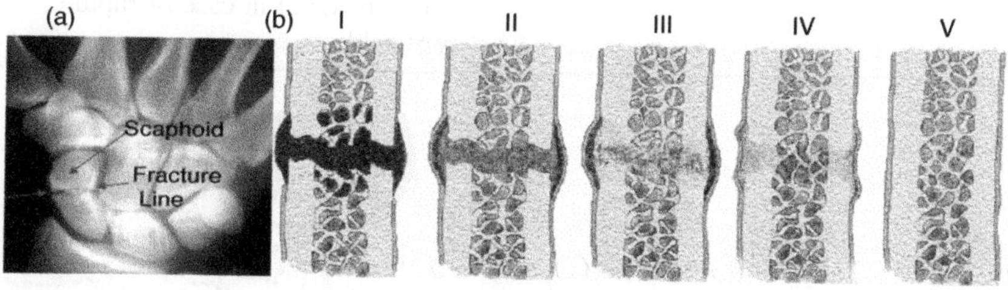

Figure 11.26 Inspiration for self-healing polymers. (a) Fracture (b) Healing process of broken bone: (I) internal bleeding, (II) unorganized fiber mesh, (III) calcification of fibrocartilage, (IV) fibrous bone, and (V) transformation into lamellar bone. (Reproduced with permission from: (a) Autonomous Materials Systems Group, Beckman Institute, University of Illinois, Image created by Eric N. Brown currently of Los Alamos National Laboratory; (b) schematic adapted with permission from Springer: E. N. Brown, N. R. Sottos, and S.R. White, *Exp. Mech*. **2002**, 42, 372.)

Wouldn't it be great if self-healing could be adapted to nonbiological systems? The relevance of the inspiration from Nature for scientists dealing with self-healing polymers is shown in Table 11.1, providing an overview of the naturally existing healing strategies and cross-referencing them against the self-healing principles. Most of the self-healing work to date has been biomimetic, which means that biological healing approaches have been used in the design of the chemical or engineering approaches by copying not the whole healing process, but the principle behind it. It is obvious that the natural healing method of "bleeding" is an important inspiration for the self-healing principles. Both capsule-based self-healing systems and hollow fibers are based on this biological healing approach.[131]

Self-healing principles can broadly be classified according to the type of polymer and also the ways of healing into three classes: (1) capsule-based, (2) vascular, and (3) intrinsic[132] (see Figures 11.27a–c).

The matrix of thermosets is itself not healable since they have covalent crosslinked molecular structures with no chain mobility. The addition of healing

TABLE 11.1 Biomimetic Self-healing Inspiration for Polymeric Self-healing Systems

Biological Inspiration	Self-healing Principle	Biomimetic Self-healing Process
Concept of self-healing	Intrinsic	Based on physical or chemical interaction, external stimulation necessary to initiate healing
Bleeding	Capsules	Healing agent stored in capsules, release of active material in case of rupture
Bleeding	Hollow fibers	Healing agent stored in hollow fibers, release of active material in case of rupture
Blood flow vascular network	Vascular	Healing agent stored in 2D or 3D network of hollow fibers, release of active material in case of rupture, repeatable

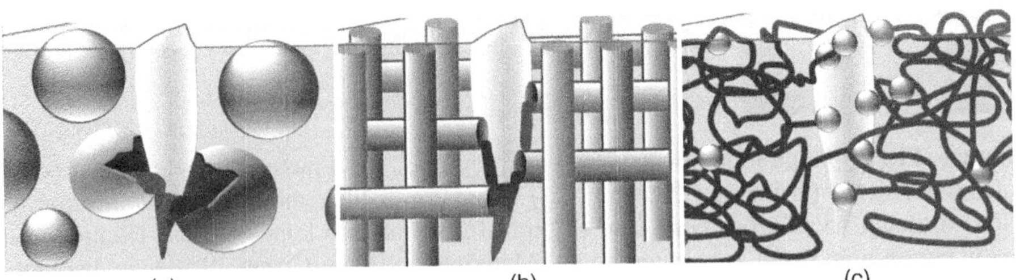

Figure 11.27 Classification of self-healing principles: (a) capsule-based, (b) vascular, and (c) intrinsic.

agents is therefore necessary, which are separately stored in microcontainers and embedded into the polymer matrix. This can be achieved via capsule-based (Figure 11.27a) and vascular (Figure 11.27b) self-healing methods.

A propagating crack ruptures the fragile vessels leading to the release of the healing agent into the crack plane by capillary action.[133] The reaction that follows yields the formation of a stable network, causing the crack to be healed and the material properties recovered. Thereby the bonding to the matrix interface plays a crucial role and can be considered to be a molecular interdiffusion mechanism.

One of the most prominent approaches is the microencapsulation of the liquid healing agent dicyclopentadiene (DCPD) for self-healing polymeric composites reacting in a ring opening metathesis polymerization (ROMP) by the aid of Grubbs's catalyst (Figure 11.28).[134]

The microencapsulated healing agent (e.g. DCPD) is embedded in a structural composite matrix (epoxy) containing a catalyst (e.g. Grubbs's catalyst) capable of polymerizing the healing agent. During the mechanical damage of the composite, the cracks are formed (a), the cracks propagate and rupture the microcapsules (b), leading to release of the healing agent (e.g. DCPD) into the crack plane through capillary action. The reactive monomer is then polymerized via ROMP when in contact with the catalyst (c), forming a crosslinked polymer (network formation) and thus healing the cracks.

Another interesting method for self-healing systems is the combination of copper(I)-catalyzed azide/alkyne-"click"-reaction, between liquid trivalent polyisobutylene (PIB) azides ($M_n = 4000$ g/mol) and liquid trivalent alkynes, with the microencapsulation approach (Figure 11.29). In response to applied stress, microcapsules filled with liquid polymers are ruptured, and encapsulated, liquid active materials are released to react under formation of a network via the azide/alkyne-"click"-reaction.[135, 136, 137]

Vascular self-healing materials (Figure 11.27b) store the active, liquid healing agent in pipelines, which may be interconnected in one (1D), two (2D), or three dimensions (3D).[132] In one-dimensional systems, hollow fibers are filled with the

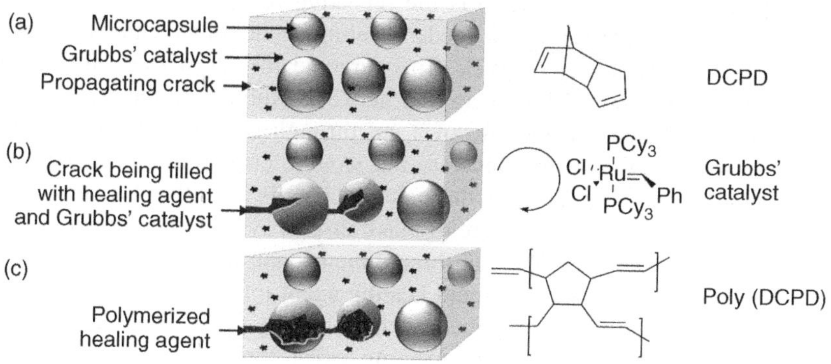

Figure 11.28 Self-healing concept via microencapsulation approach using the example of ROMP of DCPD.

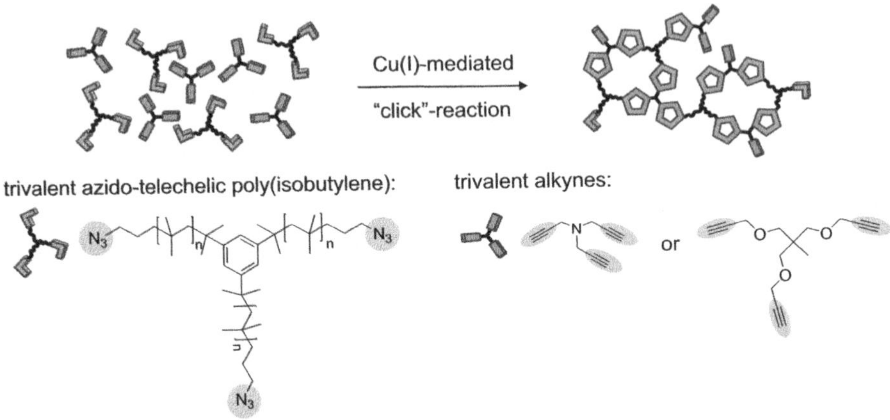

Figure 11.29 Schematic depicting the self-healing concept combining the microencapsulation approach with the azide/alkyne-"click"-reaction between liquid trivalent polyisobutylene azide and liquid trivalent alkynes.

polymerizable healing agent, which is required to be liquid at healing temperature. Therefore, it can be released from the hollow fibers to fill the damaged area and polymerize subsequently to heal the crack. All in all, three types of healing systems were developed:[133]

1. Single-part adhesive: one kind of resin (e.g., epoxy) (flowable upon heating and then curing by residual hardener) in all hollow fibers.
2. Two-part adhesive: epoxy in one hollow fiber and the hardener in neighbouring hollow fiber.
3. Two-part adhesive: first curing agent is filled in a hollow fiber, second in a microcapsule.

One major problem of one-dimensional self-healing composites (microcapsules and one-dimensional hollow fibers) is the repeating of healing. If there is a new crack in an already healed area, the needed healing agent, which cannot be refilled, is not available to heal a new crack. To overcome this drawback, material systems with three-dimensional microvascular networks mimicking the architecture of human skin are created. In the case of a cut through skin, blood flows from the capillary network in the dermal layer to the wound, causing a clot to be rapidly formed. The clot serves as a matrix through which cells and other "healing agents" migrate.[133] Due to the vascular nature of this transport system, the same area can be healed repeatedly when cut. Thus, in three-dimensional microvascular networks, a crack is healed by flowing of the liquid healing agent into the damaged area, polymerizing there to heal the crack, whereas the pipelines in this region are restored with healing agent. In the case of repeating crack formation, the healing cycle can be rerun due to subsequent loading of the three-dimensional microvascular network. However, the refilling of the pipelines, as well as the warranty that the healing agent flows into the crack, which can occur everywhere in the material, is the challenge for the synthesis of three-dimensional vascular networks.

Likewise, the impact on mechanical strength while maintaining the interconnectivity limits the application of these networks. As a compromise, two-dimensional networks combine the advantage of the repeatability of the three-dimensional vascular network and the lower impact of the mechanical strength of the unidirectional hollow fibers.[132]

Intrinsic self-healing of polymers (Figure 11.27a) describes crack healing of polymers and polymer matrices [mostly thermoplastics (no network, chain mobility)] with external stimulation (e.g. heating) by the polymers themselves. The intrinsic self-healing processes can again be divided into two modes, the physical (welding, molecular interdiffusion) and the chemical interaction (self-healing via reversible bond formation, photoinduced healing, thermally reversible crosslinking, and recombination of chain ends).[132, 138]

Welding is one of the traditional repair methods for polymeric materials. It enables the rejoining of fractured surfaces (closing cracks) or fusing new materials to the damaged region of the polymer composite due to the formation of chain entanglements between two contacting polymer surfaces.[138] *Molecular interdiffusion* occurs only above the glass transition temperature (T_g), which is often induced by heating the two samples of an amorphous polymer above the glass transition temperature. Under slight pressure the two polymer surfaces are brought into contact, because of the *molecular interdiffusion of the chains and often welding;* the establishment of mechanical strength resulted.[138]

Self-healing based on chemical interactions is focused on the recombination of broken molecules, which leads to cracks and strength decay. The *recombination of chain ends* using the example of thermoplastics prepared by condensation reactions is one method, but often high temperatures are necessary to heal the damaged region. Furthermore, chain ends capable of recombination reactions are required, limiting this repair mechanism to only a few thermoplastic polymers. Another method is self-healing via *thermally reversible crosslinking*, mostly using the Diels–Alder (DA) reaction based on highly crosslinked polymeric materials containing polyfurane and polymaleimide. At temperatures above 120 °C the linkages disconnect corresponding to the retro-DA reaction, but then reconnect during cooling as a result of the DA reaction. Advantages of this method are that no additional ingredients are needed and multiple cycles of crack healing can be executed.

New and promising intrinsic self-healing systems are mechanophores and supramolecular polymers as represented in Figure 11.30. Figures 11.30a and 11.30b are based on polymers containing mechanically active molecules, the so-called mechanophores,[139] and Figure 11.30c is based on network formation via supramolecular interactions.[140]

Mechanophores enable the transformation of energy from applied mechanical fields into productive chemical changes that can restore functionality lost by damage. When pushed or pulled with a certain force, specific *chemical reactions* are triggered in the mechanically activated molecules.

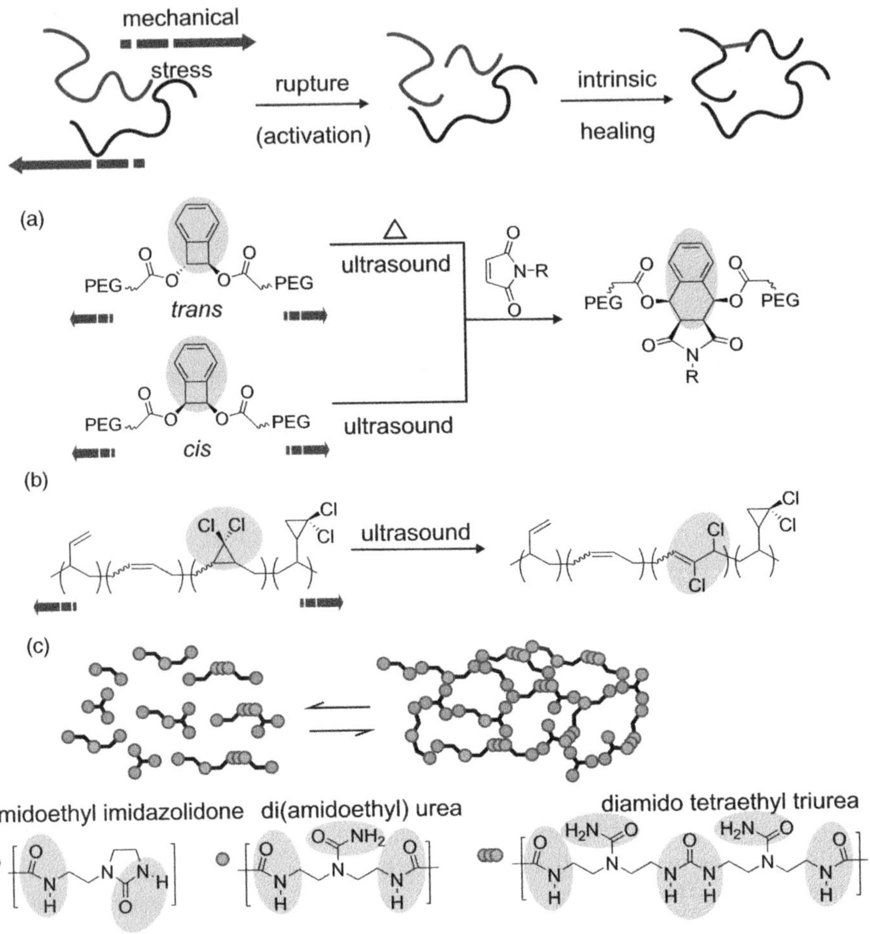

Figure 11.30 New and promising intrinsic self-healing systems: (a) and (b) mechanophores; (c) oligomers equipped with complementary hydrogen bonding groups—amidoethyl imidazolidone, di(amidoethyl) urea, and diamido tetraethyl triurea forming a supramolecular network.

Figure 11.30a shows that with mechanically sensitive chemical groups the energy of ultrasound (consistent with the effects of mechanical forces, i.e. shear force) can be used to open otherwise forbidden orbital symmetry pathways; The *trans* and *cis* isomers of 1,2-disubstituted benzocyclobutene (BCB) with long polymer strands are ring-opened in an electrocyclic reaction by ultrasound in a formally conrotatory or disrotatory process, resulting in formation of *ortho*-quinodimethide diene. This diene undergoes a fast Diels–Alder process with maleimide (proved by ^{13}C labeling experiments) without polymer chain scission.[139]

Figure 11.30b shows the mechanical active polymer based on ring-opening via ultrasound of *gem*-dichlorocyclopropanes (*g*DCCs). This shear-induced ring opening occurs several hundred times more often than chain scission, when the source of the force is coupled directly with the reaction coordinate.[139]

One special group in intrinsic self-healing via reversible bond formation are supramolecular self-healing materials. Due to multiple, reversible hydrogen bonds, a self-healing elastomeric polymer can be achieved. The synthesis of the polymers is versatile; the precursor for the hydrogen bond can be located at the end of the polymer, in the main chain and/or as a side chain, in monovalent as well as multivalent polymers.

Figure 11.30c shows the use of oligomers equipped with hydrogen bonding groups: amidoethyl imidazolidone, di(amidoethyl) urea, and diamido tetraethyl triurea forming a three-dimensional network due to supramolecular assembly.[140] In the case of applied stress, the hydrogen bonds of this rubbery self-healing material are decoupled, leading to crack formation. But bringing the fractured surfaces together reformation of the hydrogen bonds takes place at room temperature. One decisive advantage compared to other self-healing systems is the repeatability of breaking and healing. However, the mechanical properties (like creeping in response to stress) of these rubbery self-healing materials do not permit application at this moment.

Supramolecular self-healing materials by themselves are not the conclusion for autonomic healing, but they are an important step in this direction. Rather, the mixing of different self-healing systems seemed to be the right vote, for instance, supramolecular networks combined with covalent networks ensuring better mechanical properties. The supramolecular bonds should behave like mechanophores, which means breaking of these bonds instead of covalent bonds due to damage caused by stress, while the covalent network holds the basic network together and so reformation of the hydrogen bonds is possible.

How could such self-healing systems be applied in practical life? Scratches on cell phone displays, automotive coatings or propagating cracks in aerospace technology would not be a problem anymore if they were used. An important step to achieve this is a possible new coating, which can heal itself when scratched.[141] It consists of a polyurethane network blended with chitosan, a polymer found in crab and shrimp shells, which is substituted with four-member oxetane rings. Following mechanical damage of the network, the oxetane rings open creating two reactive ends. In the case of irradiation with ultraviolet light, chitosan chain scission takes place, leading to crosslinks with the reactive ends of the oxetanes and so the formation of a new network in less than an hour. Thus, it should be possible to create self-healing materials by using the power of the sun.

Therefore, it can be expected that self-healing polymers find their place in everyday life, enabling us to achieve higher lifetimes of (polymeric) (composite) materials. As the need for materials with increased life cycles and lifetimes is strongly increasing, the principles of self-healing, either by covalent forces or via supramolecular interactions, project a bright future for such polymers.

ACKNOWLEDGMENTS

We acknowledge grants DFG BI 1337/6-1 within the Forschergruppe FOR-1145, DFG 1337/7-1, DFG 1337/8-1 and DFG INST 271/249-1; INST 271/247-1, INST 271/248-1, and MINILUBES for financial support. We thank Haitham Barqawi for drawing figures.

REFERENCES

1. Wendell, D.; Patti, J.; Montemagno, C. *Small* **2006**, *2*, 1324–1329.
2. Engler, A. J.; Sen, S.; Sweeney, H. L.; Discher, D. E. *Cell* **2006**, *126*, 677–689.
3. Place, E. S.; George, J. H.; Williams, C. K.; Stevens, M. M. *Chem. Soc. Rev.* **2009**, *38*, 1139–1151.
4. Place, E. S.; George, J. H.; Williams, C. K.; Stevens, M. M. *Chem. Soc. Rev.* **2009**, *38*, 1139–1151.
5. Caruso, M. M.; Davis, D. A.; Shen, Q.; Odom, S. A.; Sottos, N. R.; White, S. R.; Moore, J. S. *Chem. Rev.* **2009**, *109*, 5755–5798.
6. Wu, D. Y.; Meure, S.; Solomon, D. *Prog. Polym. Sci.* **2008**, *33*, 479–522.
7. Antonietti, M.; Fratzl, P. *Macromol. Chem. Phys.* **2010**, *211*, 166–170.
8. Yu, S.-H.; Colfen, H.; Tauer, K.; Antonietti, M. *Nature Mater.* **2005**, *4*, 51–55.
9. Porter, D.; Vollrath, F. *Adv. Mater.* **2009**, *21*, 487–492.
10. Beachley, V.; Wen, X. *Prog. Polym. Sci.* **2010**, *35*, 868–892.
11. Shanmuganathan, K.; Capadona, J. R.; Rowan, S. J.; Weder, C. *Prog. Polym. Sci.* **2010**, *35*, 212–222.
12. Meuler, A. J.; McKinley, G. H.; Cohen, R. E. *ACS Nano* **2010**, *4*, 7048–7052.
13. Matyjaszewski, K.; Muller, A. H. E. *Prog. Polym. Sci.* **2006**, *31*, 1039–1040.
14. Matyjaszewski, K. *Macromolecules* **1993**, *26*, 1787–1788.
15. Braunecker, W. A.; Matyjaszewski, K. *Prog. Polym. Sci.* **2007**, *32*, 93–146.
16. Tang, W.; Fukuda, T.; Matyjaszewski, K. *Macromolecules* **2006**, *39*, 4332–4337.
17. Lowe, A. B.; McCormick, C. L. *Prog. Polym. Sci.* **2007**, *32*, 283–351.
18. Bielawski, C. W.; Grubbs, R. H. *Prog. Polym. Sci.* **2007**, *32*, 1–29.
19. Puskas, J. E.; Gautriaud, E.; Deffieux, A.; Kennedy, J. P. *Prog. Polym. Sci.* **2006**, *31*, 533–548.
20. Hadjichristidis, N.; Iatrou, H.; Pitsikalis, M.; Mays, J. *Prog. Polym. Sci.* **2006**, *31*, 1068–1132.
21. Polowinski, S. *Prog. Polym. Sci.* **2002**, *27*, 537–577.
22. Saito, R. *Polymer* **2008**, *49*, 2625–2631.
23. Sada, K.; Takeuchi, M.; Fujita, N.; Numata, M.; Shinkai, S. *Chem. Soc. Rev.* **2007**, *36*, 415–435.
24. Datta, B.; Schuster, G. B. *J. Am. Chem. Soc.* **2008**, *130*, 2965–2973.
25. Lutz, J.-F. *Polym. Chem.* **2010**, *1*, 55–62.

26. Watson, J. D.; Crick, F. H. C. *Nature* **1953**, *171*, 737–738.
27. Natta, G.; Pino, P.; Corradini, P.; Danusso, F.; Mantica, E.; Mazzanti, G.; Moraglio, G. *J. Am. Chem. Soc.* **1955**, *77*, 1708–1710.
28. Okamoto, Y.; Suzuki, K.; Ohta, K.; Hatada, K.; Yuki, H. *J. Am. Chem. Soc.* **1979**, *101*, 4763–4765.
29. Nakano, T.; Okamoto, Y. *Chem. Rev*. **2001**, *101*, 4013–4038.
30. Hill, D. J.; Mio, M. J.; Prince, R. B.; Hughes, T. S.; Moore, J. S. *Chem. Rev*. **2001**, *101*, 3893–4012.
31. Bassani, D. M.; Lehn, J.-M.; Baum, G.; Fenske, D. *Angew. Chem. Int. Ed*. **1997**, *36*, 1845–1847.
32. Koert, U.; Harding, M. M.; Lehn, J.-M. *Nature* **1990**, *346*, 339–342.
33. Schneider, J. P.; Kelly, J. W. *Chem. Rev*. **1995**, *95*, 2169–2187.
34. van Hest, J. C. M. *Polym. Rev*. **2007**, *47*, 63–92.
35. König, H. M.; Kilbinger, A. F. M. *Angew. Chem. Int. Ed*. **2007**, *46*, 8334–8340.
36. Ayres, L.; Adams, P. H. H. M.; Löwik, D. W. P. M.; van Hest, J. C. M. *Biomacromolecules* **2004**, *6*, 825–831.
37. Yu, T.-B.; Bai, J. Z.; Guan, Z. *Angew. Chem. Int. Ed*. **2009**, *48*, 1097–1101.
38. Kendhale, A.; Gonnade, R.; Rajamohanan, P. R.; Sanjayan, G. J. *Chem. Commun*. **2006**, 2756–2758.
39. Uemura, K.; Kumamoto, Y.; Kitagawa, S. *Chem. Eur. J*. **2008**, *14*, 9565–9576.
40. Lehn, J.-M. *Angew. Chem. Int. Ed*. **1990**, *29*, 1304–1319.
41. Berl, V.; Schmutz, M.; Krische, M. J.; Khoury, R. G.; Lehn, J.-M. *Chem. Eur. J*. **2002**, *8*, 1227–1244.
42. Binder, W.; Zirbs, R. *Adv. Polym. Sci*. **2007**, 1–78.
43. (a) Zirbs, R.; Kienberger, F.; Hinterdorfer, P.; Binder, W. H. *Langmuir* **2005**, *21*, 8414–8421; (b) Zare, P.; Stojanovic, A.; Herbst, F.; Akbarzadeh, J.; Peterlik, H.; Binder, W. H. *Macromolecules* **2012**, *45*, 2074.
44. Enders, C.; Tanner, S.; Binder, W. H. *Macromolecules* **2010**, *43*, 8436–8446.
45. Kurzhals, S.; Binder, W. H. *J. Polym. Sci. A: Polym. Chem*. **2010**, *48*, 5522–5532.
46. Herbst, F.; Schröter, K.; Gunkel, I.; Gröger, S.; Thurn-Albrecht, T.; Balbach, J.; Binder, W. H. *Macromolecules* **2010**, *43*, 10006–10016.
47. Kunz, M. J.; Hayn, G.; Saf, R.; Binder, W. H. *J. Polym. Sci. A: Polym. Chem*. **2004**, *42*, 661–674.
48. Tang, C.; Lennon, E. M.; Fredrickson, G. H.; Kramer, E. J.; Hawker, C. J. *Science* **2008**, *322*, 429–432.
49. Alarcon, C. d. l. H.; Pennadam, S.; Alexander, C. *Chem. Soc. Rev*. **2005**, *34*, 276–285.
50. Jeong, B.; Gutowska, A. *Trends Biotechnol*. **2002**, *20*, 305–311.
51. Godbey, W. T.; Wu, K. K.; Mikos, A. G. *J. Controlled Rel*. **1999**, *60*, 149–160.
52. Galaev, I. Y.; Mattiasson, B. *Trends Biotechnol*. **1999**, *17*, 335–340.
53. Hoffman, A. S.; Stayton, P. S.; Bulmus, V.; Chen, G.; Chen, J.; Cheung, C.; Chilkoti, A.; Ding, Z.; Dong, L.; Fong, R.; Lackey, C. A.; Long, C. J.; Miura, M.; Morris, J. E.; Murthy, N.; Nabeshima, Y.; Park, T. G.; Press, O. W.; Shimoboji, T.; Shoemaker, S.; Yang, H. J.; Monji, N.; Nowinski, R. C.; Cole, C. A.; Priest, J. H.; Harris, J. M.; Nakamae, K.; Nishino, T.; Miyata, T. *J. Biomed. Mater. Res*. **2000**, *52*, 577–586.

54. Blanazs, A.; Armes, S. P.; Ryan, A. J. *Macromol. Rapid Commun.* **2009**, *30*, 267–277.
55. Zhang, L.; Eisenberg, A. *Science* **1995**, *268*, 1728–1731.
56. Discher, D. E.; Eisenberg, A. *Science* **2002**, *297*, 967–973.
57. Antonietti, M.; Förster, S. *Adv. Mater.* **2003**, *15*, 1323–1333.
58. Hamley, I. W. "Self-Assembling Nanostructured Molecular Materials and Devices," in *Nanoscale Science and Technology*, John Wiley & Sons, Ltd., Chichester, UK, 2005, pp. 343–376.
59. Bermudez, H.; Brannan, A. K.; Hammer, D. A.; Bates, F. S.; Discher, D. E. *Macromolecules* **2002**, *35*, 8203–8208.
60. Binder, W. H. *Angew. Chem. Int. Ed.* **2008**, *47*, 3092–3095.
61. Binder, W. H.; Barragan, V.; Menger, F. M. *Angew. Chem. Int. Ed.* **2003**, *42*, 5802–5827.
62. Discher, B. M.; Won, Y.-Y.; Ege, D. S.; Lee, J. C.-M.; Bates, F. S.; Discher, D. E.; Hammer, D. A. *Science* **1999**, *284*, 1143–1146.
63. Dimova, R.; Seifert, U.; Pouligny, B.; Förster, S.; Döbereiner, H. G. *Eur. Phys. J. E* **2002**, *7*, 241–250.
64. Choucair, A.; Lim Soo, P.; Eisenberg, A. *Langmuir* **2005**, *21*, 9308–9313.
65. Lee, J. C. M.; Santore, M.; Bates, F. S.; Discher, D. E. *Macromolecules* **2001**, *35*, 323–326.
66. Ahmed, F.; Pakunlu, R. I.; Srinivas, G.; Brannan, A.; Bates, F.; Klein, M. L.; Minko, T.; Discher, D. E. *Mol. Pharmaceutics* **2006**, *3*, 340–350.
67. Graff, A.; Sauer, M.; Van Gelder, P.; Meier, W. *Proc. Natl. Acad. Sci. U.S.A.* **2002**, *99*, 5064–5068.
68. Lee, J. C. M.; Bermudez, H.; Discher, B. M.; Sheehan, M. A.; Won, Y.-Y.; Bates, F. S.; Discher, D. E. *Biotechnol. Bioeng.* **2001**, *73*, 135–145.
69. Meier, W.; Nardin, C. *Rev. Mol. Biotechnol.* **2002**, *90*, 17–26.
70. Bates, F. S.; Fredrickson, G. H. *Annu. Rev. Phys. Chem.* **1990**, *41*, 525–557.
71. Leibler, L. *Macromolecules* **1980**, *13*, 1602–1617.
72. Hadjichristidis, N.; Pispas, S.; Floudas, G. "Block Copolymer Morphology," in *Block Copolymers*, John Wiley & Sons, Hoboken, NJ, 2003, pp. 346–3613.
73. Yang, S. Y.; Ryu, I.; Kim, H. Y.; Kim, J. K.; Jang, S. K.; Russell, T. P. *Adv. Mater.* **2006**, *18*, 709–712.
74. Hamley, I. W. *Nanotechnology* **2003**, *14*, R39–R54.
75. Mermall, V.; Post, P. L.; Mooseker, M. S. *Science* **1998**, *279*, 527–533.
76. Goldstein, L. S. B.; Philp, A. V. *Annu. Rev. Cell Dev. Biol.* **1999**, *15*, 141–183.
77. Vale, R. D. *Cell* **2003**, *112*, 467–480.
78. Vale, R. D.; Milligan, R. A. *Science* **2000**, *288*, 88–95.
79. Osada, Y.; Gong, J.-P. *Adv. Mater.* **1998**, *10*, 827–837.
80. Ookubo, N.; Hirai, Y.; Ito, K.; Hayakawa, R. *Macromolecules* **1989**, *22*, 1359–1366.
81. Osada, Y.; Hasebe, M. *Chem. Lett.* **1985**, *14*, 1285–1288.
82. Sawahata, K.; Gong, J. P.; Osada, Y. *Macromol. Rapid Commun.* **1995**, *16*, 713–716.

83. Osada, Y.; Okuzaki, H.; Hori, H. *Nature* **1992**, *355*, 242–244.

84. Gong, J. P.; Matsumoto, S.; Uchida, M.; Isogai, N.; Osada, Y. *J. Phys. Chem.* **1996**, *100*, 11092–11097.

85. Northen, T. R.; Woodbury, N. W. *Langmuir* **2005**, *21*, 4949–4953.

86. Schild, H. G. *Prog. Polym. Sci.* **1992**, *17*, 163–249.

87. Gil, E. S.; Hudson, S. M. *Prog. Polym. Sci.* **2004**, *29*, 1173–1222.

88. Gebeshuber, I. C.; Drack, M.; Scherge, M. *Tribol. Mater. Surf. Interfaces* **2008**, *2*, 200–212.

89. Furey, M. J.; Burkhardt, B. M. *Lubr. Sci.* **1997**, *9*, 255–271.

90. Hardingham, T. E.; Muir, H. *Biochim. Biophys. Acta* **1972**, *279*, 401–405.

91. Han, L.; Dean, D.; Ortiz, C.; Grodzinsky, A. J. *Biophys. J.* **2007**, *92*, 1384–1398.

92. Klein, J.; Kumacheva, E.; Mahalu, D.; Perahia, D.; Fetters, L. J. *Nature* **1994**, *370*, 634–636.

93. Raviv, U.; Giasson, S.; Kampf, N.; Gohy, J.-F.; Jerome, R.; Klein, J. *Nature* **2003**, *425*, 163–165.

94. Kobayashi, M.; Takahara, A. *Chem. Lett.* **2005**, *34*, 1582–1583.

95. Sakata, H.; Kobayashi, M.; Otsuka, H.; Takahara, A. *Polym. J.* **2005**, *37*, 767–775.

96. Kenausis, G. L.; Voros, J.; Elbert, D. L.; Huang, N.; Hofer, R.; Ruiz-Taylor, L.; Textor, M.; Hubbell, J. A.; Spencer, N. D. *J. Phys. Chem. B* **2000**, *104*, 3298–3309.

97. Huang, N.-P.; Michel, R.; Voros, J.; Textor, M.; Hofer, R.; Rossi, A.; Elbert, D. L.; Hubbell, J. A.; Spencer, N. D. *Langmuir* **2000**, *17*, 489–498.

98. Ishihara, K.; Nomura, H.; Mihara, T.; Kurita, K.; Iwasaki, Y.; Nakabayashi, N. *J. Biomed. Mater. Res.* **1998**, *39*, 323–330.

99. Ho, S. P.; Nakabayashi, N.; Iwasaki, Y.; Boland, T.; LaBerge, M. *Biomaterials* **2003**, *24*, 5121–5129.

100. Moro, T.; Takatori, Y.; Ishihara, K.; Konno, T.; Takigawa, Y.; Matsushita, T.; Chung, U.-I.; Nakamura, K.; Kawaguchi, H. *Nature Mater.* **2004**, *3*, 829–836.

101. Kobayashi, M.; Terayama, Y.; Hosaka, N.; Kaido, M.; Suzuki, A.; Yamada, N.; Torikai, N.; Ishihara, K.; Takahara, A. *Soft Matter* **2007**, *3*, 740–746.

102. Kobayashi, M.; Takahara, A. *Chem. Rec.* **2010**, *10*, 208–216.

103. Lendlein, A.; Kelch, S. *Angew. Chem. Int. Ed.* **2002**, *41*, 2034–2057.

104. El Feninat, F.; Laroche, G.; Fiset, M.; Mantovani, D. *Adv. Eng. Mater.* **2002**, *4*, 91–104.

105. Lendlein, A.; Langer, R. *Science* **2002**, *296*, 1673–1676.

106. Maitland, D. J.; Metzger, M. F.; Schumann, D.; Lee, A.; Wilson, T. S. *Lasers Surg. Med.* **2002**, *30*, 1–11.

107. Small IV, W.; Wilson, T.; Benett, W.; Loge, J.; Maitland, D. *Opt. Express* **2005**, *13*, 8204–8213.

108. Langer, R.; Lendlein, A. *PCT Patent Application WO 2004073690 A1*, 2004.

109. Langer, R.; Lendlein, A. *PCT Patent Application WO 2006092789 A2*, 2006.

110. Hampikian, J. M.; Heaton, B. C.; Tong, F. C.; Zhang, Z.; Wong, C. P. *Mater. Sci. Eng. C* **2006**, *26*, 1373–1379.

111. Li, W.; Li, S. *Adv. Polym. Sci*. **2007**, *206*, 191–210.
112. Marty, J. D.; Mauzac, M. *Adv. Polym. Sci*. **2005**, *172*, 1–35.
113. Asanuma, H.; Hishiya, T.; Komiyama, M. *Adv. Mater*. **2000**, *12*, 1019–1030.
114. Wulff, G.; Schauhoff, S. *J. Org. Chem*. **1991**, *56*, 395–400.
115. Piletska, E. V.; Guerreiro, A. R.; Whitcombe, M. J.; Piletsky, S. A. *Macromolecules* **2009**, *42*, 4921–4928.
116. Chen, Y.; Shimizu, K. D. *Org. Lett*. **2002**, *17*, 2937–2940.
117. Haupt, K.; Mosbach, K. *Chem. Rev*. **2000**, *100*, 2495–2504.
118. Urraca, J. L.; Hall, A. J.; Moreno-Bondi, M. C.; Sellergren, B. *Angew. Chem. Int. Ed*. **2006**, *45*, 5158–5161.
119. Yamazaki, T.; Ohta, S.; Sode, K. *Electrochemistry* **2008**, *76*, 590–593.
120. Nishino, H.; Huang, C.-S.; Shea, K. J. *Angew. Chem. Int. Ed*. **2006**, *45*, 2392–2396.
121. Dickert, F. L.; Hayden, O. *Anal. Chem*. **2002**, *74*, 1302–1306.
122. Dickert, F. L.; Hayden, O.; Halikias, K. P. *Analyst* **2001**, *126*, 766–771.
123. Wulff, G. *Chem. Rev*. **2002**, *102*, 1–28.
124. Hall, A. J.; Emgenbroich, M.; Sellergren, B. *Top. Curr. Chem*. **2005**, *249*, 317–349.
125. Carboni, D.; Flavin, K.; Servant, A.; Gouverneur, V.; Resmini, M. *Chem. Eur. J*. **2008**, *14*, 7059–7065.
126. Liu, J.-Q.; Wulff, G. *J. Am. Chem. Soc*. **2008**, *130*, 8044–8054.
127. Zhang, H.; Piacham, T.; Drew, M.; Patek, M.; Mosbach, K.; Ye, L. *J. Am. Chem. Soc*. **2006**, *128*, 4178–4179.
128. Shen, X.; Zhu, L.; Li, J.; Tang, H. *Chem. Commun*. **2007**, 1163–1165.
129. Shiraishi, Y.; Suzuki, T.; Hirai, T. *New J. Chem*. **2010**, *34*, 714–717.
130. Brown, E.; Sottos, N.; White, S. *Exp. Mech*. **2002**, *42*, 372–379.
131. Trask, R. S.; et al. *Bioinspiration Biomimetics* **2007**, *2*, 1–9.
132. Blaiszik, B. J.; Kramer, S. L. B.; Olugebefola, S. C.; Moore, J. S.; Sottos, N. R.; White, S. R. *Annu. Rev. Mater. Res*. **2010**, *40*, 179–211.
133. Yuan, Y. C.; Yin, T.; Rong, M. Z.; Zhang, M. Q. *Express Polym. Lett*. **2008**, *2*, 238–250.
134. White, S. R.; Sottos, N. R.; Geubelle, P. H.; Moore, J. S.; Kessler, M. R.; Sriram, S. R.; Brown, E. N.; Viswanathan, S. *Nature* **2001**, *409*, 794–797.
135. Gragert, M.; Schunack, M.; Binder, W. H. *Macromol. Rapid Commun*. **2011**, *32*, 419–425.
136. Schunack, M.; Gragert, M.; Döhler, D.; Michael, P.; Binder, W. H. *Macromol. Chem. Phys*. **2012**, *213*, 205–214.
137. Döhler, D.; Michael, P.; Sottos, N. R.; Binder, W. H. *Macromolecules*. **2012**, ma-2012–00505v.
138. Wu, D. Y.; Meure, S.; Solomon, D. *Prog. Polym. Sci*. **2008**, *33*, 479–522.
139. Moore, J. S.; White, S. R.; Sottos, N. R.; Odom, S. A.; Shen, Q.; Davis, D. A.; Caruso, M. M. *Chem. Rev*. **2009**, *109*, 5755–5798.
140. Cordier, P.; Tournilhac, F.; Soulie-Ziakovic, C.; Leibler, L. *Nature* **2008**, *451*, 977–980.
141. Gosh, B.; Urban, M. W. *Science* **2009**, *323*, 1458–1460.

CHAPTER 12

Biomimetic Cavities and Bioinspired Receptors

STÉPHANE LE GAC

UMR CNRS 6226-Institut des Sciences Chimiques de Rennes, 263 Avenue du Général Leclerc-CS 74205, Université de Rennes 1, 35042 Rennes Cedex, France

IVAN JABIN

Laboratoire de Chimie Organique, Université Libre de Bruxelles (U.L.B.), Av. F. D. Roosevelt 50, CP160/06, B-1050 Brussels, Belgium

OLIVIA REINAUD

Laboratoire de Chimie et de Biochimie Pharmacologiques et Toxicologiques, CNRS UMR 8601, PRES Sorbonne Paris Cité, Université Paris Descartes, 45 rue des Saints Péres, 75006 Paris, France

12.1 INTRODUCTION

Molecular receptors are the heart of supramolecular chemistry and of biochemistry.[1] They allow the reversible assembly of discrete entities through the establishment of multiple weak interactions between the different components. These phenomena are fundamental in biology. In enzymes, for example, formation of the enzyme–substrate complex is the key for the selectivity and efficiency of the biocatalysis. Also, the knowledge of the major sites of interaction and their relative geometrical positioning is essential for understanding the outstanding efficiency of the recognition processes.[2] In the active site of enzymes, these processes combine nonpolar interactions for the recognition of a hydrophobic residue and electrostatic interactions such as charge–charge and charge–dipole, including, of course, hydrogen bonding. In some cases also, a metal ion is a ligating center allowing the coordination of the substrate or an intermediate in the catalytic cycle.[3] This is beautifully illustrated by Zn enzymes such as *carboxypeptidase*.[4] Hence, combining a donor or acceptor site to a hydrophobic pocket is a recurrent strategy encountered in natural systems to efficiently and selectively bind a guest. All this information from

Bioinspiration and Biomimicry in Chemistry: Reverse-Engineering Nature, First Edition.
Edited by Gerhard F. Swiegers.
© 2012 John Wiley & Sons, Inc. Published 2012 by John Wiley & Sons, Inc.

Nature has been a major source of inspiration for chemists. On the one hand, it has been widely used for the design of new inhibitors. On the other hand, it motivated chemists to design artificial hosts useful for selective recognition or catalysis.[1, 5]

Among the various molecular platforms that can be used for the construction of such receptors, macrocycles presenting multiple aromatic units have been widely used. Cavitands[6] such as resorcinarenes,[7] cyclotriveratrylenes,[8] calixarenes,[9] and cyclophanes[10] have largely been developed during the last two decades. Calix[6]arenes are particularly attractive macrocyclic systems to play the role of a host because they have a size well adapted for the selective inclusion of small organic molecules.[11] This is not the case for calix[4]arenes that have mostly been used as a platform for the preorganization of a binding site outside the cavity.[12] However, calix[6]arenes suffer from a great flexibility that must be restricted. Indeed, they must be constrained to the cone conformation in order to present a concave cavity suitable to include a guest. The second requirement is the implementation of a polarized site. Thus, we have developed different strategies, all based on the introduction of nitrogen functionalities at the small rim of the calixarene (Figure 12.1).

For biomimetic purposes in the bioinorganic field, we have first developed systems presenting three coordinating arms[13] such as imidazole groups that, upon binding to a transition metal ion (such as Cu,[14] Zn,[15] Co, or Ni[16]), close the entrance on one side and constrain the calixarene cavity in a cone open to the outside at the large rim. The *funnel* complexes proved to be remarkable receptors for neutral molecules that can bind the metal center. The second generation of calix[6]-based ligands provides a switchable functionality interacting with the metal center. These two systems are described in the first part of this chapter, with special highlights on biomimetics. The second part presents systems derived from the third generation of ligands, where the polyaza donor covalently caps the calixarene. This system opened the route to a variety of receptors, with or without metal ions, displaying highly specific properties tuned by the nature of the aza cap. Finally, self-assembled and polarized cavities, reminiscent of nonmetallic but structured proteins, illustrate other biomimetic concepts such as allosterism, for which flexibility is a key feature.

12.2 MIMICS OF THE MICHAELIS–MENTEN COMPLEXES OF ZINC(II) ENZYMES WITH POLYIMIDAZOLYL CALIXARENE-BASED LIGANDS

A biomimetic tris(imidazolyl) core for metal ion coordination associated to a cavity was constructed from the calix[6]arene scaffold. This family of cavity-based ligands mimics the polyimidazole sites found in the active site of many metalloenzymes (Figure 12.1). Synthesis of the calix[6]arene-based ligands is readily achieved through sequential and selective functionalization of the calix small rim by methyl and 2-methylimidazolyl groups in alternate positions. The large rim can be further functionalized almost at will through sequential nitration,[17] reduction, diazotation, and click chemistry.[18] These calix[6]tris(imidazole) ligands readily

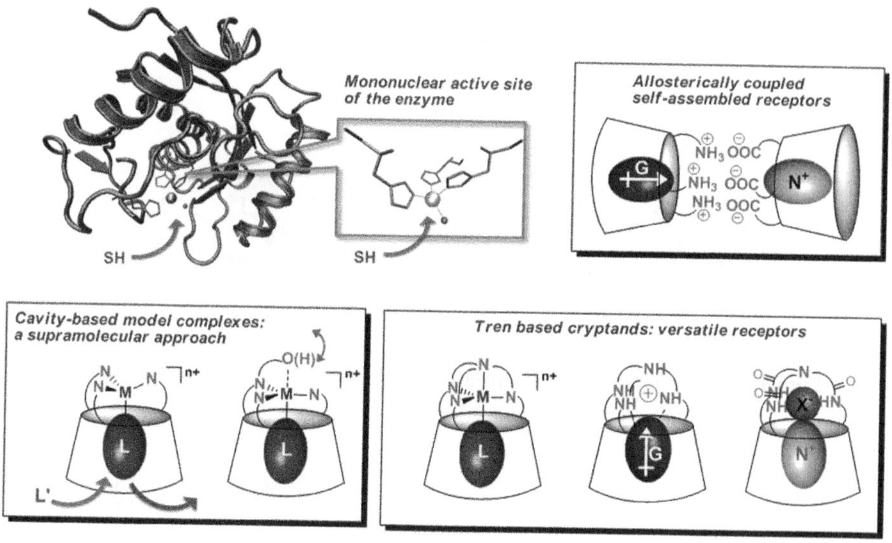

Figure 12.1 Top left: Schematic representation of the 3D structure of a mononuclear metalloenzyme with its active site in inset (adamalysine, a Zn-matrix metallopeptidase from snake venom, PDB codes: 1IAG). Schematized supramolecular models of metalloenzyme active sites based on calix[6]arenes and relevant bioinspired ditopic and self-assembled receptors with allosteric properties.

complex one Zn(II) metal ion to yield 4-coordinate mononuclear complexes where all three imidazoles wrap the metal dication in a helical way (Figure 12.2). The apical binding site is oriented toward the center of the cavity and occupied by a guest ligand. The calixarene adopts a cone conformation that is locked since the aromatic units cannot undergo flipping around the methylene bridges of the calix small rim due to the metal binding to the three nitrogen arms. Nevertheless, the aromatic units still stand alternatively in *in* and *out* positions relative to the cavity. This flattened conformation is the opposite of the one observed for the free ligand, as the anisole X substituents now adopt an *in* position relative to the three others. Altogether, they form a door controlling the cavity entrance.

12.2.1 A Bis-aqua Zn(II) Complex Modeling the Active Site of Carbonic Anhydrase

Several chemical systems have been developed by various groups to reproduce the $[Zn(His)_3(OH_2)]^{[2+]}$ coordination core encountered in many hydrolytic Zn enzymes.[4] Surprisingly, dicationic zinc aqua model complexes have proved extremely difficult to stabilize and most classical models only succeeded in stabilizing Zn-hydroxo species because of the high Lewis acidity of the Zn(II) center bound to only four neutral ligands.[19] In strong contrast, the reaction of the calix[6]tris(imidazole) ligand with $Zn(H_2O)_6(ClO_4)_2$ in THF readily yields a very stable dicationic zinc–aqua complex.[20] The unusual stability of the aqua complex

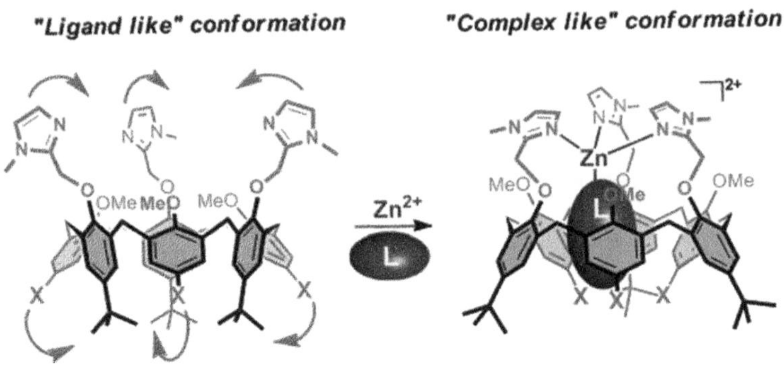

Figure 12.2 Shaping the calix[6]arene core into a Zn(II) *funnel* complex.

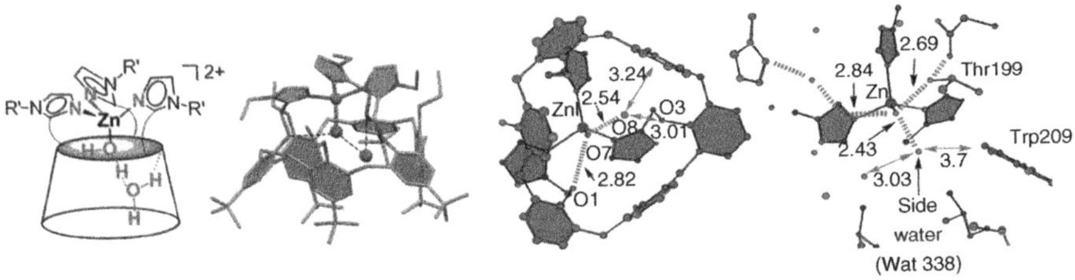

Figure 12.3 XRD structures of (from right to left) the active site of carbonic anhydrase II[21] and the calix-based model compound (bottom and side views, respectively).[20] (Reproduced with permission. Copyright the Royal Society of Chemistry: Ref. 25.)

is actually due to second coordination sphere and cavity effects: the water ligand is hydrogen-bonded to the oxygen atoms of the calix small rim that defines an electron-rich environment; it is further stabilized by a strong hydrogen bond to a second water guest suspended in the heart of the cavity, itself stabilized by OH/π interactions with the aromatic walls of the cavity. This complex shows remarkable similarities with the active site of carbonic anhydrase (Figure 12.3), which makes it the first structural model for the Zn–aqua species found in enzymes, nicely illustrating the importance of the microenvironment for the stabilization of reactive species.

In this calix[6]arene-based system, the cone constrains the metal ion in a tetrahedral geometry, precluding a second guest as small as a water molecule to coordinate the metal center in endo position. The corollary of this feature is that such a model will not allow mimicking the 5-coordinate intermediate formed during enzymatic catalysis. Indeed, no hydrolytic activity has been observed. On the other hand, this model has allowed studying, for the first time, the binding properties of such a constrained and highly Lewis acidic Zn(II) center toward a variety of exogenous ligands (vide infra).

12.2.2 Structural Key Features of the Zn(II) Funnel Complexes

The exceptional stability of this calixarene-based Zn–aqua complex is best illustrated by its reluctance to deprotonation in the presence of an amine. Instead, both water molecules are displaced by a primary amine yielding the 4-coordinate adduct depicted in Figure 12.4. Similar ternary complexes are readily formed by a wide variety of small organic coordinating molecules (L).[20, 22] In each case, X-ray diffraction analysis shows a Zn(II) center in the regular tetrahedral environment provided by the tris(imidazole) core and the guest ligand L. With protic guests such as amines, alcohols, or primary amides, hydrogen bonds always connect their acidic protons to one or two calixarene phenoxyl units as illustrated in Figure 12.4 by the XRD structure of the heptylamine complex. The guest conformation often undergoes gauche interactions for an optimized filling of the calixarene cavity with stabilizing CH/π interactions between the guest alkyl chain and the aromatic walls of the host. The ethanol[23] and acetaldehyde ternary complexes, which have been characterized by XRD as well as in solution, provide interesting models for substrate binding in *liver alcohol dehydrogenase* (LADH),[24] a zinc enzyme catalyzing the reversible dehydrogenation of alcohols to aldehydes via hydride transfer to NAD^+.

Due to their helical shape, these *funnel* complexes are chiral. The helicity, which originates from the metal binding of the three imidazolyl arms, is transmitted to the calixarene core, hence providing a chiral environment that ultimately is experienced by the guest. In solution, both enantiomers are in equilibrium. Interestingly, it has been shown that a chiral guest can control the equilibrium between the two helical forms of the complexes, thereby transmitting its own chirality to the whole calixarene-based system.[26]

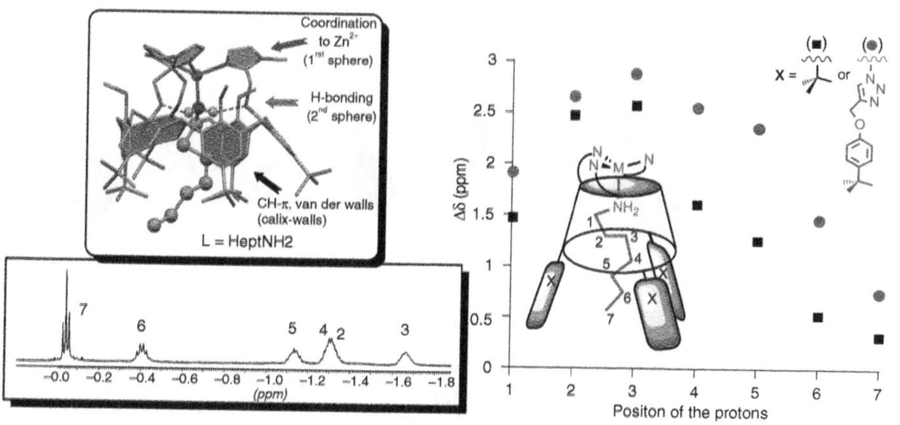

Figure 12.4 Left: XRD structure of the heptylamine dicationic Zn(II) *funnel* complex.[22] Bottom: Representative high-field shifted ^1H signature of an included guest [L = $CH_3(CH_2)_6NH_2$, $CDCl_3$, 300 K, 500 MHz].[22] Right: Mapping of the $\Delta\delta$ shifts corresponding to heptylamine complexes as a function of the large rim substitution pattern of the calix[6]arene cavity.[25]

12.2.3 Hosting Properties of the Zn(II) Funnel Complexes: Highly Selective Receptors for Neutral Molecules

^{1}H NMR spectroscopy has proved to be a powerful tool to monitor the presence of a coordinating molecule (L) inside the cavity as the in/out exchange is generally slow on the NMR analysis time scale.[15] The up-field shifts ($\Delta\delta$) measured for the guest protons, due to the shielding effect of the π electron ring current, are dependent on their spatial position in the aromatic cavity of the calixarene (Figure 12.4). The relative capacity of guest ligands to bind to the Zn(II) center was thus evaluated by ^{1}H NMR spectroscopy through competition experiments in a noncoordinating solvent (Figure 12.5). The equilibrium constants $K_{L'/EtOH}$ show that the selectivity of the inner-cavity binding is based on (1) the σ-donor property of the guest ligand, (2) the possible establishment of hydrogen bonds between the guest and the small rim, (3) the relative host/guest geometries, and (4) the charge of the ligand.

- With primary amines, coordination to the metal center is stoichiometric and quantitative at millimolar (mM) concentrations. Amides and alcohols are also excellent guests, better than nitriles. Coordination of aldehydes and carboxylic acids is much weaker, although detected. Neither ether nor ketone yields detectable coordinated species.
- Steric hindrance at the level of the coordinating atom and at its α position is a major factor of selectivity: whereas primary amines are the best ligands, secondary amines do not coordinate the metal center at all! Coordination of 1-propanol is 30 times stronger than 2-propanol.

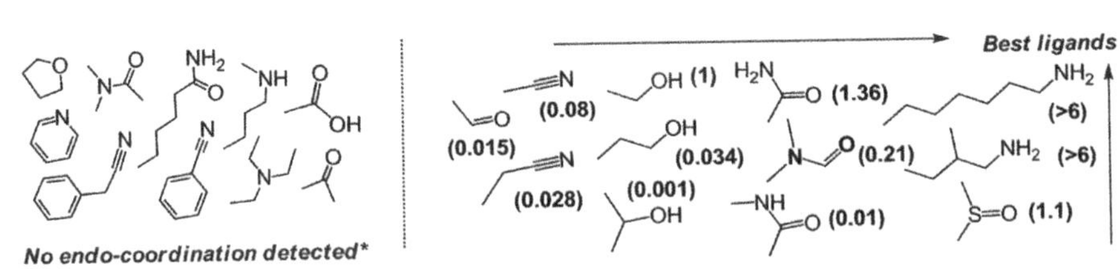

Figure 12.5 Ligand exchange at the Zn(II) center of the *funnel* complexes based on the *hexa-t*Bu-calix[6]tris(imidazole) ligand. The equilibrium constants $K_{L'/EtOH}$ (exchange of L = EtOH for L′ at 298 K in $CDCl_3$) are given in parentheses. $^*K_{EtOH/2H2O} = 0.4 mol \cdot L^{-1}$. When no endo coordination can be detected, $K_{L/2H2O} < 10^{-5}$ mol· L^{-1}.[20, 22]

- Neither a methyl substituent in 2-position nor a long alkyl chain precludes coordination at the metal center. However, benzo- and benzyl-nitrile are too sterically demanding to yield a stable adduct with the *hexa-t*Bu ligand.
- The calix cavity is surprisingly reluctant to host anions in spite of the presence of the dicationic metal center. Adding halides or hydroxides[27] in excess either leads to decoordination or induces a rearrangement of the calixarene macrocycle to give dimers and trimers with the coordinated bridging anions outside the cavity. The reason actually stems from the second coordination sphere defined by the calix oxygen-rich small rim[28]: in the conformation adopted by the tris(imidazole)-based complexes (Figure 12.2), all six oxygen atoms of the macrocycle point their lone pairs toward the cone axis, at the level of the coordinating atom. If an anion were to occupy that position, a strong electrostatic repulsion would result and lead to the destabilization of the host–guest adduct.

As shown schematically in Figure 12.2, the *para*-substituents of the anisole units are oriented in an *in* position relative to the cavity, thus constituting a door that closes the entrance of the host. Removing these three bulky *t*Bu groups[29] or replacing them by smaller ones allows the cavity to host larger guests. For example, the tris(anilino) derivative[30] (corresponding to $X = NH_2$ in Figure 12.2) accepts much larger guests, for example, dimethyldopamine, tryptamine, and benzylamine, than the parent compound ($X = t$Bu), for which endo coordination of these bulky amines has never been detected (vide infra).

12.2.4 Induced Fit: Recognition Processes Benefit from Flexibility

Highly specific enzymes generally display a relatively rigid pocket for the selective recognition of a unique substrate. In contrast, enzymes that contribute to the metabolism of drugs and xenobiotics, such as cytochromes P450, must face the efficient binding of a wide variety of substrates within the same active pocket. It has recently been recognized that this class of P450 enzymes has indeed a very flexible proteic backbone that allows the active pocket to shrink or expand depending on the substrate size.[31] The importance of such a behavior has been well recognized for other enzyme–substrate complexes as well as for drug–receptor complexes.[32] Interestingly, the calix[6]arene-based Zn(II) receptors are also capable of induced-fit behaviors for guest binding. This is well illustrated by the comparison of the XRD structures shown Figure 12.6.

Replacement of three *t*Bu groups by three small NH_2 substituents at the large rim (X in Figure 12.2) not only opens wider the entrance of the cavity but also allows spectacular induced-fit behaviors. The XRD structures of the ternary complexes obtained with dimethyldopamine and benzylamine evidence aniline walls standing almost parallel to each other to allow the large aromatic cores of the guest to fit in. Surprisingly, in the related aqua complex, the second water guest present in the parent *hexa-t*Bu system is absent in the aniline host. The stabilization of the acidic water ligand is now ensured by a direct OH–π interaction with the bent aniline unit

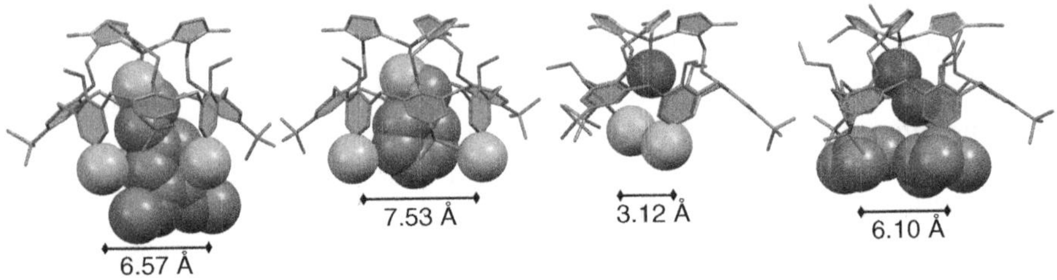

Figure 12.6 Induced-fit process undergone by the tris(aniline) derivative that allows larger and smaller guests to bind compared to the parent *t*Bu compounds, as illustrated by XRD structures. From left to right: The dimethyldopamine, benzylamine, and mono-aqua Zn(II) complexes and the bis-aqua derivative based on the *hexa-t*Bu ligand for comparison purposes.[30] (Reproduced with permission. Copyright the Royal Society of Chemistry: Ref. 25.)

that shrinks the cavity size, thus adapting it to the smallness of the guest ligand. In contrast, when stacked together, the bulky *t*Bu groups define a larger cavity space, thus requiring the hosting of a second water molecule to stabilize the structure (see Figure 12.6, right). In other words, the increased flexibility allows the optimization of noncovalent attractive interactions within the cavity, that is, hydrogen bonds and OH–π and CH–π interactions with the aromatic walls of the more or less flattened cone core with small as well as with large guests. Such a behavior stands in strong contrast to cyclodextrine[33] or resorcinarene-based[34] receptors, which, due to their rigidity, can display strong binding to organic guests, at the very condition, however, that there is a good fit between the guest size and the cavity size (described as the 55% rule by Rebek).[35] Hence, a rigid receptor presents a disadvantage as only a restricted number of guests will display a strong affinity for the receptor: only those whose size fits with the 55% rule. In the calix[6]-based system, the high, but controlled, flexibility of the host turns out to be an advantage, with a cavity that adapts to the size and nature of the guest for an optimized host–guest binding energy.

12.2.5 Multipoint Recognition

Para-substituents can also be involved in molecular recognition. The Zn(II) *funnel* complex based on the same tris(aniline) ligand, for example, displays a high propensity to interact at the level of the large rim aniline door with a variety of cations, such as a second metal ion,[36] a single proton, or an ammonium,[37] thus giving rise to stable tricationic structures. Thanks to the establishment of multiple hydrogen bonds at the level of the tris(aniline) door, this receptor is able to discriminate between mono- and polyamines; for example, it binds much better 1,3-propyldiamine than butylamine, provided, however, the former is monoprotonated. The remarkable selectivity of this multipoint recognition system is best illustrated by the regioselective binding of an unsymmetrical triamine, as illustrated in Figure 12.7.

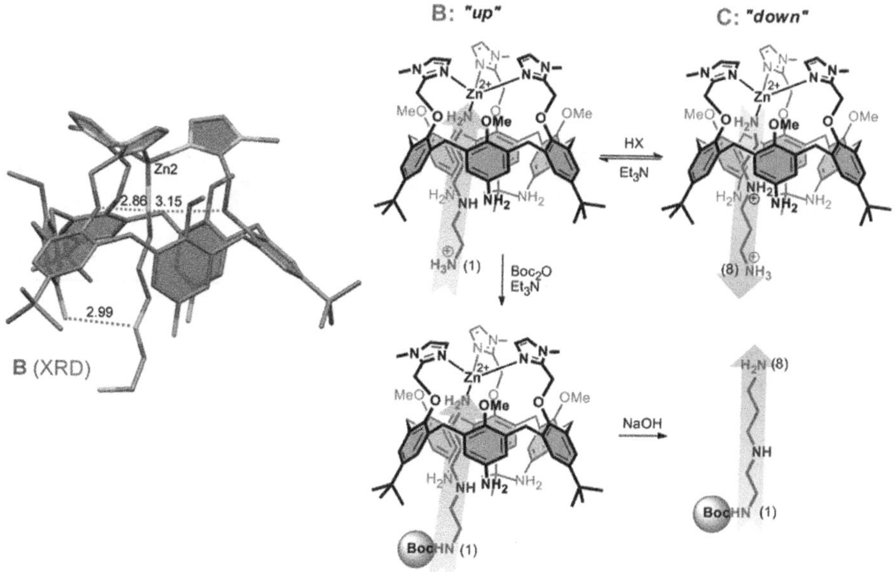

Figure 12.7 Left: Side view of the XRD refined structure[13] of tricationic complex **B**. Right: Regioselective binding of *N*-(2-aminoethyl)propane-1,3-diamine to the Zn(II) funnel complex based on the tris(aniline) ligand ($X = NH_2$ in Figure 12.2) and acid–base control of its directionality. (1) and (8) are arbitrary numbers used to differentiate the two primary N atoms of the triamino guest. Supramolecular protection allowing the regioselective carbamoylation of *N*-(2-aminoethyl)propane-1,3-diamine (Boc_2O stands for di-*t*Bu-dicarbonate).

Here, the funnel wraps and orients with a high selectivity an unsymmetrical triamine guest as a function of its protonation state. This host–guest adduct thus behaves as a bistable system: the guest reversibly changes from an "up" position to a "down" position depending on the acidity of the medium, depicting an acid–base-controlled directional switch. This remarkably discriminative recognition process has been used for orienting an electrophilic reagent (Boc_2O) at a single site (N1) with an unprecedented chemo- and regioselectivity, very much like in the SSAT enzyme,[38] an *N*-acetyl transferase that monoacetylates the triamine spermidine in vivo on N1. Indeed, in the funnel system like in the enzyme, the microenvironment not only guarantees specific binding but also positions the substrate molecule relative to the reactive species. As a result, a substrate presenting several sites susceptible to react is regio- and chemoselectively transformed. This case study also shows that a funnel-like receptor can be used as a supramolecular protecting tool allowing a transformation that is hardly feasible with conventional covalent chemistry.

12.2.6 Implementation of an Acid–Base Switch for Guest Binding

The second generation of ligands (Figure 12.1) was developed in order to introduce an additional functionality to the metal complexes.[39] The ligands present a fourth

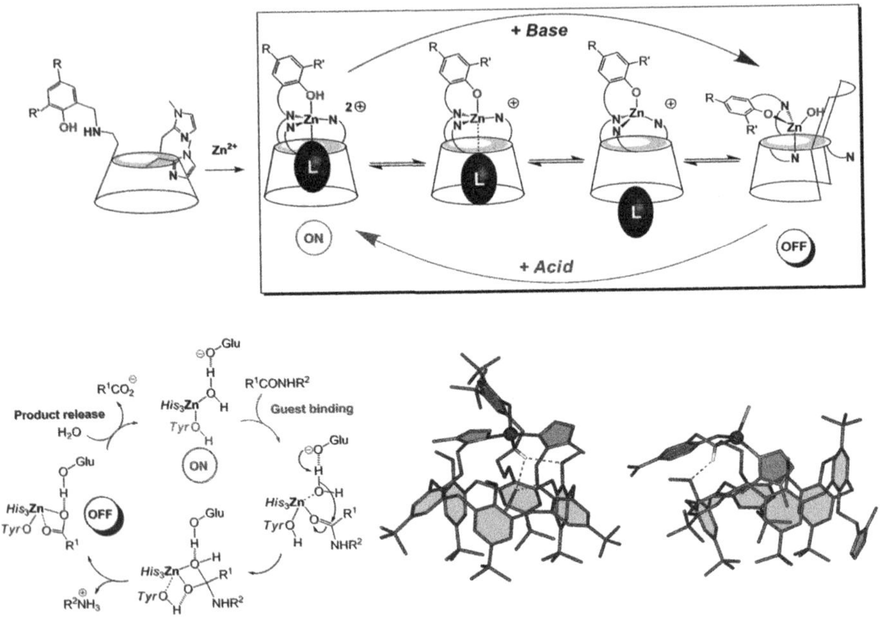

Figure 12.8 Acid–base switch for guest binding by the calix[6]N_3ArOH based Zn(II) complexes. Bottom, from left to right: Proposed mechanism for the peptidase activity of astacin and serralysin Zn-enzyme families, XRD structures of the monocationic phenoxide complex with a noncoordinated MeOH guest, and the neutral chlorophenoxide complex presenting a self-included imidazolyl arm.[41]

donor group covalently linked to one nitrogenous arm, which provides an additional cap to the system. This fourth donor can play the role of a redox-active function, such as a phenoxide, and participate in the oxidation of a substrate, hence providing a good functional model of the radical copper enzyme, galactose oxidase.[40] It can also act as a hemilabile arm and control the inner binding.[41] Indeed, with ligand calix[6]N_3ArOH (schematically illustrated in Figure 12.8), three different protonation states for the corresponding Zn(II) complexes have been characterized: [Zn(II)N_3ArOH]$^{2+}$, [Zn(II)N_3ArO]$^+$, and [Zn(II)(OH)N_3ArO]. Whereas the dicationic 5-coordinate species is very sensitive to guest binding, the monocationic complex binds a guest ligand with a lower affinity due to a decrease of the Zn(II) Lewis acidity.

The neutral species can be obtained upon reaction with a base to yield a hydroxo complex or with an anion such as a chloride that coordinates the metal center from the outside of the calixarene cavity. The simultaneous binding of two anionic donors induces an impressive conformational reorganization of the system. One imidazole arm is released by the metal center. The other one undergoes self-inclusion into the π-basic calixarene cavity, thus precluding any guest inclusion. As a result, the calix[6]N_3ArOH-based Zn(II) complexes act as an acid–base switch for guest binding. Several aspects of this system appear reminiscent of Zn-peptidases of the astacin and serralysin families.[4] For these enzymes, in addition to the three histidine residues, a side chain tyrosine coordinates the metal ion and its role has been questioned. This model system suggests that one role is to accurately control the activity of the enzymes as the pH varies, acting as an off switch upon a pH rise.

12.3 COMBINING A HYDROPHOBIC CAVITY AND A TREN-BASED UNIT: DESIGN OF TUNABLE, VERSATILE, BUT HIGHLY SELECTIVE RECEPTORS

12.3.1 Tren-Based Calix[6]arene Receptors

An alternative approach for the elaboration of biomimetic receptors consists of capping the hydrophobic cavity of the calix[6]arene skeleton by a tripodal aza subunit. Tris(2-aminoethyl)amine (tren) provides an attractive platform for the elaboration of such an aza cap. In this context, three different tren-based calix[6]arenes (i.e., calix[6]tren,[42] calix[6]trenamide,[43] and calix[6]trenurea[44]) have been synthesized (Figure 12.9). These three molecular receptors display a calix[6]arene framework constrained in a cone conformation by either a covalent trisamino, trisamido, or trisureido tren-based cap. In all cases, the grid-like nitrogenous cap closes the conic cavity at the narrow rim, leaving a single entrance controlled by the flexible *t*Bu door.

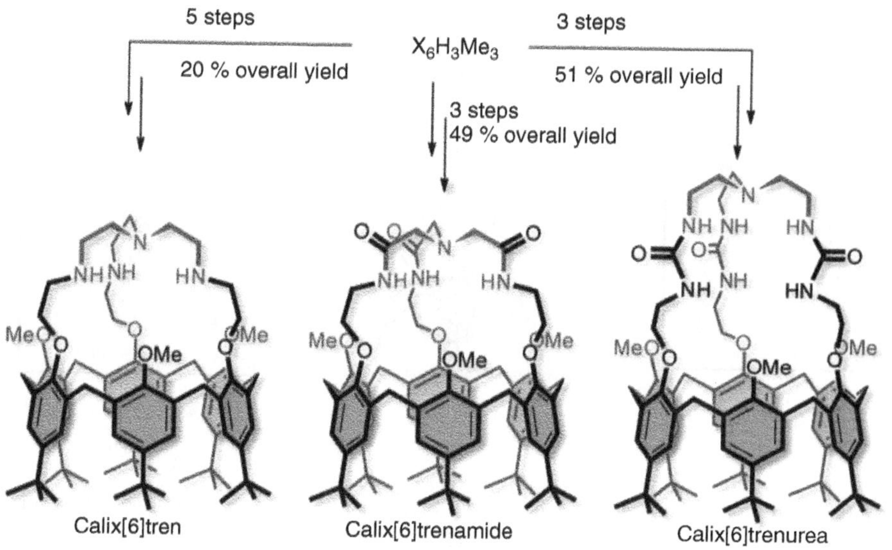

Figure 12.9 Structures of the three different tren-based calix[6]arenes. $X_6H_3Me_3$ stands for 1,3,5-trimethoxy-calix[6]arene.[42–44]

12.3.2 Versatility of a Polyamine Site[45]

12.3.2.1 Highly Resistant Funnel Complexes With Zn^{2+}, Cu^{2+}, and Cd^{2+}, the calix[6]tren ligand leads to *funnel* complexes with the metal ion coordinated by the tren cap and an exchangeable neutral molecule bound to the metal in the calixarene pocket, as depicted in Figure 12.10.[46] The XRD structure of a Zn^{2+} complex displays a 5-coordinate metal ion bound to a guest EtOH ligand and to the tren cap in an asymmetrical environment (right inset, Figure 12.10). Due to both a strong chelate effect and a full cavity-controlled access to the metal center, the Zn complexes appeared remarkably resistant and better hosts than their parent tris(imidazolyl) calix[6]-based complexes. They are resistant to bases, acids, and even electrophiles (e.g., NBu_4OH, AcOH, and MeI). Most interestingly, they proved not only capable of hosting a variety of small guest ligands such as alcohols, nitriles, amides, or primary amines, but also allowed the binding of large guests such as imidazole, benzylamine, and dodecyldiamine under experimental conditions for which the parent systems (Figure 12.5) underwent decoordination of Zn.

12.3.2.2 Complexation of Ammonium Ions In the absence of metal ion, this calix-cryptand is insensitive to neutral species such as alcohols, amides, and amines in chloroform solutions at millimolar (mM) concentrations. However,

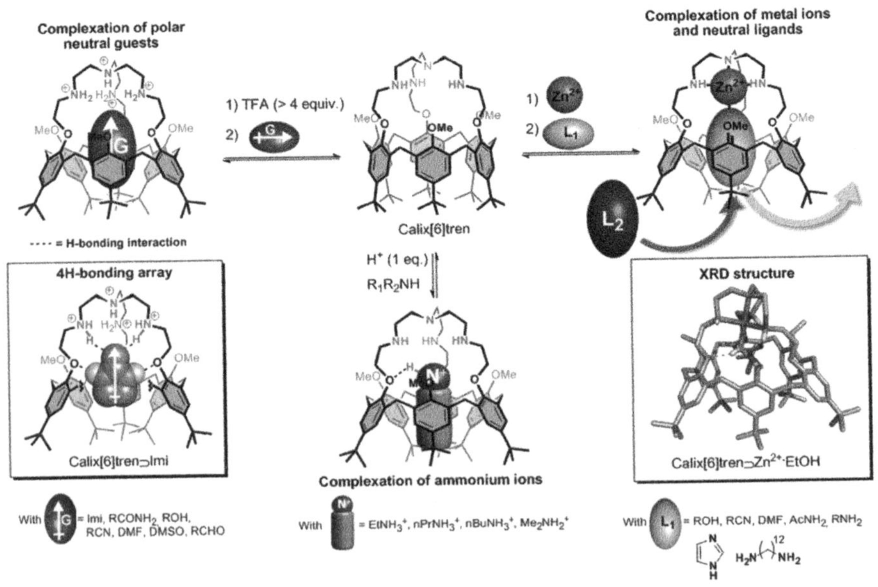

Figure 12.10 Versatility of the calix[6]tren receptor as a tetracation, a monocation, or a Zn(II) complex.[45]

the presence of primary or secondary ammonium ions leads to the competitive monoprotonation of the tren cap and formation of endo complexes with the ammonium ion accommodated into the cavity. The competitive protonation of the host is overpassed by the addition of an excess of the corresponding free amine (Figure 12.10). The endo complexation of ammonium ions is selective as the picrate salts of $EtNH_3^+$, $n PrNH_3^+$, $n BuNH_3^+$, or $Me_2NH_2^+$ are readily bound with 1/3/0.1/0.05 relative affinities, respectively, whereas inclusion of $Me_4N^+Pic^-$ is not detected under the same experimental conditions. By analogy with related calix[6]arene-based receptors,[47] the recognition is rationalized by H-bonding interactions between the guest ammonium ion and phenoxyl units of the calixarene core together with CH–π interactions within the cavity.

12.3.2.3 A Polarized Receptor for Polar Neutral Molecules In strong contrast to the neutral calix[6]tren, the tetracationic derivative, obtained through protonation with an excess of a strong acid (e.g., >4 equiv of trifluoroacetic acid, TFA), can recognize polar neutral molecules (G). Hence, intracavity complexation of alcohols, nitriles, amides, ureas, and even acetaldehydes was observed upon the addition of only a few molar equivalents of these guests G in chloroform (Figure 12.10). The driving force of the recognition event consists of a combination of electrostatic charge–dipole and H-bonding interactions between the protonated cap and the guest, as shown by the following sequence of relative affinity (rel. aff.): Imi (imidazolidin-2-one, a cyclic urea, see Figure 12.10, left inset, $\mu = 3.9$ D; rel. aff. = 500) $\gg$ DMF (3.9 D; 1) > $AcNH_2$ (3.7 D; 0.6) > EtOH (1.7 D; 0.14). Indeed, apolar molecules such as alkanes are not complexed. As shown on the XRD structure of a related system (vide infra),[48] the selectivity for Imi stems from a four H-bonding array leading to remarkable host–guest complementarities.

12.3.2.4 A Versatile Biomimetic Receptor that Needs to Be Polarized In conclusion, in the absence of a hydrophobic effect that requires water as the solvent, nonpolar interactions (van der Waals, CH–π) are not strong enough to allow the efficient binding of neutral guests by the neutral calix[6]tren receptor. However, the tren cap is highly basic and thus can be used to polarize the structure by protonation. In addition, the tren unit offers a strong chelate binding site for a metal ion. Each form of polarized receptor is capable of complexing neutral species into the hydrophobic cavity with high but different selectivities. For instance, the perprotonated calixtren strongly binds a urea, whereas the coordination of this guest was not observed with the Zn complex. All these results are reminiscent of molecular recognition processes encountered in natural systems. Indeed, enzymes are highly selective catalysts that bind their substrate in well-defined geometries and, in many cases, recognition involves cationic NH_3^+ groups belonging to Lys or Arg residues. All in all, these results validate the biomimetic approach that consists of associating in close proximity a cationic subunit and a hydrophobic cavity to build up efficient receptors for neutral species.

12.3.3 Polyamido and Polyureido Sites for Synergistic Binding of Dipolar Molecules and Anions

12.3.3.1 The Trisamido Site: A Unique Recognition Tool with Combined Dipolar and H-Bonding Interactions NMR spectroscopy studies in chloroform showed that the calix[6]trenamide host strongly binds polar neutral guests (G) displaying (1) an acceptor H-bonding group that interacts with the convergent NH groups of the trisamido cap, and (2) donor H-bonding group(s) that interact with the phenoxy groups of the calix core (Figure 12.11). Indeed, the inclusion of ureas and amides was detected, again, with high selectivities (see the rel. aff. displayed in Figure 12.11). The host–guest complex with Imi is even stable in a protic environment [$CD_3OD/CDCl_3$ (3:2), $K_a = 430\ M^{-1}$]. In strong contrast with the calix[6]tren receptor, the protonated trenamido host is insensitive to polar neutral molecules. This reluctance is rationalized by the competing formation of a stable five-membered intramolecular H-bonded ring between NH^+ and an introverted C=O (Figure 12.11). Here, the NH^+ proton induced a conformational reorganization of the trisamido recognition site into an insensitive form of the receptor.

12.3.3.2 The Trisureido Site: A Remarkable Neutral Anion Binding Site[49] In CD_3CN, the trenurea host strongly binds anions at the level of the crypturea cap thanks to (1) a conformational flip of the aromatic units, (2) the highly favorable cavity filling by the OMe groups, and (3) the spreading of the ureido arms (see structures displayed in Figure 12.12). The anion recognition proceeds through H-bonding interactions as shown by the substantial ^1H NMR downfield shift of the ureido protons (Figure 12.12). The association constants K_a displayed in Table 12.1 indicate that the binding discrimination is mostly based on the size of the anions. Very interestingly, the binding of Cl^- is efficient in a protic solvent ($K_a = 20 \pm 2\ M^{-1}$ at 298 K in CD_3OD). This remarkable result highlights the fact that neutral receptors can bind anions in protic solvents provided that they possess a highly preorganized H-bonding recognition site isolated from the solvent.

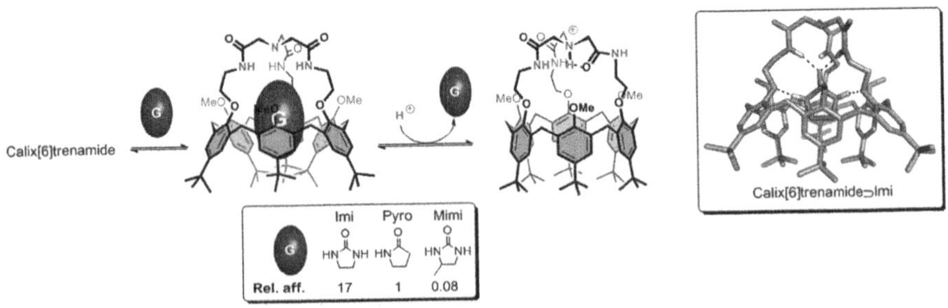

Figure 12.11 Host–guest properties of calix[6]trenamide toward neutral guests. Inset: Energy minimized structure of the inclusion complex calix[6]trenamide ⊃ Imi. Selected distances (Å): N(host)–O(Imi): 2.84 and 2.86; N(Imi)–O(host): 2.80 and 2.82.[43]

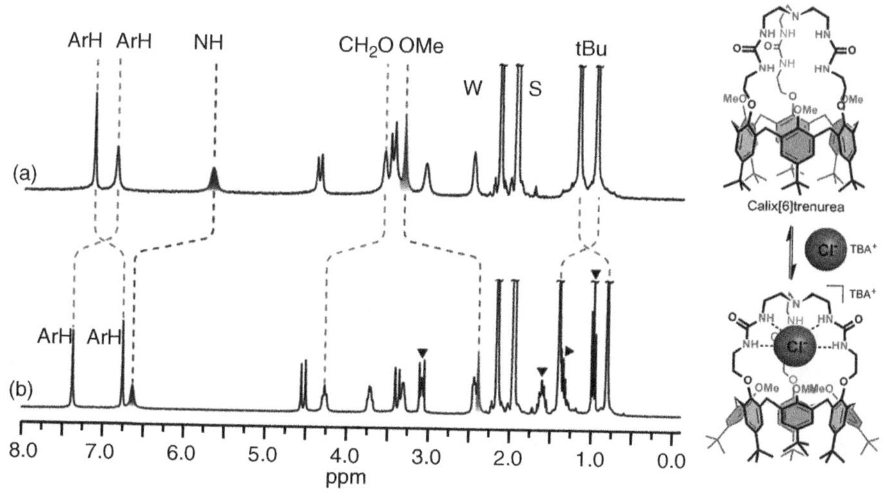

Figure 12.12 ^{1}H NMR spectra (CD_3CN, 300 MHz, 298 K) of (a) calix[6]trenurea; (b) after addition of 1 equiv of TBA^+Cl^-. ▼: TBA^+; w: water; s: residual solvent.[49]

TABLE 12.1 Association Constants K_a of Calix[6]trenurea Toward Anions X^- in CD_3CN (Ref. 49)

Anion[a] X^-	Geometry of X^-	K_a (M^{-1})[b]
Cl^-	Spherical	48300
Br^-	Spherical	1930
I^-	Spherical	nd[c]
CN^-	Linear	640
N_3^-	Linear	215[d]
AcO^-	V-shaped	160
NO_3^-	Trigonal	98

[a] TBA^+ salts.
[b] K_a determined at 243 K and defined as $K_a = [\text{host} \supset X^-]/([\text{host}]\,[X^-])$. Errors estimated ±10%.
[c] Not detected.
[d] Determined at 298 K.

12.3.3.3 Heteroditopic Receptors for Organic Contact Ion Pairs The ability of calix[6]trenamide and calix[6]trenurea to recognize an ammonium ion simultaneously to an anion was investigated by NMR spectroscopy in chloroform. In both cases, the complexation of the ammonium ion only proceeds when an anion is co-bound in the trisureido cap (Figure 12.13). This positive cooperativity is due to the close proximity between the two complexed ions and thus to their strong electrostatic interaction. These heteroditopic receptors stress the importance of binding the two ions as a contact ion pair in order to avoid their

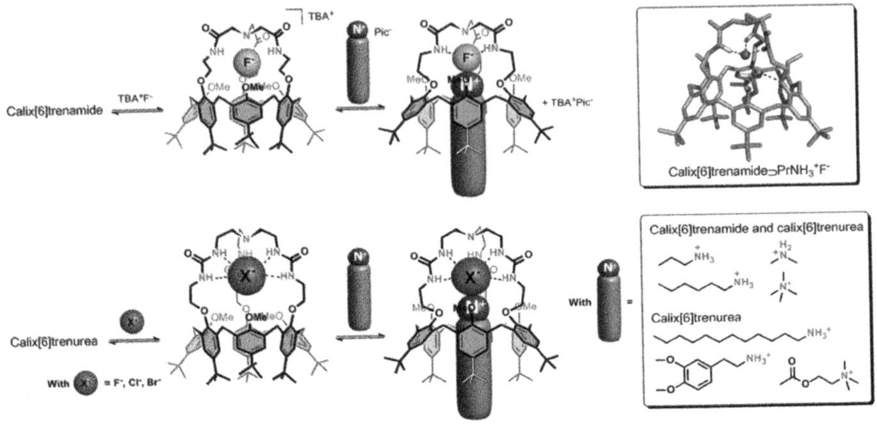

Figure 12.13 Host–guest properties of calix[6]trenamide and calix[6]trenurea toward anions and organic contact ion pairs. Top inset: Energy minimized structure of the inclusion complex calix[6]trenamide ⊃ $PrNH_3^+F^-$. Selected distances (Å): N(host)–F^-: 2.78, 2.80, and 2.99; N^+ (guest)–O(host): 2.96; N^+ (guest)–F^-: 2.08.[43, 49]

highly energetically unfavorable dissociation. In the case of the trenureido receptor, ternary complexes with various anions X^- ($X^- = F^-$, Cl^-, or Br^-) and linear ammonium ions of various length (e.g., propyl-, hexyl-, or dodecylammonium) are obtained (Figure 12.14a). High cumulative binding constants (e.g., $\beta_2 > 1.6 \times 10^9$ M^{-2} for $PrNH_3^+Cl^-$) were obtained. Calix[6]trenurea also presents a strong affinity for bulkier quaternary ammonium salts and biologically relevant ammonium salts such as the neurotransmitter acetylcholine chloride or a dopamine hydrochloride derivative ($\beta_2 > 2.3 \times 10^6$ M^{-2} in the case of TMA^+Cl^-) (Figure 12.14b). Very interestingly, most of the ternary host–guest complexes are stable in a protic

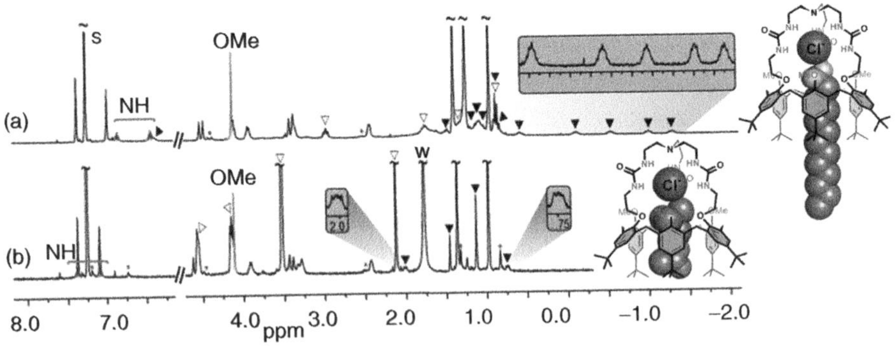

Figure 12.14 ^1H NMR spectra ($CDCl_3$, 300 MHz, 298 K) of (a) calix[6]trenurea ⊃ dodecylammonium chloride and (b) calix[6]trenurea ⊃ acetylcholine chloride. ▼: included ammonium ion; ∇: free ammonium ion; w: water; s: residual solvent.[49]

environment; for instance, the complex with acetylcholine chloride was still visible in a mixture of $CD_3OD/CDCl_3$ (4:1). In the case of the trenamido host, the binding of contact ion pairs exclusively proceeded with F^- as the anionic partner (i.e., no cation complexation was apparent with Cl^-, AcO^-, $MeSO_3^-$, NO_3^-, or SO_4^{2-}) (Figure 12.13). This remarkable selectivity is due to the smallness of the binding site provided by the convergent NH groups of the cryptamide cap. Interestingly, ^1H and ^{19}F NMR experiments revealed significant scalar couplings between the fluoride anion of the ternary complexes and the NH amido protons as well as the NH_3^+ of the included propylammonium ion (i.e., ^{1h}J scalar couplings across N–H···F^- hydrogen bonds). The energy-minimized structure of the ternary $PrNH_3^+F^-$ complex is depicted in Figure 12.13. The very short distance between the fluoride and the charged nitrogen of the ammonium ion ($d_{N+\ldots F-} = 2.08$ Å) shows the presence of a strong electrostatic interaction at the level of the ion pair. As in other complexes, the ammonium ion is further stabilized by the calixarene host through a combination of CH–π interactions with the aromatic walls and H-bonding interaction with a phenoxy oxygen. Besides its interaction with the cation, the anion is strongly bound to the convergent hydrogen-bond donor NH groups. Finally, in the case of both receptors (i.e., calix[6]trenamide and calix[6]trenurea), the protonation of the cap of the ternary complexes triggers the release of the anion and, as a consequence, of the ammonium ion.

12.3.4 Acid–Base Controllable Receptors

Another interesting feature of these tren derivatives lies in the presence of proton sensitive site(s) at the level of the polar aza cap. As shown above, the protonation of the basic cap of the hosts led to positively charged cavities whose host properties toward dipolar guests are governed by strong charge–dipole interactions. In the case of the calix[6]tren receptor, a drastic increase in binding strength was observed with the protonation of the receptor. In contrast, guest release can be triggered by addition of an acid to the trenamido receptor. This negative cooperativity is due to the conformational reorganization of the more rigid trenamide binding site that turns off the amido binding properties of the cap. These controllable binding properties make possible the reversible interconversion of different modes of recognition. For instance, a three-way supramolecular switch was obtained with the calix[6]trenurea host–guest system (Figure 12.15). The protonation of the apical nitrogen of the cap acts as an effector that allows the switch from one mode of recognition to another and a remarkable guest selection from a complex mixture. In other words, these receptors nicely illustrate how the host properties of a hydrophobic cavity can be tuned by the environment, which is reminiscent of natural systems and their propensity to pH control.

12.4 SELF-ASSEMBLED CAVITIES

Self-assembly is ubiquitous in biological systems. It allows reversible and programmable assembly of subcomponents into functional complex structures. Data

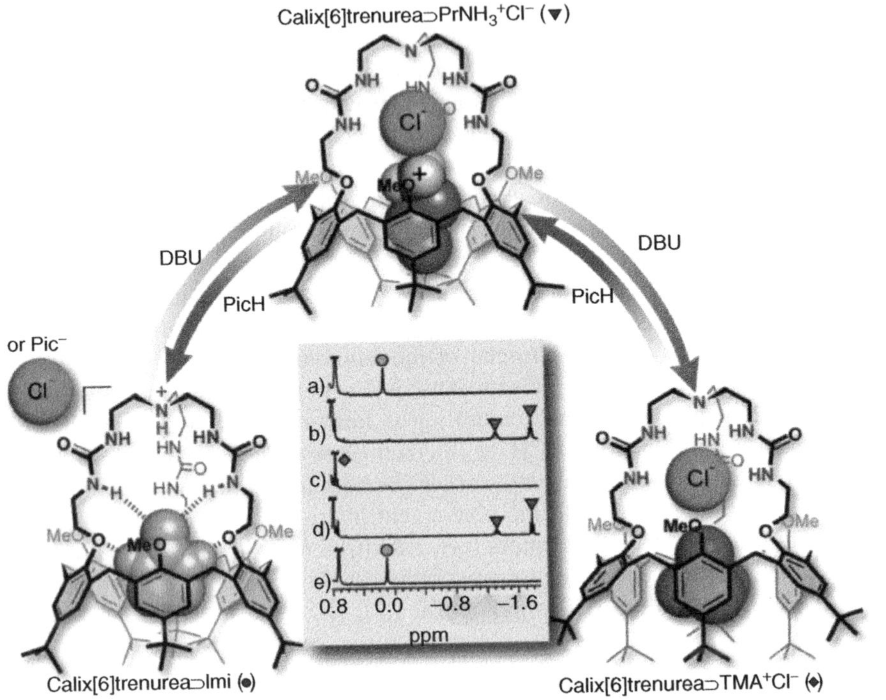

Figure 12.15 Three-way supramolecular switch triggered by the addition of acid or base to calix[6]trenurea. ^1H NMR spectra (400 MHz, 298 K, high-field region) in $CDCl_3/CD_3OD$ (98:2) of (a) mixture of calix[6]trenurea, PicH (1 equiv), TMA^+Cl^- (2.5 equiv), $PrNH_3^+$ Cl^- (15 equiv), and Imi (2.5 equiv); (b) after addition of DBU (2 equiv); (c) after addition of DBU (25 equiv); (d) after addition of PicH (12 equiv); and (e) after addition of PicH (22 equiv). •: included Imi; ▼: included $PrNH_3^+$; ◇: included TMA^+.[49]

storage, regulation and replication phenomena, catalytic transformations, membrane transport, and many other tasks essential to living systems lie on self-assembled materials. In addition to the two above described strategies (supramolecular capping via metal coordination and covalent capping), a third approach based on ionic interactions was exploited to obtain calix[6]arene-based receptors. In biological receptors, ammonium and carboxylate residues of a protein scaffold are often involved in substrate recognition. It is also a major tool for protein folding into an active state. Having this in mind, calix[6]trisamine and calix[6]trisacid (depicted in Figure 12.16) have been used for obtaining positively and negatively polarized cavities.

12.4.1 Receptors Decorated with a Triscationic or a Trisanionic Binding Site

The free base calix[6]trisamine undergoes a fast cone–cone inversion at RT and is not suitable for hosting neutral molecules. In contrast, its trifluoroacetic acid

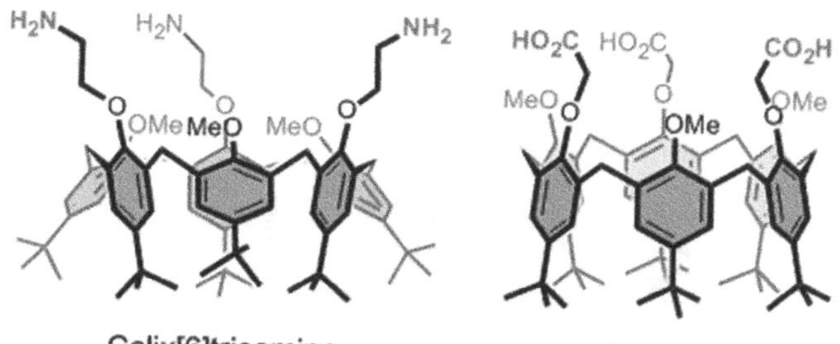

Figure 12.16 Small rim 1,3,5-trisfunctionalized calix[6]arenes with amine and carboxylic acid groups.

protonated form is locked in the cone conformation, and NMR studies in $CDCl_3$ have revealed that polar neutral molecules, such as alcohols, ureas, amides, or sulfoxydes, are efficiently endo-complexed.[50] X-ray structure of the endo complex with DMF shows that the ammonium arms are self-assembled through a network of H-bonding interactions with the trifluoroacetate counteranions (Figure 12.17). This assembly of the arms thus forms an ion paired cap that (1) freezes the cavity in the cone conformation and (2) polarizes the hydrophobic cavity with a tricationic protic site. In general, the guests orient their dipole moment along the C_3 axis of the cavity and are bound via a combination of H-bonding interactions with the ammonium arms and the phenoxyl units, CH–π interactions with the aromatic walls, and dipole–charge interactions. In adequacy with this finding, the endo-binding of neutral molecules with low polarity (ketones, ethers, halogeno-alcanes) has never been observed. Hence, guest selection based on size, shape, and electronic structure occurs in this simple system. For instance, a cyclic urea (imidazolidin-2-one) is very well recognized and, to date, behaves as the best "key" for the calix[6]trisammonium host.

The reverse situation is observed with calix[6]trisacid derivatives (Figure 12.17).[51] Upon acid–base reaction with an excess of amine, the triscarboxylate structure is frozen in the cone conformation and is able to endo-complex one of the ammonium counterion, even in pure methanol solution. NMR studies evidenced that the exo complexation of at least one ammonium ion is required for the quantitative endo-binding process to occur. The X-ray structure with $EtNH_3^+$ revealed that the two exo-bound ammonium ions are involved in ion pairing interactions with adjacent carboxylate arms, thus forming a supramolecular cap in a way similar to that observed for the protonated form of calix[6]trisamine. The third (endo) ammonium ion is stabilized via a combination of weak interactions (charge–dipole, H-bonding, and CH–π), and displays a remarkable C_3 complementarity with the triscarboxylate binding site of the host. Again, the calixarene cavity acts as a funnel and provides size and shape selection, primary ammonium ions being best recognized. Such a size selection can be used to tune the supramolecular cap. For instance, the bulky terbutylammonium ion is exclusively recognized in exo position, which allows the quantitative inclusion

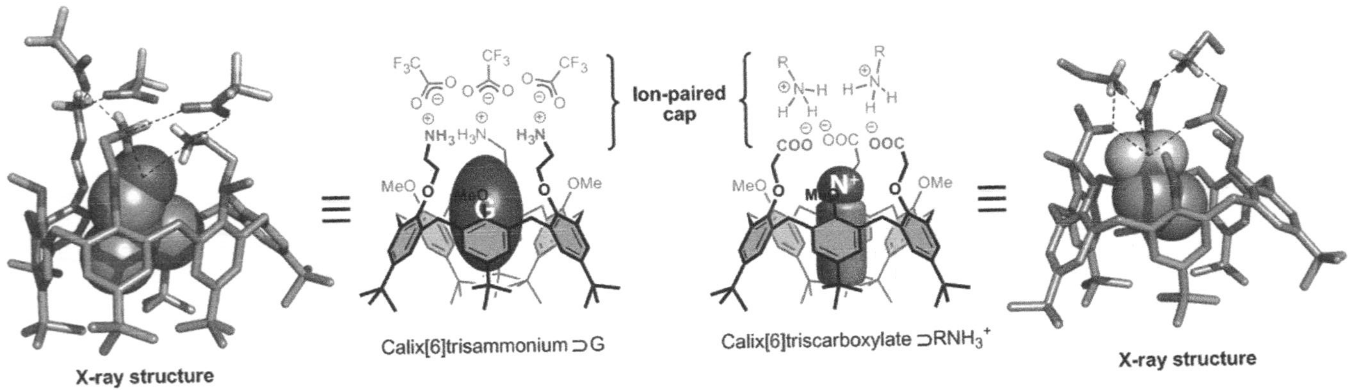

Figure 12.17 Opposite strategies for the supramolecular capping via ion pairing of the calix[6]arene core. X-ray structures of (left) calix[6]trisammonium ⊃ DMF and of (right) calix[6]triscarboxylate ⊃ $EtNH_3^+$ (dashed lines indicate H bonds).[50, 51]

of a primary ammonium ion of interest upon addition of strictly one equivalent. Inclusion of large biorelevant ammonium guests (e.g., spermine, spermidine, and organosoluble derivatives of dopamine and 6-hydroxytryptamine) has also been observed with calix[6]triacids deprived of three of the *t*Bu groups at the large rim.

12.4.2 Receptors Capped Through Assembly with a Tripodal Subunit

The self-assembly strategy not only provides a simple means to obtain efficient receptors, but can be extended to the construction of more sophisticated multitopic receptors through the coupling of two cavities. Indeed, a tripodal triscarboxylic acid can advantageously replace the excess of strong acid used for the capping of calix[6]trisamine (Figure 12.18).[52] However, in these softer acids, presence of a guest molecule was mandatory for obtaining discrete self-assembled structures, which now are ternary. In other words, the self-assembly process between the calix[6]trisamine and the triscarboxylic cap is directed by the guest molecule. Its inclusion contributes to the shaping of the cavity through an induced-fit process, which optimizes the directionality of the intersubunit interactions. The stability of the [1+1+1] complexes depends on the preorganization of the cap: with a rigid tripod they are stable in pure methanol solution (Figure 12.18). Also interestingly, when the rigid triscarboxylic platform is concave as exemplified by the bowl-shape cyclotriveratrylene scaffold, [1+1+2] self-assembled ditopic receptors are obtained (Figure 12.18). One molecule of Imi is included in each cavity, but with different recognition processes, and thus different affinities. In this system, one equivalent of guest is required to shape the calixarene receptor, whereas the second equivalent may or may not be present in the rigid bowl-shape cap.

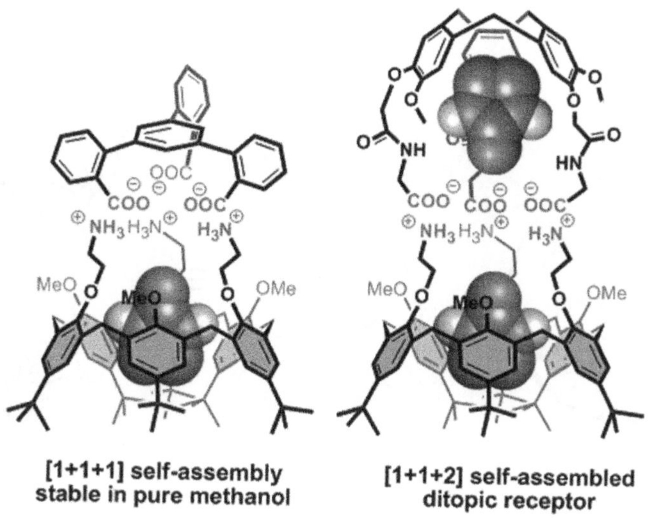

Figure 12.18 Supramolecular capping of a calix[6]trisamine with tripodal triscarboxylic acid subunits (guest molecules: Imi).[52]

12.4.3 Heteroditopic Self-Assembled Receptors with Allosteric Response

The possibility of coupling two cavities is of great interest as it may open the way to receptors with allosteric response. This is indeed what was obtained when the rigid triscarboxylic cap of the cyclotriveratrylene unit was substituted for the flexible calix[6]trisacid (Figure 12.19).[48] Mixing equimolar amounts of such partners led only to nondiscrete species. This is obviously ascribed to the high flexibility of the calix[6]arene skeletons. However, the selective formation of a discrete species can be triggered by the addition of the respective guest molecules, that is, a polar neutral molecule and an ammonium ion. The obtained supramolecular heteroditopic receptors correspond to a rare case of [1+1+1+1] quaternary self-assembly in which the calixarenes are connected in a tail-to-tail manner (diabolo-like topology). The X-ray structure with Imi and propylammonium as guests (Figure 12.19) revealed a high degree of complementarities between the four partners (size, shape, and electronic structure): triple ion pairing and up to 15 H bonds stabilize the supramolecular structure. This allows an extremely selective self-assembly process with no error and good stability as these quaternary complexes tolerate the presence of a large amount of a protic solvent at millimolar (mM) concentration.

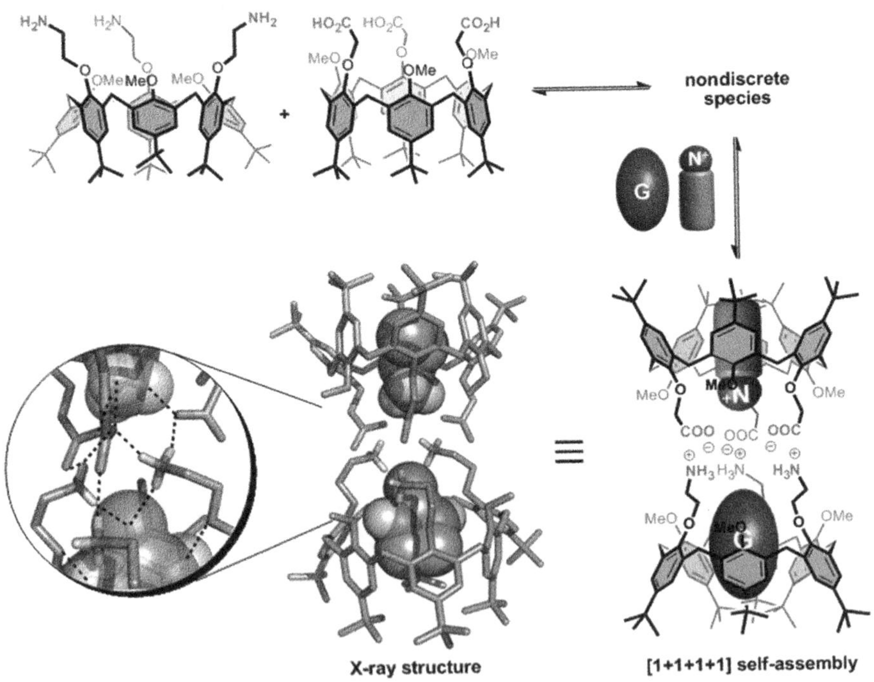

Figure 12.19 Self-assembled ditopic receptors with allosteric response. Bottom left: X-ray structure with $PrNH_3^+$ and Imi as guests. Dashed lines indicate H bonds.[48]

The whole self-assembly process corresponds to an allosterically coupled double induced-fit process and involves several cooperative events:

- The acid–base reaction provides polarized cavities with opposite binding properties.
- The inclusion of a guest rigidifies the calixarene host and increases the directionality of the intersubunit interactions.
- The two self-assembled cavities "communicate" and the guest recognition processes are mutually reinforcing.

This example nicely illustrates how the flexibility of the calixarene skeleton can be turned into an advantage for the construction of sophisticated hosts. Such self-assembled receptors displaying heterotopic allosteric response are somehow reminiscent of biological processes. For instance, a crucial step in the arginyl t-RNA synthetase activity involves a mutual double induced-fit process, in which both the enzyme and the t-RNA undergo important conformational changes.[53]

12.4.4 Interlocked Self-Assembled Receptors

Molecular recognition in biological receptors involves more than the direct host–guest interactions. Nature's strategy to reach nanomolar affinity is to use rather flexible scaffolds, such as protein backbones, into which secondary interactions are reinforced upon guest binding. Several shells of interactions might be coupled and, in turn, enhance guest association through a reduction in conformational motion of the whole proteic structure.[54] The calix[6]arene structure, because of its flexibility, is a good platform to attempt mimicking such a reinforced molecular recognition strategy.

For that purpose, calix[6]arenes with a dual recognition pattern at the small rim were tested.[55] In place of the methyl groups belonging to anisol units, the calixarenes display additional functionalities such as ureas or carboxylic acids (Figure 12.20). Hence, the assembly of the trisamino–trisureido calix[6]arene with tris- or hexacarboxylic acid substituted calix[6]arenes[56] revealed additional host–host and host–guest interactions in the final self-assembled edifices (Figure 12.20, left and right, respectively):

- When using a calix[6]hexaacid, Imi, and the free base $PrNH_2$, a neutral complex is formed (the calix[6]hexaacid is four-deprotonated, the ammonium guest is formed in situ) in which the urea groups are involved in H-bonding interactions with the carboxylate arms (host–host interactions).
- When using a calix[6]trisacid, Imi, and the chloride salt of $PrNH_3^+$, a quinternary [1+1+1+1+1] edifice is formed (also formally neutral), in which the chloride counterion of the included ammonium ion is bound to the urea groups (host–guest interactions).

These secondary intra-assembly interactions may be viewed as reminiscent of protein–protein recognition and enzyme–substrate interaction, both being

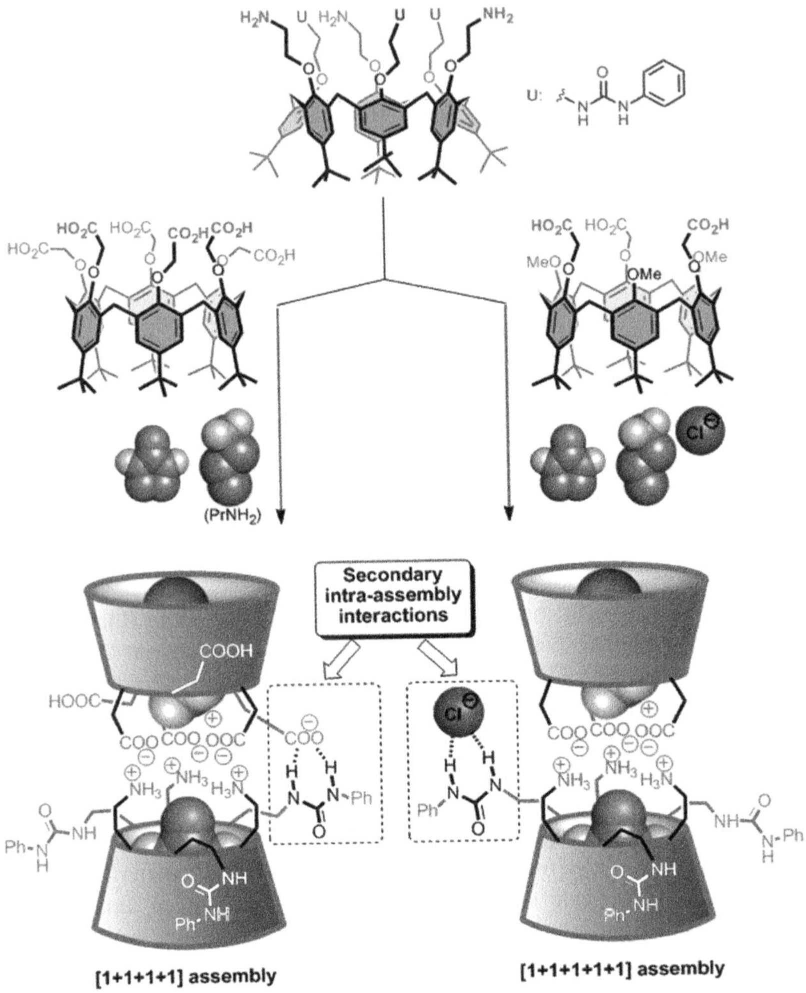

Figure 12.20 The use of a dual recognition pattern for introducing (left) host–host and (right) host–guest additional interactions within self-assembled receptors.[56]

interlocked and under allosteric control. Their achievement is possible because of the versatility of the calix[6]arene functionalization together with the controlled flexibility of its skeleton. All in all, this third approach, based on ion pairing, highlights how the organization at one site of the system due to guest recognition can be transmitted to the other site for the binding of a second guest and how interlocked the processes are (organization/recognition). The synergism leads to entropically costly assemblies of up to five different species in the above examples with a high selectivity (i.e., no error in the overall recognition pattern).

12.5 CONCLUSION

Modeling metalloenzyme active sites with cavity-embedded metal complexes led to the development of metalloreceptors that display highly specific but tunable host–guest properties. Indeed, protection of the metal center by controlling not only its first but also its second coordination sphere allowed stabilizing high Lewis acidic states and preventing its dimerization. The protected dicationic state was exploited to obtain very sensitive receptors with a selectivity defined by the calixarene cavity. In many ways, these so-called funnel complexes mimic substrate binding in metallo/enzymes. Very remarkably, this biomimetic approach led to the development of several families of receptors displaying different properties. The systems described herein have several features in common that were revealed to be keys for their host–guest properties:

- All present a π-basic cone cavity (defined by aromatic walls) closed on one side and open on the other side. This allows guest binding through partial inclusion: only part of the guest needs to be recognized by the host in order to be bound in a selective way. Due to this aspect, these *funnel* systems resemble enzyme channels, which is quite different from bowl and capsule-like receptors.
- The polarization of one end of the receptor is a prerequirement for obtaining good affinity for a guest molecule. This polarization can be achieved with simple dipoles such as amides, like in proteins. Ureas allow strong binding to anions, which, in turn, allows strong binding of an ion-paired ammonium. A polyammonium site was revealed to be very effective for the binding of a variety of neutral molecules provided they present a dipole. With dicationic metal ions, the attraction to the positive charge is further reinforced by the coordination link with the guest.
- An additional basic site can be used for triggering guest binding or guest release. Such proton switches are reminiscent of enzyme sensitivity to pH.
- Preorganization plays a key role in many ways. Covalent capping of the funnel yields better receptors in term of robustness and versatility, as exemplified by the tren-based systems compared to the tris(imidazolyl) ones for which an entropic cost needs to be paid for cone shaping. The more rigid the cap, the more selective the receptor, but the less versatile.
- Flexibility of the cavity core turns out to be an advantage as the cavity size can adapt to the guest. Such induced-fit phenomena—shaping the cone through guest binding—can be exploited for the elaboration of allosterically coupled multireceptors. Indeed, self-assembled flexible cavities can communicate thanks to the efficient transmission of the information from one side of the assembly to the other.

Hence, salient points that are common with biological system can be pointed out:

1. The tools used for molecular recognition: charge–dipole, dipole–dipole, and London interactions, the highly directional and thus discriminating H-bonding network, coordination to a metal ion.
2. The bases for selectivity: the selectivity of function (cf. point 1) due to the interaction with one side of the receptor, and selectivity of shape due to the partial inclusion in a cavity.

The motto is: polarized and flexible funnel-shaped cavities!

REFERENCES

1. (a) Lehn, J.-M. *Supramolecular Chemistry*, Wiley-VCH, Weinheim, 1995; (b) Steed, J. W.; Atwood, J. L. *Supramolecular Chemistry*; Wiley, Chichester, UK, 2000; (c) reviews on molecular recognition: *Chem. Rev*. **1997**, *97*, 1231–1734.
2. Dugas, H. *Bioorganic Chemistry: A Chemical Approach to Enzyme Action*, 3rd ed., Springer-Verlag, New York, 1996, pp. 252–387.
3. (a) Bertini, I; Gray, H. B.; Stiefel, E. I.; Valentine, J. S. *Biological Inorganic Chemistry, Structure and Reactivity*, University Science Books, Sausalito, CA, 2007; (b) Lippard, S. J.; Berg, J. M. *Principles of Bioinorganic Chemistry*, University Science Books, Mill-Valley, CA, 1994; (c) Holm, R. H.; Kennepohl, P.; Solomon, E. I. *Chem. Rev*. **1996**, *96*, 2239.
4. (a) Lipscomb, W. N.; Sträter, N. *Chem. Rev*. **1996**, *96*, 2375; (b) Auld, D. S. "Metal Sites in Proteins and Models," in *Structure and Bonding, Vol. 89* (Ed. Sadler, P. J.), Springer-Verlag, Berlin, 1997, pp. 29–50.
5. (a) Schneider, H.-J. *Angew. Chem. Int. Ed*. **1991**, *30*, 1417; (b) Davis, A. P.; Wareham, R. S. *Angew. Chem. Int. Ed*. **1999**, *38*, 2978; (c) Motherwell, W. B.; Bingham, M. J.; Six, Y. *Tetrahedron* **2001**, *57*, 4663.
6. (a) Cram, D. J.; Cram, J. M. *Container Molecules and Their Guests, Monographs in Supramolecular Chemistry*, The Royal Society of Chemistry, Cambridge, UK 1994; (b) Hof, F.; Craig, S. L.; Nuckolls, C.; Rebek, J. Jr. *Angew. Chem. Int. Ed*. **2002**, *41*, 1488; (c) Warmuth, R.; Yoon, J. *Acc. Chem. Res*. **2001**, *34*, 95; (d) Higler, I.; Timmerman, P.; Reinhoudt, D. N. *Eur. J. Org. Chem*. **1998**, 2689.
7. (a) Rudkevich, D. M.; Rebek J. Jr. *Eur. J. Org. Chem*. **1999**, *9*, 1991; (b) Jasat, A.; Sherman, J. C. *Chem. Rev*. **1999**, *99*, 931.
8. (a) Collet, A. *Tetrahedron* **1987**, *43*, 5725; (b) Collet, A.; Dutasta, J.-P.; Lozach, B.; Canceill, J. *Top. Curr. Chem*. **1993**, *165*, 103; (c) Brotin, T.; Dutasta, J.-P. *Chem. Rev*. **2009**, *109*, 88.
9. (a) Gutsche, C. D. *Calixarenes, Monographs in Supramolecular Chemistry*, The Royal Society of Chemistry, Cambridge, UK, 1989; (b) Gutsche, C. D. *Calixarenes Revisited, Monographs in Supramolecular Chemistry* (Ed. Stoddart, J. F.), The Royal Society of Chemistry, Cambridge, UK, 1998; (c) Arduini, A.; Secchi, A.; Pochini, A. *J. Org. Chem*. **2000**, *65*, 9085; (d) *Calixarenes 2001* (Eds. Asfari, Z.; Böhmer, V.;

Harrowfield, J.; Vicens, J.), Kluwer Academic Publishers, Dordrecht, Holland, 2001; (e) Mandolini, L.; Ungaro, R. (Eds.). *Calixarenes in Action*, Imperial College Press, London, 2000; (f) Ikeda, A.; Shinkai, S. *Chem. Rev*. **1997**, *97*, 1713.

10. Diederich, F. *Cyclophanes, Monographs in Supramolecular Chemistry*, The Royal Society of Chemistry, Cambridge, UK, 1994.
11. Redshaw, K. *Coord. Chem. Rev*. **2003**, *244*, 45.
12. Except in the case of molecular capsules resulting from the association of two calix[4]arene units. See, for example: (a) Shivanyuk, A.; Saadioui, M.; Broda, F.; Thondorf, I.; Vysotsky, M. O.; Rissanen, K.; Kolehmainen, E.; Böhmer V. *Chem. Eur. J*. **2004**, *10*, 2138; (b) Rebek, J. Jr. *Chem. Commun*. **2000**, 637.
13. Sénèque, O.; Rondelez, Y.; Le Clainche, L.; Inisan, C.; Rager, M.-N.; Giorgi, M.; Reinaud, O. *Eur. J. Inorg. Chem*. **2001**, 2597.
14. (a) Blanchard, S.; Le Clainche, L.; Rager, M.-N.; Chansou, B.; Tuchagues, J.-P.; Duprat, A. F.; Le Mest, Y.; Reinaud, O. *Angew. Chem. Int. Ed*. **1998**, *37*, 2732; (b) Le Clainche, L.; Giorgi, M.; Reinaud, O. *Eur. J. Inorg*. **2000**, 1931; (c) Le Clainche, L.; Rondelez, Y.; Sénèque, O.; Blanchard, S.; Campion, M.; Giorgi, M.; Duprat, A. F.; Le Mest, Y.; Reinaud, O. *Comptes Rendus Acad. Sci. Chim. Sér. IIc* **2000**, *3*, 811; (d) Rondelez, Y.; Sénèque, O.; Rager, M.-N.; Duprat, A. F.; Reinaud, O. *Chem. Eur. J*. **2000**, *6*, 4218; (e) Rondelez, Y.; Rager, M.-N.; Duprat, A.; Reinaud, O. *J. Am. Chem. Soc*. **2002**, *124*, 1334.
15. (a) Sénèque, O.; Rager, M.-N.; Giorgi, M.; Reinaud, O. *J. Am. Chem. Soc*. **2000**, *122*, 6183; (b) Sénèque, O.; Rager, M.-N.; Giorgi, M.; Reinaud, O. *J. Am. Chem. Soc*. **2001**, *123*, 8442; (c) Sénèque, O.; Giorgi, M.; Reinaud, O. *Chem. Commun*. **2001**, 984; (d) Sénèque, O.; Giorgi, M.; Reinaud, O. *Sup. Chem*. **2003**, *15*, 573.
16. Sénèque, O.; Campion, M.; Giorgi, M.; Le Mest, Y.; Reinaud, O. *Eur. J. Inorg. Chem*. **2004**, 1817.
17. Redon, S.; Li, Y.; Reinaud, O. *J. Org. Chem*. **2003**, *68*, 7004.
18. (a) Colasson, B.; Save, M.; Milko, P.; Roihová, J.; Schröder, D.; Reinaud, O. *Org. Lett*. **2007**, *9*, 4987; (b) Colasson, B.; Reinaud, O. *J. Am. Chem. Soc*. **2008**, *130*, 15226.
19. Parkin, G. *Chem. Rev*. **2004**, *104*, 699.
20. Sénèque, O.; Rager, M.-N.; Giorgi, M.; Reinaud, O. *J. Am. Chem. Soc*. **2001**, *123*, 8442.
21. HAAkansson, K.; Carlsson, M.; Svensson, L. A.; Liljas, A. *J. Mol. Biol*. **1992**, *227*, 1192.
22. Sénèque, O.; Rager, M.-N.; Giorgi, M.; Reinaud, O. *J. Am. Chem. Soc*. **2000**, *122*, 6183.
23. Sénèque, O.; Giorgi, M.; Reinaud, O. *Chem. Commun*. **2001**, 984.
24. Li, H.; Hallows, W. H.; Punzi, J. S.; Pankiewicz, K. W.; Watanabe K. A.; Goldstein, B. M. *Biochemistry* **1994**, *33*, 11734.
25. Coquière, D.; Le Gac, S.; Darbost, U.; Sénèque, O.; Jabin, I.; Reinaud, O. *Org. Biomol. Chem*. **2009**, *7*, 2485.
26. Blanchard, S.; Rager, M.-N.; Duprat A. F.; Reinaud, O. *New J. Chem*. **1998**, 1143.

27. (a) Sénèque, O.; Campion, M.; Douziech, B.; Giorgi, M.; Rivière, E.; Journaux, Y.; Le Mest, Y.; Reinaud, O. *Eur. J. Inorg. Chem.* **2002**, 2007; (b) Izzet, G.; Frapart, Y. M.; Prangé, T.; Provost, K.; Michalowicz A.; Reinaud, O. *Inorg. Chem.* **2005**, *44*, 9743; (c) Izzet, G.; Akdas, H.; Hucher, N.; Giorgi, M.; Prangé, T.; Reinaud, O. *Inorg. Chem.* **2006**, *45*, 1069.
28. Izzet, G.; Zeng, X.; Akdas, H.; Marrot, J.; Reinaud, O. *Chem Commun.* **2007**, 810.
29. Rondelez, Y.; Rager, M.-N.; Duprat, A.; Reinaud, O. *J. Am. Chem. Soc.* **2002**, *124*, 1334.
30. Coquière, D.; Marrot, J.; Reinaud, O. *Org. Biomol. Chem.* **2008**, *6*, 3930.
31. Otyepka, M.; Skopalík, J.; Anzenbacherová, E.; Anzenbacher, P. *Biochim. Biophys. Acta* **2007**, *1770*, 376.
32. Davis, A. M.; Teague, S. J. *Angew. Chem. Int. Ed.* **1999**, *38*, 736.
33. Rekharsky, M. K.; Inoue, Y. *Chem. Rev.* **1998**, *98*, 1875.
34. Sliwa, W.; Deska, M. *Chem. Heterocycl. Compd.* **2002**, *38*, 646.
35. Mecozzi S.; Rebek, J. Jr. *Chem. Eur. J.* **1998**, *4*, 1016.
36. Coquière, D.; Marrot, J.; Reinaud, O. *Chem. Commun.* **2006**, 3924.
37. Coquière, D.; de la Lande, A.; Martí, S.; Parisel, O.; Prangé, T.; Reinaud, O. *Proc. Natl. Acad. Sci. U.S.A.* **2009**, *106*, 10449.
38. Bewley, M. C. *Proc. Natl. Acad. Sci. U.S.A.* **2006**, *103*, 2063.
39. Sénèque, O.; Reinaud, O. *Tetrahedron* **2003**, *59*, 5563.
40. Sénèque, O.; Campion, M.; Douziech, B.; Giorgi, M.; Le Mest, Y.; Reinaud, O. *Dalton Trans.* **2003**, 4216.
41. Sénèque, O.; Rager, M.-N.; Giorgi, M.; Prangé, T.; Tomas, A.; Reinaud, O. *J. Am. Chem. Soc.* **2005**, *127*, 14833.
42. Jabin, I.; Reinaud, O. *J. Org. Chem.* **2003**, *68*, 3416.
43. Lascaux, A.; Le Gac, S.; Wouters, J.; Luhmer, M.; Jabin, I. *Org. Biomol. Chem.* **2010**, *8*, 4607.
44. Ménand, M.; Jabin, I. *Org. Lett.* **2009**, *11*, 673.
45. Darbost, U.; Rager, M-.N.; Petit, S.; Jabin, I.; Reinaud, O. *J. Am. Chem. Soc.* **2005**, *127*, 8517.
46. (a) Darbost, U.; Zeng, X.; Rager, M.-N.; Giorgi, M.; Jabin, I.; Reinaud, O. *Eur. J. Inorg. Chem.* **2004**, 4371; (b) Izzet, G.; Douziech, B.; Prangé, T.; Tomas, A.; Jabin, I.; Le Mest, Y.; Reinaud, O. *Proc. Natl. Acad. Sci. U.S.A.* **2005**, *102*, 6831; (c) Izzet, G.; Zeitouny, J.; Akdas-Killig, H.; Frapart, Y.; Ménage, S.; Douziech, B.; Jabin, I.; Le Mest, Y.; Reinaud, O. *J. Am. Chem. Soc.* **2008**, *130*, 9514.
47. Darbost, U.; Giorgi, M.; Reinaud, O.; Jabin, I. *J. Org. Chem.* **2004**, *69*, 4879.
48. Le Gac, S.; Marrot, J.; Reinaud, O.; Jabin, I. *Angew. Chem. Int. Ed.* **2006**, *45*, 3123.
49. Ménand, M.; Jabin, I. *Chem. Eur. J.* **2010**, *16*, 2159.
50. Darbost, U.; Giorgi, M.; Hucher, N.; Jabin, I.; Reinaud, O. *Supramol. Chem.* **2005**, *17*, 243.
51. Le Gac, S.; Giorgi, M.; Jabin, I. *Supramol. Chem.* **2007**, *19*, 185.
52. Le Gac, S.; Luhmer, M.; Reinaud, O.; Jabin, I. *Tetrahedron* **2007**, *63*, 10721.

53. Delagoutte, B.; Moras, D.; Cavarelli, J. *EMBO J.* **2000**, *19*, 5599.

54. (a) Williams, D. H.; Stephens, E.; O'Brien, D. P.; Zhou, M. *Angew. Chem. Int. Ed.* **2004**, *43*, 6596; (b) Otto, S. *Dalton Trans.* **2006**, *23*, 2861.

55. (a) Le Gac, S.; Ménand, M.; Jabin, I. *Org. Lett.* **2008**, *10*, 5195; (b) Le Gac, S.; Marrot, J.; Jabin, I. *Chem. Eur. J.* **2008**, *14*, 3316.

56. Le Gac, S.; Picron, J.-F.; Reinaud, O.; Jabin, I. *Org. Biomol. Chem.* **2011**, *9*, 2387.

CHAPTER 13

Bioinspired Dendritic Light-Harvesting Systems

ANDREA M. DELLA PELLE and SANKARAN THAYUMANAVAN

Department of Chemistry, University of Massachusetts–Amherst, Amherst, Massachusetts, USA

13.1 INTRODUCTION

Energy supply and demand has become an increasing concern in our society as we attempt to shy away from our dependence on fossil fuels. Developing new sources of energy and improving on the existing alternative energy sources will be an invaluable contribution to our future. Nature has supplied a clean and abundant energy source, the Sun, which has become a very attractive option as an alternative source of energy. The Sun provides the Earth with more energy in one day than is consumed by the entire human race in a year.[1] According to the U.S. Department of Energy, the Sun deposits 120,000 terawatts (TW) of power to the Earth's surface, far exceeding the demands of the human population, which consumes approximately 13 TW per year.[2] It is thus imperative that we find a method of harnessing and converting solar energy into electric energy.

Natural photosynthetic organisms have evolved over billions of years, resulting in extremely efficient systems for the conversion of light into chemical energy that is stored in the form of ATP. Natural light-harvesting assemblies, such as those found in purple photosynthetic bacteria, consist of a large array of chromophores surrounding a single reaction center (Figure 13.1).[3–8] This chlorophyll assembly acts as an antenna capable of absorbing photons from a broad spectral range and transferring its energy to the reaction center, where the conversion from solar energy to chemical energy is completed. Amazingly, the energy of any photon absorbed by any of the several hundreds of chlorophylls within the assembly is rapidly passed to the reaction center with energy transfer quantum yields approaching unity.[9] Therefore, it not surprising that the science community has focused on the

Bioinspiration and Biomimicry in Chemistry: Reverse-Engineering Nature, First Edition.
Edited by Gerhard F. Swiegers.
© 2012 John Wiley & Sons, Inc. Published 2012 by John Wiley & Sons, Inc.

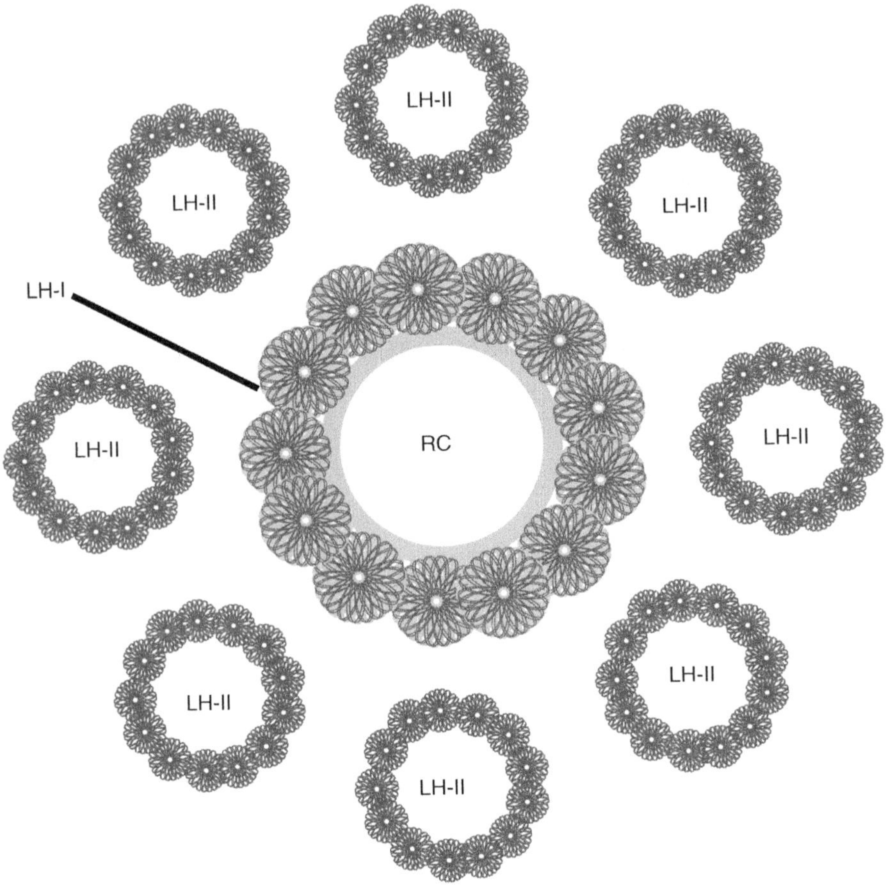

Figure 13.1 Schematic representation of bacterial light-harvesting complexes. In purple bacteria there are two types of light-harvesting complexes (LH-I and LH-II) responsible for harnessing light and transferring energy to the central reaction center.

development of artificial light-harvesting arrays, which are inspired by and mimic the highly efficient natural photosynthetic systems.[10–14]

One promising mimic of the natural systems is light-harvesting dendrimers, which demonstrate similar, although relatively more modest, properties to natural photosynthetic systems.[15–19] Dendrimers are highly branched synthetic macromolecules that, unlike their polymer counterparts, can be structurally perfect, precisely controlled, and monodisperse.[20–23] Additionally, it has been demonstrated that the dendrimer structure offers advantages over their linear polymer analogs in light harvesting.[24, 25] Their general structure consists of several branches, which all originate from a single core. For light-harvesting purposes, dendrimers can be functionalized with chromophores positioned in the core, branching units, or periphery of the dendritic structure (Figure 13.2). Functionalizing a dendrimer

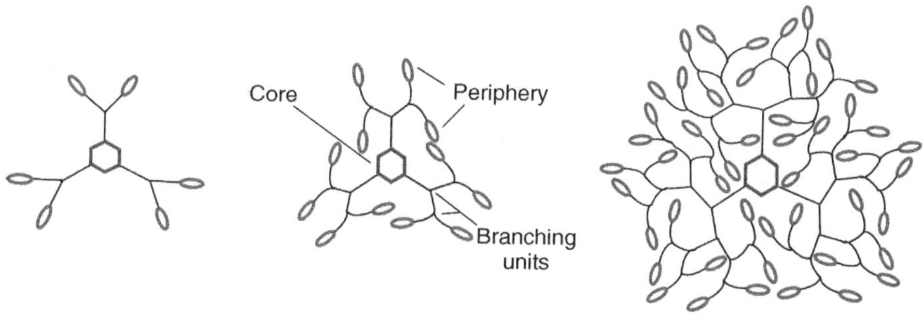

Figure 13.2 Dendrimers can be functionalized at the core, the branching units, and/or the periphery. Increasing the generation of the dendrimer not only increases the size of the molecule but also doubles the number of terminal groups.

with chromophores at the periphery positions, especially in higher generations, results in a high density of light-harvesting functionalities and increases the likelihood of capturing light (Figure 13.2). The structures of dendrimers mimic natural light-harvesting systems in that chromophores arranged around the periphery of the dendrimer act as antennae surrounding the central reactive center or core. In addition, dendrimers also possess tree-like structures, which may act as an energy gradient for the funneling of energy from the periphery to the core. Backfolding of the peripheral units results in a short spatial distance between the periphery and the core, further facilitating high efficiency of the energy transfer process.[26, 27]

Several research groups have focused on the design and synthesis of appropriate dendritic architectures as well as chromophores for efficient energy transfer.[28–36] First, we discuss the architectures of dendrimers for light-harvesting purposes. In some systems, the dendrimer backbone itself acts as a chromophore for light harvesting. In other systems, however, the dendrimer acts solely as a scaffold, and plays no photoactive role. Next, we focus on the electronic processes that occur in light-harvesting dendrimers, citing specific examples of both electronic energy transfer as well as charge transfer within the systems. Finally, we discuss specific examples in which light-harvesting dendrimers have been modified for applications in clean energy technologies.

13.2 DENDRIMER ARCHITECTURES

13.2.1 Dendrimer as a Chromophore

Dendrimers are unique macromolecules in that their parameters can be controlled and varied through functionalization at the various positions. Specifically, chromophore functionalization can be localized to just the core and the end groups or it can involve the dendrimer backbone as well. In the first case, the dendrimer structure acts solely as a scaffold and plays no photoactive role. In the second case, the dendrimer backbone itself acts as a chromophore for light harvesting. For

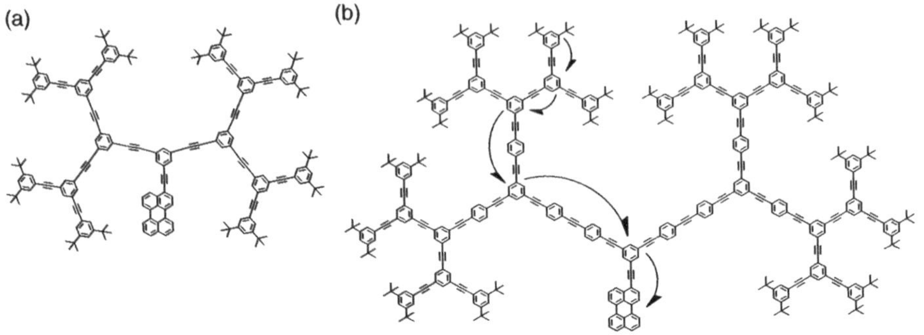

Figure 13.3 Phenylacetylene dendrimers conjugated to a perylene moiety at the focal point (a) without and (b) with an energy gradient effect.

this purpose, conjugated dendrimers are utilized, most commonly phenylacetylene dendrimers where the periphery units act as energy donors. Additionally, an energy acceptor can also be included in the core of the dendrimers to create a unidirectional energy transfer.[37]

The dendrimers, reported by Xu and Moore, consist of a phenylacetylene dendrimer conjugated to a single perylene moiety.[37–40] As seen in Figure 13.3a, each generation consists of the same repeat unit and therefore will have the same excitation energies. In the case of Figure 13.3b, an additional phenylacetylene unit is installed in each generation from the periphery to the core. Due to the increase in conjugation, there is a stepwise decrease in the excited state energy resulting in a gradient effect, funneling the energy gathered by the light-harvesting periphery to a focal point. These dendrimers show a very strong absorbance characteristic, which increases with increasing generation due to the larger number of phenylacetylene chromophores. Additionally, the absorbance spectra of the dendrimers with the energy gradient are broadened compared to those without an energy gradient because of the variations in conjugation lengths. These perylene-cored dendrimers exhibit a high efficiency intramolecular excitation energy transfer. Excitation of the dendrimer backbone results in an emission spectra, which appears to come from the perylene, with almost complete quenching of the dendrimer. The energy transfer quantum yield, Φ_{EET}, for the dendrimers are 95% and 98% for the molecules in Figures 13.3a and 13.3b, respectively. An alternate method for generating an energy gradient is for the conjugated dendrimers to consist of unsymmetrical branching.[41] An example of such a system is discussed later in Section 13.3.1.

Müllen, De Schryver, and co-workers reported polyphenylene dendrimers (Figure 13.4) that act as chromophoric backbones for light-harvesting purposes.[27, 42] In these dendrimers, there are no bridging groups connecting the phenyl rings and therefore they are twisted out of plane, limiting the conjugation of the dendrimer. Additionally, the energy collected at the periphery of the dendrimer can be funneled into an energy "sink" created by installing a biphenyl or perylenediimde core. Similar to the dendrimers reported by Xu and Moore,

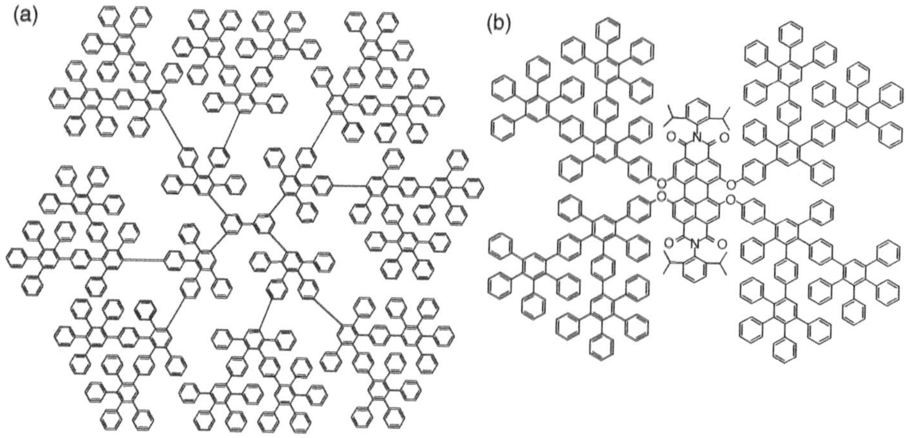

Figure 13.4 Conjugated polyphenylene dendrimers that funnel energy from the periphery into an energy "sink" created by (a) a biphenyl or (b) perylenediimide focal point.

excitation of the donor molecules results in the quenching of the polyphenylene fluorescence and a dominance of the core emission.

13.2.2 Dendrimer as a Scaffold

Dendrimers can also act solely as a scaffold for light-harvesting chromophores rather than acting as the chromophore themselves. Most commonly, benzylether linkages are utilized for the construction of dendrimers with light-harvesting moieties at the periphery or core. These dendrimers are not conjugated due to their insulating ether linkages and these dendrimers are also more flexible compared to the fully conjugated dendrimers. Here, chromophores placed at the periphery of the molecule can act as donors with an energy acceptor at the core. Due to the insulating linkers, there is a lack of electronic communication between the donors and acceptors, allowing for independent tuning of the energy levels of the chromophores.

Fréchet and co-workers have synthesized various dendrimers (Figure 13.5) where the donor and acceptor functionalities are a pair of coumarin dyes, which were specifically chosen because they fulfill the spectral overlap requirements for Förster energy transfer.[43] The lack of through-bond electronic communication in these dendrimers is further supported by the absorption spectrum of the dendrimer, which is the sum of individual absorption spectra of the donor and acceptor. Additionally, the donor absorbance doubled with increasing generation while the acceptor absorbance remained relatively constant. The selective excitation of the donor, coumarin-2, showed an emission spectra corresponding to that of the acceptor, coumarin-343, suggesting that there is an efficient energy transfer. Time-resolved experiments show that the energy migration from the periphery to the core is extremely fast, on subpicosecond time scales. Even at high generations, the energy transfer is efficient, possibly owing to the backfolding of the flexible

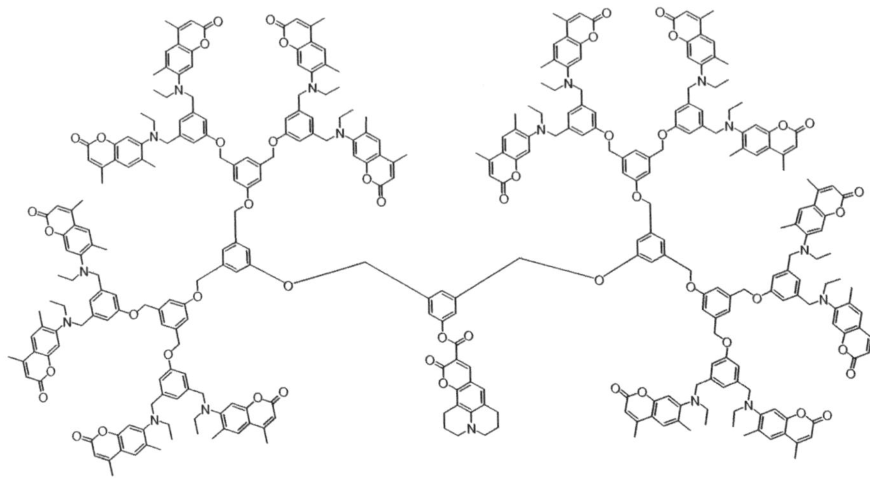

Figure 13.5 Arylether dendrimer with donor and acceptor coumarin dyes located at the periphery and core, respectively.

dendrimer, bringing the donor and acceptor functionalities into close proximity and allowing for the efficient through space energy transfer.

Fréchet and co-workers also varied the core chromophore in benzylether dendrimers with coumarin-2 donors in the periphery (Figure 13.6).[44] They installed both a pentamer and heptamer of thiophene as the core. Thiophene was chosen because its visible properties can be tuned by changing the number of repeat units. Additionally, the spectral overlap of oligothiophenes with coumarin-2 is appropriate

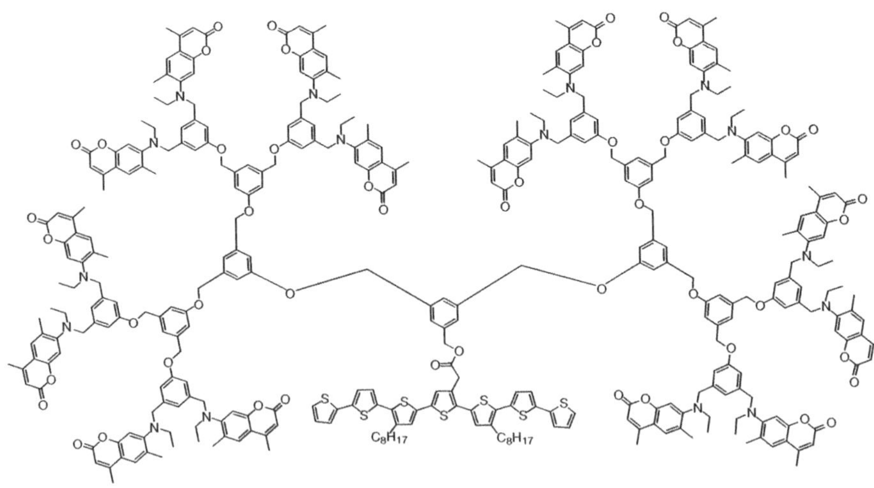

Figure 13.6 Benzylether dendrimer consisting of coumarin dyes located at the periphery and a thiophene based chromophore at the core.

for Förster energy transfer. Dendrimers up to the third generation were synthesized and it was found that the energy transfer in all cases was nearly quantitative.

13.3 ELECTRONIC PROCESSES IN LIGHT-HARVESTING DENDRIMERS

13.3.1 Energy Transfer in Dendrimers

One of the most critical processes for photosynthetic complexes is the unidirectional energy transfer process, which occurs from the chromophores to the photoactive centers. One of the successful dendritic designs for light harvesting followed by energy transfer is the use of phenyl acetylene dendrimers with a directed energy transfer described in previous sections. Here an energy gradient was synthetically included in the design resulting in efficient funneling of energy from the light-harvesting chromophores on the periphery to the core of the molecule. Alternately, designing a dendrimer with unsymmetrical branching can also create an energy gradient.[41] A majority of light-harvesting dendrimers consist of symmetrical branching, where transfer from the periphery to the core must proceed through the branching units to the core as seen in Figure 13.7a. However, in the dendrimers with unsymmetrical branching seen in Figure 13.7b, there exist "shortcuts" from the periphery units directly to the core of the dendrimer. Additionally, unsymmetrical branching produces branches with various conjugation lengths leading to an energy gradient directed toward the core of the molecule.

Peng and co-workers have synthesized a class of unsymmetrical branched phenyl acetylene dendrimers (Figure 13.8), with branching at the ortho and para positions.[45, 46] As a result of the unsymmetrical branching, there is not only a variation in the conjugation length of the branches but also a broad absorption spectra. In addition, as the size of the dendrimer increases, its spectrum red-shifts and the amount of light that is absorbed doubles with increasing generations. To determine the energy transfer efficiencies of the dendrimers, a perylene acceptor functionality was installed into the core. The absorption spectra of the perylene-cored dendrimer, Figure 13.8b, is similar to that of the dendrimer

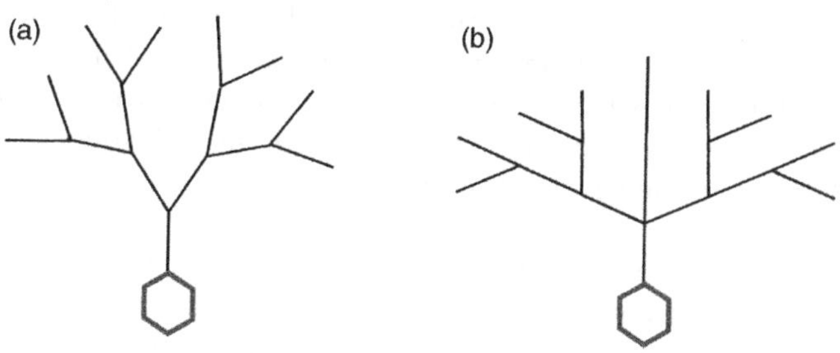

Figure 13.7 Schematic representation of dendrimers with (a) symmetrical and (b) unsymmetrical branching.

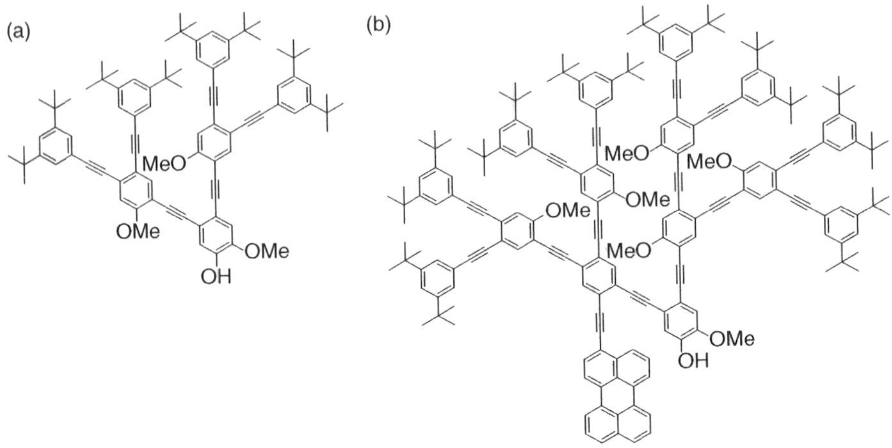

Figure 13.8 Unsymmetrically branched dendrimers containing a unidirectional EET (a) without and (b) with a perylene core.

itself, Figure 13.8a, suggesting that the ground state properties of the dendrimer remains the same when the acceptor is added. Steady state fluorescence of the perylene-cored dendrimer was taken, and when excited at the absorption maximum of the phenyl acetylene backbone, a majority of the fluorescence originated from the perylene core. The energy transfer efficiencies are greater than 98% for generations one through three and 95% for generation four. This system efficiently transfers energy from the periphery to the core and the energy transfer efficiency does not appear to significantly decrease with increasing generation.

In addition to rigid scaffolds, dendrimers with flexible scaffolds, such as benzyl ether dendrimers, can also be used in energy transfer processes. Fréchet and co-workers have synthesized benzyl ether dendrimers that consist of several chromophores and result in a tandem energy transfer from the periphery to the core.[47] The dendrimers contain a perylenebis(dicarboximide) (PDI) moiety as an energy acceptor at the core as well as fluorol-7 GA and coumarin-2 as the donor functionalities at the second and third branching points, respectively (Figure 13.9). This dendrimer was specifically designed to fulfill the spectral requirements necessary for a Förster resonance energy transfer (FRET) process. Here the coumarin-2 molecules placed at the third branching point will be used to harness light energy. The coumarin-2 molecules have appropriate spectra overlap with the fluorol-7 GA molecules, which are located closer to the core, resulting in favored energy transfer toward the core of the molecule. The energy can then be transferred to the PDI by an additional FRET process. While the absorption spectrum of this molecule demonstrates a composite of all three individual absorbance spectra, the emission spectrum shows almost complete quenching of the coumarin-2 and fluorol-7 GA chromophores. The efficiency of energy transfer from coumarin-2 to fluorol-7 GA was found to be 99% and the efficiency of transfer from fluorol-7 GA to PDI was found to be 96%. The direct energy transfer from coumarin-2 to PDI was

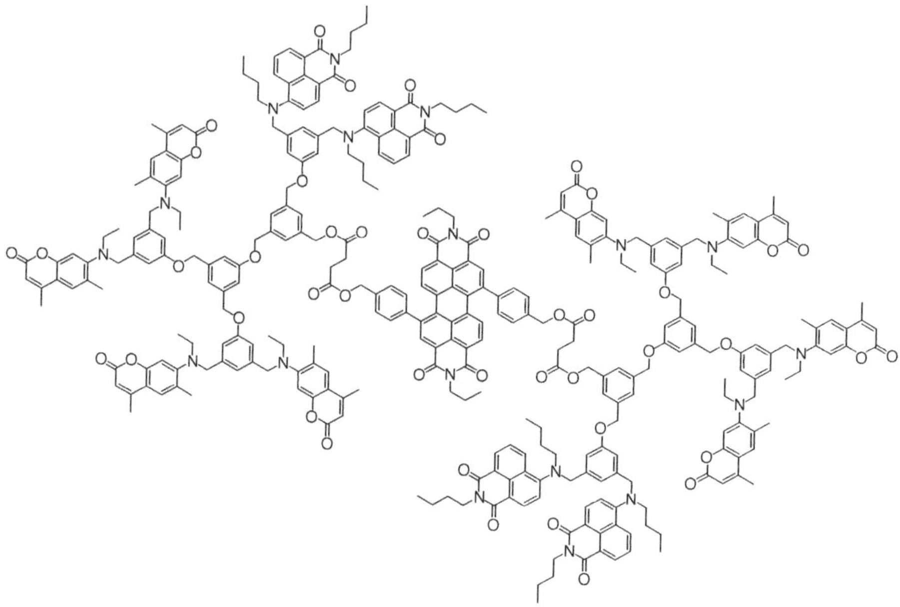

Figure 13.9 Benzyl ether dendrimer containing fluorol-7GA and coumarin-2 at the second and third branching positions to create a tandem EET from the periphery to the core.

less favorable at 79% efficiency due to the significantly lower amount of spectral overlap between the two chromophores as well as the increased distance between them.

13.3.2 Charge Transfer in Dendrimers

Electron transfer plays a crucial role in most natural photosynthetic systems. Stewart and Fox first reported electron transfer in dendrimers with benzyl ether backbones containing aryl groups in the periphery and tertiary amines tethered to the focal point.[48] Selective excitation of the peripheral chromophores results in an efficient intramolecular electron transfer from the donor at the core, which quenches the fluorescence of the electron acceptors at the periphery.

Efficient charge transfer can also occur in rigid conjugated dendrimers, as demonstrated by Guldi and co-workers, through the incorporation of a fullerene acceptor at the core of a phenylenevinylene dendrimer.[49] Müllen and co-workers have demonstrated photoinduced charge transfer in polyphenylene dendrimers with a PDI core and peripheral triphenylamine (TPA) units (Figure 13.10).[50] The steady state and transient spectroscopy data suggest that this system undergoes an energy transfer process as well as an electron transfer process from the periphery to the core, mimicking the complete photosynthetic process. Solvent polarity plays a major factor on which process is preferred and it is noted that complete energy transfer is observed in nonpolar media and electron transfer efficiency increases

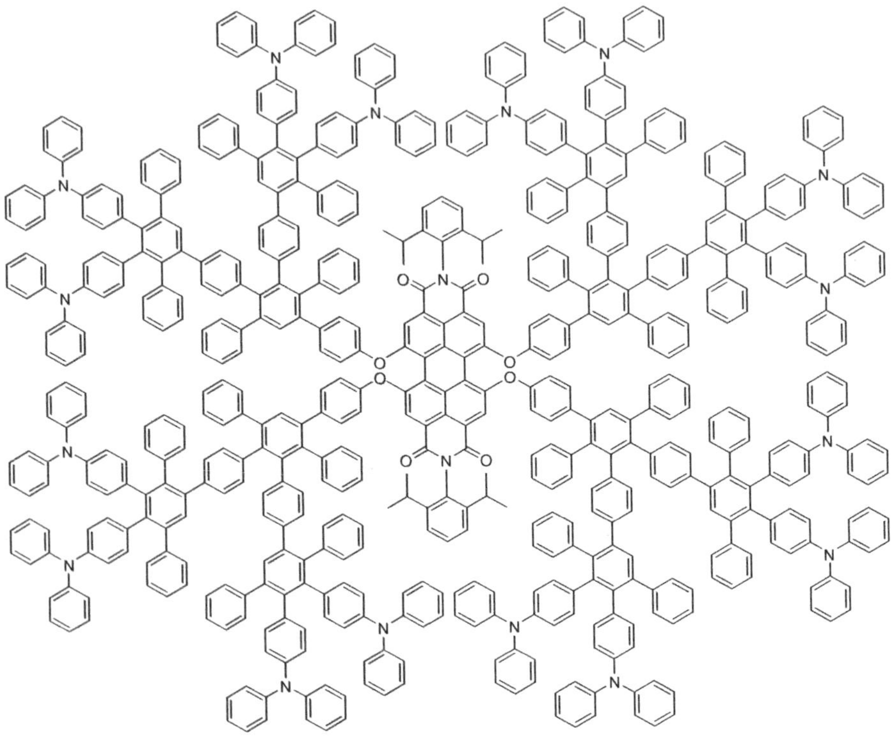

Figure 13.10 Conjugated dendrimer that undergoes both energy and electron transfer mimicking the complete photosynthetic process.

with solvent polarity. This charge transfer decreases with increasing generation most likely due to the increased distance between the donor and acceptor molecules.

Thayumanavan and co-workers have reported dendrimers based on flexible benzyl ether dendrimers, which also mimic the complete photosynthetic event in that they are capable of undergoing both energy and electron transfer (Figure 13.11).[51] These dendrimers contain an energy and electron accepting core that consists of a benzothiadiazole-based conjugated oligomer and diarylaminopyrene units at the periphery, which act as energy and electron donors. The spectral overlap of the donor and acceptor moieties is appropriate for FRET, implying that there will be energy transfer. Additionally, the oxidation potential of the benzothiadiazole core, found from cyclic voltammetry, is above that of the diarylaminopyrene units, suggesting that it is possible for the excited state of the core to be reduced by the peripheral units. Efficient energy transfer, even in high generations, occurs as a result of the excitation of the peripheral chromophores, yielding a fast rise in fluorescence of the acceptor functionality. Solvent-dependent fluorescent quenching and transient absorption spectroscopy data suggest that there is a charged-intermediate species implying the presence of a charge transfer process. The charge transfer efficiency in these dendrimers is as high as 70% and the efficiency of a photon

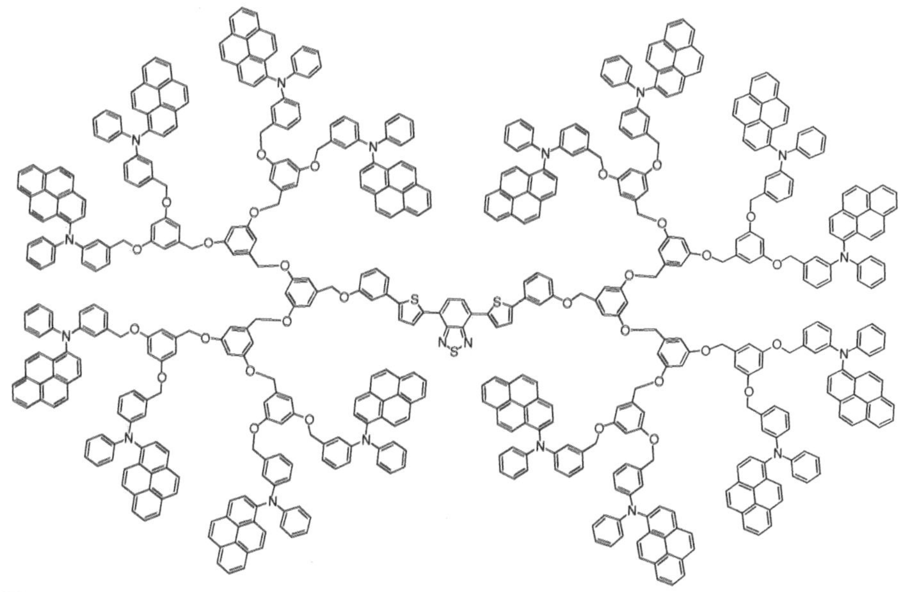

Figure 13.11 Benzyl ether dendrimer capable of undergoing both energy and electron transfer.

to charge-separated state process was 50%. Unlike the systems studied by Müllen, these dendrimers do not show a decrease in charge transfer with increasing generation. This can be explained by the flexibility in the dendrimer scaffold. The rate of electron transfer depends on the distance between the donor and acceptor functionalities, and in the case of the benzyl ether dendrimers, the flexible backbone allows for backfolding, which brings the donor functionalities in close proximity to the acceptor core in high generations.[26]

13.4 LIGHT-HARVESTING DENDRIMERS IN CLEAN ENERGY TECHNOLOGIES

Cost effective and clean hydrogen evolution from water has been recognized as a major challenge limiting the use of hydrogen as fuel in several clean energy technologies. Not surprisingly, there have been several attempts to develop new processes for generating hydrogen from water.[52–54] Among these processes, light-driven hydrogen evolution has been recognized as a truly clean technology where light from the Sun would be used to convert water to hydrogen.[55–58] In order to realize photosensitized hydrogen evolution, early attempts focused on the use of organic dyes, such as ruthenium bipyridyl complexes.[59–61] Only later did people begin to focus on the use of extended pi-conjugated organic polymers. These polymers offer major benefits owing to their large absorption cross sections; however, they are often water insoluble, thereby limiting their use in hydrogen production.

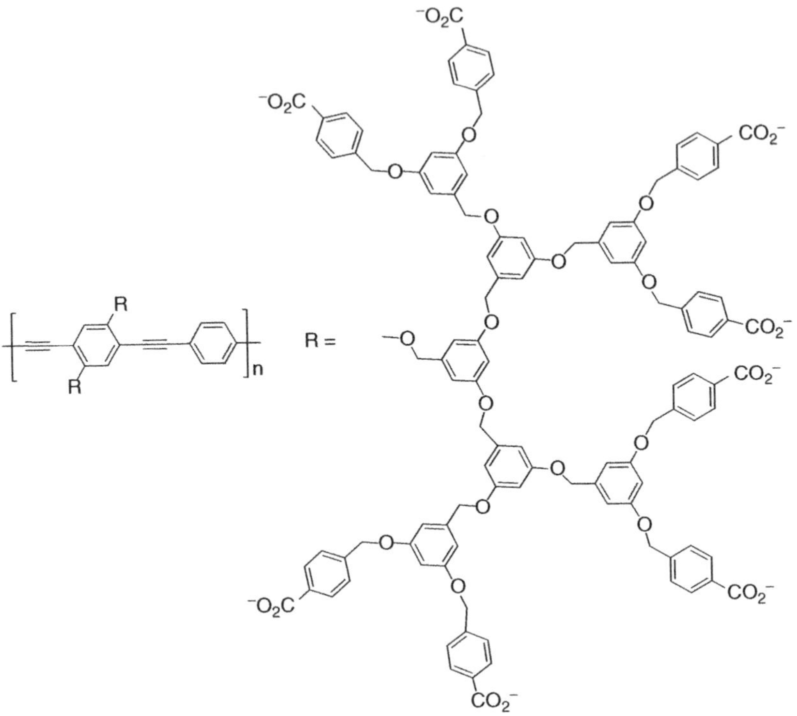

Figure 13.12 Conjugated polymer backbone with negatively charged dendrimer shell.

Aida and co-workers have developed a design that uses dendrimers to "wrap up" conjugated polymers allowing for water solubility.[62, 63] Their approach utilizes a large negatively or positively charged dendritic shell that is capable of encapsulating a poly(phenylene ethynylene) conjugated backbone (Figure 13.12). They find that the larger dendrimers, which create a 2 nm thick shell, are effective in isolating the conjugated backbone, preventing the loss of excitation energy that results from the formation on an excimer.

In order to investigate the photoinduced electron transfer, methyl viologen (MV^{2+}) was electrostatically incorporated onto the surface of the dendrimer as the electron acceptor. Upon excitation of the conjugated polymer–dendrimer complex, a charge transfer takes place, significantly decreasing the fluorescence intensity of the complex. As a result of this electron transfer, a hole is left behind in the conjugated backbone and MV^{2+} is reduced to $MV^{+\bullet}$. They suggest that this molecule may readily exchange with free MV^{2+} in solution liberating the $MV^{+\bullet}$ into solution. High accumulation of $MV^{+\bullet}$ in solution can result in hydrogen evolution from water in the presence of colloidal platinum according to Eq. 13.1:

$$2\,MV^{+\bullet} + H_2O \xrightarrow{\text{colloidal Pt}} H_2 + 2OH^- + 2MV^{2+} \qquad (13.1)$$

They found that there was a steady generation of hydrogen (1.3×10^{-6} mol h^{-1}) over the course of 5 hours without a decrease in activity. Additionally, the absorbance of the backbone did not change, indicating that the system is stable against photobleaching. The overall quantum efficiency for hydrogen evolution, that is, the number of H_2 generated per number of photon absorbed, was 13%.

Recently, light-harvesting dendrimers have been of interest for solar energy conversion devices such as organic photovoltaics.[64–67] Polymer-based organic photovoltaic devices offer a low-cost, large-scale, and solution processable alternative to the traditional silicon-based photovoltaics.[68–70] One of the major limitations of polymer photovoltaics is a lack in the control of the active layer morphology, which is dependent on the molecular weight of the polymer.[71, 72] As mentioned previously, unlike their polymer equivalents, dendrimers are completely controlled and monodisperse, resulting in a well-defined morphology. Additionally, the use of dendrimers offers a large array of chromophores for light absorption and efficient energy transfer to the core. P-type, or hole transporting, dendrimer semiconductors can be blended with fullerene acceptors such as [6,6]-phenyl-C_{61}-butyric methyl ester (PCBM), creating bulk heterojunction photovoltaics.

Early attempts at dendrimer-based heterojunction solar cells, first published in 2004 by Crossley, Imahori, Kamat, Fukuzumi, and co-workers, were focused on the use of dendrimers with porphyrin moieties, which closely resemble natural photosynthetic arrays (Figure 13.13), blended with fullerene.[73] It was found that the dendrimers form clusters with the fullerene molecules forming a π-complex and resulting in absorbance in the 700–800 nm region. These solar cells showed an overall power conversion efficiency 0.32%.

Rance, Kopidakis, and co-workers reported a dendrimer consisting of an electron-withdrawing core and electron-rich dendrons, based on thiophene, for bulk heterojunction solar cells (Figure 13.14).[74] The combination of electron-rich and electron-poor constituents in conjugated molecules can be used to lower the bandgap of the material, causing a red shift in absorbance and allowing for improved photon harvesting.[75–77] The dendrimers were found to be planar, allowing for well-ordered morphology in both neat and blended films. When compared to a control, in which there are no electron-withdrawing groups in the core, the dendrimer shows higher photoconductivity and longer photocarrier lifetimes. These improvements, as well as the ordered morphology, result in a device efficiency of 1.12%.

Wong and co-workers have also synthesized thiophene dendrimers with a hexa-peri-hexabenzocoronene (HBC) core (Figure 13.15), which is a planar aromatic hydrocarbon well known to self-assemble into columns, yielding very well-organized films.[78] The absorption intensity and bandgap of these dendrimers can easily be tuned by changing the generation of the dendrimer resulting from the change in the number of thiophene units. Solar cells were constructed from all generations by blending with both $PC_{61}BM$ and $PC_{71}BM$. The highest efficiency was found in the second-generation dendrimer blended with $PC_{71}BM$ and was found to have an efficiency of 2.5%.

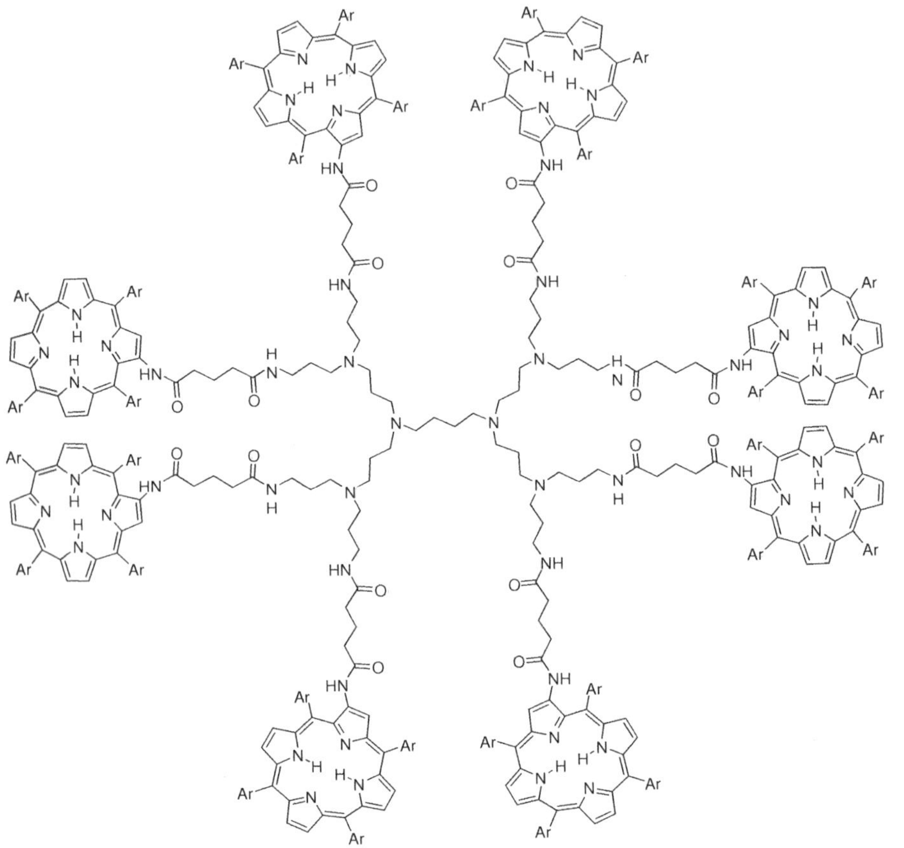

Figure 13.13 Porphyrin-based light-harvesting dendrimers that were blended with fullerene to create a dendrimer bulk heterojunction solar cell.

Thayumanavan and co-workers have reported a molecular design based on their previous work reported in Section 13.3.2, which would eliminate the need for blending in a fullerene-based acceptor functionality.[79] In the charge-separated state of light-harvesting dendrimers, one of the charges is localized in the periphery and the other is encapsulated in the core and thus trapped. In a photovoltaic device, both efficient charge separation as well as migration are necessary and therefore trapping one of the charges is not beneficial. Thayumanavan and co-workers envisage a hybrid material that combines the advantages of a dendritic architecture for light harvesting and charge transfer as well as a linear polymer for intermolecular transport of the charges. The design reported here consists of a dendron–rod–coil architecture with a hole-transporting dendron, low bandgap sensitizer that also acts as the core, and an electron-transporting coil polymer (Figure 13.16a). Additionally, dendron–rod–coil architectures have been shown to provide phase-separated nanoscale domains.

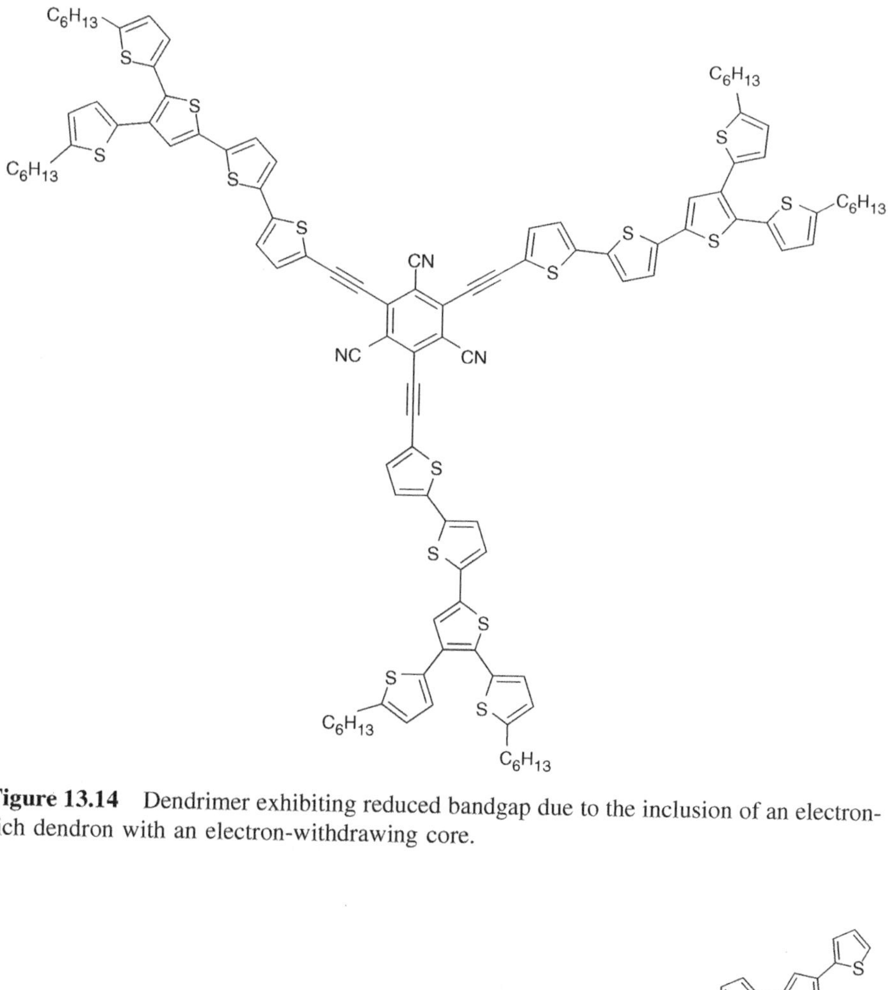

Figure 13.14 Dendrimer exhibiting reduced bandgap due to the inclusion of an electron-rich dendron with an electron-withdrawing core.

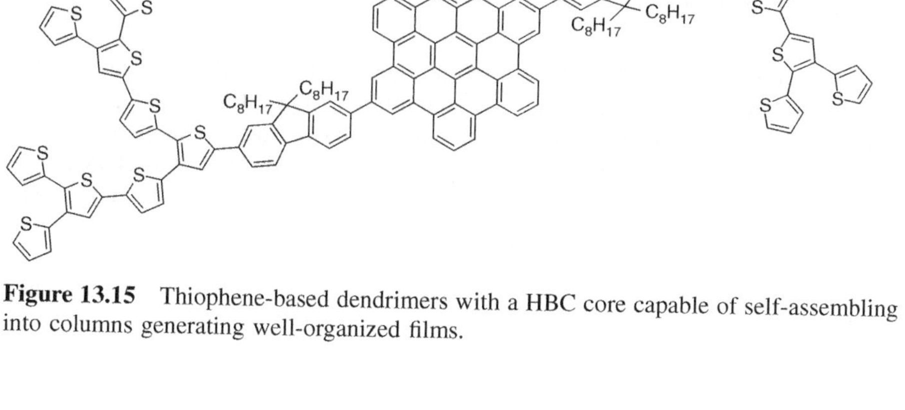

Figure 13.15 Thiophene-based dendrimers with a HBC core capable of self-assembling into columns generating well-organized films.

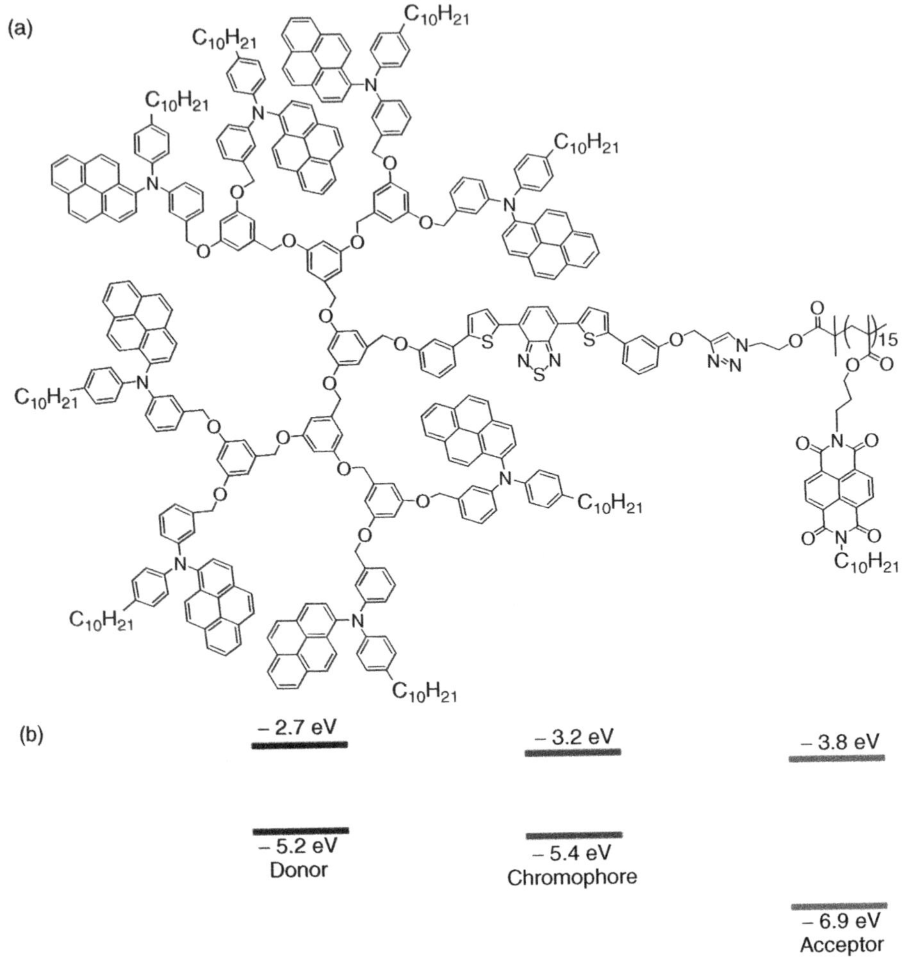

Figure 13.16 (a) Dendron–rod–coil architecture for an all-organic photovoltaic device. (b) The energy levels of the three constituents are appropriate for CT from the donor dendrimer to the excited state of the chromophore and from the excited state of the chromophore to the acceptor polymer

It has already been demonstrated (see Section 13.3.2) that a benzyl ether dendrimer with diarylaminopyrene functionalities at the periphery is capable of reducing the excited state of the benzothiadiazole-based rod chromophore. Cyclic voltammetry and absorption spectroscopy were used to estimate the HOMO and LUMO energy levels of all three functionalities in the molecule, and it was found that the energy levels are appropriate for electron transfer not only from the dendrimer to the excited state of the chromophore but also from the excited state of the chromophore to the coil polymer (Figure 13.16b). Fluorescence quenching experiments were done to determine if the molecule would undergo charge transfer. It was found that upon selective excitation of the core there was an efficient electron transfer resulting in complete quenching of the fluorescence of the core. Interestingly, the charge transfer efficiencies of the dendron–rod–coil are much higher than those of the dendrimers previously studied.

Stern–Volmer quenching experiments were used to determine the ability of the charge transport functionalities, in both the dendron and the coil polymer, to quench the chromophore. It was found that although the NDI moiety in the coil polymer is much more effective than the diarylaminopyrene moiety in the dendron at quenching the excited state of the chromophore, the difference in quenching between the dendron–rod and rod–coil functionalities is relatively minor. This suggests that the dendrimer provides some architectural advantage over the coil polymer. These findings, along with literature precedence for the formation of self-assembling nanostructures from dendron–rod–coils,[80, 81] suggest that this material may have promise in an organic photovoltaic device.

13.5 CONCLUSION

Owing to their unique structures and properties, dendrimers have been of great interest to the scientific community in a rather diverse set of research fields.[82–84] They offer many structural benefits including high functional group density, controlled molecular weight, and complete control over specific functionalization at the core, branching units, and periphery. For light-harvesting purposes, dendrimers offer a great synthetic analogy to natural photosynthetic systems. We have illustrated here that, through synthetic manipulation, the properties of these dendrimers can be tuned to generate effective photosynthetic mimics. Although dendrimers offer excellent platforms for the fundamental understanding of structure–property relationships, their difficult, lengthy, and often expensive syntheses may prevent dendrimers from being utilized in the mass production of light-harvesting devices. Future generations of light-harvesting dendrimers should be based on easy and cost-effective syntheses of well-controlled dendrimers, or hyperbranched polymers, with control over varying the functionalities at specific positions. In order to realize this, much attention must be paid to the development of new routes for dendrimer synthesis. With these new and easier synthetic approaches, light-harvesting dendrimers can become competitive with polymers for use as active materials in organic devices.

ACKNOWLEDGMENTS

The authors acknowledge partial support from the U.S. Department of Energy through the EFRC at UMass Amherst (DE-SC0001087) and the U.S. Army Research Office (54635CH) through the Green Energy Center.

REFERENCES

1. Morton, O. *Nature* **2006**, *443*, 19.
2. Basic Research Needs for Solar Energy Utilization; U.S. Department of Energy Office of Science: **2005**, `http://www.sc.doe.gov/bes/reports/files/SEU_rpt.pdf`.
3. Deisenhofer, J.; Epp, O.; Miki, K.; Huber, R.; Michel, H. *Nature* **1985**, *318*, 618.
4. Deisenhofer, J.; Michel, H. *Science* **1989**, *245*, 1463.
5. Barber, J.; Andersson, B. *Nature* **1994**, *370*, 31.
6. McDermott, G.; Prince, S. M.; Freer, A. A.; Hawthornthwaite-Lawless, A. M.; Papiz, M. Z.; Cogdell, R. J.; Isaacs, N. W. *Nature* **1995**, *374*, 517.
7. Pullerits, T.; Sundström, V. *Acc. Chem. Res*. **1996**, *29*, 381.
8. Balzani, V.; Credi, A.; Venturi, M. *Curr. Opin. Chem. Biol*. **1997**, *1*, 506.
9. Hu, X.; Damjanovic, A.; Ritz, T.; Schulten, K. *Proc. Natl. Acad. Sci. U.S.A.* **1998**, *95*, 5935.
10. Wasielewski, M. R. *Chem. Rev*. **1992**, *92*, 435.
11. Gust, D.; Moore, T. A.; Moore, A. L. *Acc. Chem. Res*. **1993**, *26*, 198.
12. Gust, D.; Moore, T. A.; Moore, A. L. *Acc. Chem. Res*. **2001**, *34*, 40.
13. Webber, S. E. *Chem. Rev*. **1990**, *90*, 1469.
14. Fox, M. A. *Acc. Chem. Res*. **1992**, *25*, 569.
15. Adronov, A.; Fréchet, J. M. J. *Chem. Commun*. **2000**, 1701.
16. Nantalaksakul, A.; Reddy, D. R.; Bardeen, C. J.; Thayumanavan, S. *Photosynthesis Res*. **2006**, *87*, 133.
17. Zeng, Y.; Li, Y.-Y.; Chen, J.; Yang, G.; Li, Y. *Chem. Asian J*. **2010**, *5*, 992.
18. Balzani, V.; Ceroni, P.; Maestri, M.; Vicinelli, V. *Curr. Opin. Chem. Biol*. **2003**, *7*, 657.
19. D'Ambruoso, G. D.; McGrath, D. V. *Adv. Polym. Sci*. **2008**, *214*, 87.
20. Fréchet, J. M. J. *Science* **1994**, *263*, 1710.
21. Tomalia, D. A.; Baker, H.; Dewald, J.; Hall, M.; Kallos, G.; Martin, S.; Roeck, J.; Ryder, J.; Smith, P. *Polym. J*. **1985**, *17*, 117.
22. Newkome, G. R.; Yao, Z.; Baker, G. R.; Gupta, V. K. *J. Org. Chem*. **1985**, *50*, 2003.
23. Hawker, C. J.; Fréchet, J. M. J. *J. Am. Chem. Soc*. **1990**, *112*, 7638.
24. Nantalaksakul, A.; Dasari, R. R.; Ahn, T.-S.; Al-Kaysi, R.; Bardeen, C. J.; Thayumanavan, S. *Org. Lett*. **2006**, *8*, 2981.
25. Ahn, T.-S.; Nantalaksakul, A.; Dasari, R. R.; Al-Kaysi, R. O.; Müller, A. M.; Thayumanavan, S.; Bardeen, C. J. *J. Phys. Chem. B* **2006**, *110*, 24331.
26. Krishnamoorthy, K.; Dasari, R. R.; Nantalaksakul, A.; Thayumanavan, S. *Chem. Commun*. **2007**, 739.
27. Liu, D.; De Feyter, S.; Cotlet, M.; Stefan, A.; Wiesler, U.-M.; Herrmann, A.; Grebel-Koehler, D.; Qu, J.; Müllen, K.; De Schryver, F. C. *Macromolecules* **2003**, *36*, 5918.

28. Denti, G.; Campagna, S.; Serroni, S.; Ciano, M.; Balzani, V. *J. Am. Chem. Soc.* **1992**, *114*, 2944.
29. Jiang, D. L.; Aida, T. *J. Am. Chem. Soc.* **1998**, *120*, 10895.
30. Du, P.; Zhu, W.-H.; Xie, Y.-Q.; Zhao, F.; Ku, C.-F.; Cao, Y.; Chang, C.-P.; Tian, H. *Macromolecules* **2004**, *37*, 4387.
31. Yuan, M.; Yin, X.; Zheng, H.; Ouyang, C.; Zuo, Z.; Liu, H.; Li, Y. *Chem. Asian J.* **2009**, *4*, 707.
32. Yeow, E. K. L.; Ghiggino, K. P.; Reek, J. N. H.; Crossley, M. J.; Bosman, A. W.; Schenning, P. H. J.; Meijer, E. W. *J. Phys. Chem. B* **2000**, *104*, 2596.
33. Bar-Haim, A.; Klafter, J.; Kopelman, R. *J. Am. Chem. Soc.* **1997**, *119*, 6197.
34. Gilat, S. L.; Adronov, A.; Fréchet, J. M. J. *Angew. Chem. Int. Ed.* **1999**, *38*, 1422.
35. Dichtel, W. R.; Hecht, S.; Fréchet, J. M. J. *Org. Lett.* **2005**, *7*, 4451.
36. Xu, T. H.; Lu, R.; Qiu, X. P.; Liu, X. L.; Xue, P. C.; Tan, C. H.; Bao, C. Y.; Zhao, Y. Y. *Eur. J. Org. Chem.* **2006**, *2006*, 4014.
37. Xu, Z.; Moore, J. S. *Acta Polymerica* **1994**, *45*, 83.
38. Devadoss, C.; Bharathi, P.; Moore, J. S. *J. Am. Chem. Soc.* **1996**, *118*, 9635.
39. Shortreed, M. R.; Swallen, S. F.; Shi, Z.-Y.; Tan, W.; Xu, Z.; Devadoss, C.; Moore, J. S.; Kopelman, R. *J. Phys. Chem. B* **1997**, *101*, 6318.
40. Swallen, S. F.; Shi, Z.-Y.; Tan, W.; Xu, Z.; Moore, J. S.; Kopelman, R. *J. Lumin.* **1998**, *76–77*, 193.
41. Peng, Z.; Melinger, J. S.; Kleiman, V. *Photosynthesis Res.* **2006**, *87*, 115.
42. Qu, J.; Liu, D.; De Feyter, S.; Zhang, J.; De Schryver, F. C.; Müllen, K. *J. Org. Chem.* **2003**, *68*, 9802.
43. Adronov, A.; Gilat, S. L.; Fréchet, J. M. J.; Ohta, K.; Neuwahl, F. V. R.; Fleming, G. R. *J. Am. Chem. Soc.* **2000**, *122*, 1175.
44. Adronov, A.; Malenfant, P. R. L.; Fréchet, J. M. J. *Chem. Mater.* **2000**, *12*, 1463.
45. Peng, Z.; Pan, Y.; Xu, B.; Zhang, J. *J. Am. Chem. Soc.* **2000**, *122*, 6619.
46. Melinger, J. S.; Pan, Y.; Kleiman, V. D.; Peng, Z.; Davis, B. L.; McMorrow, D.; Lu, M. *J. Am. Chem. Soc.* **2002**, *124*, 12002.
47. Serin, J. M.; Brousmiche, D. W.; Fréchet, J. M. J. *Chem. Commun.* **2002**, 2605.
48. Stewart, G. M.; Fox, M. A. *J. Am. Chem. Soc.* **1996**, *118*, 4354.
49. Guldi, D. M.; Swartz, A.; Luo, C.; Gómez, R.; Segura, J. L.; Martín, N. *J. Am. Chem. Soc.* **2002**, *124*, 10875.
50. Qu, J.; Pschirer, N. G.; Liu, D.; Stefan, A.; De Schryver, F. C.; Müllen, K. *Chem. Eur. J.* **2004**, *10*, 528.
51. Thomas, K. R. J.; Thompson, A. L.; Sivakumar, A. V.; Bardeen, C. J.; Thayumanavan, S. *J. Am. Chem. Soc.* **2005**, *127*, 373.
52. Navarro, R. M.; Peña, M. A.; Fierro, J. L. G. *Chem. Rev.* **2007**, *107*, 3952.
53. Wagner, F. T.; Lakshmanan, B.; Mathias, M. F. *J. Phys. Chem. Lett.* **2010**, *1*, 2204.
54. Armaroli, N.; Balzani, V. *Chem Sus Chem* **2011**, *4*, 21.
55. Kiwi, J.; Grätzel, M. *Nature* **1979**, *281*, 657.

56. Tabata, M.; Maeda, K.; Ishihara, T.; Minegishi, T.; Takata, T.; Domen, K. *J. Phys. Chem. C* **2010**, *114*, 11215.
57. Allakhverdiev, S. I.; Thavasi, V.; Kreslavski, V. D.; Zharmukhamedov, S. K.; Klimov, V. V.; Ramakrishna, S.; Los, D. A.; Mimuro, M.; Nishihara, H.; Carpentier, R. *J. Photochem. Photobiol. C Photochem. Rev*. **2010**, *11*, 101.
58. Ajayakumar, G.; Kobayashi, M.; Masaoka, S.; Sakai, K. *Dalton Trans*. **2011**, *40*, 3955.
59. Richoux, M.-C.; Harriman, A. *J. Chem. Soc. Faraday Trans*. **1982**, *78*, 1873.
60. Islam, S. D.-M.; Yoshikawa, Y.; Fujitsuka, M.; Ito, O.; Komatsu, S.; Usui, Y. *J. Photochem. Photobiol. A Chem*. **2000**, *134*, 155.
61. Kalyanasundaram, K. *Coord. Chem. Rev*. **1982**, *46*, 159.
62. Jiang, D.-L.; Choi, C.-K.; Honda, K.; Li, W.-S.; Yuzawa, T.; Aida, T. *J. Am. Chem. Soc*. **2004**, *126*, 12084.
63. Tomioka, N.; Takasu, D.; Takahashi, T.; Aida, T. *Angew. Chem. Int. Ed*. **1998**, *37*, 1531.
64. Lo, S.-C.; Burn, P. L. *Chem. Rev*. **2007**, *107*, 1097.
65. Mitchell, W.J.; Kopidakis, N.; Rumbles, G.; Ginley, D. S.; Shaheen, S. E. *J. Mater. Chem*. **2005**, *15*, 4518.
66. Kopidakis, N.; Mitchell, W. J.; Van de Lagemaat, J.; Ginley, D. S.; Rumbles, G.; Shaheen, S. E.; Rance, W. L. *Appl. Phys. Lett*. **2006**, *89*, 103524.
67. Roquet, S.; Cravino, A.; Leriche, P.; Alévêque, O.; Frère, P.; Roncali, J. *J. Am. Chem. Soc*. **2006**, *128*, 3459.
68. Günes, S.; Neugebauer, H.; Sariciftci, N. S. *Chem. Rev*. **2007**, *107*, 1324.
69. Spanggaard, H.; Krebs, F. C. *Solar Energy Mater. Solar Cells* **2004**, *83*, 125.
70. Po, R.; Maggini, M.; Camaioni, N. *J. Phys. Chem. C* **2010**, *114*, 695.
71. Yang, X.; Loos, J. *Macromolecules* **2007**, *40*, 1353.
72. Ma, W.; Kim, J. Y.; Lee, K.; Heeger, A. J. *Macromol. Rapid Commun*. **2007**, *28*, 1776.
73. Hasobe, T.; Kashiwagi, Y.; Absalom, M. A.; Sly, J.; Hosomizu, K.; Crossley, M. J.; Imahori, H.; Kamat, P. V.; Fukuzumi, S. *Adv. Mater*. **2004**, *16*, 975.
74. Rance, W. L.; Rupert, B. L.; Mitchell, W. J.; Köse, M. E.; Ginley, D. S.; Shaheen, S. E.; Rumbles, G.; Kopidakis, N. *J. Phys. Chem. C* **2010**, *114*, 22269.
75. Bundgaard, E.; Krebs, F. C. *Solar Energy Mater. Solar Cells* **2007**, *91*, 954.
76. Chen, J.; Cao, Y. *Acc. Chem. Res*. **2009**, *42*, 1709.
77. Havinga, E. E.; Hoeve, W.; Wynberg, H. *Polym. Bull*. **1992**, *29*, 119.
78. Wong, W. W. H.; Ma, C.-Q.; Pisula, W.; Yan, C.; Feng, X.; Jones, D. J.; Müllen, K.; Janssen, R. A. J.; Bäuerle, P.; Holmes, A. B. *Chem. Mater*. **2010**, *22*, 457.
79. Nantalaksakul, A.; Mueller, A.; Klaikherd, A.; Bardeen, C. J.; Thayumanavan, S. *J. Am. Chem. Soc*. **2009**, *131*, 2727.
80. Palmer, L. C.; Stupp, S. I. *Acc. Chem. Res*. **2008**, *41*, 1674.

81. Zubarev, E. R.; Sone, E. D.; Stupp, S. I. *Chem. Eur. J*. **2006**, *12*, 7313.

82. Azagarsamy, M. A.; Gomez-Escudero, A.; Yesilyurt, V.; Vachet, R. W.; Thayumanavan, S. *Analyst* **2009**, *134*, 635.

83. Patri, A. K.; Kukowska-Latallo, J. F.; Baker Jr., J. R. J. *Adv. Drug Deliv. Rev*. **2005**, *57*, 2203.

84. Dufés, C. Uchegbu, I. F. Schätzlein, A. G. *Adv. Drug Deliv. Rev*. **2005**, *57*, 2177.

CHAPTER 14

Biomimicry in Organic Synthesis

REINHARD W. HOFFMANN

Fachbereich Chemie der Philipps Universität, D-35032 Marburg, Germany

14.1 INTRODUCTION

Since the dawn of organic chemistry, synthesis of new compounds or those that prevail in Nature has been a major task of organic chemistry. Over the last two centuries the complexity of the synthesis targets has steadily increased. Only with a parallel development of new synthetic methods had it become possible by the end of the last century to complete syntheses of molecules such as palytoxin[1] (Scheme 14.1). This synthesis—of which not all details are yet published—involved approximately 180 individual steps, demonstrating the enormous effort in personnel and materials to attain this goal.

When the targets of natural product synthesis become even more complex in the 21st century, it is evident that the strategies and methods used in the last century reach their limits. Hence, organic chemistry is faced in the 21st century with the necessity to substantially increase the efficiency of syntheses by turning to new strategies. Combined with better synthesis methods, this should reduce the number of steps necessary to reach complex target structures. The same can be said for the synthesis of active pharmaceutical ingredients in the pharmaceutical industry, where the need for more efficient synthesis strategies has also become more pressing in the 21st century.

Natural products are synthesized by Nature in the living cells from simple starting materials. And, many pharmaceuticals are patterned according to a similarity to certain natural products. When new strategies for synthesis of such compounds are needed, it is obvious and advantageous to ask how Nature synthesizes such molecules in the process of biosynthesis. This raises the hope that Nature has found, through the process of evolution, an efficient route for the synthesis of a particular natural product, a route that could serve as a model for in vitro synthesis. Thus, knowledge of a biosynthetic pathway for a natural product of interest could

Bioinspiration and Biomimicry in Chemistry: Reverse-Engineering Nature, First Edition.
Edited by Gerhard F. Swiegers.
© 2012 John Wiley & Sons, Inc. Published 2012 by John Wiley & Sons, Inc.

Palytoxin

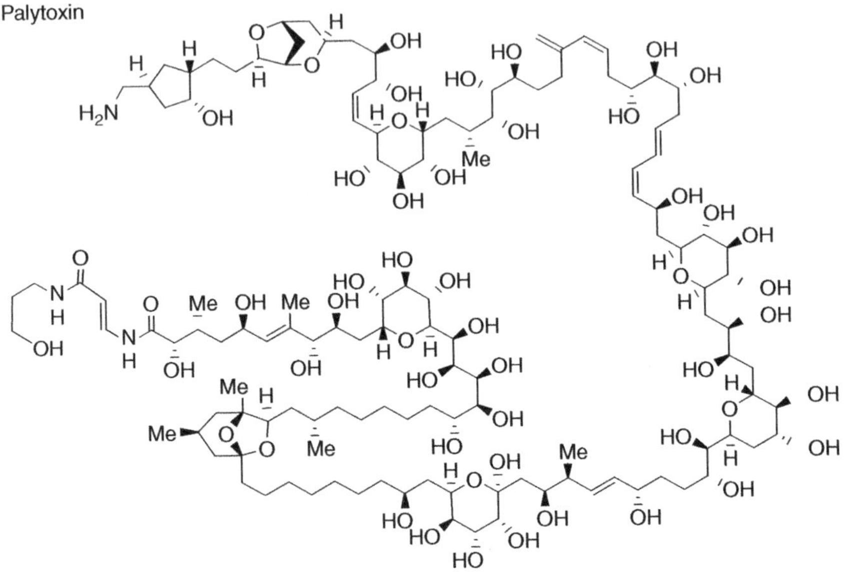

Scheme 14.1

serve as a guideline to develop a "biomimetic" synthesis. This line of thought could be expected to open reasonable approaches to the synthesis of a natural product, or at least provide a much better synthetic route than used before.

14.2 BIOMIMETIC SYNTHESIS OF NATURAL PRODUCTS

Biomimetic synthesis of a natural product requires knowledge of the biosynthesis pathway. This, however, becomes a limiting condition, because this information is available at the outset of a synthesis project in a only few cases. This fortunate situation held for the biomimetic synthesis of progesterone by the A. W. Johnson group.[2] in 1968. At the time that Johnson's work had commenced, major aspects of the biosynthesis of progesterone had been clarified[3] by the seminal studies of Bloch, Cornforth, Eschenmoser, Lynen, Popjack, and Stork as shown in Scheme 14.2. Later, further details of the cholesterol biosynthesis were established.[4]

It is characteristic that the Johnson group in their synthesis of progesterone did not try to reproduce the whole sequence of the progesterone biosynthesis. Rather, they focused on the key step in the biosynthesis, the one that led to the largest increase in complexity, the cyclization of squalene into lanosterol. This step involves an enzyme-mediated cationic polyene cyclization cascade of squalene-epoxide. Instead of copying Nature and starting from squalene, Johnson chose a different polyene **1**, which after cyclization would allow him a more ready conversion into progesterone than the route Nature uses. In doing so the Johnson group succeeded in the series of transformations to 16,17-dehydro-progesterone[2]

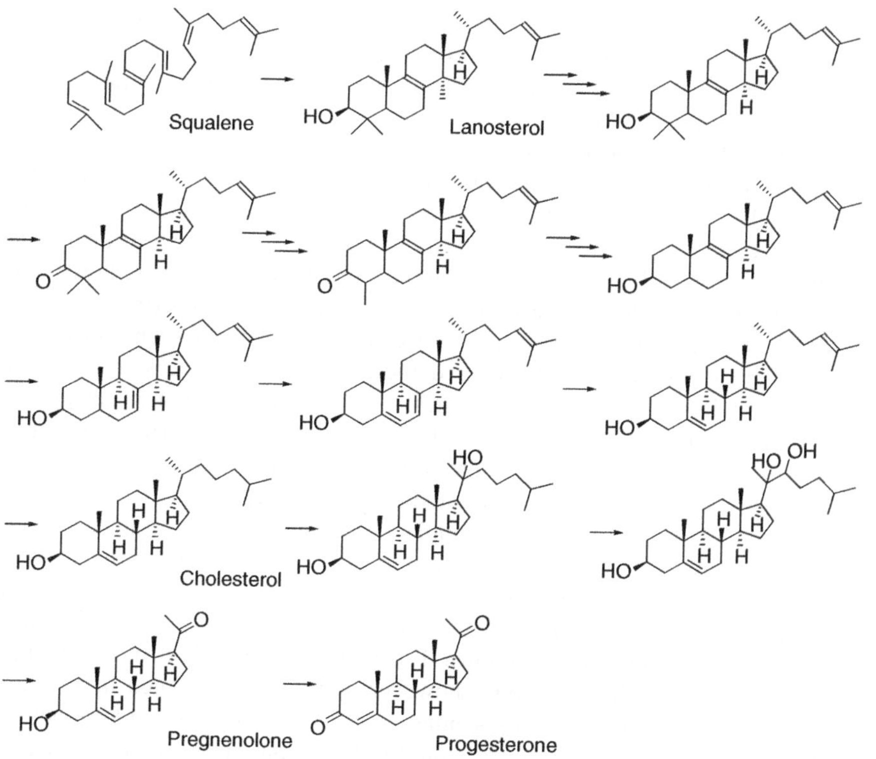

Scheme 14.2

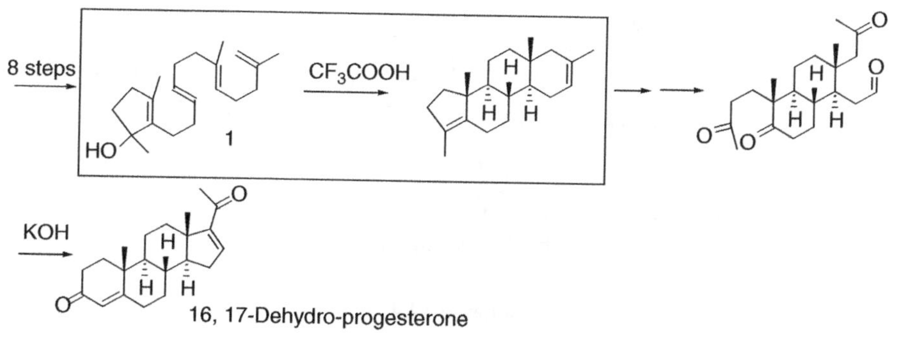

Scheme 14.3

shown in Scheme 14.3. Since the biomimetic transformation was carried out in a nonenzymatic manner, the product of the reaction is a racemate. It is, moreover, characteristic that the steps before and after the "biomimetic transformation" were carried out by "conventional" organic chemistry. This synthesis thus demonstrated that a rapid increase in complexity is possible by following in a biomimetic manner Nature's precedent of polyene cyclization. However, with regard to reach a highly efficient synthesis of progesterone, the eight steps before and three steps after the biomimetic step render this synthesis not efficient enough to surpass the generation of progesterone by partial synthesis from the readily available natural product diosgenine [5] shown in Scheme 14.4.

Johnson in his biomimetic progesterone synthesis followed Nature's precedent in principle, the polyene cyclization cascade, whereas the way to carry out this reaction in the flask—trifluoroacetic acid—was far from the way Nature uses in vivo. A restrictive view at biomimetic synthesis has been postulated by Mannich[6] and exemplified by Schöpf[7] in numerous examples: biomimetic synthesis should allow only in vitro reactions that proceed under physiological conditions (pH 3–8, water, room temperature) that prevail in a living cell. The benefit of such endeavors is the insight that Nature could have used these reaction sequences in biosynthesis, actually without the help of enzymes![8] As R. Robinson put it aptly,[9] such demonstrations are "an achievement, but not necessarily a contribution to plant physiology." In the end, proof or disproof of such concepts, as convincing as they may be, rests on detailed biosynthesis studies to be carried out.

When one looks for the earliest example of such a biomimetic synthesis, one is frequently referred[8, 10] to the classical tropinone synthesis by Robinson[11] shown in Scheme 14.5. However, Robinson gives in his report[11] no indication that this synthesis was carried out with the aim to follow Nature's route to the tropane alkaloids, which at that time was not known. It is therefore a myth[12] that this synthesis was a biomimetic synthesis. In turn, however, Robinson speculated in an immediately following paper[13] that this synthesis proceeded under so mild conditions that it would seem likely that Nature could use this route in biosynthesis. Hence,

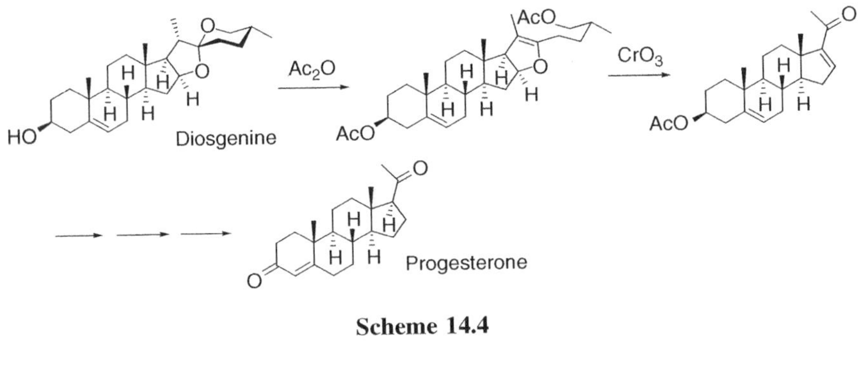

Scheme 14.4

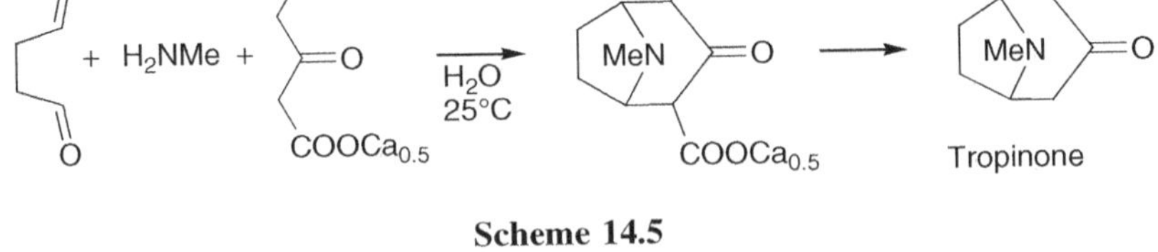

Scheme 14.5

Robinson's tropinone synthesis is an early example of a "*potentially* biomimetic synthesis." Later studies on the biosynthesis of tropane alkaloids showed[14] that Robinson was by and large right with his prediction.

14.2.1 Potentially Biomimetic Synthesis

Literally, biomimetic synthesis requires solid knowledge of the biosynthesis pathway of the natural product of interest. This is clearly distinct from a *potentially* biomimetic synthesis, which speculates about the biosynthetic pathway and tries to benefit from this concept in an actual synthesis to be carried out in the lab. This term has been suggested to the author by Albert Eschenmoser and appears in one of his recent publications.[15]

Serious studies of biosynthesis pathways became possible only after isotopically labeled precursors became available for feeding studies under in vivo conditions. These studies focused mainly on the primary metabolic pathways and addressed rarely the generation of secondary metabolites, the diversity of which formed and forms the targets of natural product synthesis. As a consequence of this situation, the field of natural product synthesis advanced much more rapidly than the studies of the biogenetic pathways. Hence, biomimetic considerations in organic synthesis were forced to be predominantly of the "potentially biomimetic" type. As will be shown in the following, in terms of its scientific value, potentially biomimetic synthesis is not at all inferior to true biomimetic synthesis. First of all, a synthesis of the former type is a test on the feasibility of a proposed biosynthesis route and, if successful, gives credence to the biosynthesis proposal.

Take usnic acid as an example. Once its structure was secured,[16] it was recognized that usnic acid could be considered as a composite of two identical building

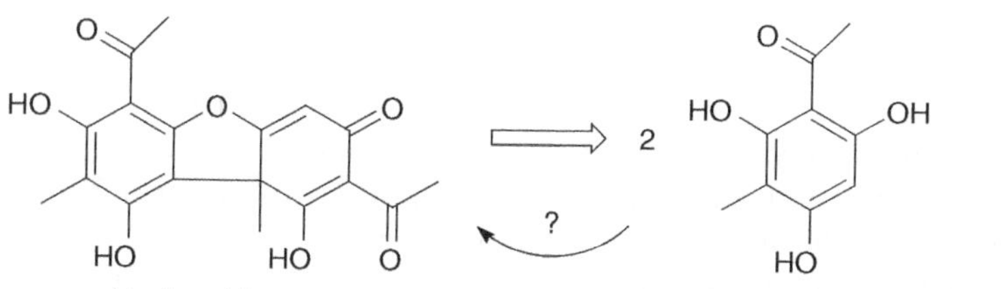

Scheme 14.6

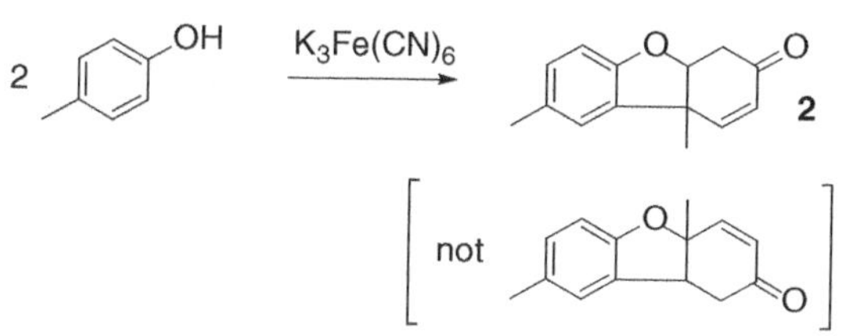

Scheme 14.7

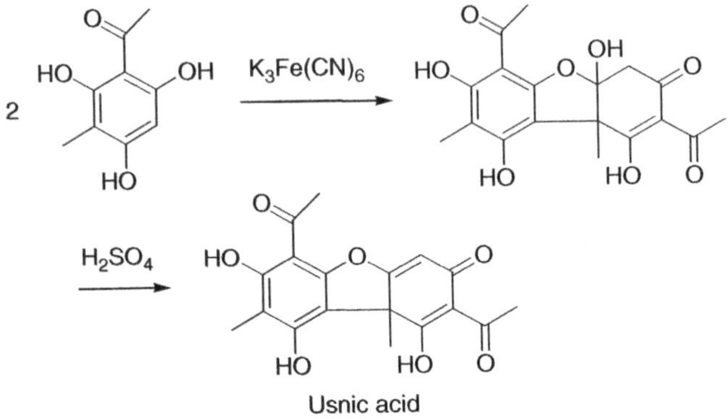

Scheme 14.8

blocks; see Scheme 14.6. A few years later, Schöpf and Roß[17] discussed the possibility that usnic acid could indeed be formed in Nature by an oxidation of methyl-phloroacetophenone. Finally, Barton et al.[18] realized that Pummerer and coworkers had in 1925[19] carried out a reaction, which in essence is a model reaction for this process (cf. Scheme 14.7) but failed in the correct assignment of the structure of the product.

With the correct structure of **2** proved, Barton was in a position to synthesize usnic acid along the proposed biogenetic pathway, Scheme 14.8, thus completing its potentially biomimetic synthesis.

The significance of this potentially biomimetic synthesis becomes apparent when one considers usnic acid as a member of a family of structurally related natural products. Could they all be formed in Nature by oxidation of phenolic precursor molecules? This insight led Barton and Cohen to postulate[20] a general (potentially) biosynthetic scheme for the formation of such natural products. This would apply as well to galanthamine **3**, for example, which might arise from the bisphenol **5** by oxidation and subsequent intramolecular oxa-Michael addition to narwedine **4** followed by eventual reduction of the carbonyl group (cf. Scheme 14.9).

To test this proposal Barton and Kirby subjected the bisphenol **5** to oxidation and substantiated the formation of narwedine, albeit in very low yield.[21] After galanthamine gained interest as a means to treat Alzheimer's disease, highly

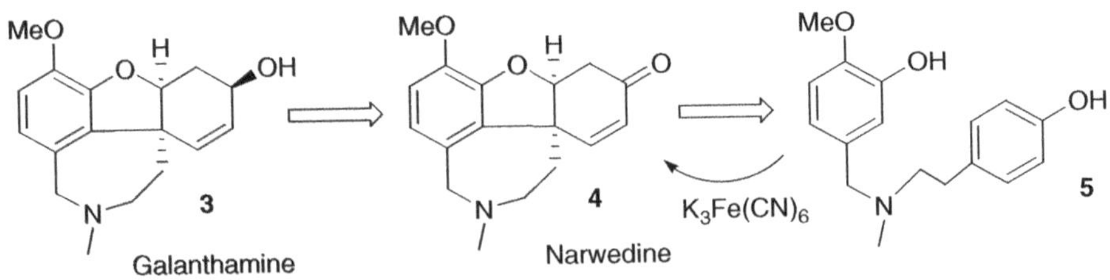

Scheme 14.9

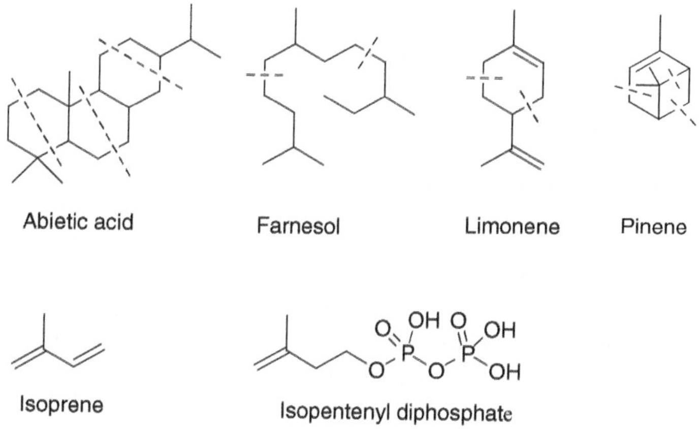

Scheme 14.10

efficient synthesis routes to **3** have been developed along these lines, using modified substrates and other oxidizing agents.[22]

At this point it does not actually matter whether Nature uses phenolic oxidation in the synthesis of galanthamine or not. It is the concept of a potentially biomimetic pathway that led to an efficient synthesis.

Not any speculation about conceivable biogenetic connections between the members of a class of natural products[23] leads to a meaningful proposal of a biosynthetic pathway. Rather, a number of restrictions have to be applied until a meaningful

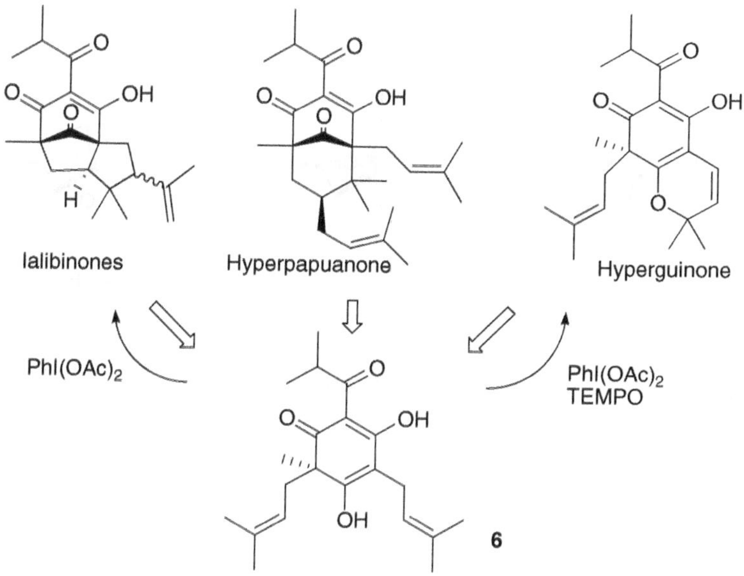

Scheme 14.11

proposal of a biosynthetic pathway can be put forward.[24] The most common way to start is to look at the structures of a class of natural products and to inspect their carbon backbones for common features. C. Schöpf[25] likened this procedure to "comparative anatomy." R. Robinson, with his book *The Structural Relations of Natural Products*,[9] gives an example of the chances and pitfalls of this approach to reach insights into the biogenesis of natural products.

The most well-known case for identifying common structural features in a class of compounds is the "isoprene rule" for terpenes[26] in Scheme 14.10. Inspection of terpene structures showed that essentially all terpenes can be considered as being composites of isoprene building blocks. This led to the hypothesis that the terpenes might be formed in Nature by an actual combination of isoprene building blocks, leaving, however, open what the exact building block is that is used in Nature. The actual isoprene building block, isopentenyl diphosphate, and the correctness of this hypothesis were only established later.[27] While the biosynthetic hypothesis thus was indeed correct, this has not yet led to a biomimetic synthesis of terpenes using isopentenyl building blocks.

Terpene substructures and phenolic skeleta meet in the class of natural products called polyprenylated acylphloroglucinols.[28] A potentially biomimetic synthesis of some members of the set in Scheme 14.11 was carried out to support the notion of a biosynthesis scheme that involved the diprenylated 1-methyl-3-acyl-phloroglycinol **6** as a key biosynthetic intermediate. After **6** had been synthesized in conventional manner, when subjected to one-electron phenolic oxidation it cyclized smoothly to both stereoisomers of the racemic ialibinones[29] and, with a different oxidation regime, a smooth cyclization to racemic hyperguinone could be effected. These cyclizations occur in vivo probably by different enzymes that control the cyclization mode and lead to enantiomerically pure products. Overall, this study strengthens the

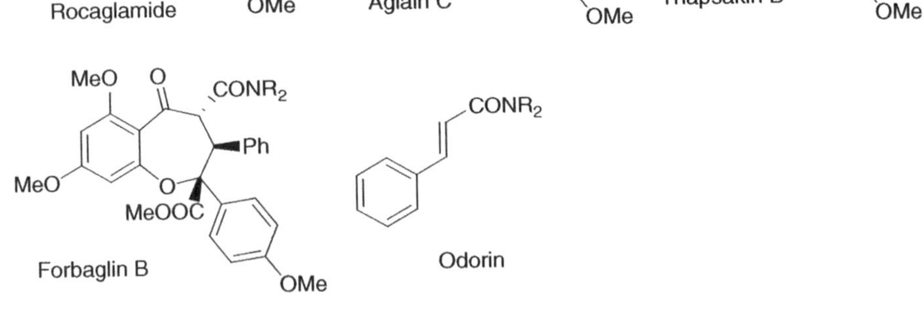

Scheme 14.12

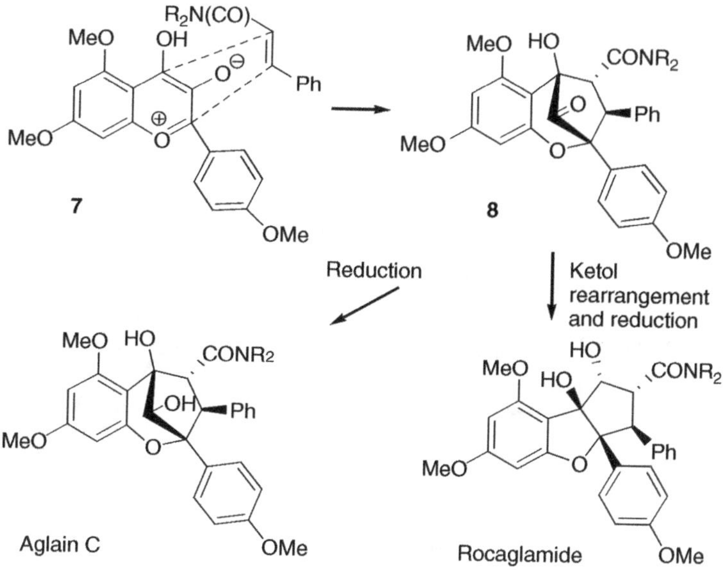

Scheme 14.13

notion that the ialibinones and hyperguinone have a common biogenetic precursor, **6**, and that this compound ought to be a natural product that awaits isolation from the leaves of *Hypericum papuanum*: that is, such a potentially biomimetic synthesis points to the presence of distinct, not yet isolated natural products.

An arbitrarily chosen, yet nevertheless typical, example of a potentially biomimetic synthesis is given by the rocaglamides as representatives of the cyclopenta[*b*]tetrahydrobenzofuran natural products isolated from the plant genus *Aglaia*.[30] From these plants similar products such as aglain C, thapsakin B, or forbaglin B have been isolated as well[31, 32] (Scheme 14.12). Scrutinizing these structures by "comparative anatomy" reveals among others a cinnamic amide as a common structural element. And since odorin has been isolated along with the other products,[31] this may well be a biogenetic precursor of the cyclopenta[*b*]tetrahydrobenzofuran natural products. Chemical reasoning, a 1,3-dipolar cycloaddition, suggests then an oxidopyrylium intermediate **7** as the other component in biosynthesis. Thus, the unifying biosynthesis scheme given in Scheme 14.13 could be proposed[33] with **8** as key intermediate, from which reduction of the carbonyl group would result in the aglain series.

A well precedented α-ketol rearrangement should give the rocaglamides, and oxidative cleavage of **8** should lead to the forbaglins. It is the hallmark of such a biosynthesis proposal that it can be subjected to in vitro experimental verification. In due course, the oxidopyrylium intermediate was generated in the presence of methyl cinnamate by shining light on compound **9**, a well-known natural product, resulting in the formation of the expected cycloadduct **10**, from which base treatment led to the rocaglamide derivative **11**[30] (Scheme 14.14). While the feasibility of the proposed biosynthesis scheme could thus be verified, this synthesis

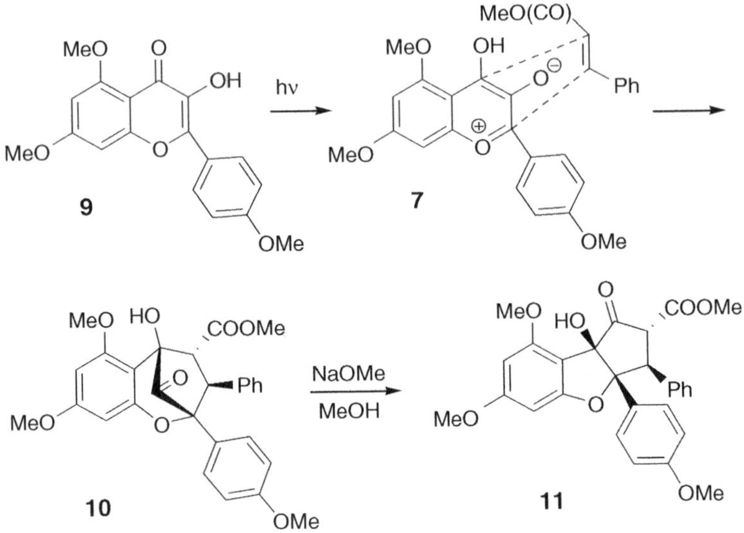

Scheme 14.14

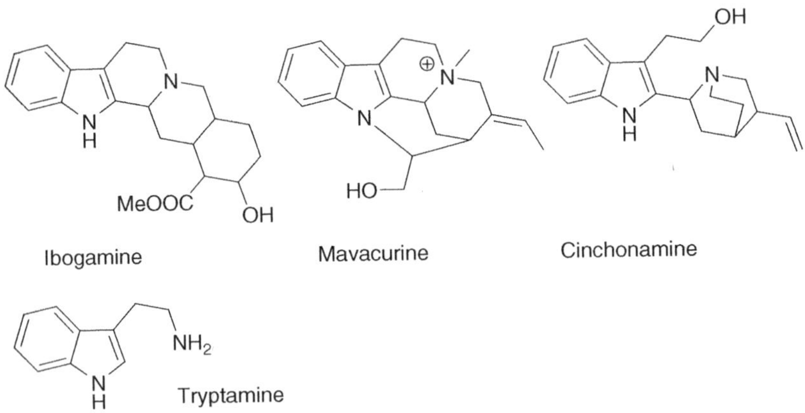

Scheme 14.15

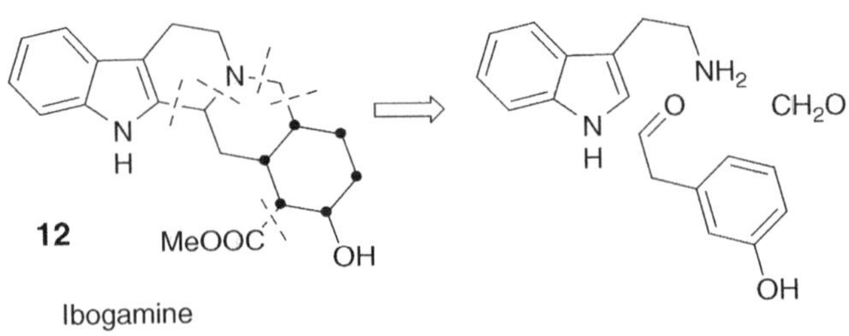

Scheme 14.16

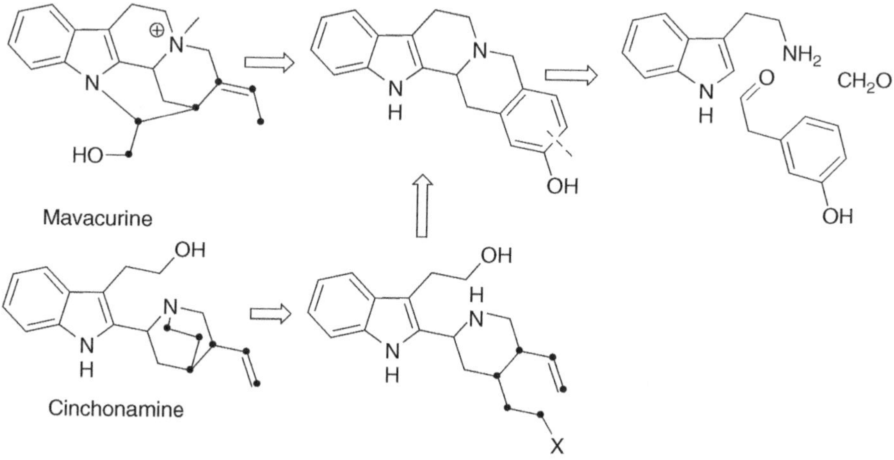

Scheme 14.17

remains a potentially biomimetic synthesis, as long as the in vivo proof of such a synthesis sequence has not been established. Regarding the latter, it has not yet been addressed whether an oxidopyrylium intermediate **7**, if it is generated from **9** in Nature, arises by irradiation with light or in a dark enzymatic process.

What is typical in the above example? The biogenetic hypothesis is derived by inspection of a set of compounds that co-occur in the same genus or family of organisms (plants). This suggests that they are not only biogenetically related, but that they may be members of a biochemical sequence of reactions or have a common ancestor structure. The size or restriction of this set of compounds obviously affects the appropriateness and quality of the conclusions to be drawn. The resulting biogenetic hypothesis is based only on structures that are known to occur in Nature and implies only mechanistically sound and well precedented chemical reactions. The merit of such a biogenetic hypothesis lies in the fact that it gives biochemists a lead—which labeled compounds should be tested in feeding experiments in order to establish the biosynthesis pathway.

Another focal point of biosynthesis consideration has been the biogenesis of the indole alkaloids. Comparative anatomy of even a very small set of representatives in Scheme 14.15 points to tryptamine as one possible common constituent and biochemical precursor.

Whereas the source of the remaining parts of these alkaloids remained in the dark, a proposal for the biosynthesis of ibogamine could be made on the grounds of well-precedented chemistry[34] (Scheme 14.16). This proposal stipulated the six marked skeletal atoms in **12** to be derived from an aromatic ring in a *m*-hydroxyphenylacetaldehyde. It was then Woodward[34] who recognized that also in mavacurine and cinchonamine the disposition of six skeletal atoms could be accounted for from a similar phenylacetaldehyde precursor provided one postulates a fission of the aromatic ring as shown in Scheme 14.17.

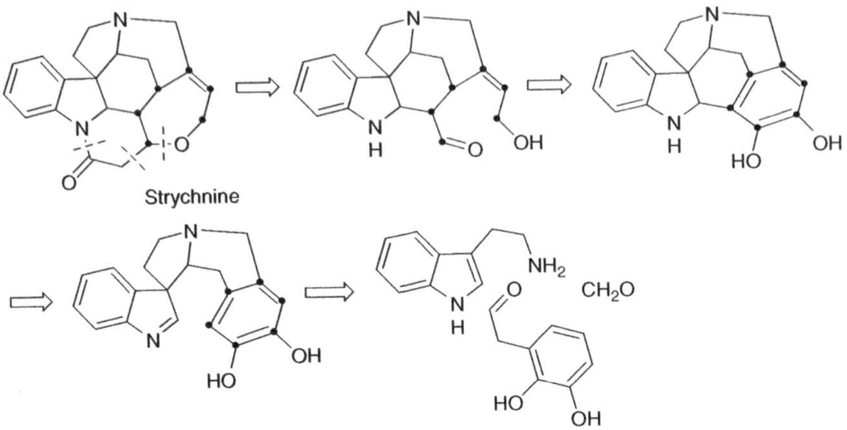

Scheme 14.18

Animated by these insights, Woodward[35] envisioned that even the much more complex indole alkaloid strychnine might have a similar biosynthetic background as shown in Scheme 14.18.

The oxidative fission of a phenol or catechol nucleus (the so-called Woodward fission) became thus a key part of the biosynthesis schemes postulated for the indole alkaloids. For Woodward it was beyond doubt that these alkaloids have been formed in Nature by this simple fundamental biogenetic route.[34] He thus embarked on a synthesis of strychnine along these lines with the oxidative fission of a veratrole ring as a key feature. Yet the fission of the veratrole ring was effected at another bond than that which is postulated to be cleaved in biosynthesis! Thus, Woodward's synthesis of strychnine shown in Scheme 14.19 is a bioinspired synthesis, the completion[36] of which marked a major achievement in natural product synthesis.

However, already during the execution of this synthesis, doubts as to the validity of the Woodward-fission hypothesis in the biosynthesis of indole alkaloids arose, mainly because of some stereochemical issues.[37] Hence, alternative precursors for the six marked skeletal atoms in the indole alkaloids were considered. Eventually, feeding studies with isotope-labeled compounds established that the precursor to that part of the indole alkaloids is the terpene-derived secologanin[38] (cf. Scheme 14.20).

At this point Woodward's biogenetic hypothesis of the fission of an aromatic ring was obsolete. Hence, his strychnine synthesis, which was initiated as a potentially biomimetic one, missed this attribute. It nevertheless remains what it is, a brilliant synthesis that was inspired by biosynthetic considerations.

When one is looking for classics in potentially biomimetic synthesis, one has to list the endiandric acids. For simplicity just one member, **14** (R = H, endiandric acid A), of this family is shown. Shortly after isolation and structure assignment of this tetracyclic compound,[39] a daring biosynthesis proposal was put forward[40] that

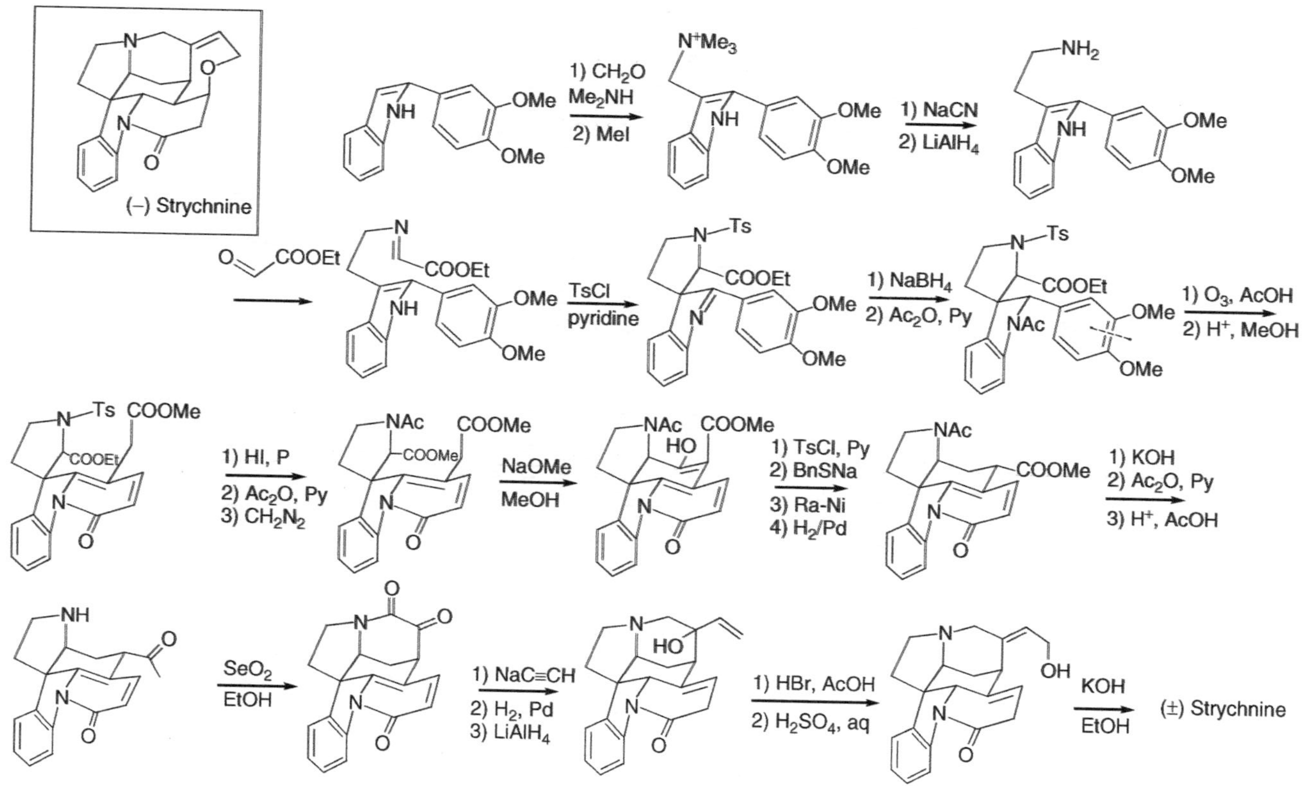

Figure 14.0 (With kind permission from R. W. Hoffmann, *Elements of Synthesis Planning*, Springer Science & Business Media, 2009, p.167)

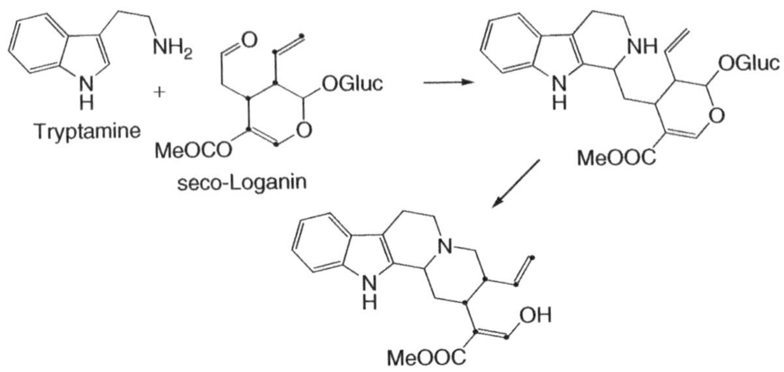

Scheme 14.20

takes into account the unusual phenomenon that endiandric acid and its congeners are found in racemic form in Nature. This proposal was based on a thorough understanding of the mechanisms and stereochemistry of electrocyclic and pericyclic reactions (cf. Scheme 14.21). This scheme needed experimental verification and it did not take long until the group of K. C. Nicolaou [41] showed that the open-chain precursor **13** (R = Me), on briefly heating to 100 °C, reacted smoothly to the hoped for tetracyclic product **14**. The biogenetic scheme allows for several parallel stereoisomeric pathways. Accordingly the in vitro synthesis provided not only endiandric acid A, but several of its constitutional and stereoisomers. Some of them have indeed been found in endiandra species. Others remain to be identified as natural products, whereby synthetic samples may be especially helpful. For other examples of classics in potentially biomimetic synthesis, see Bulger et al.[42]

When discussing the impact of potentially biomimetic synthesis, there is the classical study of Heathcock[43, 44] on the synthesis of the daphniphyllium alkaloids (cf. Scheme 14.22). Initially, Heathcock had completed an impressive synthesis of methyl homodaphniphyllate in a conventional synthetic manner. Being confronted

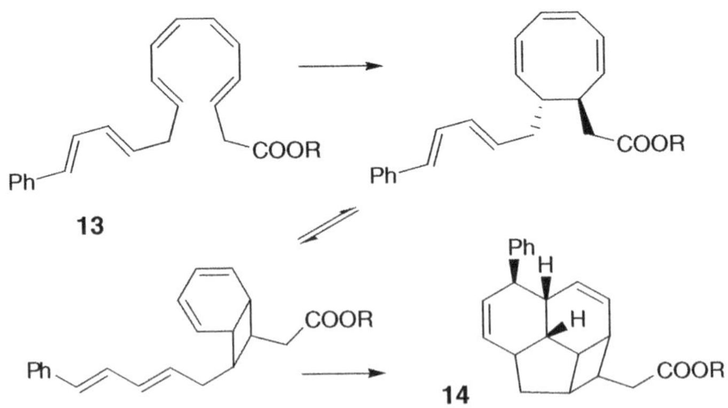

Scheme 14.21

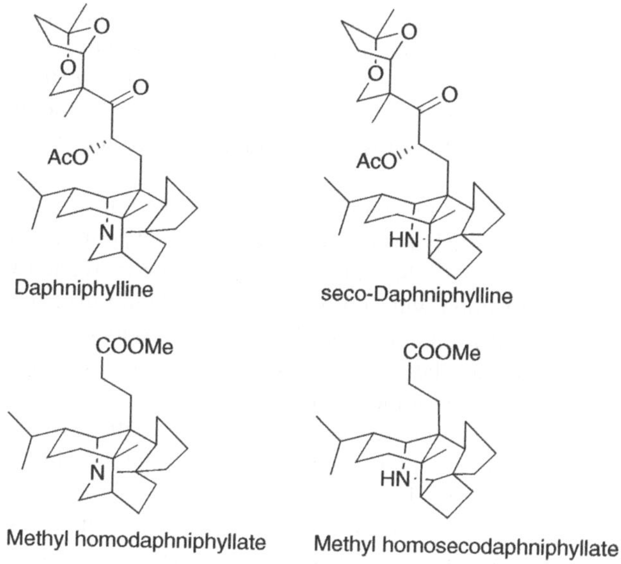

Scheme 14.22

with many difficulties during this synthesis, Heathcock started to wonder how Nature could arrive at such a diabolically intertwined pentacyclic skeleton. In due course, he realized that the carbon skeleton of secodaphniphylline contained an uninterrupted terpene chain of squalene, whereas such a correlation was not possible for daphniphylline. Heathcock concluded that the daphniphyllium alkaloids should

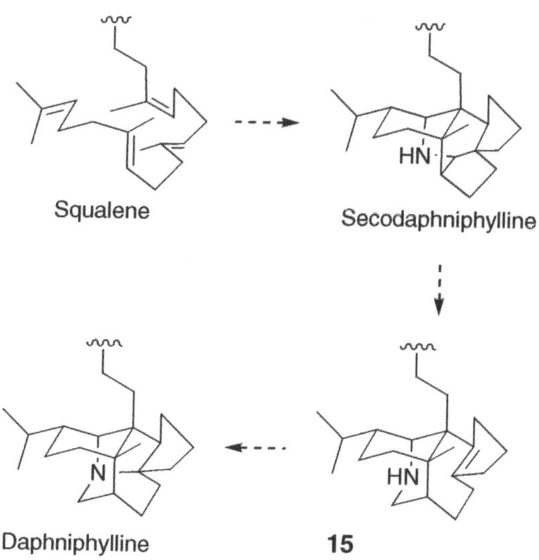

Scheme 14.23

be derived from squalene and that secodaphniphylline would be earlier on the reaction path, that is, a biochemical progenitor of daphniphylline. As a corollary, he speculated that there must be a reasonable chemical pathway (that Nature uses) to convert the secodaphniphylline skeleton into that of daphniphylline. He then envisioned the amine **15** as the biosynthetic link between the two skeletons (cf. Scheme 14.23).

But the true secret of this biosynthesis is the conversion of squalene into the secodaphniphylline skeleton. It took a lot of chemical experience to propose the sequence shown in Scheme 14.24. Squalene ought to be converted to the dialdehyde **16**, which upon reaction with a nitrogen source should give the enamine **17**. In the course of this transformation one double bond equivalent of the squalene skeleton should be reduced. The enamine is then converted over some precedented steps to the bicyclic dihydropyridine **18**. This dihydropyridine is perfectly set up to undergo an acid catalyzed intramolecular Diels–Alder addition to give **19**. An aza-Prins-reaction to **20** would complete the pentacyclic skeleton, which requires only reduction of the propenyl side chain to furnish the secodaphniphylline derivatives. Daring as it may be, such a hypothesis is subject to experimental test. After some model studies, Heathcock was able to demonstrate that dihydrosqualene-dialdehyde **21** cyclized by the action of ammonia and acetic acid to the pentacyclic amine **22**, which he dubbed protodaphniphylline (cf. Scheme 14.25). The ease by which six

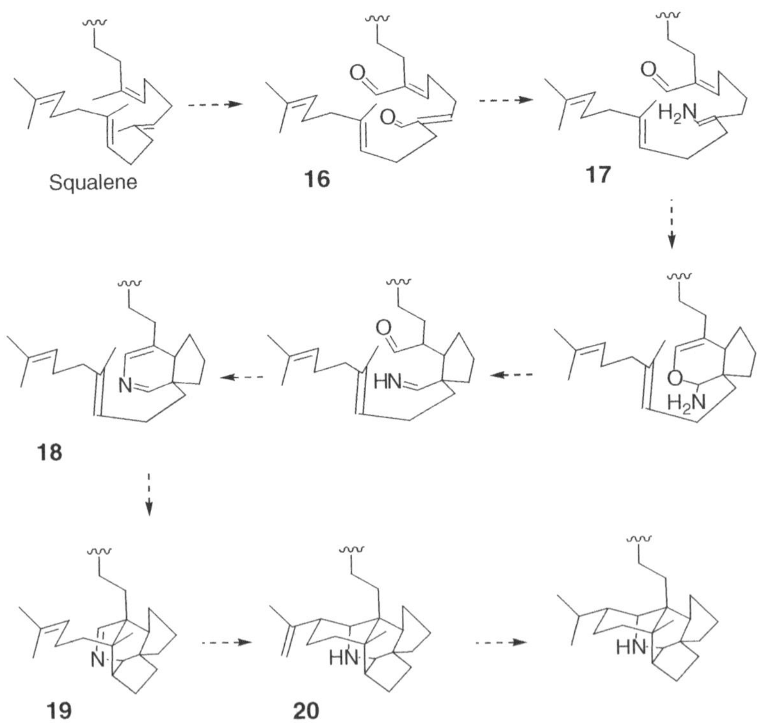

Scheme 14.24

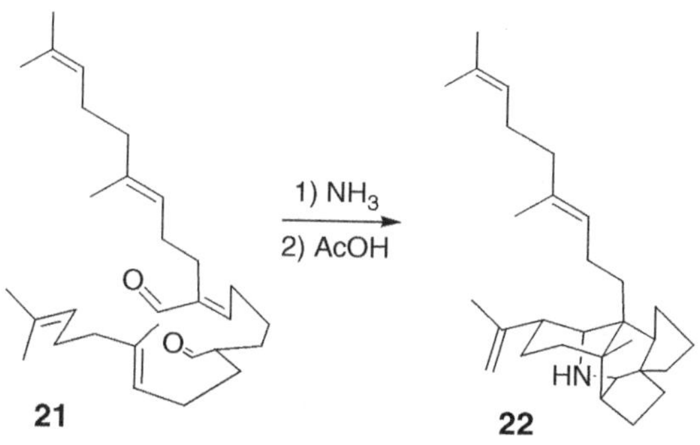

Scheme 14.25

sigma bonds and five rings were closed in one magnificent reaction casacade is truly suggestive that Nature uses this efficient pathway in the synthesis of the secodaphniphyllines. Yet one step of the postulated biogenesis of daphniphylline still lacked precedent, the rearrangement of the secodaphniphylline to the daphniphylline skeleton alluded to above. The following model study revealed that there is indeed a chemically reasonable way to go from the secodaphniphylline to the daphniphylline skeleton (cf. Scheme 14.26). Based on this information, Heathcock was able to complete a second, significantly more efficient synthesis of methyl homodaphniphyllate,[43, 44] a synthesis that in most of its steps is a potentially biomimetic synthesis.

This overall study is exemplary for the interplay between the search for simple reactions Nature could have and might have used in the biosynthesis of complex natural products, and their realization in the laboratory and utilization in efficient laboratory synthesis of such complex natural products. In the end, as far as organic synthesis is concerned, it does not matter whether the projected scheme is the actual pathway of biosynthesis or not. The consideration of potentially biosynthetic pathways is, by itself, what leads to the leap forward in synthesis strategy.

In the foregoing example one could follow how, by studies directed at the chemical behavior of a natural product, its potential biogenesis became progressively clearer until a truly convincing scheme remained, which culminated in a highly efficient in vitro synthesis. An even more penetrating example is given by a potentially biomimetic synthesis of agelastatin A[45] (cf. Scheme 14.27).

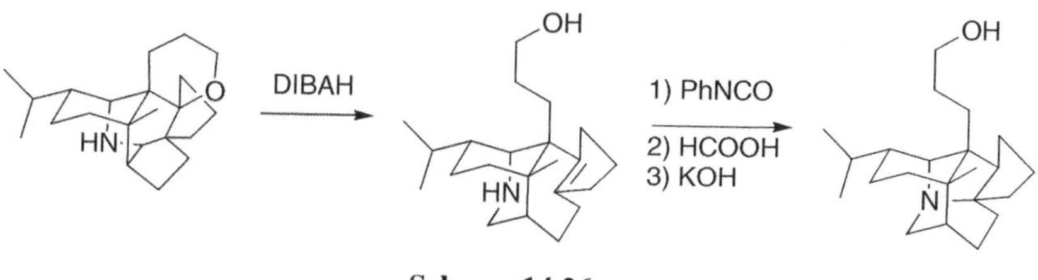

Scheme 14.26

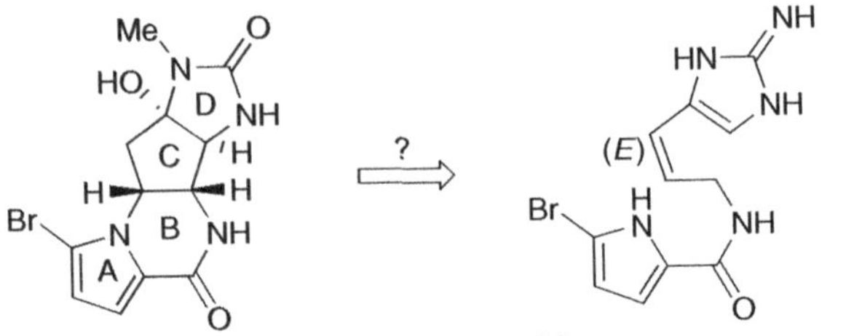

Agelastatin A

23 Debromo-oroidin

Scheme 14.27

Right after the isolation and characterization of agelastatin A,[46] it was speculated that this natural product could arise by cyclization and refunctionalization of an oroidine derivative **23**, occurring in the same family of marine organisms. This biotransformation was envisaged by Mourabit and Potier[47] to involve first a dehydrogenation of **23**, then closure of ring C to **24**, followed by closure of ring B and refunctionalization in ring D (cf. Scheme 14.28). In this scheme C-4 served as an electrophile in the closure of ring C. Movassaghi entertained a different sequence of events for the biosynthesis of agelastatin.[45] He saw the formation of ring B from **25** as the initial step followed by hydrolysis of the guanidine to a urea function. Oxidation of **26** at C-8 would generate an electrophile **27** that could combine with C-4 as a nucleophilic component to form ring C. Addition of water at C-5 would conclude the route to agelastatin A (cf. Scheme 14.29). Intermediate **27** is the hallmark of this biosynthesis pathway. Movassaghi synthesized the equivalent compound **28** by conventional means and demonstrated its ready

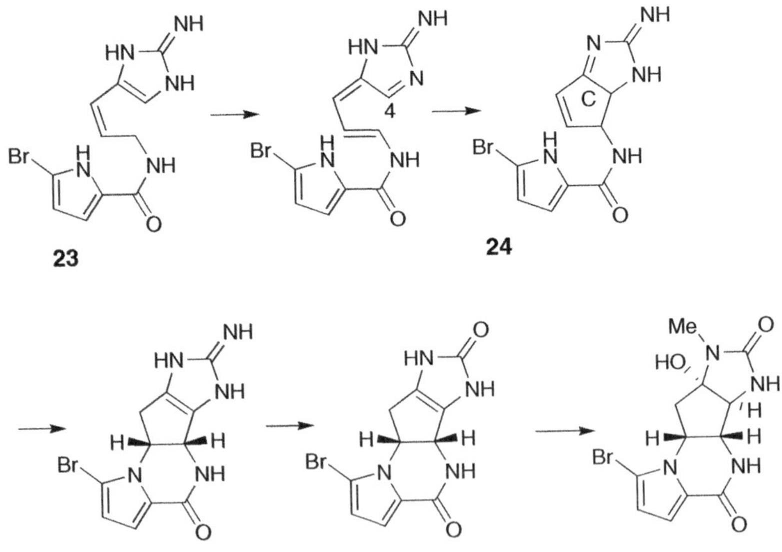

23

24

Scheme 14.28

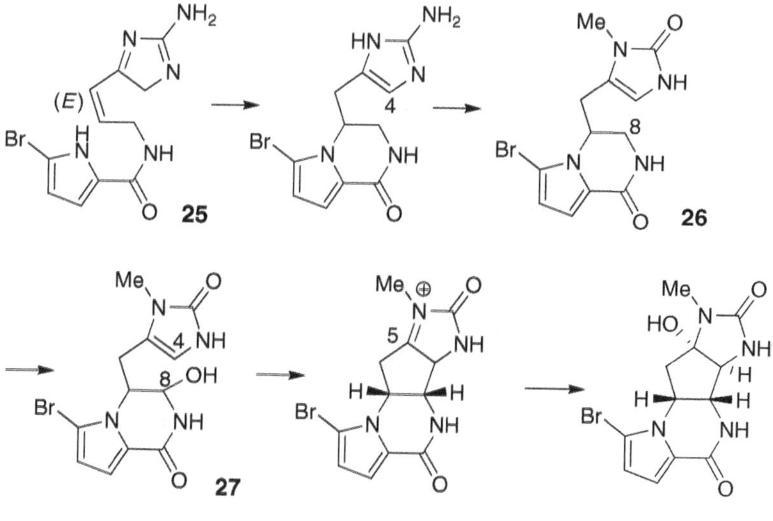

Scheme 14.29

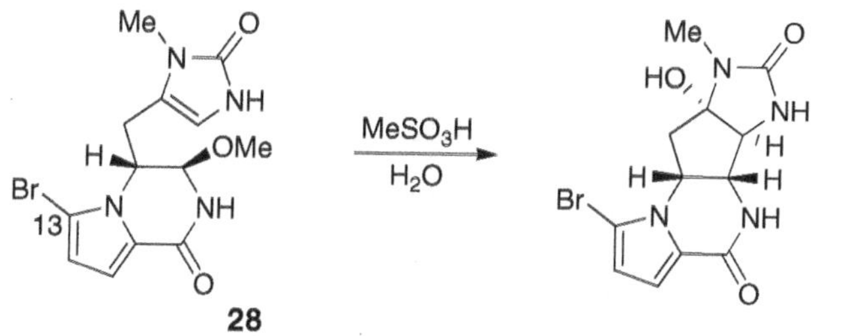

Scheme 14.30

cyclization to agelastatin A under acidic conditions as shown in Scheme 14.30. He thus concluded the so far shortest and most concise synthesis of agelastatin A. Movassaghi's studies moreover revealed important features of this process, for example, that the bromine atom in position 13 is essential for the access to and configurational stability of **28**, and that the guanidine function in **25** has to be converted to a urea before the cyclization of ring C may be effected. What this amounts to is that a deeper insight into the sequence of events in the potentially biosynthetic scheme has been gained, whereas in the more rudimentary biosynthesis proposals commonly encountered, there is only an indication of the bonds to be formed, but no informed guess as to the sequence of these steps.

14.3 BIOMIMETIC REACTIONS IN ORGANIC SYNTHESIS

So far considerations of literal or potential biomimesis in organic synthesis were concerned with the synthesis of an individual member of or a class of natural products. Biomimesis could relate, however, also to a type or class of reactions for which Nature has found an optimal way to carry them out. Breslow[48] referred to biomimetic chemistry as "imitating the style of enzyme catalyzed processes in an effort to achieve some of the advantages, which Nature has realized by the use of enzymes" regarding selectivity and high speed of reactions. This statement is an invitation to screen the active centers of enzymes and to see whether the functional entities present would act as catalysts by themselves without the intervention of a protein backbone. Obviously, matters are not that simple. For example, the manner in which class I aldolases affect the addition of ketones to aldehydes has been elucidated[49] (cf. Scheme 14.31). The key function is the primary amine of a lysine residue of the enzyme. Yet primary amines by themselves are poor catalysts—if at all—for an intramolecular aldol addition.[50] However, once the mechanism of the enzymatic aldol addition was understood, it was a small logical step to move to secondary amines and then to secondary amino acids as potential catalysts. In this way, it was discovered that proline is a superb catalyst for enantioselective intermolecular aldol additions[50] (Scheme 14.32).

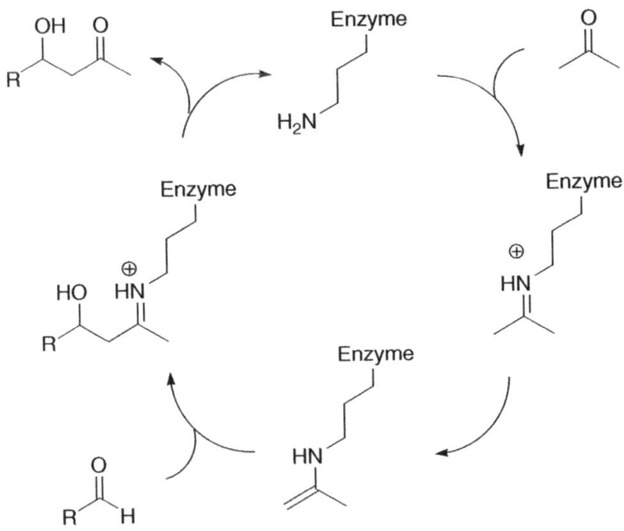

Scheme 14.31

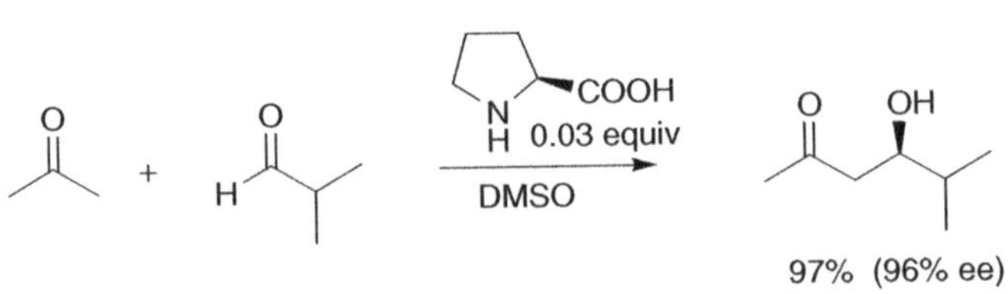

Scheme 14.32

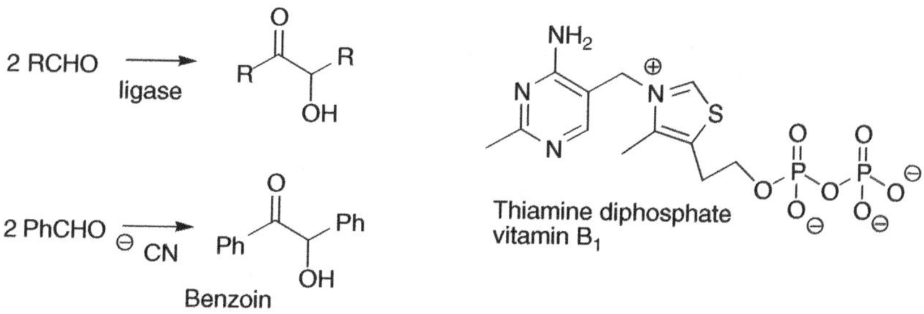

Scheme 14.33

This finding of a biomimetic catalysis marked the beginning of an exponential development of the field of "organocatalysis." Within a few years the results fill voluminous monographies.[51] Here is not the place to go deeper into these achievements. Rather, we will consider the biomimicry of other reactions, for which Nature has developed specific cofactors.

An example is given by the thiamine diphosphate (vitamin B_1) dependent enzymatic reactions. Typical among them are carboligase reactions that effect the formation of an α-ketol by ligation of two aldehydes, Scheme 14.33. This ligation corresponds to the venerable in vitro benzoin addition, a reaction known

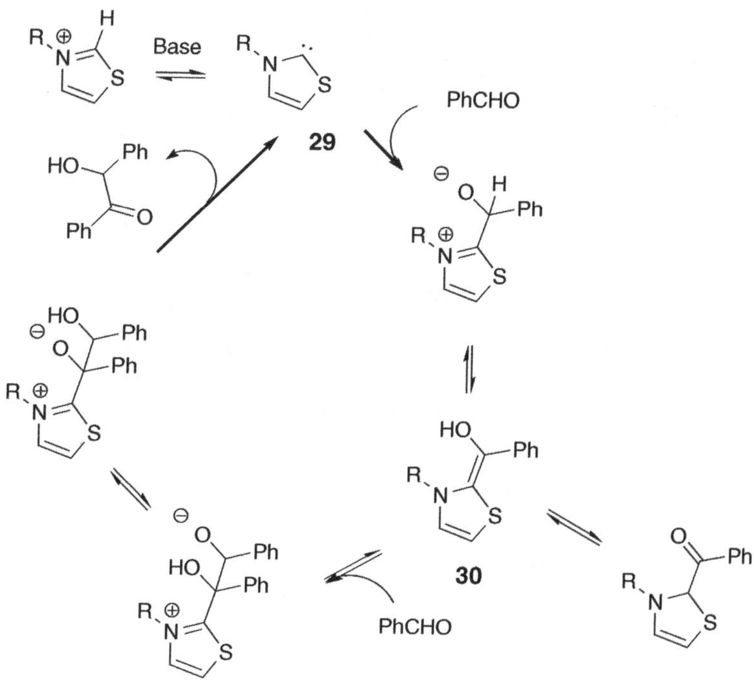

Scheme 14.34

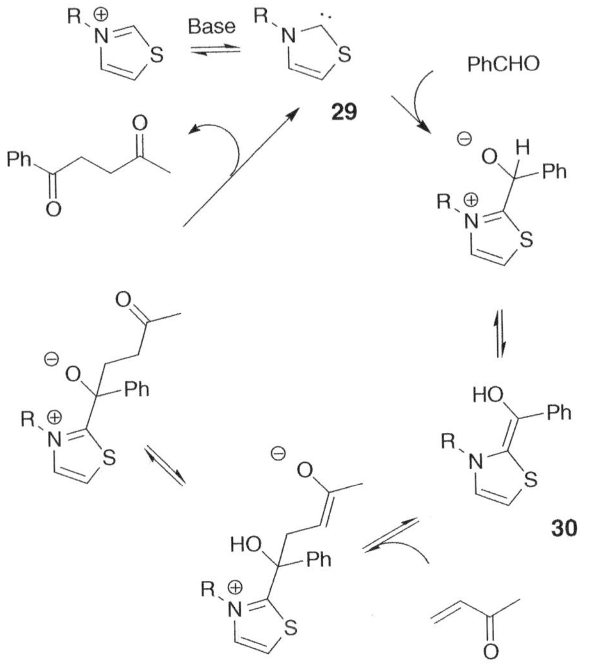

Scheme 14.35

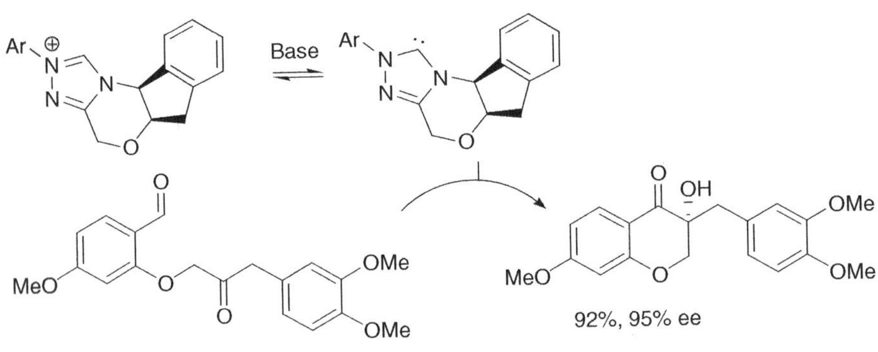

Scheme 14.36

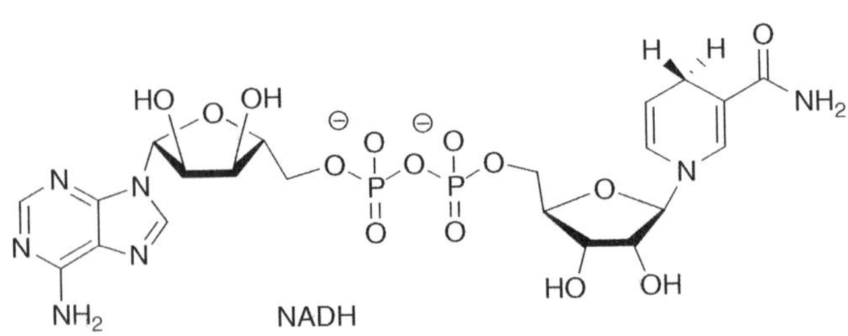

Scheme 14.37

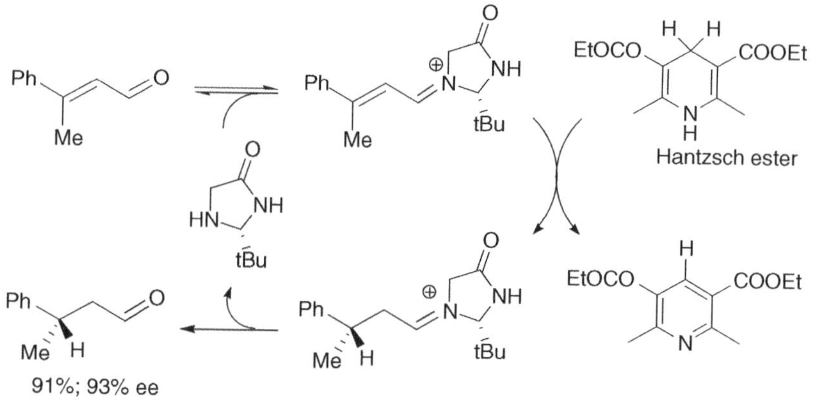

Scheme 14.38

for almost two centuries.[52] The similarity of these two reactions was underscored when it was found,[53] that vitamin B_1 catalyzes the benzoin condensation without any enzyme. Likewise, this reaction could be catalyzed by a variety of thiazolium salts, identifying this heterocyclic portion of vitamin B_1 as the warhead to effect the benzoin condensation. In due course, the mechanism of this reaction was recognized by Breslow (cf. Scheme 14.34); that is, the conjugate base of thiamine, the nucleophilic carbene **29**, is the actual catalytic species,[54] much like cyanide in the original Lapworth mechanism[55] of the benzoin addition.

With this insight it became clear that Nature uses a nucleophilic carbene **29** for the "umpolung" of an aldehyde. It took some time until chemists realized that this precedent is a valuable lead for the development of new synthetic methodology by imaginative interception of the intermediate **30** in the catalytic cycle. An early example is the interception of **30** by an α, β-unsaturated ketone, a process that led to the development of the Stetter reaction[56] (cf. Scheme 14.35).

A few years later followed an ever increasing wave of new reactions based on N-heterocyclic carbenes as organocatalysts,[57] as illustrated by the intramolecular enantioselective benzoin addition in Scheme 14.36.[58] The success story[59] of this now highly active research field goes clearly back to the biochemical precedent of Nature's thiamine catalysis.

Another cofactor in primary metabolism is NADH, Scheme 14.37, which was early recognized as Nature's reducing agent, corresponding to $NaBH_4$ in synthetic organic chemistry. It is the dihydropyridine moiety of NADH that was identified as being the hydride source of the cofactor. Hence, once the constitution and structure of NADH had been established,[60] chemists were in a position to take advantage of the reactivity pattern akin to NADH's reactivity by using dihydropyridines in general as NADH mimetics.[61] However, having $NaBH_4$ at their disposal, there was no pressure on organic chemists to develop NADH mimetics, until more chemoselective reducing agents than $NaBH_4$ were required. This was the case on reduction of acidic immonium salts, where the reducing agent had to be stable

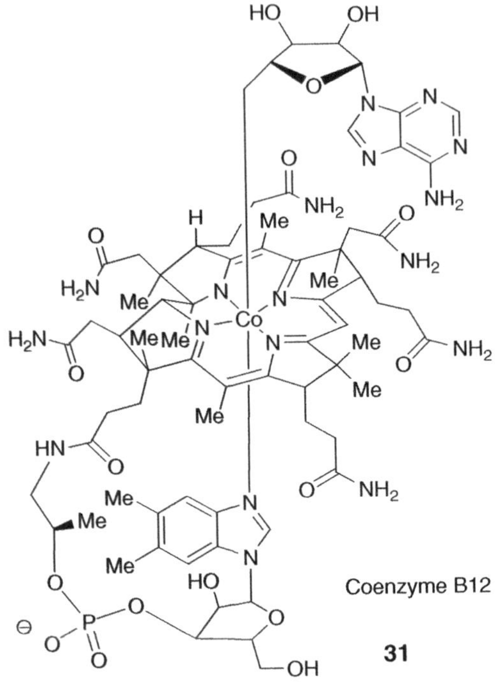

Coenzyme B12

31

Scheme 14.39

under the acidic conditions[62] on the one side, but yet reactive enough to effect the reduction. The best balance in reducing power[63] was met by the Hantzsch ester, which allowed, for example, for an efficient enantioselective organocatalytic reduction of α, β-unsaturated aldehydes and ketones[64, 65] shown in Scheme 14.38.

These studies were guided by the aim to "use the conceptual blueprint of biochemical reduction, wherein enzymes and NADH are replaced by small molecule amine catalysts and Hantzsch ester pyridines"[65] and triggered off a massive series of efforts to use Hantzsch esters in biomimetic reductions of a variety of unsaturated nitrogen compounds.[66]

Among the cofactors that Nature uses is coenzyme B_{12} (**31**), which is a key factor to carry out rather unusual transformations that have little precedent in organic chemistry.[67] These reactions are mainly skeletal rearrangements, in which two neighboring groups swap their places. Accordingly, the structure and the mechanism of reactions of coenzyme B_{12} met with great attention. The key feature is an organometallic cobalt–carbon bond in **31**, Scheme 14.39.

The reactions of coenzyme B_{12} can be summarized as shown in Scheme 14.40. The reaction sequence is initiated by cobalt–carbon bond homolysis in **31** to give an adenosyl radical **32**. The adenosyl radical may abstract a hydrogen atom from a substrate R–H giving adenosine (**33**). The resulting substrate radical R• will rearrange to a product radical R′•, which abstracts a hydrogen atom from adenosine (**33**) to give the product R′H and to regenerate the coenzyme. Wherever radicals are

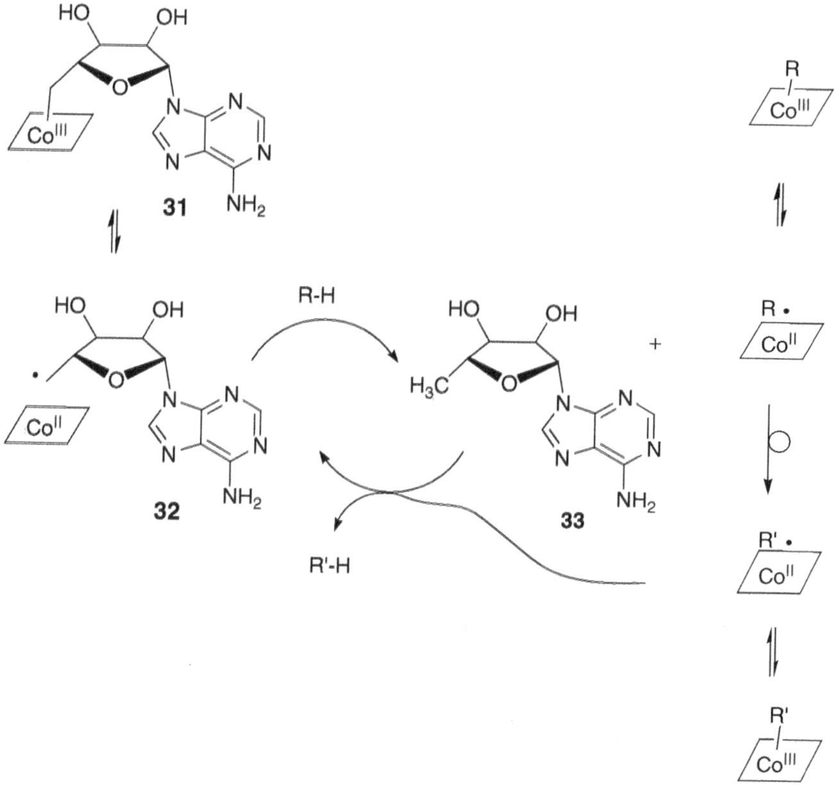

Scheme 14.40

formulated in this scheme, they arise in the proximity of cobalt^{II} and, hence, are in equilibrium with the corresponding alkyl-cobalt^{III} species due to the persistent radical effect.[68] Mechanistically, it is not yet fully settled whether the rearrangements induced by coenzyme B_{12} involve free radicals—as formulated above—or the corresponding alkyl-cobalt^{III} species.[69] But from a more distant point of view the situation may be summarized as follows:[70] "coenzyme B_{12} fulfills its biochemical role by serving as a *free radical reservoir* from which 5′-deoxyadenosyl radicals (**32**) are reversibly released under mild conditions." In the eyes of an organic chemist this statement points to a mild way to generate free radicals. Accordingly, coenzyme B_{12} model compounds have been studied with the aim to generate and utilize free radicals in a biomimetic fashion.[71, 72] In the following example a ring closure is attained by cyclization of a radical, which was generated from an alkyl-cobalt^{III} species **34**. The direct product is again an alkyl-cobalt^{III} species **36**. This was ultimately refunctionalized in a subsequent oxidation step to give the alcohol **35**,[71] as shown in Scheme 14.41.

All in all, these and other examples of coenzyme B_{12} mimetic reactions show that Nature's use of organometallic reagents can readily be adapted into preparative organic chemistry.

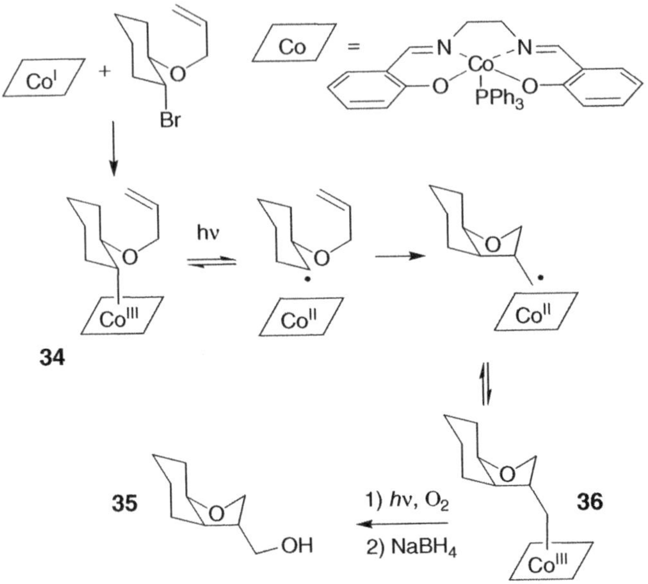

Scheme 14.41

A very favorable constellation for the development of biomimetic reactions arises when Nature handles a certain type of transformation perfectly, for which chemists have no way, or no good way, to achieve. This holds, for instance, for the oxyfunctionalization of unactivated C–H bonds. Nature does this with the cytochrome P450 enzymes,[73] which all have a protoporphyrin IX iron complex **37** as a cofactor,[74] Scheme 14.42.

This iron(III) species is initially oxidized to an oxo-iron(IV) intermediate **38**, which is the actual agent that oxidizes C–C double bonds, or inserts into carbon–hydrogen bonds. In mimicking these reactions, simple iron–porphyrin complexes such as **39** have been utilized as catalysts in oxidation reactions,[75] as shown in Scheme 14.43.

Yet from a practical point of view, this system still had too many deficiencies. It became necessary to deviate further from Nature's precedent in turning

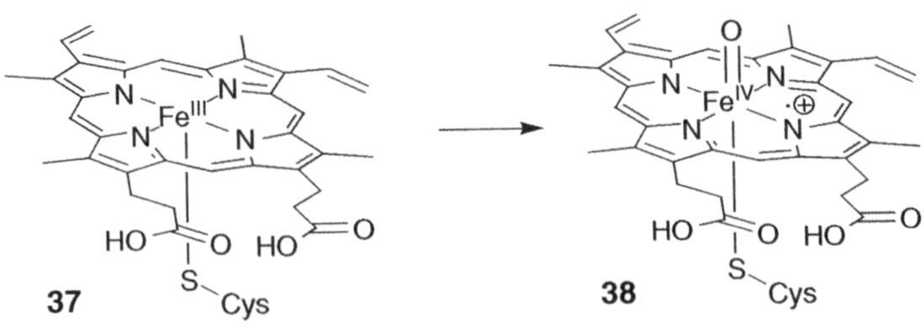

Scheme 14.42

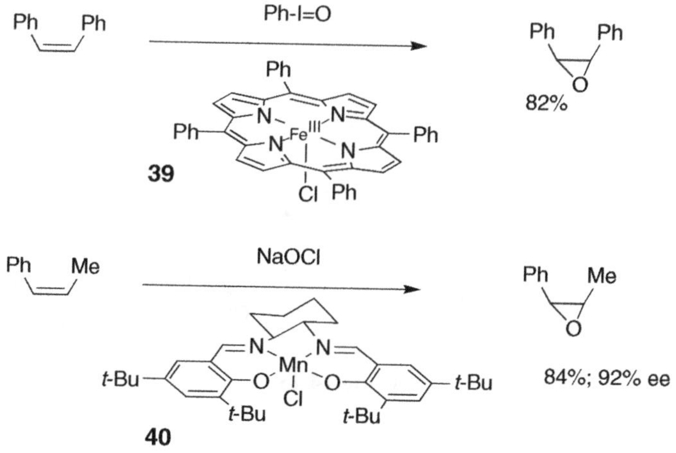

Scheme 14.43

to analogous manganese complexes **40** to reach a generally applicable reaction scheme for enantioselective olefin epoxidation.[76]

However, the real challenge is not epoxidation of alkenes, rather it is the position specific hydroxylation of unactivated CH_2 groups in complex molecules. Nature

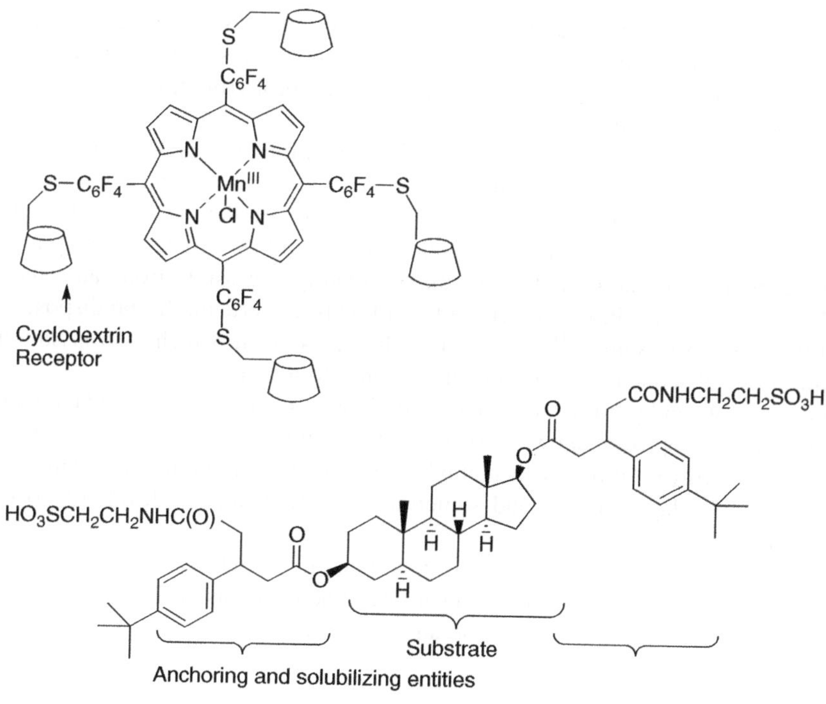

Scheme 14.44

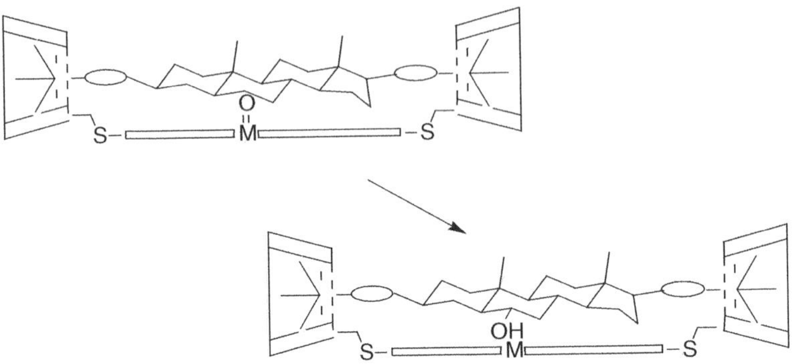

Scheme 14.45

is perfect in attaining such transformations with her cytochrome P450 enzymes,[73] inviting chemists to copy this precedent. This implies two features to be mimicked: first, a porphyrin Fe(III) or Mn(III) system to reproduce the action of protoporphyrin IX (**37**); and second, a molecular vice to hold the substrate in place relative to the metal porphyrin oxidant, something the protein part of the cytochrome P450 enzyme does in vivo. Both features have been attained by Breslow and co-workers in the system[77] shown in Scheme 14.44.

To the manganese porphyrin catalyst were attached four cyclodextrin receptor groups, into which nonpolar hydrocarbon residues such as *t*-butyl groups would bind. Two such nonpolar anchor groups were attached to a steroid substrate. These nonpolar groups should fix the substrate in the porphyrin device as indicated schematically in Scheme 14.45.

With 1 mole percent of the catalyst upon addition of stoichiometric PhI=O oxidant, the substrate was quantitatively hydroxylated in the indicated position. While thus a truly biomimetic hydroxylation could be demonstrated, the effort involved renders this approach impractical as a synthesis route. More practical approaches to biomimetic hydroxylation use readily accessible iron catalysts and apply it on substrates, in which certain C–H bonds are structurally predisposed for hydroxylation. An example[78] is given in Scheme 14.46, in which hydroxylation is followed by subsequent oxidation of the alcohol to a ketone.

However, when hydroxylations and oxidations of C–H bonds are to be attained, the biomimetic approach shown above is not the only option. Nonbiomimetic oxidations and hydroxylations with methyl(trifluoromethyl)dioxirane are really competitive and display similar—and sometimes complementary—levels of position selectivity.[79]

These examples—all referring to reactions in which Nature uses specific cofactors—show that biomimicry regarding reactions has truly stimulated and enriched synthetic methodology in organic chemistry.

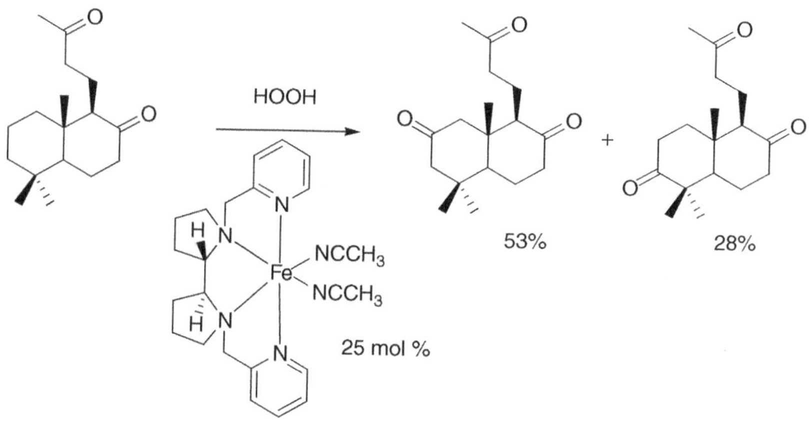

Scheme 14.46

14.4 BIOMIMETIC CONSIDERATIONS AS AN AID IN STRUCTURAL ASSIGNMENT

Considerations about biosynthesis pathways had still another impact in the first half of the last century. This was the time when the structures of natural products were arrived at solely by degradation studies without recourse to present-day spectroscopic methods. Structure assignments then were, by necessity, tentative and still are today.[80] In the same period, considerations about biosynthetic pathways were flourishing[9] and by the 1950s the conviction that these speculations must be correct had become a dominant trend in organic chemistry. For instance, with regard to alkaloids, Woodward wrote in 1956: [34] "We may go at present so far as

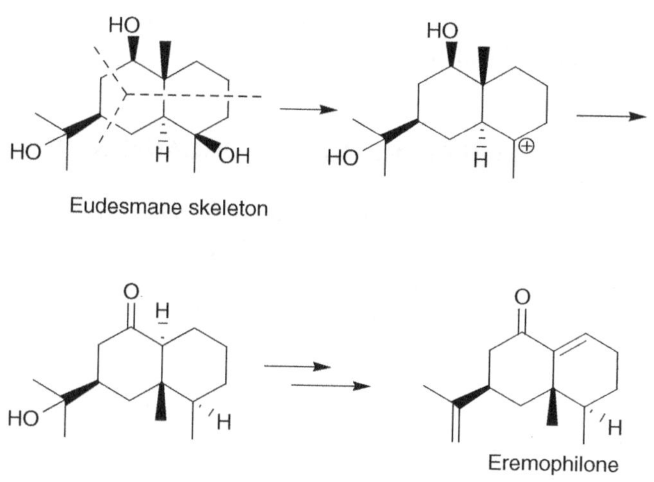

Scheme 14.47

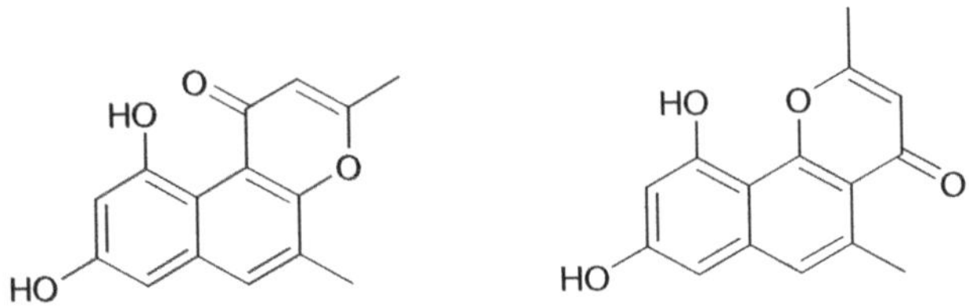

Eleutherinol
proposed structure 1952

Eleutherinol
corrected structure 1953

Scheme 14.48

to state that any structural assignment of an alkaloid that is not compatible with the biogenesis principles must be considered as dubious" (translation by the current author). In the terpene field, structural assignments that did not comply with the isoprene rule could—correctly or not—be reconciled with reference to skeletal rearrangements that occur during the cationic polycylization cascade.[9] For instance, whereas the eudesmane skeleton corresponds to the empirical isoprene rule,[26] that of eremophilone and of the related α-vetivone or nootkatone does not. However, it complies with the biogenetic isoprene rule.[81] Accordingly, the proposed biogenesis scheme of the latter (Scheme 14.47) was assumed to feature a deep reaching molecular rearrangement.[82]

Biogenetic schemes for polyacetate derived natural products are much more rigid and do not allow for exceptions. Hence, when the structure of eleutherinol was published,[83] it was evident that this did not correspond to the usual pattern of acetate building blocks being connected head to tail.[84] Either eleutherinol did not belong to the class of polyacetate derived natural products as do its congeners,[85] or its structure needed revision (cf. Scheme 14.48).

That situation was reason enough to synthesize a key degradation product of eleutherinol and thereby establish its alternate revised structure.[84, 86] This demonstrates that at times consideration of biosynthetic pathways became an important aspect in the structural elucidation of natural products.

14.5 REFLECTIONS ON BIOMIMICRY IN ORGANIC SYNTHESIS

For most of the natural products of interest to organic synthesis, biosynthesis studies have not been carried out yet. Therefore, most considerations of the biosynthesis pathway for a natural product are speculations. These speculations are based on a limited set of structural features common to a family of related and often co-occurring natural products, and these speculations build on well-established reactivity patterns and reaction types of classes of organic compounds rather than on specific intermediates and specific reactions. At this stage of biogenesis speculations, these details are not yet necessary, because Nature may use enzymes to facilitate and speed up specific reactions to occur under physiological conditions, whereas the chemist would have to make recourse to reagents and reaction conditions not available to Nature. However, in these speculations one should keep in mind that Nature's processes in the living cell are bound to the invariable laws of chemistry and physics. Enzymes can facilitate only those processes that have a thermodynamic driving force, whereas they cannot render impossible reactions viable! When this is kept in mind, qualified speculations about a biosynthesis pathway may be arrived at, speculations that merit becoming a blueprint for an actual synthesis in the lab.

When the resulting potentially biomimetic synthesis turns out to be more efficient than previous synthesis attempts, one major goal of biomimicry has been realized: an increase in synthesis efficiency.[8] Ironically, this result does not depend in any way on the correctness of the speculation. For organic synthesis, it does not matter at this point whether Nature uses the contrived biogenetic path or not! Nevertheless, thoughts about biogenesis are just *one* of the obligatory considerations in retrosynthesis planning. Therefore, such biogenetic speculations should not prevent chemists to look as well at other synthesis routes not precedented by Nature!

Nature, on the one side, is quite apt in using chemically meaningful reactions, such as the Diels–Alder addition, 1,3-dipolar cycloadditions, or electrocyclic processes, reactions that proceed without the need of enzymes. But on the other side, Nature is restricted to a limited set of starting materials. That means that chemists who follow a biomimicry paradigm are using a restricted set of synthesis options, much like a vegetarian, who uses a restricted set of food options. Both are hoping for an imaginary benefit by doing so. However, a chemist in doing synthesis is not a priori subject to such a restriction; rather he/she has many more options, both regarding starting materials and reactions, such as transition metal catalyzed coupling reactions, to build up a molecular skeleton. Therefore, the following statement of Heathcock[43] can serve as a beacon for future development: "Although our approaches to problems have matured, we need even more mature strategies of synthesis. There is no reason that organic chemists should not be able to surpass Nature's virtuosity in the synthesis of complex organic structures."

Even if we have to admit that biomimetic synthesis has no merit by itself, its execution has quite a multifarious impact on organic chemistry and on biochemistry. In practically all potentially biomimetic syntheses that have been carried out, it is the *concept* of a potentially biogenetic pathway that has an impact on the planning and strategy of the synthesis, with the consequence that "biogenetic type syntheses are often neater, shorter and more efficient than normal routes in which no attention is paid to natural processes".[8] The impact on biosynthesis studies seems obvious (but is not so much taken advantage of), namely, that the concepts of a biogenetic pathway can be subjected to experimental test in vitro and thus provide strong hints as to which compounds should be tested in vivo as precursors and intermediates in a biosynthetic pathway.[47] The longest lasting impact of biomimetic synthesis on the chemical sciences is the information on the inherent reactivity of postulated and tested intermediates in a biogenetic pathway.[87] Chemists recognize in this manner that even highly complex reactions of these intermediates proceed almost spontaneously,[43] to the point that enzyme intervention in the living cell may at best just increase selectivity and reaction rate. It is then left to the synthetic chemist to take advantage of such inherent reactivities in order to pattern a highly efficient synthetic route based on this information.[45] At this point the cycle closes, that biomimesis is one—but not the only one—route to increase the efficiency of natural product synthesis.

REFERENCES

1. (a) Armstrong, R. W.; Beau, J.-M.; Cheon, S. H.; Christ, W. J.; Fujioka, H.; Ham, W.-H.; Hawkins, L. D.; Jin, H.; Kang, S. H.; Kishi, Y.; Martinelli, M. J.; McWhorther Jr., W. W. ; Mizuno, M.; Nakata, M.; Stutz, A. E.; Talamas, F. X.; Taniguchi, M.; Tino, J. A.; Ueda, K.; Uenishi, J.; White, J. B.; Yonaga, M. *J. Am. Chem. Soc*. **1989**, *111*, 7530–7533; (b) Suh, E. M.; Kishi, Y. *J. Am. Chem. Soc*. **1994**, *116*, 11205–11206.
2. Johnson, W. S.; Semmelhack, M. F.; Sultanbawa, M. U. S.; Dolak, L. A. *J. Am. Chem. Soc*. **1968**, *90*, 2994–2996.
3. Clayton, R. B. *Q. Rev. Chem. Soc. London* **1965**, *19*, 168–230.
4. Gaylor, J. L. *Biochem. Biophys. Res. Commun*. **2002**, *292*, 1139–1146.
5. Mann, J. *Chem. World* **2010**, 56–60.
6. Mannich, C. *Mitteil. Dtsch. Pharmaz. Ges*. **1928**, *5*, 93–96.
7. Schöpf, C.; Lehmann, G. *Liebigs Ann. Chem*. **1935**, *518*, 1–37.
8. van Tamelen, E. E. *Fortschr. Chem. Org. Naturstoffe* **1961**, *19*, 242–290.
9. Robinson, R. *The Structural Relations of Natural Products*, Oxford University Press, Oxford, UK, 1955.
10. Scholz, U.; Winterfeldt, E. *Nat. Prod. Rep*. **2000**, *17*, 349–366.
11. Robinson, R. *J. Chem. Soc. London* **1917**, *111*, 762–768.
12. Birch, A. J. *Notes Rec. R. Soc. London* **1993**, *47*, 277–296.
13. Robinson, R. *J. Chem. Soc. London* **1917**, *111*, 876–899.
14. Henscheidt, T. *Top. Curr. Chem*. **2000**, *209*, 175–206.
15. Eschenmoser, A. *Helv. Chim. Acta* **2010**, *93*, 1439–1499; see Abb. 24.
16. Curd, F. H.; Robertson, A. *J. Chem. Soc. London* **1937**, 894–902.
17. Schöpf, C.; Roß, F. *Liebigs Ann. Chem*. **1941**, *546*, 1–40.
18. Barton, D. H. R.; Deflorin, A. M.; Edwards, O. E. *J. Chem. Soc. London* **1956**, 530–534.
19. Pummerer, R.; Puttfarcken, H.; Schopflocher P. *Ber. Dtsch. Chem. Ges*. **1925**, *58*, 1808–1820.
20. Barton, D. H. R.; Cohen, T. *Festschrift Arthur Stoll*, Birkhäuser, Basel, 1957, pp. 117–143.
21. Barton, D. H. R.; Kirby, G. W. *Proc. Chem. Soc. London* **1960**, 392–393.
22. (a) Node, M.; Kodama, S.; Hamashima, Y.; Baba, T.; Hamamichi, N.; Nishide, K. *Angew. Chem*. **2001**, *113*, 3150–3152; *Angew. Chem. Int. Ed*. **2001**, *40*, 3060–3062; (b) de la Torre, M. C.; Sierra, M. A. *Angew. Chem*. **2004**, *116*, 162–184; *Angew. Chem. Int. Ed*. **2004**, *43*, 160–181.
23. Hall, J. A. *Chem. Rev*. **1937**, *20*, 305–344.

24. Birch, A. J. *Interdisciplinary Sci. Rev*. **1976**, *1*, 215–233.
25. Schöpf, C. *Liebigs Ann. Chem*. **1932**, *497*, 1–6.
26. Ruzicka, L. *Experientia* **1953**, *9*, 357–367.
27. Lynen, F. *Angew. Chem*. **1965**, *77*, 929–944.
28. Singh, I. P.; Bharate, S. B. *Nat. Prod. Rep*. **2006**, *23*, 558–591.
29. George, J. H.; Hesse, M. D.; Baldwin, J. E.; Adlington, R. M. *Org. Lett*. **2010**, *12*, 3532–3535.
30. Gerard, B.; Jones II, G.; Porco Jr., J. A. *J. Am. Chem. Soc*. **2004**, *126*, 13620–13621.
31. Dumontet, V.; Thoison, O.; Omobuwajo, R.; Martin, M.-T.; Perromat, G.; Chiaroni, A.; Riche, C.; Pais, M.; Sévenet, T.; Hadi, A. H. A. *Tetrahedron* **1996**, *52*, 6931–6942.
32. Bacher, M.; Hofer, O.; Brader, G.; Vajrodaya, S.; Greger, H. *Phytochemistry* **1999**, *52*, 253–263.
33. Proksch, P.; Edrada, R. A.; Ebel, R.; Bohnenstengel, F. I.; Nugroho, B. W. *Curr. Org. Chem*. **2001**, *5*, 923–938.
34. Woodward, R. B. *Angew. Chem*. **1956**, *68*, 13–20.
35. Woodward, R. B. *Nature* **1948**, *162*, 155–156.
36. Woodward, R. B.; Cava, M. P.; Ollis, W. D.; Hunger, A.; Daeniker, H. U.; Schenker, K. *Tetrahedron* **1963**, *19*, 247–288.
37. Wenkert, E.; Bringi, N. V. *J. Am. Chem. Soc*. **1959**, *81*, 1474–1481.
38. Battersby, A. R. *Pure Appl. Chem*. **1967**, *14*, 117–136.
39. Bandaranayake, W. M.; Banfield, J. E.; Black, D. S. C.; Fallon, G. D.; Gatehouse, B. M. *J. Chem. Soc. Chem. Commun*. **1980**, 162–163.
40. Bandaranayake, W. M.; Banfield, J. E.; Black, D. S. C. *J. Chem. Soc. Chem. Commun*. **1980**, 902–903.
41. Nicolaou, K. C.; Petasis, N. A.; Zipkin, R. E. *J. Am. Chem. Soc*. **1982**, *104*, 5560–5562.
42. Bulger, P. G.; Bagal, S. K.; Marquez, R. *Nat. Prod. Rep*. **2008**, *25*, 254–297.
43. Heathcock, C. H. *Angew. Chem*. **1992**, *104*, 675–691; *Angew. Chem. Int. Ed. Engl*. **1992**, *31*, 665–681.
44. Heathcock, C. H. *Proc. Natl. Acad. Sci. U.S.A*. **1996**, *93*, 14323–14327.
45. Movassaghi, M.; Siegel, D. S.; Han, S. *Chem. Sci*. **2010**, 561–566.
46. D'Ambrosio, M.; Guerriero, A.; Debitus, C.; Ribes, O; Pusset, J.; Leroi, S.; Pietra, F. *J. Chem. Soc. Chem. Commun*. **1993**, 1305–1306.
47. Mourabit, A. A.; Potier, P. *Eur. J. Org. Chem*. **2001**, 237–243.
48. Breslow, R. *Chem. Soc. Rev*. **1972**, *1*, 553–580.
49. Rutter, W. J. *Fed. Proc*. **1964**, *23*, 1248–1257.
50. List, B.; Lerner, R. A.; Barbas III, C. F. *J. Am. Chem. Soc*. **2000**, *122*, 2395–2396.
51. (a) List, B. *Top. Curr. Chem*. **2010**, *291*, 1–460; (b) Berkessel, A.; Gröger, H. *Asymmetric Organocatalysis*, Wiley-VCH, Weinheim, Germany, 2005, pp. 1–440.
52. Wöhler, F.; Liebig, J. *Ann. Pharm*. **1832**, *3*, 249–282; see p. 276.

53. (a) Ukai, T.; Tanaka, R.; Dokawa, T. *J. Pharm. Chem. Soc. Jpn*. **1943**, *63*, 296–300; *Chem. Abstr*. **1951**, *45*, 5148; (b) Mizuhara, S.; Handler, P. *J. Am. Chem. Soc*. **1954**, *76*, 571–573.

54. (a) Breslow, R. *J. Am. Chem. Soc*. **1958**, *80*, 3719–3726; (b) Breslow, R.; Schmuck, C. *Tetrahedron Lett*. **1996**, *37*, 8241–8242; (c) Berkessel, A.; Elfert, S.; Etzenbach-Effers, K.; Teles, J. H. *Angew. Chem*. **2010**, *122*, 7275–7279; *Angew. Chem. Int. Ed*. **2010**, *49*, 7120–7124.

55. Lapworth, A. J. *J. Chem. Soc. London* **1903**, 995–1005.

56. Stetter, H. *Angew. Chem*. **1976**, *88*, 695–704; *Angew. Chem. Int. Ed. Engl*. **1976**, *15*, 639–647.

57. Enders, D.; Balensiefer, T. *Acc. Chem. Res*. **2004**, *37*, 534–541.

58. Takikawa, H.; Suzuki, K. *Org. Lett*. **2007**, *9*, 2713–2716.

59. (a) *N-Heterocyclic Carbenes in Synthesis* (Ed. Nolan, S. P.), Wiley-VCH, Weinheim, Germany, 2006; (b) Enders, D.; Niemeier, O.; Henseler, A. *Chem. Rev*. **2007**, *107*, 5606–5655.

60. (a) Schlenk, F. *J. Biol. Chem*. **1942**, *146*, 619–625; (b) Pullman, M. E.; San Pietro, A.; Colowick, S. P. *J. Biol. Chem*. **1954**, *206*, 129–141.

61. (a) Pandit, U. K.; Mas Cabré, F. R. *J. Chem. Soc. Chem. Commun*. **1971**, 552; (b) Stout, D. M.; Meyers, A. I. *Chem. Rev*. **1982**, *82*, 223–243.

62. Itoh, T.; Nagata, K; Miyazaki, M.; Ishikawa, H.; Kurihara, A.; Ohsawa, A. *Tetrahedron* **2004**, *60*, 6649–6655.

63. Richter, D.; Mayr, H. *Angew. Chem*. **2009**, *121*, 1992–1995; *Angew. Chem. Int. Ed*. **2009**, *48*, 1958–1961.

64. (a) Pandit, U. K.; Mas Cabré, F. R.; Gase, R. A.; de Nie-Sarink, M. J. *J. Chem. Soc. Chem. Commun*. **1974**, 627–628; (b) Yang, W. Y.; Hechavarria Fonseca, M. T.; List, B. *Angew. Chem*. **2004**, *116*, 6829–6832; *Angew. Chem. Int. Ed*. **2004**, *43*, 6660–6662; (c) Tuttle, J. B.; Ouellet, S. G.; MacMillan, D. W. C. *J. Am. Chem. Soc*. **2006**, *128*, 12662–12663.

65. Ouellet, S. G.; Tuttle, J. B.; MacMillan, D. W. C. *J. Am. Chem. Soc*. **2005**, *127*, 32–33.

66. (a) Connon, S. J. *Org. Biomol. Chem*. **2007**, *5*, 3407–3417; (b) Ouellet, S. G.; Walgi, A. M.; MacMillan, D. W. C. *Acc. Chem. Res*. **2007**, *40*, 1327–1339.

67. Marsh, E. G. N.; Patterson, D. P.; Li, L. *Chem. Bio. Chem*. **2010**, *11*, 604–621.

68. Studer, A. *Chem. Eur. J*. **2001**, *7*, 1159–1164.

69. He, M.; Dowd, P. *J. Am. Chem. Soc*. **1998**, *120*, 1133–1137.

70. Halpern, J. *Science* **1985**, *227*, 869–875.

71. Pattenden, G. *Chem. Soc. Rev*. **1988**, *17*, 361–382.

72. (a) Scheffold, R. *Chimia* **1985**, *39*, 203–211; (b) Ghosez, A.; Göbel, T.; Giese, B. *Chem. Ber*. **1988**, *121*, 1807–1811.

73. *Cytochrome P450, Structure, Mechanism, and Biochemistry*, 3rd ed. (Ed. Ortiz de Montellano, P. R.), Kluwer Academic, New York, 2005.

74. Groves, J. T. *Inorg. Biochem*. **2006**, *100*, 434–447.

75. Groves, J. T.; Nemo, T. E.; Myers, R. S. *J. Am. Chem. Soc*. **1979**, *101*, 1032–1033.

76. Jacobsen, E. N.; Zhang, W.; Muci, A. R.; Ecker, J. R.; Deng, L. *J. Am. Chem. Soc.* **1991**, *113*, 7063–7064.

77. Yang, G.; Gabriele, B.; Belvedere, S.; Huang, Y.; Breslow, R. *J. Org. Chem.* **2002**, *67*, 5057–5067.

78. Chen, M. S.; White, M. C. *Science* **2010**, *327*, 566–571.

79. Curci, R.; D'Accolti, L.; Fusco, C. *Acc. Chem. Res.* **2006**, *39*, 1–9.

80. Nicolaou, K. C.; Snyder, S. A. *Angew. Chem.* **2005**, *117*, 1036–1069; *Angew. Chem. Int. Ed.* **2005**, *44*, 1013–1045.

81. Eschenmoser, A.; Arigoni, D. *Helv. Chim. Acta* **2005**, *88*, 3011–3050.

82. Hendrickson, J. B. *Tetrahedron* **1959**, *7*, 82–89.

83. Ebnöther, A.; Meijer, T. M.; Schmid, H. *Helv. Chim. Acta* **1952**, *35*, 910–928.

84. Birch, A. J.; Donovan, F. W. *Aust. J. Chem.* **1953**, *6*, 373–378.

85. Birch, A. J.; Donovan, F. W. *Aust. J. Chem.* **1953**, *6*, 360–368.

86. Frei, H.; Schmid, H. *Liebigs Ann. Chem.* **1957**, *603*, 169–177.

87. Gravel, E.; Poupon, E. *Eur. J. Org. Chem.* **2008**, 27–42.

CHAPTER 15

Conclusion and Future Perspectives: Drawing Inspiration from the Complex System that Is Nature

CLYDE W. CADY, DAVID M. ROBINSON, and PAUL F. SMITH

Department of Chemistry and Chemical Biology, Rutgers The State University of New Jersey, Piscataway, New Jersey, USA

GERHARD F. SWIEGERS

Intelligent Polymer Research Institute and ARC Centre of Excellence for Electromaterials Science, University of Wollongong, Wollongong, NSW 2522, Australia

15.1 INTRODUCTION: NATURE AS A COMPLEX SYSTEM

Perhaps the most important feature of Nature is the sheer and overwhelming complexity of biological systems. Life as we know it involves a staggering array of biochemical entities that interact with each other in an enormous and extraordinarily complicated web. Despite its bewildering complexity, Nature not only sustains life in all its facets, but systematically amplifies it over time. How does that happen? Indeed, how did it all come about in the first place? Can we unravel and start to understand the processes at work—the processes of life itself?

A new and emerging field of study draws inspiration from the web of synergistic biochemical interactions that give rise to life. While disparate aspects of this inquiry have been studied for many years, it is only recently that the field has coalesced into a single coherent body. The coalescence was arguably prompted by the realization that biological systems are examples of complex systems.

The field of complex systems science, which has formed and established itself in the last 40 years,[1] examines the way in which multiple independent elements within systems interact with each other to create self-propagating chains of action and reaction. Numerous examples of complex systems exist, including weather systems (chains of interacting weather events), traffic patterns on intersecting roads

Bioinspiration and Biomimicry in Chemistry: Reverse-Engineering Nature, First Edition.
Edited by Gerhard F. Swiegers.
© 2012 John Wiley & Sons, Inc. Published 2012 by John Wiley & Sons, Inc.

(chains of automobile movements), stock markets (chains of mutually beneficial financial transactions), and family trees (chains of procreation events), to name but a few. The common feature in all of these systems is that they change their state based on the accumulated action–reaction interactions over time, between the individual elements in the system. Each individual action–reaction event is typically relatively simple and easily understood, but the accumulation of them is extraordinarily complicated.

In a sense, complex systems can be envisioned to be like intersecting lines of dominos. When each individual domino in any one line falls, we can know with near certainty that the next one will be knocked over and will also fall. We can further know how the following one will fall and when it will fall. But, when two lines of falling dominos intersect each other, it becomes more difficult to know what will happen at the intersection points. One line may prematurely set off the other line. Alternatively, the two lines may mutually halt each other. A further possibility is that the two lines will pass through each other without affecting each other at all. Several other options may arise. Now multiply that uncertainty by hundreds, or thousands, or millions of intersecting lines of dominos, and it becomes essentially impossible to predict the overall state of the system with any degree of certainty even over short intervals into the future.

It was, perhaps, weather forecasting that gave the greatest impetus to the foundation of complex systems science. By the 1970s, it was starting to dawn on scientists that no amount of mathematical modeling could accurately predict weather for more than a period of a few days. This was because weather events display an extreme sensitivity to their initial conditions, an effect commonly known as the *butterfly effect*, after a paper presented by Edward Lorenz to the AAAS in Washington D.C. on December 29, 1979, entitled "Predictability: Does the Flap of a Butterfly's Wings in Brazil Set off a Tornado in Texas?"

The problem is that weather is comprised, effectively, of innumerable, interacting, highly localized chains of action–reaction weather events. While each individual event and the chain it sets off (e.g., warming of a particular location by the sun) can be modeled in not inconsiderable detail, it is more difficult to model and predict how two local chains interact with and influence each other. As more such local chains are added to the model, it very quickly and nonlinearly becomes an increasingly impossible task.

One of the reasons is that complex systems display a property known as amplification. When there are many chains present and interacting with each other, a small event in even one of the remote chains can quickly be transmitted and amplified through the other chains to create a powerful overall effect somewhere entirely different. Thus, the butterfly flapping its wings in Brazil could potentially set off a tornado in Texas. We will discuss the origin and character of amplification in later sections.

Biological systems and life itself have a very similar character. They comprise of innumerable biochemical entities interacting with each other in action–reaction chains to thereby create something that is more than the simple sum of their parts. Most of the individual interactions are, in themself, not overly complicated. The individual chain that is created or changed by the interaction is similarly amenable to understanding. However, when two or more chains interact and influence each other, then things quickly get complicated. When many, many chains are present and interacting, then the system rapidly becomes an impenetrable maze. For this reason, complex systems have also been termed *chaotic* (i.e., described by chaos theory). In a sense they can also be considered to be cascade processes; that is, the system cascades along a seemingly inevitable path driven by the timing and character of the individual action–reaction events between the system elements.

Biology has recently started examining these webs of biochemical interactions that comprise life, in a subdiscipline known as systems biology. The field that draws inspiration from such biological systems has therefore come to be called *systems chemistry*. It is around this concept that efforts devoted to biomimicry and bioinspiration involving the concept of life itself have coalesced. A new journal, *Journal of Systems Chemistry*, was established recently to cater to studies of this type.[1]

15.2 COMMON FEATURES OF COMPLEX SYSTEMS AND THE AIMS OF SYSTEMS CHEMISTRY

Complex systems typically display a range of common features that arise out of their unique character, involving, as it does, interacting, localized, chains of action–reaction. We do not discuss all of these common features here, but will mention a few of the most pertinent. These include:

- Amplification
- Self-replication
- Emergence
- Evolution
- Feedback
- Autonomy/autonomous agents
- Nonequilibrium processes
- Time and path dependence

A key feature of complex systems is that they often systematically expand to fill the space allotted to them. For example, what we call biology has expanded from a prebiotic primordial state to, effectively, cover the Earth. A threatening virus that arises in some very particular location will generally, over time, tend to expand its reach to more and more locations.

This property is very different from what is common for human-made creations. Our buildings, machines, and the like inevitably get old, degrade, become impossible to maintain, and ultimately disappear. In a sense, extinction is inherent in human-made devices. But Nature's way of making things is different. Nature makes things, which make new things, and so on. This "machines-making-machines" property of biology is called *self-replication*, and it involves the incorporation within the system of a capacity for renewal. Such a capacity is absent in human-made manufacturing.

Thus, the DNA molecule of biology is unique in that it replicates itself. It does so over and over, creating the outwardly moving cascade that we call biology.

Self-replication is likely a form of the property of amplification noted earlier. Somehow, when the individual elements in a complex system engage in action–reaction events, they can end up creating something that is more than the simple sum of their parts. How they do that is a key topic in systems biology. Studying, understanding, and creating amplication and self-replication is similarly a central theme of systems chemistry.[1]

In other complex systems, amplification often seems to arise out of individual events occurring in a synchronous fashion. That is, the timing of the individual events tends to make the events mutually reinforce one another, to thereby achieve a larger overall effect than might otherwise have been the case. For example, extreme weather tends to derive from synchronous confluences of a collection of more moderate weather events that end up combining to creating a "perfect storm." There seems little doubt that amplification in biological systems and life itself has similar origins. In later sections in this chapter, we discuss work aimed at elucidating amplification effects in chemistry.

We should note here also that, when we speak of amplification, we refer not only to unexpected, nonlinear enlargement effects, but also to nonlinear contraction effects. Nature can shrink back just as rapidly and vigorously as it can expand.

Emergence is the phenomenon where amplification leads to new patterns of behavior by the system. These patterns, which have typically not existed before, emerge spontaneously from the background. Emergence commonly derives from another property of complex systems: *evolution*.

Because they are so dynamic, complex systems tend to undergo continuous change in response to changes in the external conditions. In effect, they adapt to the changing environment. This process of adaption is known as evolution and one of its physical manifestations is emergence. Understanding emergence is a key topic in systems chemistry and biology research.[1] For example, how did Nature come to adopt a left-handed chirality in all of its amino acids? Can we, through chemistry, come to better understand that process or maybe even simulate it? Understanding, or better still, creating and harnessing systems that *evolve* is a prime target of systems chemistry.

Feedback is the process where an outcome in one part of the system dramatically influences—amplifies or diminishes—a process that takes place in another part of the system. Thus, one typically speaks of a *feedback loop*, where a system generates something that dramatically influences the system that has created it. Examples include the allosteric responses of biology discussed in an earlier chapter. Another famous example is that of the bacterial growth curve of an isolated population of microbes on a new, but static food source.[2a] Initially, the microbial population grows exponentially as there is an apparently unlimited source of food (the "exponential phase"). But then, once the bacterial population has exploded, the population first stabilizes (the "stationary phase") and then declines in an extreme, nonlinear way as the decreasing food availability and the toxic products generated by the microbes during their consumption of the food create a vigorously destructive feedback loop (the "death phase").

Because of the amplification involved, feedback effects are also of interest in systems chemistry. How do they work? Can we create such effects?

Many complex systems are known to contain within them mini, self-contained systems that display elements of complexity. These "systems-within-systems" may act totally autonomously from the rest of the system, giving them the property of autonomy. Some researchers also speak of *autonomous agents*.

An example of an autonomous agent is you. Your body is a complex system that exists autonomously within Nature, which is also a complex system, albeit much larger. Indeed, Nature contains many autonomous agents, including all of the microbes, organisms, and other known biological entities.

But the concept of autonomy does not extend only to living things, it goes much further than that. For example, while you are an autonomous agent, there are subsystems within your body that are themselves autonomous. These systems go all the way down to the molecular level. For example, it could be argued that proteins are autonomous agents in the way that they fold. Protein folding has been and is the subject of a large volume of research. The problem that arises is that protein folding is a relatively quick process, occurring on a time scale of seconds. However, studies show that if a protein were to search for its global energy minimum by sampling every conformation available to it, it would typically take innumerable years to fold.[2b] How, then, do proteins fold so quickly? This question is known as the *Levinthal paradox*[2b] and the answer may lie in systems chemistry. Proteins likely fold by a cascade process in which each local conformational change sets off another local conformational change and so on, until there are no more conformational changes available. At that stage, the protein is fully folded. In other words, the individual residues in the protein act autonomously and interact with the other residues in a series of action–reaction events to set up a one-way sequence of conformational changes that cause the protein to fold.

This explanation, if it is generally correct, illustrates another more microscopic feature of complex systems; namely, an apparent prevalence of nonequilibrium (or "one-way") processes at the most basic, fundamental level of the system. Many (although certainly not all) complex systems seem to display a "one-way" character at their most elemental level. There is often no searching for, nor achieving, an equilibrium. For example, the complex system of a family tree propagates exclusively in a one-way manner; you cannot unmake the birth of a child or obtain a new child from someone already dead. This is not to say that all action–reaction steps in complex systems involve nonequilibrium processes. Some may occur as equilibrium processes under thermodynamic control. In a later section we describe examples of complex systems driven by equilibrium and nonequilibrium molecular events.

The apparent prevalence of nonequilibrium processes has much to do with the fact that in many complex systems the individual action–reaction events are pathway and time dependent. As noted above, timing and synchronicity appear to play a large part in the phenomenon of amplification. But timing can only ever be important if pathway is also important. In other words, an element can only ever be at the right place at the right time. It cannot be only at the *right time*.

Understanding the fundamental character of nonequilibrium processes, as well as path and time dependence in chemical reactions and processes, is another, highly fundamental avenue of research in systems chemistry. The field of chemistry has a well-developed and rich foundation in chemical reactions and processes driven by their thermodynamics under equilibrium conditions. Our understanding of chemical processes under nonequilibrium conditions is far less developed.[3]

We should note here too that nonequilibrium processes are sometimes termed *kinetic processes*; that is, processes driven by the kinetics of reaction, not by its thermodynamics. In the case of chemical reactions, an alternative terminology is *mechanical processes*; that is, processes driven by the mechanics of reactant collisions.[3] Other terms may also be used. In the sections that follow, we use the original terms employed by the researchers involved. Where applicable, we have added the expression nonequilibrium for improved clarity.

15.3 EXAMPLES OF RESEARCH IN SYSTEMS CHEMISTRY

In the following sections we discuss several examples of research in systems chemistry. This research is mainly drawn from recent scientific publications in chemistry journals and should be considered to offer only a representative slice of this new field.

15.3.1 Self-Replication, Amplification, and Feedback

Many areas of interest in systems chemistry are developed around the concepts of self-replication and amplification. A possible method of self-replication in chemistry and biology is *autocatalysis*, or "making molecules that make themselves."[4] The two primary methods employed to foster autocatalysis are:[5]

1. Using the reaction product as a templating molecule that goes on to facilitate the synthesis of further reactants (Figure 15.1a).
2. Cross-catalytic systems in which products of multiple reactions catalyze the counterpart reactions (Figure 15.1b).

In cross-catalyzed systems, two reaction trajectories produce separate products A and B, with product A then catalyzing the formation of B and vice versa. Figure 15.2 depicts a reaction scheme exhibiting both autocatalytic and cross-catalytic mechanisms.[5]

Autocatalytic reactions are of fundamental interest. They potentially also have implications for our understanding of the origin of life and its molecular building blocks. We discuss these topics in a later section.

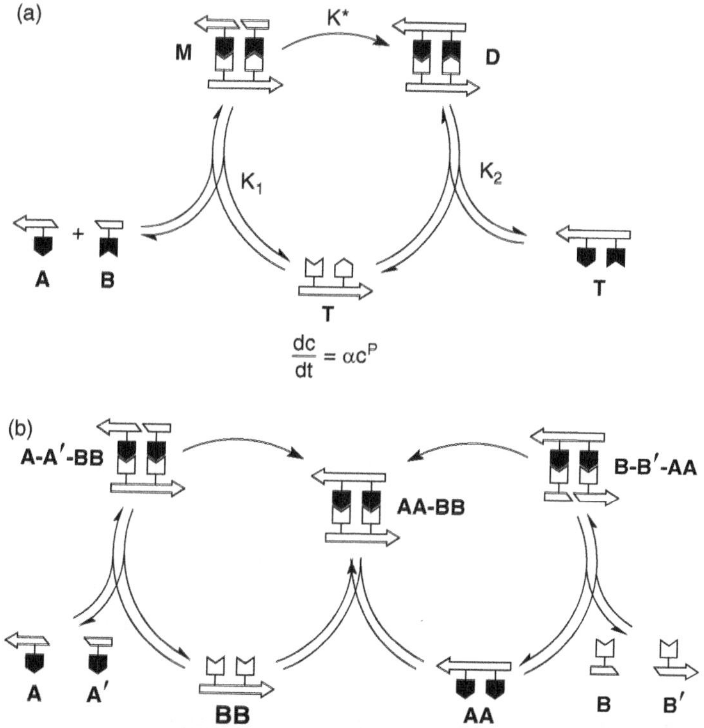

Figure 15.1 Schematic illustrations by von Kiedrowski of two common mechanisms for self-replication: (a) representation of template directed self-replication, and (b) representation of cross-catalytic self-replication. (Reprinted as an open-source image from ARKIVOC: Ref. 5.)

15.3.1.1 Self-Replication While many self-replicating systems employ autocatalytic reactions using one of the above models, there are exceptions including self-replicating systems promoted by product-induced solvation of reactants.[6] One example of such a system involves lipophilic amine and hydrophilic aldehydes reversibly generating amphiphilic imines that aggregate and form micelles. The micelles increase the concentration of the dissolved lipophilic reactant allowing for a product-induced increase in reactants. In its use of a feedback loop, this system was said to offer a chemical model of homeostasis.[7]

In further work, multiple variations of lipophilic amines were introduced into the system allowing for self-replicating chemical "selection" of the "fittest" products.[6] One example in this respect involved the use of a mix of both templated autocatalysis and cross-catalysis for the condensation of 3-aminobenzamidine and phenoxy acids.[5] By varying aromatic substituent groups of the reactants, the mechanism of autocatalysis could be selected (Figure 15.2). With NO_2 substituted phenoxy acids, the product exhibited the classic square-root dependence of autocatalytic systems in which the templates interact to form a stable complex inhibiting the reaction.

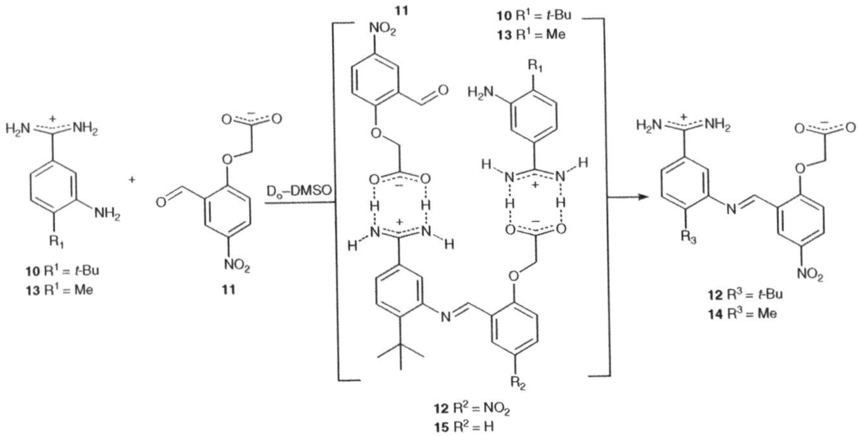

Figure 15.2 Reaction scheme exhibiting both *autocatalytic* and *cross-catalytic* mechanism. (Reprinted with permission. Copyright Wiley-VCH: Ref. 4.)

When the NO_2 was replaced with H, the resulting template displayed first order dependence, indicating a cross-catalytic system in which the template formed a less stable complex with itself.

15.3.1.2 Amplification: Asymmetric Autocatalysis Another important concept in autocatalytic systems chemistry is the potential for enantioselectivity in the self-replication of chiral products, or asymmetric autocatalysis. One early system focused on the addition of dialkylzinc to pyridine-3-carbaldehyde to which a 20 mol % enantiomeric excess (ee) of chiral product was added. Seeding with chiral product had the effect of autocatalytically generating enantioselectivity, which matched the seeded product in an otherwise racemic reaction.[8]

This work was later extended to produce consecutive autocatalytic reactions capable of producing almost enantiomerically pure product upon addition of minimal amounts (<0.0001% ee) of enantio-enriched product. Through the use of consecutive reactions, the enantioselectivity imparted by a slight excess was, effectively, amplified to produce enantiomerically pure, chiral pyrimidyl alkanol.

In a first example of organocatalytic asymmetric autocatalysis, autoamplification of enantiomeric excess was observed for the Mannich reaction of acetone and an iminoethyl gloxylate when a small amount of chiral product was added (Figure 15.3).[9] Interestingly, both chiral products can be selected for, depending only on the choice of initial product chirality. Comparable unseeded reactions produced racemic mixtures.

Further work studying the origin of homochirality revealed that stochastic fluctuations in the distribution of enantiomers may account for significant enantio enrichment of products, generating, at random, an excess of (*S*) or (*R*) form.[10] While many efforts in asymmetric autocatalysis have focused on autoamplification of existing chiralty, this work provided insight into the origin of chirality and its

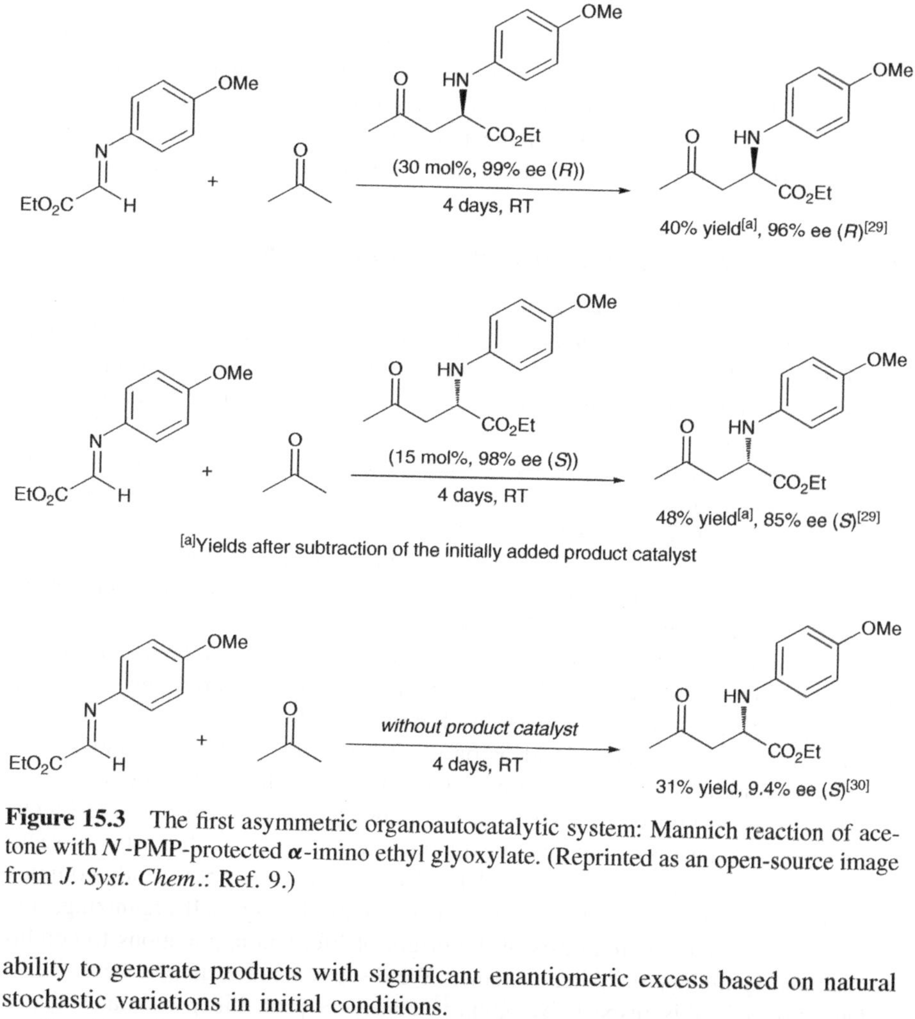

Figure 15.3 The first asymmetric organoautocatalytic system: Mannich reaction of acetone with *N*-PMP-protected **α**-imino ethyl glyoxylate. (Reprinted as an open-source image from *J. Syst. Chem.*: Ref. 9.)

ability to generate products with significant enantiomeric excess based on natural stochastic variations in initial conditions.

The biological context of symmetry-breaking reactions has been studied by Eschenmoser and colleagues,[11] who demonstrated that copolymerization of heterochiral pyranosyl tetramers into pyranosyl-RNA resulted in a ligation process that was highly stereoselective. Shortly thereafter, Feringa and van Delden highlighted the origin of chirality in relation to several stereoselective systems.[12]

While broad in scope, the field of autocatalysis and self-replication has historically focused on the implications for our understanding of the origin of life; in particular, the generation and replication of the information encoded RNA and DNA molecules necessary for the propagation of life.

Autocatalytic replication of nucleic acids was first reported using complementary trimers linking to form a templating hexamer.[13] A simpler system was later discovered in which single complementary adenosine base pairs were autocatalyzed by the resulting product.[14, 15] In elaboration of this work, an experiment was devised in which a bimolecular reaction (A + B) was used to generate a product (A + B → T), where the product (T) then templated and catalyzed further synthesis.[16] Self-replication of larger peptide chains from fragments was also demonstrated using a similar bimolecular templating approach.[17] Using cross-catalytic systems, 10–23 deoxyribozymes were activated from inactive precursors using complementary pairs of cyclized deoxyribozymes to produce their autocatalytic linear versions.[18]

15.3.2 Emergence, Evolution, and the Origin of Life

While not necessarily autocatalytic, efforts in systems chemistry have been made to generate fundamental biological molecules from prebiotically plausible conditions. Synthesis of pyrimidine ribonucleotides has been shown from simple precursers—cyanamide, cyanoacetylene, glycolaldehyde, glyceraldehyde, and inorganic phosphate.[19]

As mentioned, studies of autocatalysis are often directed at elucidating the origin of life and its molecular building blocks. Enzymes are nature's agents for replication of complex and simple biomolecules. How, though, were the first organisms able to replicate the molecules (nucleotides) with which enzymes are encoded?

Both Pross[20] and Eschenmoser[21] have proposed that replication and metabolism developed concurrently not separately. Regardless of the order in which they came about, it remains true that while racemization is the most thermodynamically stable state of a heterochiral mixture, there are many systems, including enzymes, that perform processes with often remarkably high stereoselectivity.

The level to which Darwin's ideas on evolution influenced subsequent studies is rarely seen in any field of experimental science. But while we know how complex life is, the origin of life—that is, the transfer of inanimate building blocks into animate organisms—remains unexplained. Many systems chemists are seeking to explain this transition using models. These models attempt to study complex molecular networks and their phenomena, self-reproducing, self-organizing, and autocatalytic systems with a basis of the origin of life, and applications to combinatorial and symmetry-breaking systems.[1] We discuss some examples here.

One model in this respect, by Kuhn,[22] considers probability estimates that a homochiral oligomer of adequate length will form from a heterochiral pool and successfully replicate, given that the physical interactions rely on chance. Citing information theory, this probability was interpreted with respect to a "knowledge" term, which indicated how much information is required to form the replicating oligomer. Knowledge increases with time due to growth in chain lengths or if environmental conditions either kill the oligomer or force it to adapt otherwise, via stabilization effects. The model, which was applied to the hairpin structures of RNA and DNA, suggested that chemical entities that successfully replicated and adapted were essential to the formation of life.

In another model,[20] Pross refers to the work performed by Joyce and colleagues,[23] which showed that two RNA enzymes could not coexist over time in the presence of a substrate that was essential to both. Pross[20] argued that while there is value in being able to monitor "survival of the fittest" experimentally, the appearance of competitive behavior should not be considered the important observation. While it could be interpreted that the more adaptive enzyme drove the lesser into extinction, the observation is not so much that "molecules exhibit Darwinian behavior" but rather that "biology can be explained by chemistry." That is, the actions and evolutionary paths of organisms should be attributed to actions on the atomic scale, and not vice versa. The behavior seen in the RNA enzymes is therefore not analogous to evolution but rather a potential cause of evolution.

Pross proposed, in effect, that evolution is based on increasing "dynamic kinetic stability." By this he means that the system as a whole moves to a state of kinetic (nonequilibrium) stability. It does so independent of thermodynamic barriers experienced by its individual components.

An example of such a system in Nature is photosystem II, the universal protein that contains the active sites of photosynthesis, which turns over and rebuilds itself approximately every 30 minutes. The analogy to cell division and death, regardless of the complexity of the organism, follows similarly. Thus, knowledge of the conditions that maintain this form of autocatalysis can, if repeated in laboratory settings, yield elegant chemistry. By understanding how complex systems work, coupled with a background in evolutionary study, we may therefore not only learn from, but also be able to ultimately mimic evolution, at least in principle.

15.3.3 Autonomy and Autonomous Agents: Examples of Equilibrium and Nonequilibrium Systems

15.3.3.1 Equilibrium Systems: Dynamic Combinatorial Libraries (DCLs)

Dynamic combinatorial chemistry involves the creation of a *dynamic combinatorial library* (DCL) under thermodynamic control and at equilibrium.[25] DCLs constitute an important basis of systems chemistry in the field of screening and sensing.[16] Where standard screening techniques involve the interaction of one sensor compound with its corresponding molecule, systems chemistry involves the interaction of three or more molecules to form a unique complex identifiable by simple spectroscopy. Creating understandable results from complex mixtures is one of the hallmarks of systems chemistry and lends itself to screening large numbers of molecules simultaneously in, for example, mass screenings and sensor applications.[16, 24, 26]

To illustrate the concept of a DCL, consider the following example. In a system where reactants A, B, and C are present, along with a template T, which favors binding trimers, complexity quickly becomes evident. The binding constant of T to each of the monomer and dimer combinations must be considered, as well as the trimer combinations. The trimer combinations alone are affected by concentration, which dictates that for A_3T (Figure 15.4) to accumulate, it must be a considerably better binder than ABCT if A, B, and C are in equal concentrations.[27] The possibility of larger oligomers must also be factored even if they do not bind to the template, in order for the proper concentrations of oligomers to be modeled. All these factors and more must be considered when examining such DCL systems chemistry. It can lead to nonintuitive results.

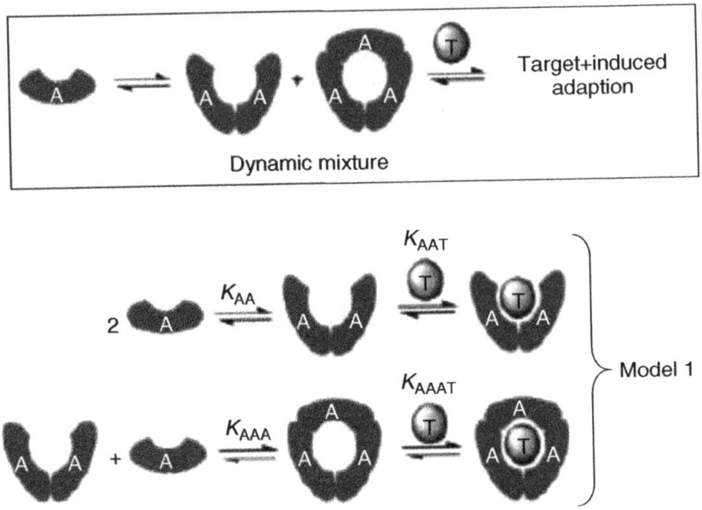

Figure 15.4 A simple interaction model for a DCL containing only one building block, A, and a template,T. (Reprinted with permission. Copyright Wiley-VCH: Ref. 27.)

In this field, the relation between biology and systems chemistry is clear. Classical chemistry would dictate that each oligomer be isolated and tested against T to find its unique binding constant, slowly building up a web of interactions. Isolation and step-by-step examination is impossible in biology, so methods were developed to examine systems as a whole.

A proof-of-principle study showing how DCLs are capable of sensing different molecules in a complex mixture was reported by Severin and co-workers (Figure 15.5).[27, 28] A solution, containing two metal ions and three coordinating dye molecules, was created as an initial DCL. The UV–visible spectrum of this solution served as the baseline of the system. Addition of a mixture of dipeptides to the baseline DCL resulted in a perturbation of the metal–dye interactions, resulting in unique UV–Vis shifts. This complex system, containing three dye molecules and two metals, allowed for the distinction of closely related dipeptides (Figure 15.6). In a similar study, a sensor was developed to distinguish Gly-Gly-His from His-Gly-Gly and Gly-His-Gly.[29]

15.3.3.2 Nonequilibrium Systems Moving beyond thermodynamic control, similar principles can be used to create kinetically controlled, or nonequilibrium networks. Such systems are of interest because, in both chemistry and biology, what reacts fastest is often just as important as what binds strongest. Two examples of such nonequilibrium DCL-like systems are discussed briefly here.

In the thermodynamically controlled examples above, the selectivity allowed by the system was rarely higher than that provided by the affinity between the template and the binding molecule. In nonequilibrium DCL-like systems, it becomes possible to enhance the selectivity as demonstrated by Gleason, Kazlauskas, and

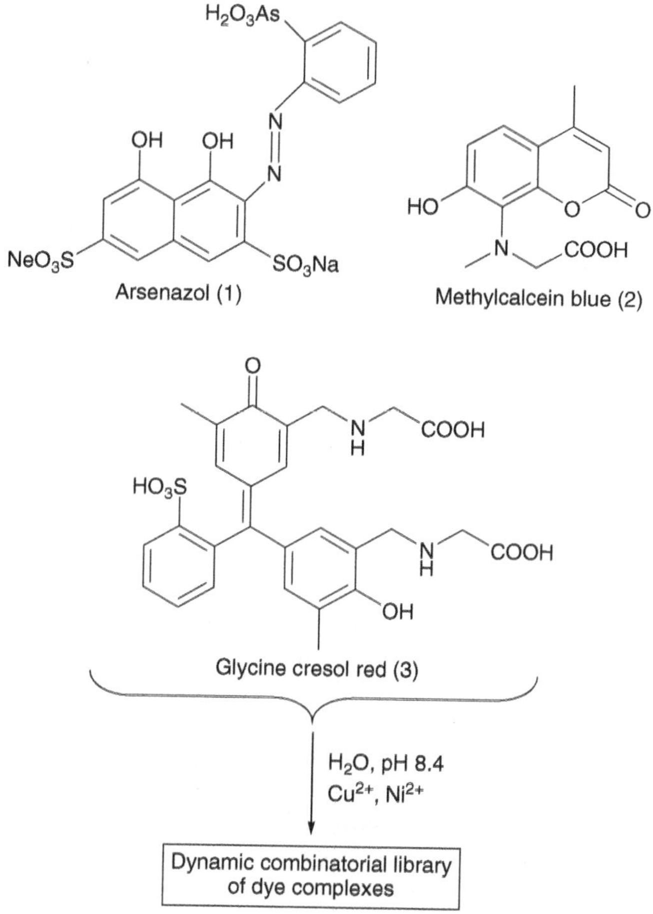

Figure 15.5 The mixture of dyes and metals used to create a *dynamic combinatorial library* (DCL) by Severin and co-workers. (Reprinted with permission. Copyright Wiley-VCH: Ref. 28.)

co-workers.[30, 31] Their experiment involved adding peptide samples to a container with both a *protease enzyme* and *carbonic anhydrase* as illustrated in Figure 15.7. The protease and carbonic anhydrase were separated by a membrane that allowed free passage of the peptide analytes. The peptide samples that were bound by carbonic anhydrase were found to spend less time in the presence of the protease. In the end, they constituted the majority of the peptide samples in solution because the peptides that spent less time bound to carbonic anhydrase were destroyed by the protease on the other side of the membrane. This simple system was elaborated upon in a later publication to include a third chamber that contained a synthesis compartment to regenerate the decomposed peptides. The addition of the third chamber allowed a 100:1 selection for the best carbonic anhydride binders despite an affinity ratio of only 2.2:1 (Figure 15.8).[31]

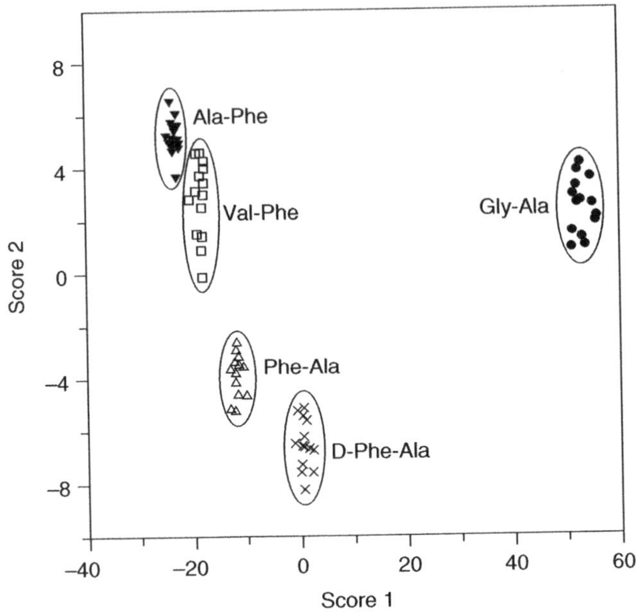

Figure 15.6 A linear discriminant analysis score plot showing clear separation of a series of closely related dipeptide analytes. (Reprinted with permission. Copyright The Royal Society of Chemistry: Ref. 16.)

Finally, we should note an example of mechanosensitive self-replication. Otto and co-workers recently reported two self-replicating peptide-derived macrocycles that emerged from a dynamic combinatorial library and competed for a common feedstock.[32] In that experiment, replication was driven by nanostructure formation that arose when the peptides assembled into fibers held together by β-sheets. The dominant replicator was determined by whether the sample was shaken or stirred. Thus, mechanical forces may generate a selection pressure in the competition between replicators and can determine the outcome of a covalent synthesis.

15.4 CONCLUSION: SYSTEMS CHEMISTRY MAY HAVE IMPLICATIONS IN OTHER FIELDS

There is an invigorating optimism in the sheer and monumental scale of the task that the young field of systems chemistry sets for itself. The questions it seeks to elucidate are among the biggest of the big questions in not only chemistry but, perhaps, also in the human experience itself. Some may argue that they are too big to be pertinently addressed at this point in time. There may well be some merit to that argument. Systems chemists certainly risk diffusion and overreach in their assertion that their field of study is "unbounded." Some limits always seem advisable, even if only to improve focus and to avoid falling into arcane, unprovable, conjecture.

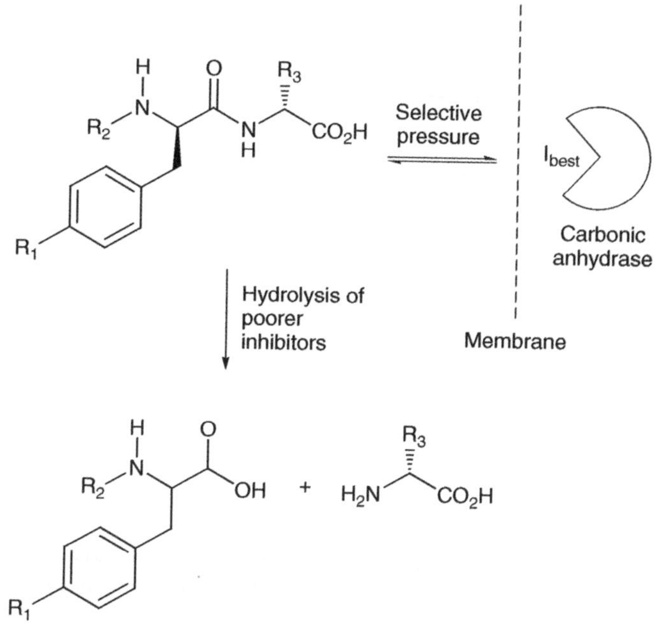

1 $R_1 = SO_2NH_2$ $R_2 = H$

Figure 15.7 An illustration of how carbonic anhydrase selectively binds and protects the peptides from exposure to a protease. (Reprinted with permission. Copyright American Chemical Society: Ref. 30.)

However, research is by its nature optimistic and research in new, untested fields is necessarily the most optimistic of all. As we have shown above, regardless of whether systems chemistry is able to start answering the "big questions," it already holds significant, practical promise in niche areas, like sensor technology, autocatalysis, chiral symmetry breaking, complex behaviors in molecular systems, and chemical self-organization, to name but a few.[22]

For these reasons, we believe that it does and will, in future, prove to be an increasingly important aspect of biomimicry and bioinspiration. The suggestion of, for example, Darwinian behavior on the molecular scale is, frankly, breathtaking and extraordinary.[20] Who could not consider it to be worthy of attention?

A very important point about studies in systems chemistry is that they may potentially generate information and implications that go beyond mere chemistry and influence a host of other fields. For example, studies in self-replication and amplification may yield valuable information in developing "self-improving" computer programs in information technology. Lehn has recently equated self-assembly in biology and chemistry with an "interactional algorithm" that drives the formation of novel chemical entities.[33] The implication is that algorithms in information technology could one day learn something from the way that biological or chemical

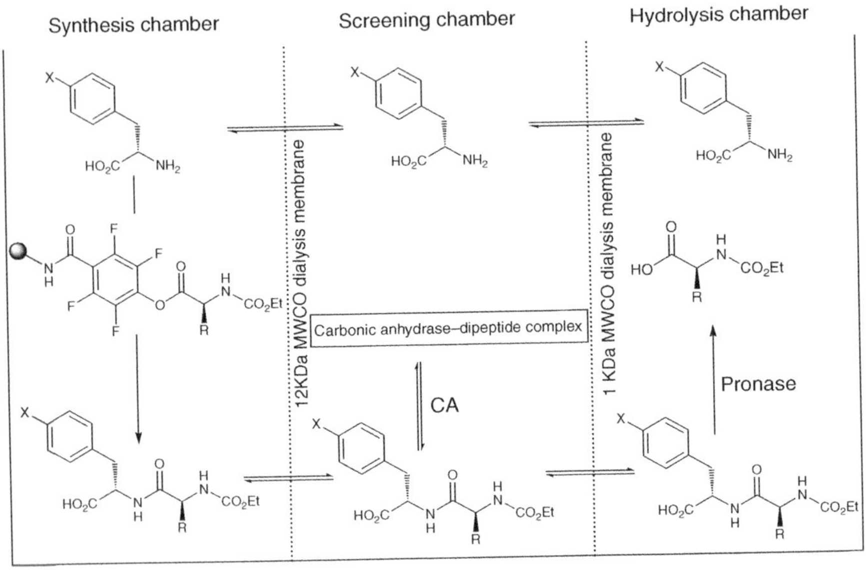

Figure 15.8 The three chamber experimental setup used by Gleason and Kazlauskas in their demonstration of a nonequilibrium-like DCL. (Reprinted with permission. Copyright Wiley-VCH: Ref. 31.)

self-assembly runs its course. Several other complex systems could perhaps eventually draw inspiration or insights from chemical and biological complex systems, including social interactions and human behavior (e.g., criminology, sociology, ethics) and economics (the phenomenon of "economic growth").[1] As such, systems chemistry offers the prospect of playing a role in unifying science and improving the human experience generally.[1] And that could well prove the best reason to study it in the future.

REFERENCES

1. (a) von Kiedrowski, G.; Otto, S.; Herdewijn, P. *J. Syst. Chem.* **2010**, *1,* 1; (b) Peyralans, J.J. P.; Otto, S. *Curr. Opin. Chem. Biol*. **2009**, *13,* 705.
2. (a) `http://en.wikipedia.org/wiki/Bacterial_growth`; (b) Levinthal, C. *J. Chim. Phys*. **1968**, *65,* 44.
3. *Mechanical Catalysis* (Ed. Swiegers, G. F.), John Wiley & Sons, Hoboken, NJ, 2008.
4. Vidonne, A.; Philp, D. *Eur. J. Org. Chem*. **2009**, *5,* 583.
5. Patzke, V.; von Kiedrowski, G. *ARKIVOC* **2007**, (*v*), 293–310.
6. Nguyen, R.; Allouche, L.; Buhler, E.; Giuseppone, N. *Angew. Chem. Int. Ed*. **2009**, *48,* 1093.
7. Zepik, H.H.; Blöchliger, E.; Luisi, P. L. *Angew. Chem. Int. Ed. Engl*. **2001**, *113,* 205.

8. Suzuki, K., Hatase, K.; Nishiyama, D.; Kawasaki, T.; Soai, K. *J. Syst. Chem*. **2010**, *1,* 5.
9. Tsogoeva, S. *J. Syst. Chem*. **2010**, *1,* 8.
10. Mauksch, M.; Tsogoeva, S. B.; Martynova, I. M.; Wei, S. *Angew. Chem. Int. Ed*. **2007**, *119,* 397.
11. Bolli, M.; Micura, R.; Eschenmoser, A. *Chem. Biol*. **1997**, *4,* 309.
12. Feringa, B. L.; van Delden, R. A. *Angew. Chem. Int. Ed*. **1999**, *38,* 3418.
13. von Kiedrowski, G. *Angew. Chem. Int. Ed. Engl*. **1986**, *25,* 932.
14. Nowick, J. S.; Feng, Q.; Tjivikua, T.; Ballester, P.; Rebek J. *J. Am. Chem. Soc*. **1991**, *113,* 8831.
15. Tjivikua, T.; Ballester, P.; Rebek, J. *J. Am. Chem. Soc*. **1990**, *112,* 1249.
16. Ludlow, R. F.; Otto, S. *Chem. Soc. Rev*. **2008**, *37,* 101.
17. Issac, R.; Chmielewski, J. *J. Am. Chem. Soc*. **2002**, *124,* 6808.
18. Levy, M.; Ellington, A. D. *Proc. Natl. Acad. Sci. U.S.A.* **2003**, *100,* 6416.
19. Powner, M. W.; Gerland, B.; Sutherland, J. D. *Nature* **2009**, *459,* 239.
20. Pross, A. *J. Syst. Chem*. **2011**, *2,* 1.
21. Eschenmoser, A. *Orig. Life Evol. Biosphere* **2007**, *37,* 309.
22. Kuhn, H. *J. Syst. Chem*. **2010**, *1,* 3.
23. Joyce, G. F. *Nature* **1989**, *338,* 217.
24. Whitesides, G. M.; Ismagilov, R. M. *Science* **1999**, *284,* 89.
25. Corbett, P. T.; Leclaire, J.; Vial, L.; West, K. R.; Wietor, J.-L.; Sanders, J. K. M.; Otto, S. *Chem. Rev*. **2006**, *106,* 3652.
26. Newth, D.; Finnigan, J. *Aust. J. Chem*. **2006**, *59,* 841.
27. Severin, K. *Chem. Eur. J*. **2004**, *10,* 2565.
28. Buryak, A.; Severin, K. *Angew. Chem. Int. Ed*. **2005**, *44,* 7935.
29. Buryak, A.; Severin, K. *J. Comb. Chem*. ***2006***, 8, 540.
30. Cheeseman, J. D.; Corbett, A. D.; Shu, R.; Croteau, J.; Gleason, J. L.; Kazlauskas, R. J. *J. Am. Chem. Soc*. **2002**, *124,* 5692.
31. Corbett, A. D.; Cheeseman, J. D.; Kazlauskas, R. J.; Gleason, J. L. *Angew. Chem. Int. Ed*. **2004**, *43,* 2432.
32. Carnall, J. M. A.; Waudby, C. A.; Belenguer, C. A.; Stuart, M. C. A.; Peyralans, J. J.-P.; Otto, S. *Science* **2010**, *327,* 1502.
33. Lehn, J.-M. *J. Chem. Sci*. **1994**, *106,* 915.

INDEX

Bioinspiration and Biomimicry in Chemistry: Reverse-Engineering Nature, First Edition.
Edited by Gerhard F. Swiegers.

© 2012 John Wiley & Sons, Inc. Published 2012 by John Wiley & Sons, Inc.

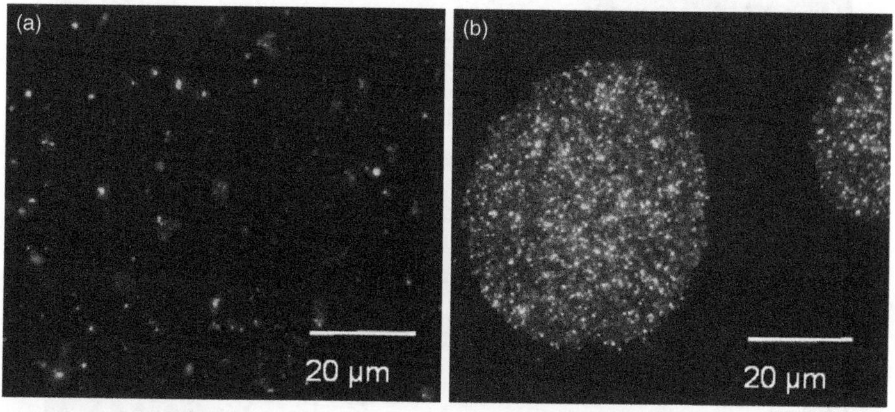

Figure 8.8 Fluorescence micrographs of vesicles containing 5 mol % of amphiphilic histidine (red) mixed with vesicles containing 5 mol % of Cu(IDA) (blue) in (a) DMPC and (b) DMPC/cholesterol. (Reproduced with permission. Copyright American Chemical Society: Ref. 38.)

Bioinspiration and Biomimicry in Chemistry: Reverse-Engineering Nature, First Edition.
Edited by Gerhard F. Swiegers.
© 2012 John Wiley & Sons, Inc. Published 2012 by John Wiley & Sons, Inc.

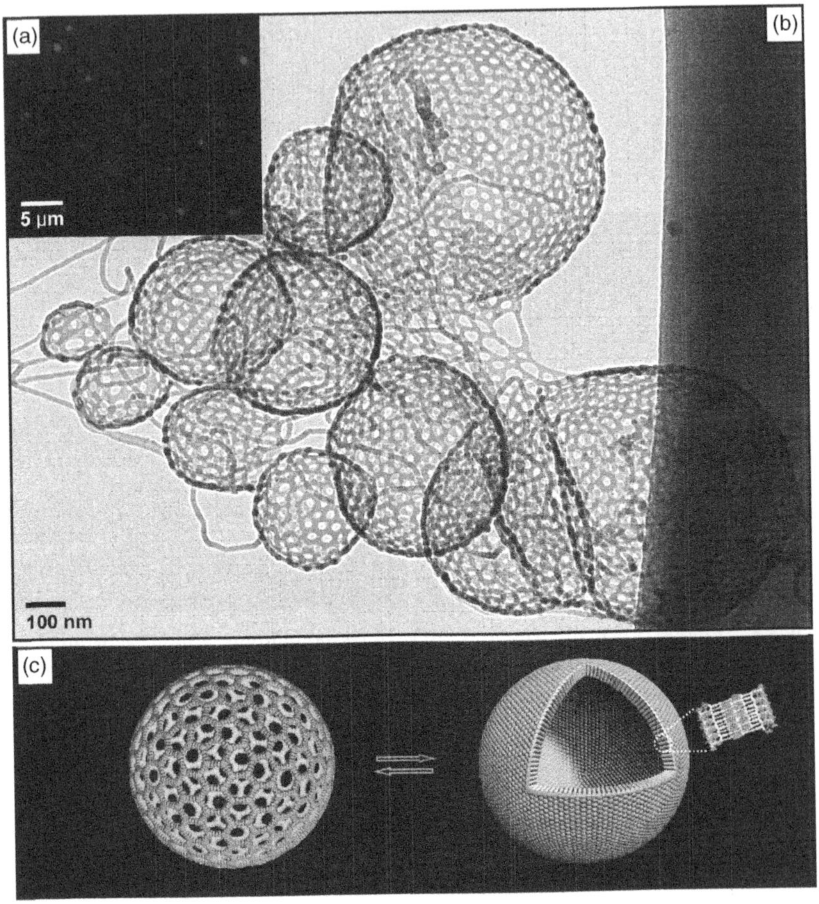

Figure 8.16 (a) Fluorescence micrograph and (b) cryo-TEM images of hollow spheres with a lateral nanoporous shell formed by self-assembly of rigid rode amphiphiles in aqueous solution (0.01 wt %). (c) Schematic representation of a reversible open/closed gating motion in the lateral nanopores of the capsules (green, polyether dendrons; yellow, aromatic segments; blue, hydrophobic branches). (Reproduced with permission. Copyright Wiley-VCH: Ref. 59.)

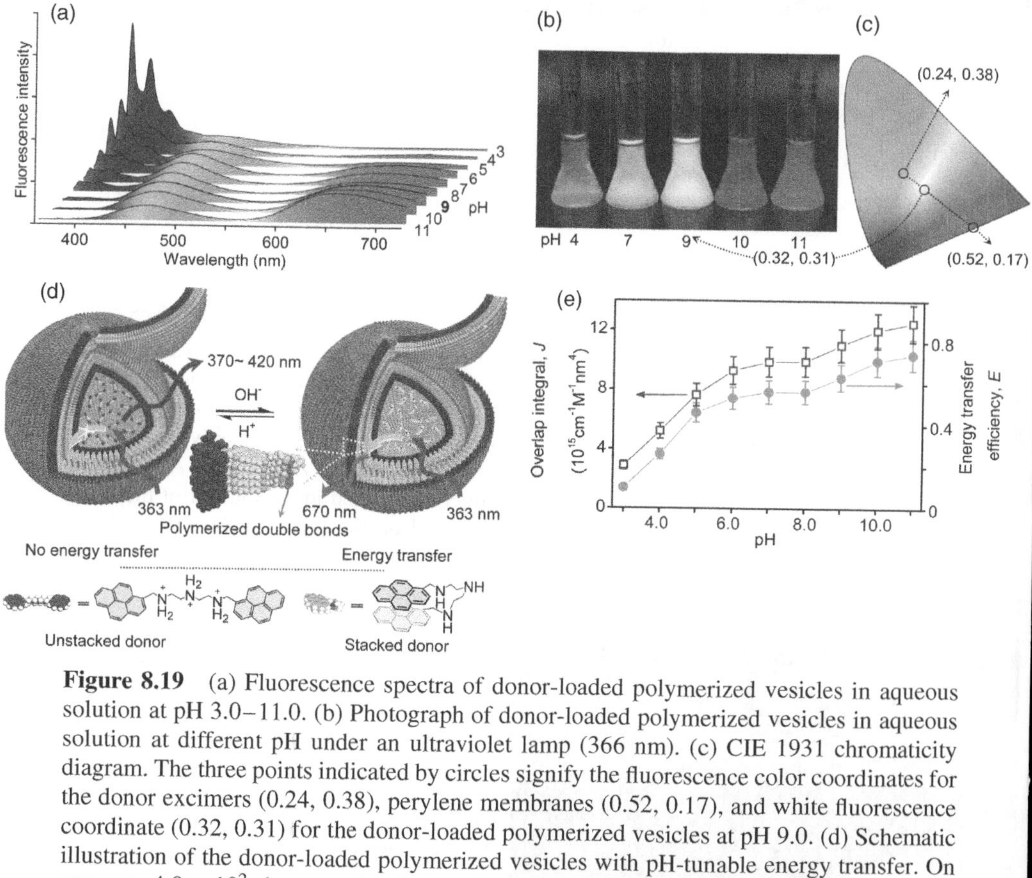

Figure 8.19 (a) Fluorescence spectra of donor-loaded polymerized vesicles in aqueous solution at pH 3.0–11.0. (b) Photograph of donor-loaded polymerized vesicles in aqueous solution at different pH under an ultraviolet lamp (366 nm). (c) CIE 1931 chromaticity diagram. The three points indicated by circles signify the fluorescence color coordinates for the donor excimers (0.24, 0.38), perylene membranes (0.52, 0.17), and white fluorescence coordinate (0.32, 0.31) for the donor-loaded polymerized vesicles at pH 9.0. (d) Schematic illustration of the donor-loaded polymerized vesicles with pH-tunable energy transfer. On average, 4.0×10^2 donor molecules are loaded into one perylene bisimide vesicle. The inner and outer layers of the vesicle consist of 5.2×10^3 and 8.4×10^3 perylene acceptor molecules, respectively. Their hydrophilic chains (blue) are exposed to water, with the hydrophobic part (orange) packed together and stabilized by polymerized double bonds (red). (e) pH-dependent energy-transfer efficiency (*E*, orange line) and overlap integral (*J*, blue line) of donor-loaded polymerized vesicles at pH 3.0–11.0. (Reproduced with permission. Copyright Nature Publishing Group: Ref. 68.)

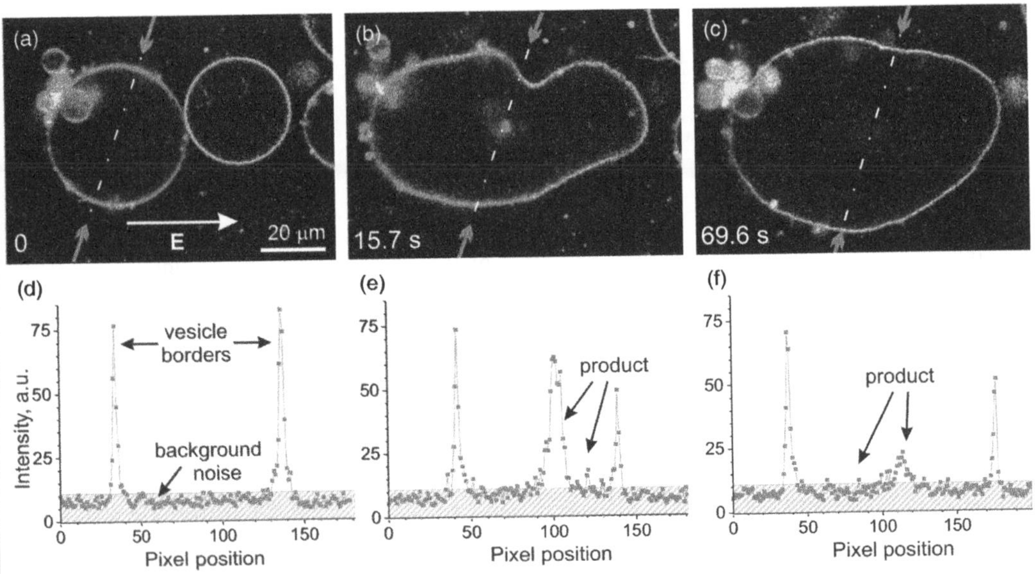

Figure 8.21 (a)–(c) Confocal scans of vesicles loaded with 0.3 mM Na_2S (red) and 0.3 mM $CdCl_2$ (green) undergoing fusion. (d)–(f) Intensity line profiles along the dash-dotted lines indicated by red arrows in (a)–(c), respectively. The direction of the field is indicated in (a). Before fusion [(a) and (d)], the vesicle interior shows only background noise similar to the external solution as indicated by the shaded zone in (d). After fusion [(b), (c), (e), and (f)], fluorescence from the product is detected in the interior of the fused vesicle. The time after applying the pulse is indicated on the micrographs. (Reproduced with permission. Copyright Wiley-VCH: Ref. 71.)

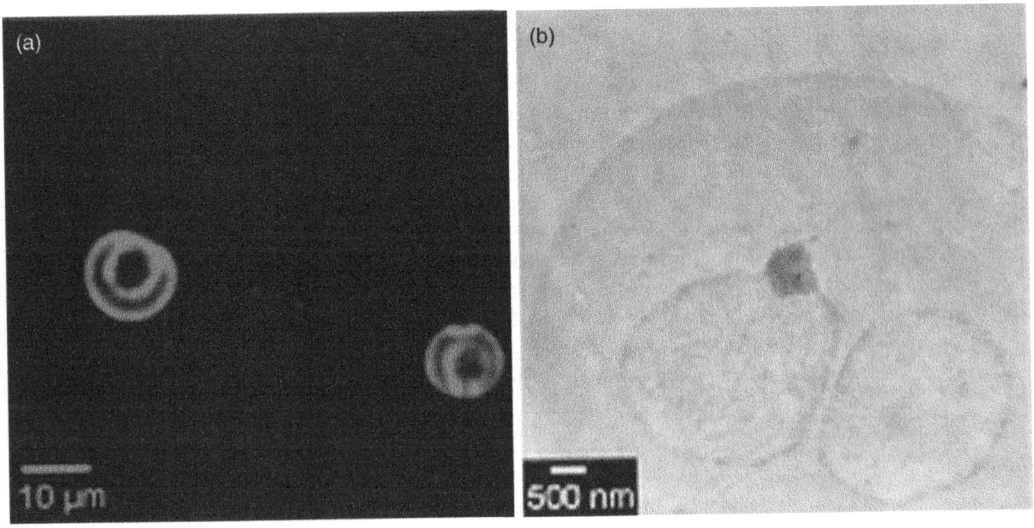

Figure 8.24 (a) Multivesicle assemblies (LSCM) and (b) a sectioned specimen (ca. 80 nm thickness) of the multivesicle assembly (TEM). (Reproduced with permission. Copyright Wiley-VCH: Ref. 73.)

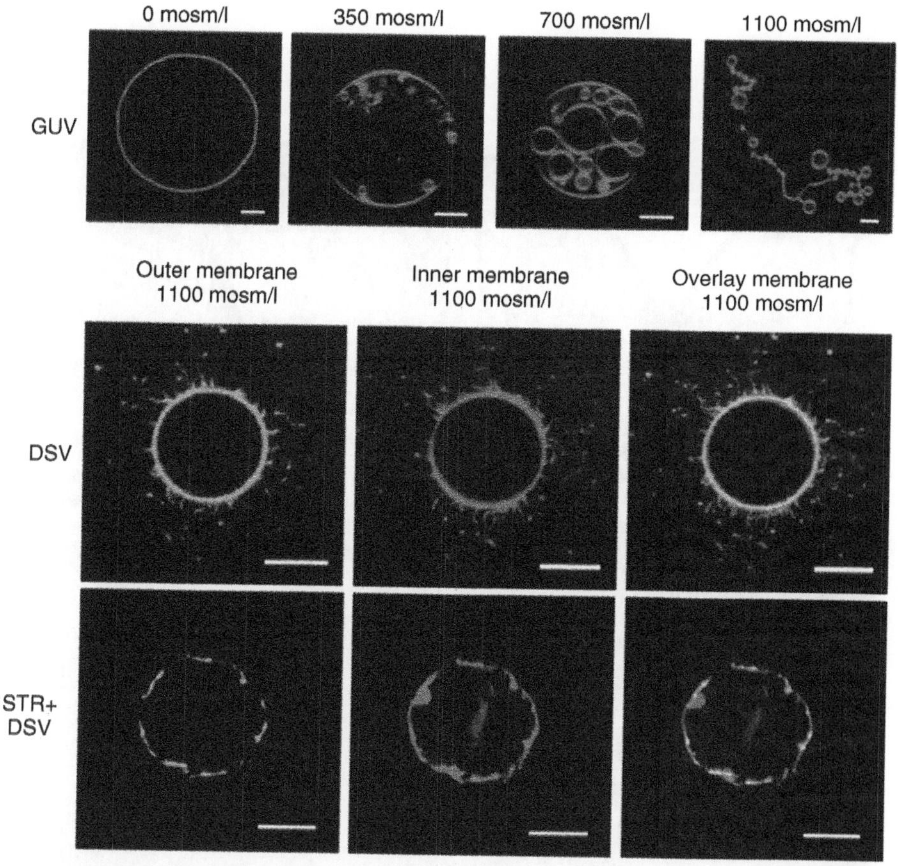

Figure 8.25 Upper row: Effect of hyperosmotic pressure on GUVs labeled with red dye. All images represent different vesicles. Middle row: Double-shell giant vesicle at 1100 mosm/L hyperosmotic pressure. Double-shell vesicles (DSVs) without surface proteins underwent an outside budding process under hyperosmotic conditions. The two membranes stuck together in newly formed nanotubes (yellow signal), or they formed separately new buds and tubes (green and red signals). Lower row: Vesicle coated with a crystalline streptavidin layer on the outer membrane surface (STR+DSV) showed a slight asymmetrical shape deformation without membrane budding (green signal), while the inner membrane released the osmotic pressure, separately forming daughter vesicles (red signal). Scale bars = 5 μm. (Reproduced with permission. Copyright American Chemical Society: Ref. 74.)

Figure 10.1 Various colorations provided by living creatures in Nature. (a) Blue *Morpho* butterfly. (b) Peacock. (c) Longhorn beetles *Tmesisternus isabellae*. (Reproduced with permission from Ref. 81. Copyright © 2009, the Optical Society of America.) (d) Myxomycetes *Diachea leucopoda*. (Reproduced with permission. Copyright the Optical Society of America: Ref. 3.)

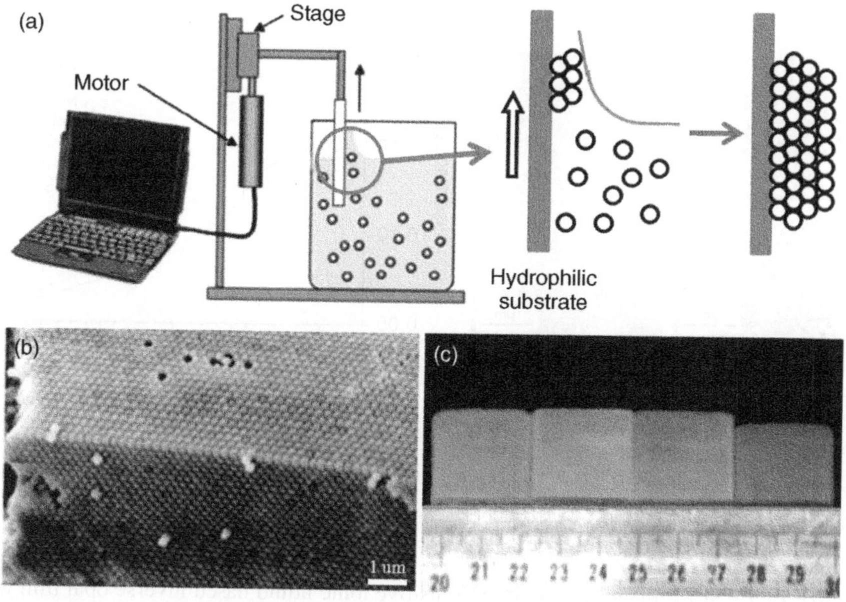

Figure 10.4 (a) Schematic illustration of the lifting method. During the lifting process, the self-assembly of colloidal crystals takes place at the air–liquid interface, due to the capillary force and the evaporation of solvent. The lifting speed can be precisely controlled by the computer. (b) SEM images of the obtained highly ordered PC films. (c) PC films with various distinct brilliant structural color fabricated by using colloidal crystals with different diameters. (Reproduced with permission. Copyright the American Chemical Society: Ref. 31.)

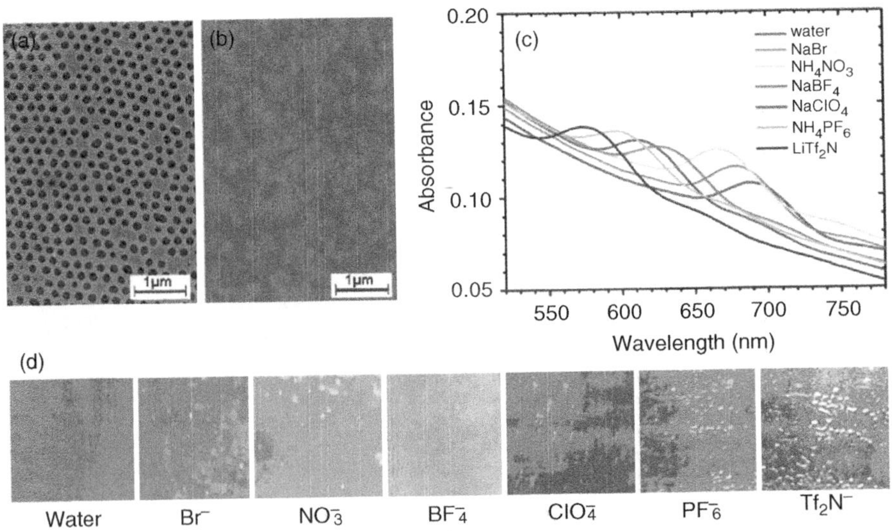

Figure 10.9 SEM images of the anion-sensitive ionic liquid based inverse opal film with (a) opened pore structures and (b) closed pore structures. (c) Stop band shift of the film in response to various kinds of anion aqueous solutions. (d) Color presented by the film when soaking into different kinds of anion aqueous solutions. Different colors will be exhibited in response to the anions due to the change of the solubility and the refraction index, which can be directly recognized by the naked eye. (Reproduced with permission. Copyright Wiley-VCH: Ref. 73.)

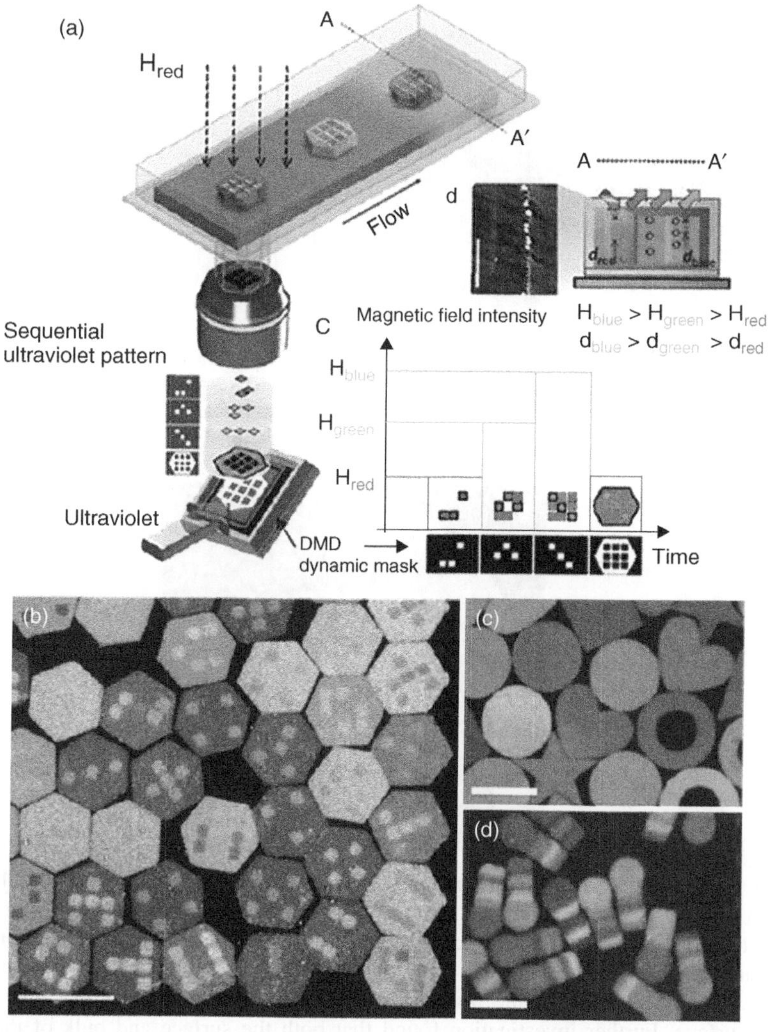

Figure 10.10 (a) Schematic illustration of high-throughput bioassays generated from M-ink with the help of an external magnetic field and a computer-controlled spatial light modulator as mask. Taking advantage of both spectral encoding and graphical encoding, various barcodes can be obtained (b–d). (b) Hexagon-type 2D color-barcoded microparticles; (c) microparticles with various shapes and colors; (d) bar-type 1D color-barcoded microparticles. (Reproduced with permission. Copyright Nature Publishing Group: Ref. 79.)

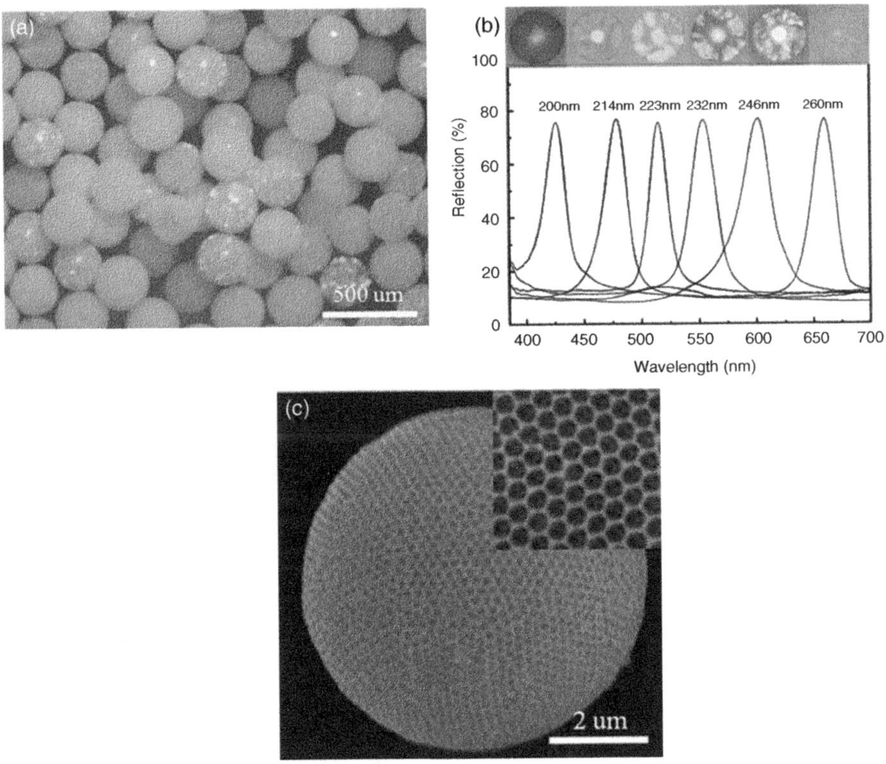

Figure 10.11 (a) Microscope image of various kinds of inverse opal beads fabricated through microfluidic technology by using polystyrene spheres with different size as sacrificial templates. (b) Reflection spectra of these beads. These optical signals are quite stable and have close relationships with the size of colloidal crystals used for the fabrication. (c) SEM images of the inverse opal beads. The insert shows that they have a porous surface. Further investigation found that both the surface and bulk of the beads are composed of well-ordered 3D periodical structures, providing a rather high specific surface area for the potential application in the bioassays. (Reproduced with permission. Copyright Wiley-VCH: Ref. 78.)

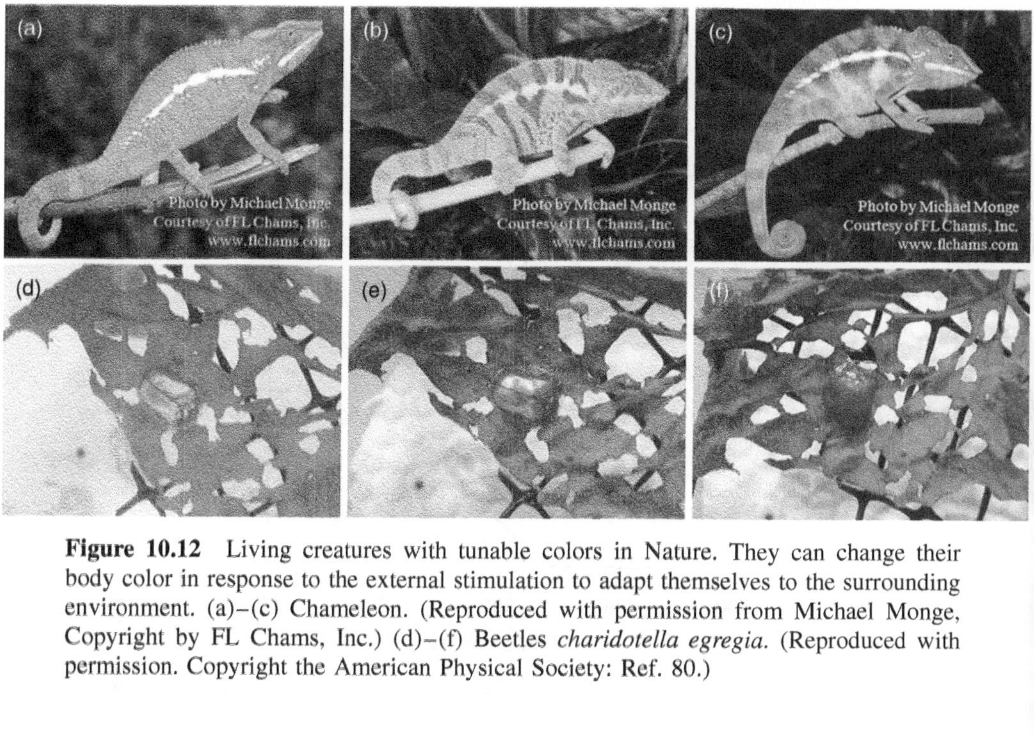

Figure 10.12 Living creatures with tunable colors in Nature. They can change their body color in response to the external stimulation to adapt themselves to the surrounding environment. (a)–(c) Chameleon. (Reproduced with permission from Michael Monge, Copyright by FL Chams, Inc.) (d)–(f) Beetles *charidotella egregia.* (Reproduced with permission. Copyright the American Physical Society: Ref. 80.)

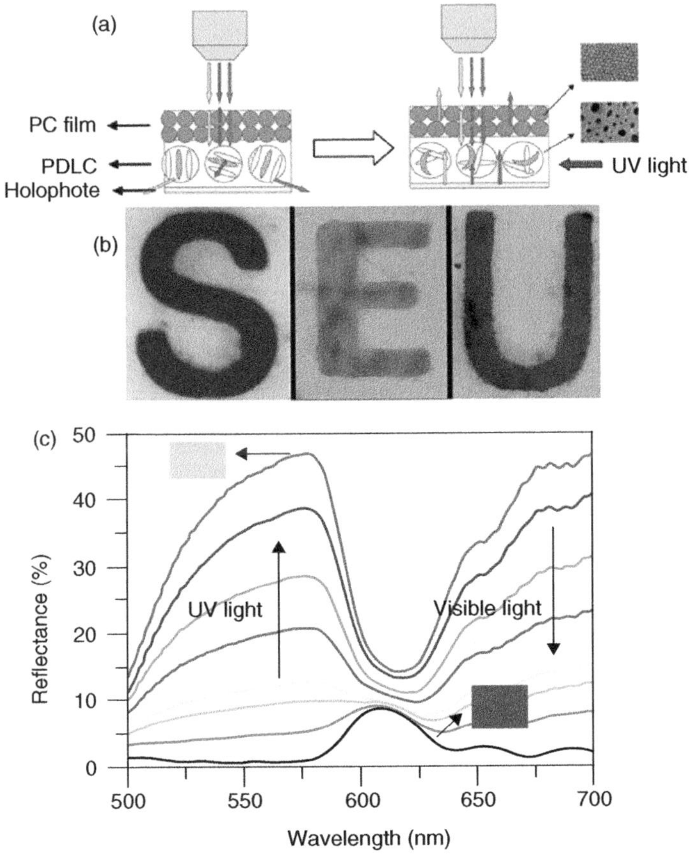

Figure 10.13 (a) Schematic illustration of a reversible photonic device composed of PC film and liquid crystals, which can conveniently be modulated by UV and visible light; (b) characteristics presented by the photonic devices; (c) reflection spectra change of the film under the irradiation of UV light. These changes can be reversed to the original state under the irradiation of visible light. The insets show the colors before and after UV irradiation. (Reproduced with permission. Copyright the American Institute of Physics: Ref. 84.)

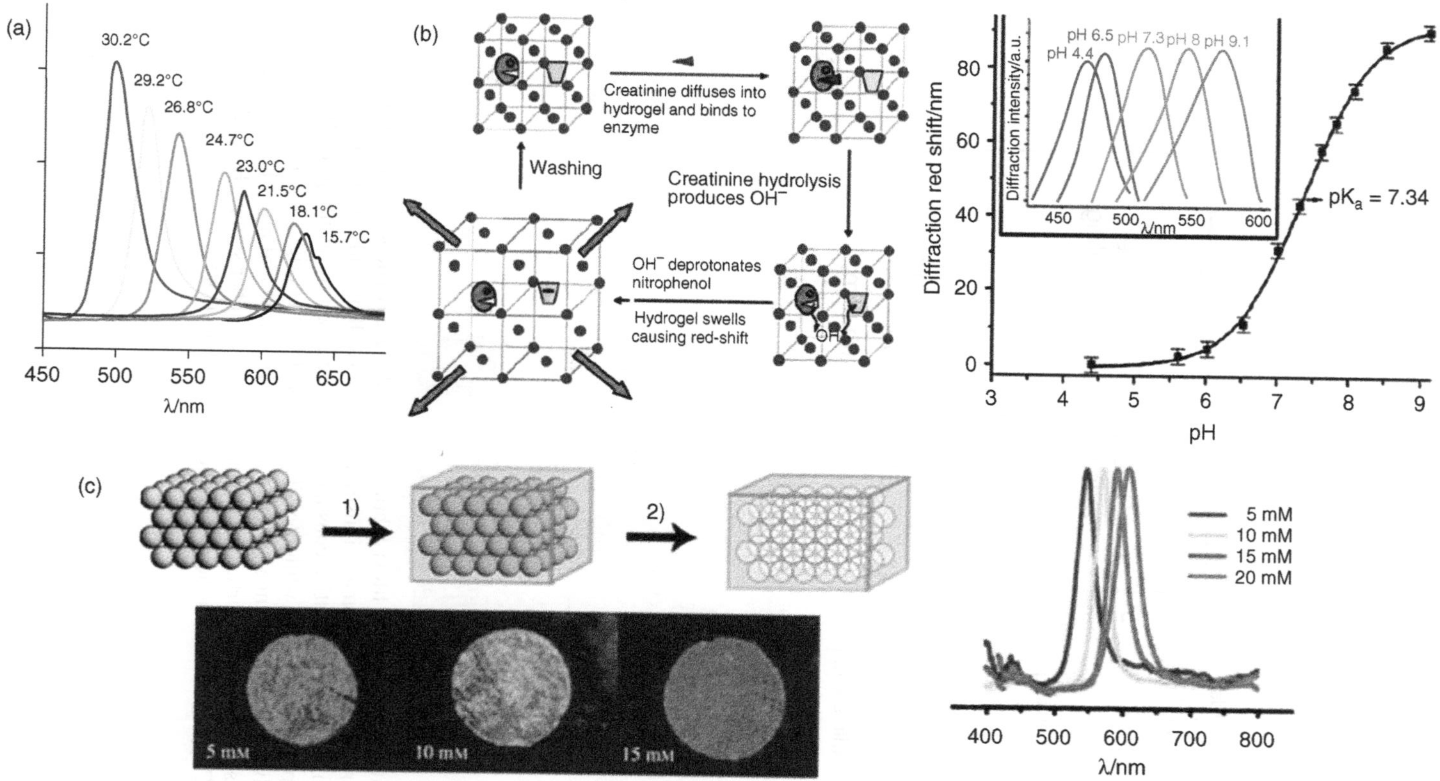

Figure 10.14 Responsive photonic materials-based PCs and their responsiveness to specific external stimulation such as temperature, pH, and chemicals. (a) Temperature dependence of the reflection spectra of the porous NIPA gel made using close-packed silica colloidal crystals as template. (Reproduced with permission. Copyright the American Chemical Society: Ref. 86.) (b) Schematic illustration of pH-responsive polymerized crystalline colloidal array (left) and the pH dependence of reflection spectra of this intelligent sensing array (right). (Reproduced with permission. Copyright the American Chemical Society: Ref. 88.) (c) Photographs (left) and reflection spectra (right) of the periodically ordered interconnecting porous poly(NIPA-*co*-AAPBA) gel in response to different concentrations of glucose. (Reproduced with permission. Copyright Wiley-VCH: Ref. 87.)

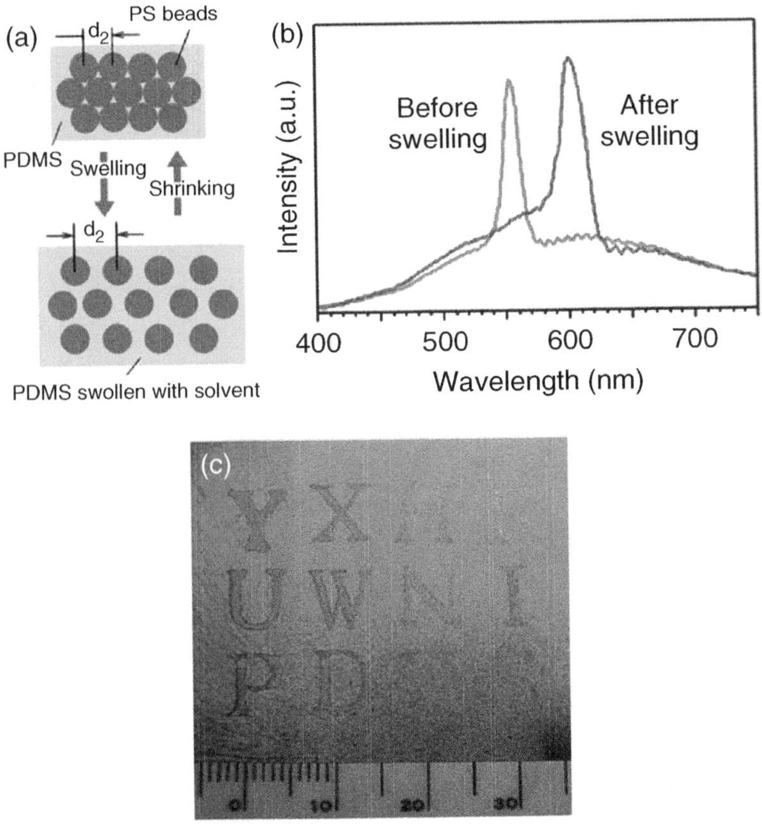

Figure 10.15 Patterned photonic films fabricated by taking advantage of the swelling process of polymer matrix. (a) Schematic illustration of the swelling and shrinking process of tunable colloidal crystals. The lattice constant will be increased by swelling the PDMS matrix with an appropriate solvent, while it will shrink back to the original state after the evaporation of solvent. (b) The change of reflection spectra before and after the swelling process. (c) Letters printed on the surface of the matrix using a rubber stamp. (Reproduced with permission. Copyright the American Chemical Society: Ref. 89.)

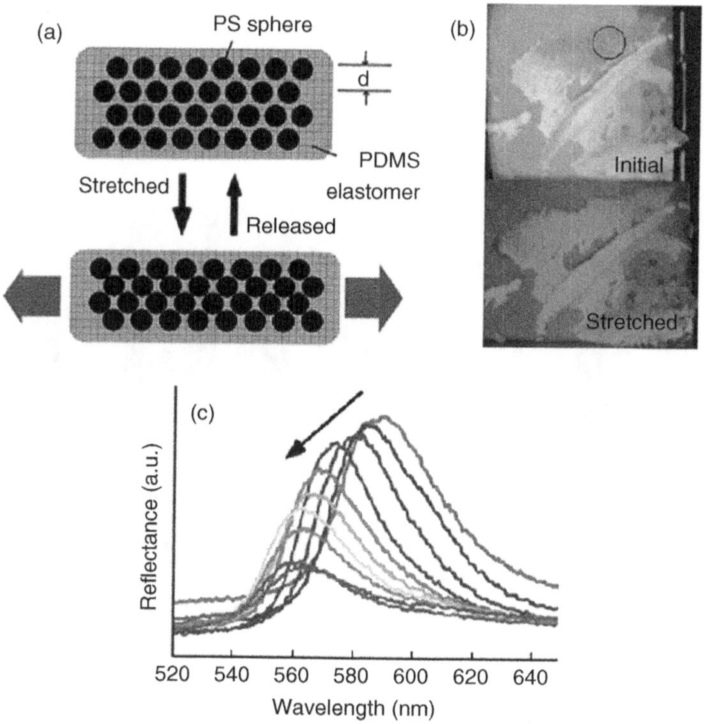

Figure 10.16 External mechanical-force-responsive photonic films fabricated by embedding colloidal crystals in PDMS matrix. (a) Schematic illustration of the reversible elastic deformation of the composite colloidal crystal film. The lattice constant can easily be modulated by the stretch rate of the film. (b) Digital photographs of the composite colloidal crystal film before (top) and after (down) stretch. (c) The change of reflection spectra during the stretching process. (Reproduced with permission. Copyright the American Chemical Society: Ref. 91.)

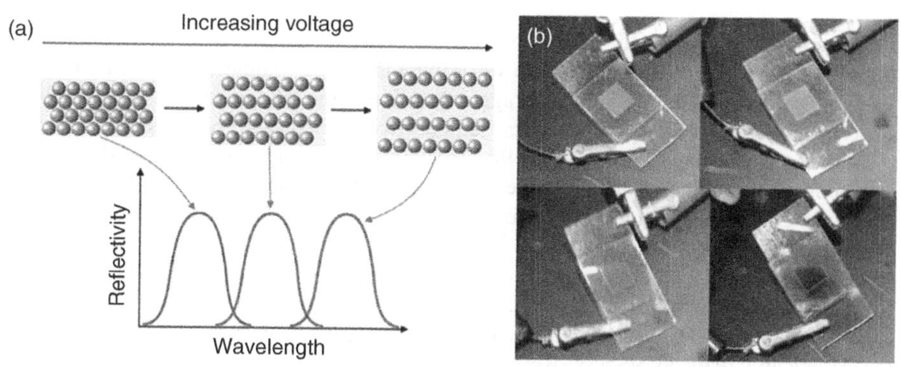

Figure 10.17 Electric-field-induced tunable photonic materials. (a) Schematic illustration of the operation of voltage-tunable full-color opal film. As the film was obtained by embedding the silica colloidal crystals into the matrix of polyferrocenylsilane gel, modulation of the voltage causes a change in the lattice constant. (b) Photographs of the tunable opal film in response to the voltage change. This kind of film has a tuning range covering the whole visible region. (Reproduced with permission. Copyright Elsevier: Ref. 37.)

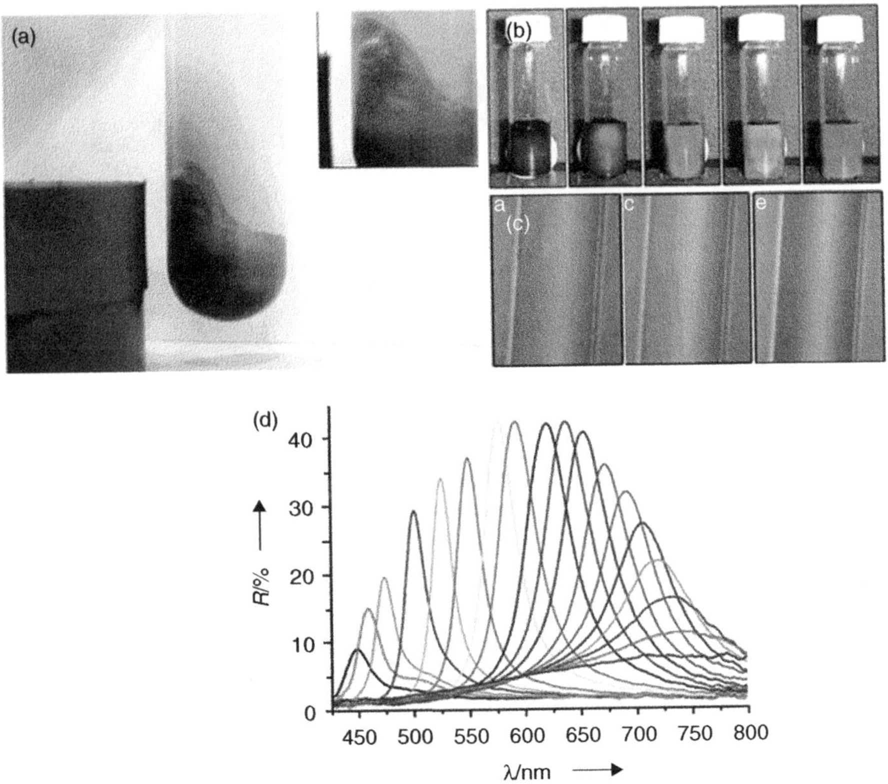

Figure 10.18 Magnetically tunable photonic crystals. With the help of an external magnetic field, they have a tuning range covering the whole visible region. (a) Magnetic colloidal crystals exhibiting various Bragg colors due to the inhomogeneous field gradient, which compresses or relaxes the crystal lattice. (Reproduced with permission. Copyright the American Chemical Society: Ref. 96.) (b) Photographs of magnetic colloidal crystals in response to an external magnetic field. The magnetic crystals will exhibit various colors while gradually altering the magnet–sample distance. (c) Optical microscope images of magnetic colloidal crystal solution enclosed in a glass capillary under an increasing magnetic field. (d) External magnetic field intensity dependence of the reflection spectra of the magnetic colloidal crystals. (Panels (b) and (d) reproduced with permission. Copyright 2007 Wiley-VCH: Ref. 98. Panel (c) reproduced with permission. Copyright the American Chemical Society: Ref. 99.)

Printed and bound by CPI Group (UK) Ltd, Croydon, CR0 4YY
16/09/2026
14963407-0003